ICSA Book Series in Statistics

Series editors

The ICSA Book Series in Statistics showcases research from the International Chinese Statistical Association that has an international reach. It publishes books in statistical theory, applications, and statistical education. All books are associated with the ICSA or are authored by invited contributors. Books may be monographs, edited volumes, textbooks and proceedings.

More information about this series at http://www.springer.com/series/13402

Yichuan Zhao • Ding-Geng Chen
Editors

New Frontiers of Biostatistics and Bioinformatics

Springer

Editors
Yichuan Zhao
Department of Mathematics and Statistics
Georgia State University
Atlanta, GA, USA

Ding-Geng Chen
Department of Biostatistics
Gillings School of Global Public Health
University of North Carolina at Chapel Hill
Chapel Hill, NC, USA

Department of Statistics
University of Pretoria
Pretoria, South Africa

ISSN 2199-0980 ISSN 2199-0999 (electronic)
ICSA Book Series in Statistics
ISBN 978-3-319-99388-1 ISBN 978-3-319-99389-8 (eBook)
https://doi.org/10.1007/978-3-319-99389-8

Library of Congress Control Number: 2018962164

This Springer imprint is published by the registered company Springer Nature Switzerland AG
The registered company address is: Gewerbestrasse 11, 6330 Cham, Switzerland

Preface

This book is mainly comprised of excellent presentations delivered in the 5th Workshop on Biostatistics and Bioinformatics held in Atlanta on May 5–7, 2017. Biostatistics and bioinformatics have been playing a key role in statistics and other scientific research fields in recent years. The aim of the 5th Workshop on Biostatistics and Bioinformatics was to stimulate research, foster interaction among researchers in field, and offer opportunities for learning and facilitating research collaborations in the era of big data. From this successful workshop, the two editors selected excellent presentations for this book. All the 22 chapters are peer-reviewed and revised multiple times before the final acceptance. This book provides the most recent advances in the field, presenting new methods and case applications at the frontiers of biostatistics and bioinformatics research and interdisciplinary areas. This timely book makes invaluable contributions to biostatistics and bioinformatics and offers insights for researchers, students, and industry practitioners.

The 22 chapters are organized into 5 parts. Part I includes five chapters that present a review of the theoretical framework in biostatistics. Part II consists of four chapters on wavelet-based approach for complex data. Part III is composed of six chapters that present clinical trials and statistical modeling. Part IV outlines high-dimensional gene expression data analysis. Part V consists of four chapters on survival analysis. We organize the chapters as self-contained units, and the references of the chapter are at the end of the each chapter so that readers can refer to the cited sources easily. To better understand the proposed procedures in the book, the readers can readily request the data sets and computer programs from the two editors. Therefore, the readers can apply these new statistical methods of the book for their own research.

Part I: Review and Theoretical Framework in Biostatistics (Chaps. 1–4)

The chapter "Optimal Weighted Wilcoxon–Mann–Whitney Test for Prioritized Outcomes" reviews concepts of prioritized outcomes in a two-group randomized clinical trial of multiple outcomes, where mortality affects the assessment of the other follow-up outcomes. In this chapter, Matsouaka, Singhal, and Betensky develop a weighted Wilcoxon–Mann–Whitney test procedure to analyze the data and determine the optimal weights that maximize its power. The authors obtain the analytical power formula for the test statistic and compare its results with those obtained via simulation studies using a range of treatment effects on the outcomes.

In the chapter "A Selective Overview of Semiparametric Mixture of Regression Models," Xiang and Yao conduct a systematic overview of new semiparametric mixture of regression models, which have been popularly used in many applications. Recent advances and open questions are also discussed.

In the chapter "Rank-Based Empirical Likelihood for Regression Models with Responses Missing at Random," Bindele and Zhao consider a general regression model with responses missing at random. From an imputed rank-based objective function, the authors derive a rank-based estimator, and its asymptotic distribution is established. An empirical likelihood approach is proposed based on the rank-based objective function, from which its asymptotic distribution is established.

In the chapter "Bayesian Nonparametric Spatially Smoothed Density Estimation," Hanson, Zhou, and de Carvalho develop a Bayesian nonparametric density estimator, which changes smoothly in space. The estimator is built using the predictive rule from a marginalized Polya tree so that observations are spatially weighted by their distance from the location of interest. The authors propose a simple refinement to accommodate arbitrarily censored data and develop a test for whether the density is spatially varying.

Part II: Wavelet-Based Approach for Complex Data (Chaps. 5–8)

The chapter "Mammogram Diagnostics Using Robust Wavelet-Based Estimator of Hurst Exponent" presents the robust estimation of Hurst exponent in two-dimensional images based on non-decimated wavelet transforms. The properties of the proposed estimators are studied both theoretically and numerically. In this chapter, Feng, Mei, and Vidakovic show how to apply proposed methods to digitized mammogram images, estimate Hurst exponent, and then use it as a discriminatory descriptor to classify mammograms to benign and malignant.

In the chapter "Wavelet-Based Profile Monitoring Using Order-Thresholding Recursive CUSUM Schemes," Zhang, Mei, and Shi propose a novel wavelet-based profile monitoring procedure, which is based on the order-thresholding transformation of recursive CUSUM statistics of multiple wavelet coefficients. The authors carry out extensive simulation studies and a case study of tonnage profile data, which show that proposed procedure is efficient for detecting the unknown local changes on the profile.

In the chapter "Estimating the Confidence Interval of Evolutionary Stochastic Process Mean from Wavelet-Based Bootstrapping," de Medeiros and de Souza propose to estimate the uncertainty for the evolutionary mean of a stochastic process based on bootstrapping of wavelet coefficients. By discrete wavelet transform, the authors apply bootstrap to estimate the confidence interval of the autocorrelation for a time series. Moreover, these methods with few modifications are implemented.

In the chapter "A New Wavelet-Based Approach for Mass Spectrometry Data Classification," Cohen, Messaoudi, and Badir propose a statistical methodology of a reliable diagnosis for classifying mass spectrometry data with a type of cancer. The authors go over wavelets, principal component analysis, and support vector machines, and perform a study on low-mass SELDI spectra from patients with breast cancer and from normal controls. The performance is evaluated with a k-fold cross validation technique and simulation study. The performance of the proposed method is excellent with an accurate classification of mass spectrometry.

Part III: Clinical Trials and Statistical Modeling (Chaps. 9–14)

In the chapter "Statistical Power and Bayesian Assurance in Clinical Trial Design," Chen and Chen propose a Bayesian assurance as an alternative to the conventional statistical power to incorporate the uncertainties of this observed treatment effect. In this chapter, the authors review the transition from conventional statistical power to Bayesian assurance and discuss the computations of Bayesian assurance using a Monte Carlo simulation-based method.

The chapter "Equivalence Tests in Subgroup Analyses" proposes that the consistency of the treatment effect in two subgroups should be assessed using an equivalence test called *consistency* test. In this chapter, Ring, Scharpenberg, Grill, Schall, and Brannath present tests for both quantitative and binary outcome variables and review the basic properties of these consistency tests using simulation studies. The authors also indicate that equivalence tests can be used both to assess the consistency of treatment effects across subgroups and to detect medically relevant heterogeneity in treatment effects across subgroups.

In the chapter "Predicting Confidence Interval for the Proportion at the Time of Study Planning in Small Clinical Trials," Yu and Vexler discuss "future" confidence interval prediction with binomial outcomes for small clinical trials and sample size calculation, where the "future" confidence interval emphasizes the confidence interval as a function of a random sample that is not observed at the planning stage of a study. The authors propose three probabilistic approaches to future confidence interval prediction when the sample size is small. The approach based on the expectation of the boundaries has the most desirable properties and is easy to implement.

The chapter "Importance of Adjusting for Multi-Stage Design When Analyzing Data from Complex Surveys" illustrates possible discrepancies in point estimates and standard errors using 2014–2015 TUS data. In this chapter, Ha and Soulakova show the importance of using the guidelines when analyzing complex surveys. The authors discuss three methods: method I ignores any weighting, method II incorporates the main weight only, and method III utilizes the main weight and balanced repeated replications with specified replicate weights.

In the chapter "Analysis of the High School Longitudinal Study to Evaluate Associations Among Mathematics Achievement, Mentorship and Student Participation in STEM Programs," Murillo, Tiwari, and Affuso analyze a subsample of the High School Longitudinal Study (2009–2013) dataset (HSLS:09). Regression models are applied to evaluate mathematics achievement and student enrollment in STEM major/careers based on their individual participation. Differences based on sex, race/ethnicity, and socioeconomic status are assessed.

The chapter "Statistical Modeling for the Heart Disease Diagnosis via Multiple Imputation" addresses a common challenge of missing data during statistical analysis of clinic data. Missing data causes severe problems in statistical analysis and leads to invalid conclusions. Multiple imputation is a useful strategy for handling missing data. In the chapter, Li and Zhao apply the multiple imputation to a public accessible heart disease dataset, which has a high missing rate, and build a prediction model for the heart disease diagnosis.

Part IV: High-Dimensional Gene Expression Data Analysis (Chaps. 15–18)

In the chapter "Learning Gene Regulatory Networks with High-Dimensional Heterogeneous Data," Jia and Liang propose to model the heterogeneous data using a mixture Gaussian graphical model and apply the imputation-consistency algorithm to estimate the parameters of the mixture model and cluster the samples to different subgroups. The proposed method is compared with an existing method for learning mixture Gaussian graphical models as well as a few other methods for homogeneous data, such as graphical Lasso, etc.

The chapter "Performance Evaluation of Normalization Approaches for Metagenomic Compositional Data on Differential Abundance Analysis" assesses normalization methods for metagenomic sequence data. In this chapter, Du, An, and Fang further study the impact of normalization on subsequent differential abundance analysis. The authors suggest the selection of a normalization method for metagenomic compositional data should be made on a case-by-case basis.

The chapter "Identification of Pathway-Modulating Genes Using the Biomedical Literature Mining" centers on an effective use of biomedical literature for the identification of the relationships among genes. A Bayesian hierarchical model was proposed, which allows to identify indirect relationship between genes by linking them using the gene ontology terms. In this chapter, Yu, Nam, Couch, Lawson, and Chung illustrate this method using the web interface GAIL. It provides the PubMed literature mining results, along with the R package by the Bayesian hierarchical model.

The chapter "Discriminant Analysis and Normalization Methods for Next-Generation Sequencing Data" studies discriminating and normalization methods for gene expression analysis with the development of high-throughput techniques. A number of new discriminant analysis methods have been proposed to discriminate next-generation sequencing data. In this chapter, Zhou, Wang, Zhao, and Tong introduce three methods including the Poisson linear discriminant analysis, the zero-inflated Poisson logistic discriminant analysis, and the negative binomial linear discriminant analysis and further introduce several normalization methods for processing next-generation sequencing data.

Part V: Survival Analysis (Chaps. 19–22)

In the chapter "On the Landmark Survival Model for Dynamic Prediction of Event Occurrence Using Longitudinal Data," Zhu, Li, and Huang demonstrate that a joint distribution of longitudinal and survival data exists that satisfy the modeling assumptions without additional restrictions. In addition, the authors propose an algorithm to generate data from this joint distribution and generalize the results to the more flexible landmark linear transformation models, which include the landmark Cox model.

In the chapter "Nonparametric Estimation of a Cumulative Hazard Function with Right Truncated Data," Zhang, Jiang, Zhao, and Akcin develop the nonparametric inference for the forward-time hazard. The authors study a family of weighted tests for comparing the hazard function between two independent samples. The authors analyze the data set about AIDS incubation time to illustrate the proposed procedures.

In the chapter "Empirical Study on High-Dimensional Variable Selection and Prediction Under Competing Risks," Hou and Xu consider competing risk analysis and explore statistical properties in the presence of high-dimensional predictors. The authors study the accuracy of prediction and variable selection of existing statistical learning methods using extensive simulation studies, including different approaches to choosing penalty parameters in each method.

In the chapter "Nonparametric Estimation of a Cumulative Hazard Function with Right Truncated Data," Akcin, Zhang, and Zhao study the nonparametric inference for the hazard rate function with right truncated data. Kernel smoothing techniques are used to get smoothed estimates of hazard rates. Three commonly used kernels, uniform, Epanechnikov, and biweight kernels are applied on the AIDS data to illustrate the proposed methods.

We are very grateful to all of people, who have supported the creation of this book with Springer. First, we thank the authors of each chapter for their wonderful contributions. Second, our deep gratitude goes to all the reviewers for their dedicated reviews, which improved the quality of the book significantly. Third, we would like to acknowledge the leadership of the organizing committee and all the volunteers of the 5th Workshop on Biostatistics and Bioinformatics because this book would be impossible without this workshop. Last but not least, our sincere appreciations go to the professional support and great assistance of Nicholas Philipson (Springer/ICSA Book Series coordinator and editorial director, Business/Economics & Statistics), Nitza Jones-Sepulveda (associate editor) from Springer New York, and Sindhuraj Thulasingam (Project Coordinator of Books) from Springer Nature, which made this book published. We welcome readers' comments on typos, errors, and improvements about the book. If there is an exchange, please send comments and suggestions to Dr. Yichuan Zhao (email: yichuan@gsu.edu) and Dr. Ding-Geng Chen (email: dinchen@email.unc.edu).

Atlanta, GA, USA — Yichuan Zhao
Chapel Hill, NC, USA — Ding-Geng Chen

Contents

List of Contributors

Olivia Affuso Department of Epidemiology, University of Alabama at Birmingham, Birmingham, AL, USA

Haci Akcin Department of Risk Management and Insurance, Georgia State University, Atlanta, GA, USA

Lingling An Department of Agricultural and Biosystems Engineering, University of Arizona, Tucson, AZ, USA

Hassan Badir National School of Applied Sciences-Tangier, ENSAT, Abdelmalek Essaadi University, Tangier, Morocco

Rebecca A. Betensky Department of Biostatistics, Harvard T.H. Chan School of Public Health, and Harvard NeuroDiscovery Center, Harvard Medical School, Boston, MA, USA

Huybrechts F. Bindele Department of Mathematics and Statistics, University of South Alabama, Mobile, AL, USA

W. Brannath Faculty of Mathematics/Computer Sciences, Competence Center for Clinical Trials Bremen, University of Bremen, Bremen, Germany

Ding-Geng Chen Department of Biostatistics, Gillings School of Global Public Health, University of North Carolina at Chapel Hill, Chapel Hill, NC, USA

Department of Statistics, University of Pretoria, Pretoria, South Africa

Jenny K. Chen Department of Biostatistics, Gillings School of Global Public Health, University of North Carolina at Chapel Hill, Chapel Hill, NC, USA

Dongjun Chung Department of Public Health Sciences, Medical University of South Carolina, Charleston, SC, USA

Achraf Cohen Department of Mathematics and Statistics, University of West Florida, Pensacola, FL, USA

Daniel Couch Department of Public Health Sciences, Medical University of South Carolina, Charleston, SC, USA

Vanda Inácio de Carvalho University of Edinburgh, Edinburgh, Scotland, UK

Aline Edlaine de Medeiros Graduate Program in Biostatistics, State University of Maringa, Maringa, Brazil

Eniuce Menezes de Souza Department of Statistics, State University of Maringa, Maringa, Brazil

Ruofei Du Biostatistics Shared Resource, University of New Mexico Comprehensive Cancer Center, Albuquerque, NM, USA

Zhide Fang Biostatistics Program, School of Public Health, Louisiana State University Health Sciences Center, New Orleans, LA, USA

Chen Feng H. Milton Stewart School of Industrial and Systems Engineering, Georgia Institute of Technology, Atlanta, GA, USA

S. Grill Faculty of Mathematics/Computer Science, Competence Center for Clinical Trials Bremen, University of Bremen, Bremen, Germany

Leibniz Institute for Prevention Research and Epidemiology - BIPS, Bremen, Germany

Trung Ha Burnett School of Biomedical Sciences, College of Medicine, University of Central Florida, Orlando, FL, USA

Timothy Hanson Medtronic Inc., Minneapolis, MN, USA

Jiayi Hou Altman Clinical and Translational Research Institute, University of California, San Diego, La Jolla, CA, USA

Xuelin Huang Department of Biostatistics, MD Anderson Cancer Center, Houston, TX, USA

Bochao Jia Eli Lilly and Company, Lilly Corporate Center, IN, USA

Yong Jiang MetLife Inc., Whippany, NJ, USA

Andrew Lawson Department of Public Health Sciences, Medical University of South Carolina, Charleston, SC, USA

Lian Li Department of Epidemiology, Rollins School of Public Health, Emory University, Atlanta, GA, USA

Liang Li Department of Biostatistics, MD Anderson Cancer Center, Houston, TX, USA

Faming Liang Department of Statistics, Purdue University, West Lafayette, IN, USA

Roland A. Matsouaka Department of Biostatistics and Bioinformatics, Duke University, Durham, NC, USA

Program for Comparative Effectiveness Methodology, Duke Clinical Research Institute, Duke University, Durham, NC, USA

Yajun Mei H. Milton Stewart School of Industrial and Systems Engineering, Georgia Institute of Technology, Atlanta, GA, USA

Chaimaa Messaoudi National School of Applied Sciences-Tangier, ENSAT, Abdelmalek Essaadi University, Tangier, Morocco

Anarina L. Murillo Department of Biostatistics, University of Alabama at Birmingham, Birmingham, AL, USA

Jin Hyun Nam Department of Public Health Sciences, Medical University of South Carolina, Charleston, SC, USA

Arne Ring Department of Mathematical Statistics and Actuarial Science, University of the Free State, Bloemfontein, South Africa

medac GmbH, Wedel, Germany

R. Schall Department of Mathematical Statistics and Actuarial Science, University of the Free State, Bloemfontein, South Africa

IQVIA Biostatistics, Bloemfontein, South Africa

M. Scharpenberg Faculty of Mathematics/Computer Sciences, Competence Center for Clinical Trials Bremen, University of Bremen, Bremen, Germany

Jianjun Shi H. Milton Stewart School of Industrial and Systems Engineering, Georgia Institute of Technology, Atlanta, GA, USA

Aneesh B. Singhal Department of Neurology, Massachusetts General Hospital, Boston, MA, USA

Julia N. Soulakova Burnett School of Biomedical Sciences, College of Medicine, University of Central Florida, Orlando, FL, USA

Hemant K. Tiwari Department of Biostatistics, University of Alabama at Birmingham, Birmingham, AL, USA

Tiejun Tong Department of Mathematics, Hong Kong Baptist University, Kowloon, Hong Kong

Albert Vexler Department of Biostatistics, University at Buffalo, State University of New York, Buffalo, NY, USA

Brani Vidakovic H. Milton Stewart School of Industrial and Systems Engineering and Wallace H. Coulter Department of Biomedical Engineering, Georgia Institute of Technology, Atlanta, GA, USA

Junhui Wang School of Data Science, City University of Hong Kong, Kowloon, Hong Kong

Sijia Xiang School of Data Sciences, Zhejiang University of Finance & Economics, Hangzhou, China

Ronghui Xu Department of Family Medicine and Public Health, University of California, San Diego, La Jolla, CA, USA

Department of Mathematics, University of California, San Diego, La Jolla, CA, USA

Weixin Yao Department of Statistics, University of California, Riverside, CA, USA

Jihnhee Yu Department of Biostatistics, University at Buffalo, State University of New York, Buffalo, NY, USA

Zhenning Yu Department of Public Health Sciences, Medical University of South Carolina, Charleston, SC, USA

Ruizhi Zhang H. Milton Stewart School of Industrial and Systems Engineering, Georgia Institute of Technology, Atlanta, GA, USA

Xu Zhang Center for Clinical and Translational Sciences, University of Texas Health Science Center, Houston, TX, USA

Yichuan Zhao Department of Mathematics and Statistics, Georgia State University, Atlanta, GA, USA

Haiming Zhou Northern Illinois University, DeKalb, IL, USA

Yan Zhou College of Mathematics and Statistics, Institute of Statistical Sciences, Shenzhen University, Shenzhen, China

Yayuan Zhu Department of Biostatistics, MD Anderson Cancer Center, Houston, TX, USA

List of Chapter Reviewers

David Benkeser Department of Biostatistics and Bioinformatics, Emory University, Atlanta, GA, USA

Frazier Bindele Department of Mathematics and Statistics, University of South Alabama, Mobile, AL, USA

Qingpo Cai Department of Biostatistics and Bioinformatics, Emory University, Atlanta, GA, USA

Hsin-wen Chang Institute of Statistical Science, Academia Sinica, Taipei, Taiwan

Ding-Geng Chen School of Social Work, University of North Carolina, Chapel Hill, NC, USA

Yunxiao Chen Department of Psychology and Institute for Quantitative Theory and Methods, Emory University, Atlanta, GA, USA

Eric Chicken Department of Statistics, Florida State University, Tallahassee, FL, USA

Xin Dang Department of Mathematics, University of Mississippi, University, MS, USA

Yang Feng Department of Statistics, Columbia University, New York, NY, USA

Cindy Fu Department of Mathematics and Statistics, York University, Toronto, ON, Canada

Matt Hayat Division of Epidemiology and Biostatistics, School of Public Health, Georgia State University, Atlanta, GA, USA

Zhiguang Huo Department of Biostatistics, University of Florida, Gainesville, FL, USA

Hsin-Hsiung Bill Huang Department of Statistics, University of Central Florida, Orlando, FL, USA

Linyuan Li Department of Mathematics and Statistics, University of New Hampshire, Durham, NH, USA

Pengfei Li Department of Statistics and Actuarial Science, University of Waterloo, Waterloo, ON, Canada

Jason Liao Nonclinical Statistics, Merck & Co., Inc., Whitehouse Station, NJ, USA

Antonio Linero Department of Statistics, Florida State University, Tallahassee, FL, USA

Yang Liu Statistics, Programming, and Economics Branch, Division of Analysis Research and Practice Integration, National Center for Injury Prevention and Control, Atlanta, GA, USA

Xuewen Lu Department of Mathematics and Statistics, University of Calgary, Calgary, AB, Canada

Sheng Luo Department of Biostatistics and Bioinformatics, Duke University School of Medicine, Durham, NC, USA

Xiaoyi Min Department of Mathematics and Statistics, Georgia State University, Atlanta, GA, USA

Timothy O'Brien Department of Mathematics and Statistics, Loyola University Chicago, Chicago, IL, USA

Li-Xuan Qin Department of Epidemiology and Biostatistics, Memorial Sloan Kettering Cancer Center, New York, NY, USA

Lixin Shen Department of Mathematics, Syracuse University, Syracuse, NY, USA

Chuck Song Division of Biostatistics, College of Public Health, Ohio State University, Columbus, OH, USA

Xin Tian Office of Biostatistics Research, National Heart, Lung, and Blood Institute, Bethesda, MD, USA

Brani Vidakovic H. Milton Stewart School of Industrial and Systems Engineering, and Wallace H. Coulter Department of Biomedical Engineering, Georgia Institute of Technology, Atlanta, GA, USA

Antai Wang Department of Mathematical Sciences, New Jersey Institute of Technology, Newark, NJ, USA

Dongliang Wang Department of Public Health and Preventive Medicine, SUNY Upstate Medical University, Syracuse, NY, USA

Lianming Wang Department of Statistics, University of South Carolina, Columbia, SC, USA

Weizheng Wang Department of Mathematics and Statistics, Wright State University, Dayton, OH, USA

Wei (Peter) Yang Department of Biostatistics and Epidemiology, University of Pennsylvania School of Medicine, Philadelphia, PA, USA

Lili Yu Department of Biostatistics, Georgia Southern University, Statesboro, GA, USA

Jiajia Zhang Department of Statistics, University of South Carolina, Columbia, SC, USA

Mei-Jie Zhang Division of Biostatistics, Medical College of Wisconsin, Milwaukee, WI, USA

Min Zhang Department of Statistics, Purdue University, West Lafayette, IN, USA

Xu Zhang Division of Clinical and Translational Sciences, Department of Internal Medicine, Medical School, University of Texas Health Science Center at Houston, Houston, TX, USA

Ying-Ying Zhang Department of Statistics and Actuarial Science, College of Mathematics and Statistics, Chongqing University, Chongqing, China

Yichuan Zhao Department of Mathematics and Statistics, Georgia State University, Atlanta, GA, USA

Yunpeng Zhao Department of Statistics, Volgenau School of Engineering, George Mason University, Fairfax, VA, USA

Hongjian Zhu Department of Biostatistics and Data Science, University of Texas Health Science Center at Houston, Houston, TX, USA

Changliang Zou Institute of Statistics, Nankai University, Tianjin, China

About the Editors

Yichuan Zhao is a professor of statistics at Georgia State University in Atlanta. He has a joint appointment as associate member of the Neuroscience Institute, and he is also an affiliated faculty member of School of Public Health at Georgia State University. His current research interest focuses on survival analysis, empirical likelihood method, nonparametric statistics, analysis of ROC curves, bioinformatics, Monte Carlo methods, and statistical modeling of fuzzy systems. He has published over 80 research articles in statistics; has co-edited two books on statistics, biostatistics, and data science; and has been invited to deliver more than 170 research talks nationally and internationally. Dr. Zhao has organized the Workshop Series on Biostatistics and Bioinformatics since its initiation in 2012. He also organized the 25th ICSA Applied Statistics Symposium in Atlanta as a chair of the organizing committee to great success. He is currently serving as editor, or on the editorial board, for several statistical journals. Dr. Zhao is an elected member of the International Statistical Institute.

(Din) Ding-Geng Chen is a fellow of the American Statistical Association and currently the Wallace H. Kuralt distinguished professor at the University of North Carolina at Chapel Hill, USA, and an extraordinary professor at the University of Pretoria, South Africa. He was a professor at the University of Rochester and the Karl E. Peace endowed eminent scholar chair in biostatistics at Georgia Southern University. He is also a senior consultant for biopharmaceuticals and government agencies with extensive expertise in clinical trial biostatistics and public health statistics with multimillion dollars' federal-funded research projects. Professor Chen has written more than 150 refereed publications and co-authored/co-edited 23 books on clinical trial methodology, meta-analysis, causal-inference, and public health statistics.

Part I
Review of Theoretical Framework in Biostatistics

Chapter 1
Optimal Weighted Wilcoxon–Mann–Whitney Test for Prioritized Outcomes

Roland A. Matsouaka, Aneesh B. Singhal, and Rebecca A. Betensky

This chapter reviews keys concepts of prioritized outcomes in a two-group randomized clinical trial of multiple outcomes, where mortality affects the assessment of the other follow-up outcomes. The main concepts related to prioritized endpoints along with the different terminologies used in the literature are discussed. Then, statistical tenets of worst-rank composite endpoints are reviewed using a combined endpoint of mortality and a continuous outcome.

We motivate the approach using a randomized clinical trial of normobaric oxygen therapy on patients who underwent an acute ischemic stroke where we combine a continuous outcome with mortality into a single composite endpoint using the worst-rank framework. We develop a weighted Wilcoxon–Mann–Whitney test statistic to analyze the data and determine the optimal weights that maximize its power. We provide the rationale for the weights and their relative importance in data analysis. In addition, we derive the analytical power formula for the test statistic. To demonstrate that the proposed power formula produces valid power estimations, we compare its results with those obtained empirically via Monte-Carlo simulations using a range of treatment effects on the outcomes. Finally, we illustrate the method using data from the clinical trial of normobaric oxygen therapy.

R. A. Matsouaka (✉)
Department of Biostatistics and Bioinformatics, Duke University, Durham, NC, USA
e-mail: roland.matsouaka@duke.edu

Program for Comparative Effectiveness Methodology, Duke Clinical Research Institute, Duke University, Durham, NC, USA

A. B. Singhal
Department of Neurology, Massachusetts General Hospital, Boston, MA, USA

R. A. Betensky
Department of Biostatistics, Harvard T.H. Chan School of Public Health and Harvard NeuroDiscovery Center, Harvard Medical School, Boston, MA, USA
e-mail: betensky@hsph.harvard.edu

Y. Zhao, D.-G. Chen (eds.), *New Frontiers of Biostatistics and Bioinformatics*, ICSA Book Series in Statistics, https://doi.org/10.1007/978-3-319-99389-8_1

1.1 Introduction

In clinical trials of multifaceted diseases, multiple outcomes are usually evaluated to estimate and compare the effects of a new active treatment over a control treatment. Although these outcomes can be analyzed separately, they are usually combined into a single composite endpoint to take into account the complexity of the disease manifestations and capture different aspects of the treatment effects. Combining outcomes has several advantages: it increases statistical precision and efficacy, reduces considerably the number of patients needed to enroll for a given expected treatment effect or to reach a specific statistical power, circumvents the needs for multiple testing, and provides an overall assessment of the treatment effect.

One of the most commonly used methods for combining multiple outcomes is the time-to-first event. For this method, only a patient's initial event during the trial is considered in the analysis while all of the subsequent events are ignored. However, such a composite endpoint have serious practical limitations that often result in misleading interpretations and poor medical decisions which are of greater concerns. Usually, component outcomes of a composite endpoint are not equally important or clinically relevant; they do not occur at the same frequency and are not similarly impacted by the treatment. More than often, treatment effects and significant statistical analyses are driven by components of lesser importance. As such, they do not provide a more comprehensive perspective of the disease burden that is realistic, congruent with clinical judgment or aligned with the perceptions and expectations of patients and their caregivers.

This is illustrated in many cardiovascular disease trials where mortality remains the major outcome of interest which, fortunately, is often less frequent and tends to occur later in a trial (see, for instance, the relative perceived clinical severity of typical components of composite endpoints considered in recent cardiovascular trials given in Fig. 1.1). In a clinical trial of "death or heart failure hospitalization" (whichever comes first), for example, a patient may experience multiple heart failure hospitalizations and eventually die. Clearly, a patient who has a minor heart attack after 1 week of follow-up but remain event-free subsequently for several consecutive years should not be considered as having a worse outcome compared to another patient in the trial who dies after 2 months of follow-up.

Therefore, standard statistical analyses based on these time-to-first outcome event where subsequent events are ignored may skew the assessment of the treatment effect, lead to biased results, and poorly reflect the true burden of the patient's disease experience (Anker and Mcmurray 2012; Anker et al. 2016; Ferreira-González et al. 2007b; Freemantle et al. 2003; Lubsen and Kirwan 2002; Prieto-Merino et al. 2013; Freemantle et al. 2003; Heddle and Cook 2011; Claggett et al. 2013; Brown et al. 2016). Moreover, the composite endpoint of time-to-first event are not applicable when the component outcomes are on different scales, such as a mixture of discrete, continuous, time-to-event, and quality-of-life outcomes (Felker et al. 2008; Tyler et al. 2011; Bebu and Lachin 2015).

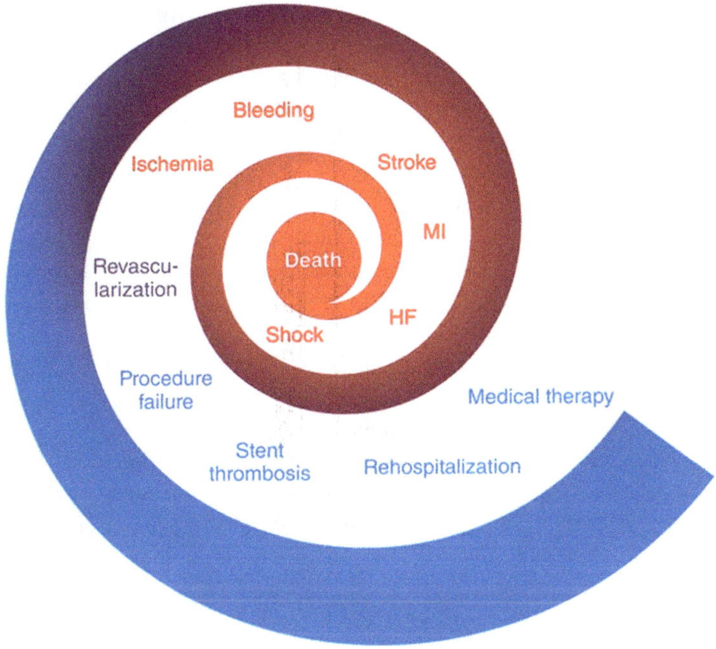

Fig. 1.1 Relative severity of cardiovascular disease outcomes (from death onward out of the spiral). *Acronyms:* MI: myocardial infarction; HF: heart failure (used with permission from Armstrong and Westerhout (2017))

Despite these serious limitations, analyses of composite endpoints are ubiquitous in a large number of clinical research areas including cardiovascular disease (Lisa and James 1997; Bakal et al. 2012b,a, 2015; Neaton et al. 2005; Follmann et al. 1992; Brittain et al. 1997; Felker et al. 2008), infectious diseases (Neaton et al. 1994; Finkelstein and Schoenfeld 1999; Follmann et al. 2007), oncology (Freemantle et al. 2003), nephrology (Hariharan et al. 2003; Li et al. 2001), neurology and psychiatry (Davis et al. 2003), health services, autoimmune disease, dermatology (Kaufman et al. 1998), respiratory (Spencer et al. 2007), rheumatoid arthritis, limb ischemia (Subherwal et al. 2012), orthopedics (DeCoster et al. 1999), urology, anesthesia, migraines, obstetrics, and gynecology (Ross 2007; Wen et al. 2017)—even though their limitations and unsatisfactory characteristics are widely recognized and genuinely mentioned in most publications (Manja et al. 2017; Zhang et al. 1997; Anker et al. 2016; Tyler et al. 2011; Cordoba et al. 2010; Rowan et al. 2008; Prieto-Merino et al. 2013).

Several alternative methods have been proposed to combine multiple outcomes while taking into account their clinical priority (Lisa and James 1997; Bakal et al. 2015; Neaton et al. 2005; Follmann et al. 1992; Brittain et al. 1997; Felker et al. 2008). Among them are the methods based on prioritized outcomes where component outcomes are prioritized and ordered—following a specific, prespecified hierarchy and with respect to their clinical importance—from the most severe (e.g., mortality) to the least severe one (or more favorable). Usually, the clinical questions of interest dictate the choice and order of the prioritized endpoints. Treatment comparison requires pairwise comparisons of patients' outcomes, where each pair comprise one patient from one treatment group (e.g., active treatment) and another patient from the alternative treatment group (e.g., control treatment). The statistical underpinnings of these methods are based on ranks. These ranks are used to draw inference as to whether a randomly selected patient in the active treatment will have, on average, a better overall composite endpoint compared to a randomly selected patient in the control treatment group by using the Wilcoxon–Mann–Whitney (WMW) test statistic. These methods, which are considered as part of the global rank approaches (Huang et al. 2008; Ramchandani et al. 2016), can be classified into two distinct categories based on the decision rules that dictate how to proceed from one outcome to a subsequent outcome on the hierarchy of outcomes.

On the one hand, we have the *proportion in favor of treatment* (PFT) of Buyse (2010) (also known as the win difference Luo et al. 2017) and the *win ratio* (WR) introduced by Pocock et al. (2011), which follow the ideas from Moyé et al. (1992) and Finkelstein and Schoenfeld (1999). In these methods, pairwise outcome comparisons between patients from the active and control treatment groups are conducted, starting from the most severe outcome. For each pairwise comparison, the patient with a better outcome is declared a winner. If it is not possible to determine the winner (e.g., comparison inconclusive or indeterminate) on the most severe outcome, the two patients are then compared on the second most severe outcome, and so forth. Finally, each patient score is recorded as a win (better outcome in the pairwise comparison), a loss (worse outcome), or a tie (when unable to declare a winner after exhausting all available outcomes).

The PFT is defined as the difference between the proportions of wins in the active and control treatment groups. The null hypothesis of no difference between the treatment groups corresponds to a PFT that is equal to 0, while a positive (resp., negative) value demonstrates that the active treatment is better (worse) than the control treatment. Similarly, the WR is the ratio of the proportion of wins in the active treatment over the proportion of wins in the control treatment. Under the null hypothesis, the WR is equal to 1. It is greater (resp., less) than 1 when the active treatment is beneficial (disadvantageous) compared to the control treatment.

On the other hand, we have the *worst-rank score* analysis—based on the original idea of Gould (1980) and O'Brien (1984). For this method, patients are placed into "buckets" (to use the analogy from Subherwal et al. 2012) on the hierarchy of component outcomes. In other words, each patient is categorized based on her or his worst personally experienced outcome. All the patients who have experienced the worst outcome (e.g., those who died) are assigned to the lowest-ranked bucket, patients who did not experience the worst outcome, but the second worst outcome are placed in the second lowest-ranked bucket, and so forth. Finally, depending on the predetermined choice of the component outcomes, patients with the less severe outcome or who did not experience any of the component outcomes are assigned to the highest-ranked bucket (Lachin 1999; Matsouaka and Betensky 2015; Matsouaka et al. 2016). Then, every patient in the active treatment group is compared to every patient in the control treatment group to determine whether the actively treated patient's outcome is better than or the same as the outcome of the patient in the control treatment.

The final result is determined by the buckets the compared patients belong to and by their respective outcomes. If the pair of patients is from the same bucket, they are compared by the magnitude of their outcome measures or by their first times to the event (whichever characterizes the bucket), where the longer the time-to-event the better (e.g., later death will be considered better compared to earlier death). If the two patients belong to two different buckets, the patient in the higher-ranked bucket is considered to have a better outcome than the patient in the lower-ranked bucket. Therefore, at the end of the process, all patients are ranked.

Despite the seemingly resemblance between the WR (or the PFT) and the worst-rank score analysis, there are stark clinical and statistical methodological differences between them. Therefore, the choice of one method versus the other must be motivated by the clinical questions of interest and should be predetermined before any analysis. This choice must not be merely dictated by the convenience to pick a method that provides the most significant results. Unlike the win ratio where the focus is put first on the worst outcome and where the next consecutive ranked outcomes (or events experienced by patients) are leveraged only to break ties, with the worst-rank score analysis the first most important step is to place patients in buckets, depending on the worst outcome or event they have personally experienced. Pairwise comparison of patients in one group versus the other is done within and between buckets. When the outcomes of patients from the same bucket are tied, the patients are declared similar and are ranked accordingly. No further comparison is needed. Likewise, when patients are from two different buckets, the patient in the higher-ranked bucket is always considered to have a better outcome.

In practice, the win ratio (or the proportion in favor of the treatment) is used in randomized trials where the most severe outcome is the main outcome of interest. In those trials, it is anticipated that a good percentage of patients will have the most severe outcome, which justify the a priori set to such an outcome. For instance, Pocock et al. (2011) reanalyzed the EMPHASIS-HF data to compare eplerenone against placebo in 2737 patients with NYHA class II heart failure and an ejection fraction less than 35% who were recruited at 278 centers in 29 countries. 1364 patients were randomly assigned to eplerenone and 1373 to placebo and the median follow-up time was 21 months. Pairs of patients from eplerenone and placebo were compared first on cardiovascular (CV) death and, if it was not possible to determine who had a CV death before the other, it was then determined who had a heart failure hospitalization first. Overall, there were a total of 147 deaths (10.8%) in the eplerenone group and 185 (13.5%) in the placebo group attributed to cardiovascular causes. Of the patients receiving eplerenone, 164 (12.0%) were hospitalized for heart failure, as compared with 253 patients (18.4%) receiving placebo.

The worst-rank score analysis is mostly used in trials where the most severe outcome is not the primary outcome. Usually, it is expected that a small percentage of patients will experience the most severe outcome. Therefore, it is mostly used in settings where the primary interest lies on a nonterminal (nonfatal) outcome, but for which analyses of the observed data are complicated due to the presence of missing observations due to death.

Felker and Maisel (2010) proposed the use of worst-rank score analysis in a hypothetical study of a phase II acute heart failure trial. They suggested a global rank score analysis of 200 patients with 101 patients in the active treatment group and 99 in the placebo group. Patients were compared for in-hospital mortality (4% patients in each group), lack of dyspnea improvement at 24 h (44% patients in active treatment group and 54% in the placebo group), detectable troponin or an increase in troponin by 25% during index hospitalization (7 and 5%, respectively), creatinine increase by more than 0.3 mg/dl (7% and 10%), and finally on change in pro-BNP from randomization to discharge. In another example, Lachin (1999) reexamined a clinical trial of the effect of vesnarinone versus placebo on patients with congestive heart failure and used a worse-rank score analysis of exercise time after 12 weeks of treatment after treatment and death (Feldman et al. 1991). Of the 80 patients randomized (40 in each group), six died before week 12 with five of them in the placebo group.

In this chapter, we consider the worst-rank score analysis and present a framework that allows us to weight the components of a worst-rank (composite) endpoint by relying uniquely on the data at hand. Matsouaka and Betensky studied the statistical properties of the worst-rank analyses based on the (ordinary) Wilcoxon–Mann–Whitney (WMW) test. They considered both tied worst-rank scores (all patients who died are assigned a fixed score) and untied worst-rank scores (where patients who died are ranked based on their time to the death, with the longer time to death the better) in the ranking of the components of the composite outcomes (Matsouaka and Betensky 2015).

For this chapter, we focus on the untied worst-rank score analyses. We assume that we have a data set where we can identify approximately well the time-to-death for each patient who died during the follow-up time. Although, one can easily adapt our method and result in the context of a tied worst-rank analysis. The current framework extends the worst-rank analysis of Matsouaka and Betensky by providing a weighted test statistic where its corresponding weights are optimal in the sense that they maximize the power of the test under a particular alternative hypothesis. We explore the statistical properties of the optimal weighted WMW test on a worst-rank composite endpoint, looking at the null hypothesis of no difference between treatment against a unidirectional alternative hypothesis that the treatment has a favorable effect on the components of the worst-rank composite endpoints or it is at least as effective as the control treatment.

To anchor the framework in the context of worst-rank score analysis, we use, as an example, a randomized clinical trial of acute ischemic stroke conducted at the Massachusetts General Hospital in Boston, Massachusetts. In this trial, a total of 85 patients who had acute ischemic stroke were randomly assigned to either room air (control therapy) or normobaric oxygen therapy (NBO), administered for 8 h. Then, the patients were assessed serially for clinical function scores including the National Institutes of Health stroke scale (NIHSS) score—a function rating scale (range from 0 to 42) used to quantify neurological deficit due to stroke—and MRI imaging (Singhal et al. 2005; Singhal 2007). The primary and efficacy endpoints were, respectively, the mean change in NIHSS scores from baseline to 4 h (during therapy) and 24 h (after therapy). To illustrate the method, we focus on the secondary endpoint to examine the change in NIHSS scores from baseline to 3 months or at discharge.

Of the 85 patients enrolled in the study, 43 were assigned to the NBO group. A total of 53 patients were discharged prior to the 3-month follow-up period and out of the 24 patients who died during the follow-up period, 17 of them were from the NBO group. Early deaths of patients precluded the measurement of their NIHSS scores at 3 months. As both death from stroke and poor 3-month NIHSS score were indicative of disease worsening, patients with missing follow-up NIHSS scores must be included in the analysis. Therefore, we consider a worst-rank composite endpoint of both death and NIHSS scores; death is considered as the worse outcome on the same scale as any measured NIHSS score.

The rest of this chapter of the book is organized as follows. In Sect. 1.2, we generalize the worst-rank endpoints where we introduce first the test proposed by Matsouaka and Betensky (2015). Then, we provide the rationale for a weighted test in the context of worst-rank endpoints. In Sect. 1.3, we determine the optimal weights for such worst-rank endpoints, while accounting for the possible presence of ties. We describe different algorithms necessary to estimate these weights using the data at hand. We present the simulation results in Sect. 1.4 and illustrate the method using the NBO trial data in Sect. 1.5. Finally, we close this chapter by discussing the merits and limitations of using weighted worst-rank endpoints and how to interpret the results.

1.2 Wilcoxon–Mann–Whitney Test for Prioritized Endpoints

1.2.1 Notation

We consider a clinical trial involving N independent patients randomized into two treatment groups of patients, with N_i patients in each group ($i = 1, 2$), and followed over a period of time T_{max}.

Let (T_{ij}, X_{ij}) denote the survival time and change in NIHSS scores for a patient j in the control ($i = 1$) or the treatment ($i = 2$) group over a follow-up time T_{max}. The observed data consist of $(T, X, \delta) = \{(T_{ij}, X_{ij}, \delta_{ij}),\ i = 1, 2; j = 1, \ldots, N_i\}$, where $\delta_{ij} = I(T_{ij} \leq T_{max})$ indicates whether the patient died before the end of follow-up time T_{max}. Since lower values of the change in NIHSS scores are worse, to define the composite worst-rank endpoints, we consider two constants $\zeta = min(X) - 1$ and $\eta = \zeta - T_{max}$, such that $\eta + T_{ij} < \zeta < X_{ij}$. The worst-rank endpoint of each patient is then defined as:

$$\widetilde{X}_{ij} = \delta_{ij}(\eta + T_{ij}) + (1 - \delta_{ij})X_{ij},\ i = 1, 2 \text{ and } j = 1, \ldots, N_i \tag{1.1}$$

The choice of η and ζ is not unique as long as when we replace T_{ij} by $\eta + T_{ij}$, we can compare observed outcomes where patients who died receive the lowest ranks compared to the survivors. In addition, the corresponding ranks of patients will reflect the relative ordering of their respective event times T_{ij} or NIHSS scores X_{ij}.

Consider F_i and G_i the cumulative distributions of the informative survival times and the NIHSS for such a patient, i.e., $F_i(v) = P(T_{ij} \leq v|0 < T_{ij} \leq T_{max})$ and $G_i(x) = P(X_{ij} \leq x|T_{ij} > T_{max})$. The distribution of the random variable $\widetilde{X} = (\widetilde{X}_{i1}, \ldots, \widetilde{X}_{iN_i})_{\{i=1,2\}}$ is given by:

$$\widetilde{G}_i(x) = p_i F_i(x - \eta) I(x < \zeta) + (1 - p_i) G_i(x) I(x \geq \zeta), \tag{1.2}$$

with $p_i = E(\delta_{ij}) = P(T_{ij} \leq T_{max})$ the probability of death before T_{max}.

We would like to test the null hypothesis of no difference between the treatment and the placebo,

$$H_o : (F_1 = F_2 \text{ and } G_1 = G_2)$$

against the unidirectional alternative hypothesis that the treatment is at least as effective as the control treatment for both mortality and the change in NIHSS scores and that it is not harmful for either treatment

$$H_1 : (F_1 \prec F_2 \text{ and } G_1 \prec G_2) \text{ or } (F_1{=}F_2 \text{ and } G_1 \prec G_2) \text{ or } (F_1 \prec F_2 \text{ and } G_1{=}G_2).$$

The symbol $\prec$ denotes the stochastic ordering of the cumulative distributions (Lachin 1999). Thus, the notation $G_1 \prec G_2$ means that $G_1(x)$ is shifted to the left of $G_2(x)$, or that the NIHSS scores for patients in the control group tend to be less than those of patients in the treatment group. In other words, there is a difference in favor of the treatment group since higher values of the change in NIHSS scores X are better.

1.2.2 *Wilcoxon–Mann–Whitney Test*

We define the Wilcoxon–Mann–Whitney U-statistic by

$$U = (N_1 N_2)^{-1} \sum_{k=1}^{N_1} \sum_{l=1}^{N_2} \left[I(\widetilde{X}_{1k} < \widetilde{X}_{2l}) + \frac{1}{2} I(\widetilde{X}_{1k} = \widetilde{X}_{2l}) \right]. \tag{1.3}$$

Theorem 1.1 *Using the worst-rank endpoints $\widetilde{X}_{ij}$ from Eq. (1.1) and $q_i = 1 - p_i$, $i = 1, 2$ and $j = 1, \ldots, N_i$, we have the following results:*

(i) the mean and variance of U under the alternative hypothesis H_1 are given by:

$$\begin{aligned}\mu &= E(U) = \pi_{U1}, \\ \sigma^2 &= Var(U) \\ &= (N_1N_2)^{-1}\Big[\pi_{U1}(1-\pi_{U1}) + (N_1-1)(\pi_{U2}-\pi_{U1}^2) \\ &\qquad + (N_2-1)(\pi_{U3}-\pi_{U1}^2)\Big]\end{aligned} \tag{1.4}$$

where

$$\begin{aligned}\pi_{U1} &= p_1p_2\pi_{t1} + p_1q_2 + q_1q_2\pi_{x1}, \\ \pi_{U2} &= p_1^2q_2 + p_1^2p_2\pi_{t2} + 2p_1q_1q_2\pi_{x1} + q_1^2q_2\pi_{x2}, \\ \pi_{U3} &= p_1q_2^2 + p_1p_2^2\pi_{t3} + 2p_1p_2q_2\pi_{t1} + q_1q_2^2\pi_{x3}, \\ \pi_{t1} &= P(T_{1k} < T_{2l} | t_{1k} \le T_{max},\ t_{2l} \le T_{max}), \\ \pi_{t2} &= P(T_{1k} < T_{2l}, T_{1k'} < T_{2l} | t_{1k} \le T_{max},\ t_{1k'} \le T_{max},\ \ t_{2l} \le T_{max}), \\ \pi_{t3} &= P(T_{1k} < T_{2l}, t_{1k} < t_{2l'} | t_{1k} \le T_{max},\ t_{2l} \le T_{max},\ \ t_{2l'} \le T_{max}), \\ \pi_{x1} &= P(X_{1k} < X_{2l}),\ \pi_{x2} = P(X_{1k} < X_{2l}, X_{1k'} < X_{2l}), \\ \pi_{x3} &= P(X_{1k} < X_{2l}, X_{1k} < X_{2l'}).\end{aligned}$$

(ii) Under the null hypothesis H_0 of no difference between the treatment groups, the mean and variance become

$$\begin{aligned}\mu_0 &= E_0(U) = \frac{1}{2}, \\ \sigma_0^2 &= Var_0(U) = \frac{N_1+N_2+1}{12N_1N_2}\end{aligned} \tag{1.5}$$

The proofs of Theorem 1.1 (i) and (ii) can be found in Appendix 1.

The asymptotic distribution of the WMW test statistic

$$Z = \frac{U - E_0(U)}{\sqrt{Var_0(U)}} \tag{1.6}$$

converges to the standard normal distribution $N(0, 1)$ as N_1 and N_2 tend to infinity, and $N_1/N_2 \longrightarrow \rho$, $0 < \rho < 1$. Its power is given by:

$$\Phi\left(\frac{\sigma_0}{\sigma}z_{\frac{\alpha}{2}} + \frac{\mu-\mu_0}{\sigma}\right) + \Phi\left(\frac{\sigma_0}{\sigma}z_{\frac{\alpha}{2}} - \frac{\mu-\mu_0}{\sigma}\right) \approx \Phi\left(\frac{\sigma_0}{\sigma}z_{\frac{\alpha}{2}} + \frac{|\mu-\mu_0|}{\sigma}\right). \tag{1.7}$$

See the proof in Matsouaka and Betensky (2015).

1.2.3 Weighted Wilcoxon–Mann–Whitney Test

In this section, we motivate a weighted WMW test by writing the WMW U-statistic (1.3) applied to the worst-rank scores (1.1) as a sum of three dependent WMW U-statistics. This allows us to demonstrate that to optimally compare two treatment groups using worst-rank scores, we need to use a weighted statistic that takes into account the dependence that exists among the three statistics.

Assume that there exist weights $\mathbf{w} = (w_1, w_2)$, $w_1 + w_2 = 1$, such that (1.1) becomes

$$\widetilde{X}_{ij} = w_1\delta_{ij}(\eta + T_{ij}) + w_2(1-\delta_{ij})X_{ij},\ i = 1,2 \text{ and } j = 1,\ldots,N. \quad (1.8)$$

The U-statistic (1.3) then becomes $U_w = w_1^2 U_t + w_1 w_2 U_{tx} + w_2^2 U_x$, where U_t, U_{tx}, and U_x are defined by:

$$\begin{aligned}
U_t &= (N_1N_2)^{-1}\sum_{k=1}^{N_1}\sum_{l=1}^{N_2}\delta_{1k}\delta_{2l}\left[I(T_{1k} < T_{2l}) + \frac{1}{2}I(T_{1k} = T_{2l})\right],\\
&\quad T_{1k} \le T_{max},\ T_{2l} \le T_{max}\\
U_{tx} &= (N_1N_2)^{-1}\sum_{k=1}^{N_1}\sum_{l=1}^{N_2}\delta_{1k}(1-\delta_{2l}) \qquad (1.9)\\
U_x &= (N_1N_2)^{-1}\sum_{k=1}^{N_1}\sum_{l=1}^{N_2}(1-\delta_{1k})(1-\delta_{2l})\left[I(X_{1k} < X_{2l}) + \frac{1}{2}I(X_{1k} = X_{2l})\right].
\end{aligned}$$

Using vector notation, we can write weighted WMW U-statistic U_w as $U_w = \mathbf{c}'\mathbf{U}$ where we define $\mathbf{U}' = (U_t, U_{tx}, U_x)$ and $\mathbf{c}' = (c_1, c_2, c_3) = (w_1^2, w_1w_2, w_2^2)$. Notice that $c_1 + 2c_2 + c_3 = (w_1 + w_2)^2 = 1$.

Theorem 1.2 *Based on the worst-rank endpoints $\widetilde{X}_{ij}$, we have*

$$\begin{aligned}
\mu_w &= E(U_w) = \mathbf{c}'\,(p_1p_2\pi_{t1}, p_1q_2, q_1q_2\pi_{x1})'\\
\sigma_w &= Var(U_w) = \mathbf{c}'\Sigma\mathbf{c},
\end{aligned}$$

where $\Sigma = Var(\mathbf{U})$ is a 3×3 matrix defined in Appendix 2.
Under the null hypothesis,

$$\begin{aligned}
\mu_{0w} &= E_0(U_w) = \frac{1}{2}\mathbf{c}'\left(p^2, 2pq, q^2\right)'\\
&= \frac{1}{2}\left[w_1^2p^2 + 2w_1w_2pq + w_2^2q^2\right] = \frac{1}{2}\left[w_1p + w_2q\right]^2\\
\sigma_{0w} &= Var_0(U_w) = \mathbf{c}'\Sigma_0\mathbf{c},
\end{aligned}$$

where $\Sigma_0 = Var_0(\mathbf{U})$ is a 3×3 matrix given in Appendix 2.
(See proof of Theorem 1.2 in Appendix 2.)
As previously, Theorem 2 allows us to define both the weighted WMW test statistic

$$Z_w = \frac{U_w - E_0(U_w)}{\sqrt{Var_0(U_w)}}. \tag{1.10}$$

It can be shown that Z_w converges to the standard normal distribution $N(0, 1)$ as N_1 and N_2 tend to infinity, and $N_1/N_2 \longrightarrow \rho, 0 < \rho < 1$.
Its corresponding power is given by:

$$\Phi\left(\frac{\sigma_{0w}}{\sigma_w} z_{\frac{\alpha}{2}} + \frac{\mu_w - \mu_{0w}}{\sigma_w}\right) + \Phi\left(\frac{\sigma_{0w}}{\sigma_w} z_{\frac{\alpha}{2}} - \frac{\mu_w - \mu_{0w}}{\sigma_w}\right)$$
$$\approx \Phi\left(\frac{\sigma_{0w}}{\sigma_w} z_{\frac{\alpha}{2}} + \frac{|\mu_w - \mu_{0w}|}{\sigma_w}\right). \tag{1.11}$$

Note that when the weights w_1 and w_2 are equal, i.e., $c_1 = c_2 = c_3 = w_1^2$, the test statistic Z_w coincides with the (ordinary) Wilcoxon–Mann–Whitney test statistic Z given in (1.6). Indeed, in that case, $\mathbf{c}'\mathbf{U} = w_1^2[U_t + U_{tx} + U_x] = w_1^2 U$ with U given by the Eq. (1.3). Thus, $\mathbf{c}' E_0(\mathbf{U}) = w_1^2 E_0(U)$ and $Var_0(\mathbf{c}'\mathbf{U}) = w_1^4 Var_0(U)$, which implies that $Z = Z_w$.

1.2.3.1 Prespecified Weights

When there are prespecified weights, usually determined as to reflect the relative importance or the severity of component outcomes, they can be used to calculate the weighted WMW test statistic Z_w. For instance, after surveying a panel of clinical investigators, Bakal et al. (2012a) used prespecified weights in a study that used composite endpoints of death, cardiogenic shock (Shock), congestive heart failure (CHF), and recurrent myocardial infarction (RE-MI). The weights were 1 for death, 0.5 for Shock, 0.3 for hospitalization for CHF, and 0.2 for RE-MI, i.e., in this context $\mathbf{w} = \frac{1}{2}(1, 0.5, 0.3, 0.2)$.

In another example Sampson et al. (2010), the composite outcome consisted of events weighted according to their severity: recurrent myocardial infarction (weight $w_1 = 0.415$), congestive heart failure that required the use of open-label angiotensin-converting enzyme (ACE) inhibitors (weight $w_2 = 0.17$), and hospitalization to treat congestive heart failure (weight $w_3 = 0.415$).

Although the use of prespecified weights provides a more nuanced approach to the importance of individual endpoints of a composite outcome, recognizes the potential underlying differences that exist among them, and facilitates the results interpretation compared to traditional composite endpoints, the selection of appropriate weights is not straightforward since inherently subjective (Ahmad et al. 2015; Sampson et al. 2010; Wilson and Berger 2011). However, when they exist, failing to use such utility (or severity) weights to highlight clinical importance of the component outcomes of a composite endpoint implies that we assume equal weights, which is sometimes even worse (Ahmad et al. 2015; Armstrong et al. 2011; Wilson and Berger 2011; Bakal et al. 2012b; Armstrong and Westerhout 2013).

It has been advocated that the method of assigning utility weights can be relatively reproducible and even refined (Armstrong et al. 2011; Bakal et al. 2015, 2012a), such refinement has been compared to a process used in baseball where a more refined analysis of a player's types of hits is taken into account to assess the value of the player beyond the simple batting average (Anstrom and Eisenstein 2011).

1.2.3.2 Optimal Weights

Now, we want to estimate the optimal weights w for the weighted WMW test statistic

$$Z_{\mathbf{c}} = \frac{\mathbf{c}'\left(\mathbf{U} - E_0(\mathbf{U})\right)}{\sqrt{Var_0(\mathbf{c}'\mathbf{U})}} = \frac{\mathbf{c}'\left(\mathbf{U} - E_0(\mathbf{U})\right)}{\sqrt{\mathbf{c}'Var_0(\mathbf{U})\mathbf{c}}}, \tag{1.12}$$

that maximizes its power, with $\mathbf{U}' = (U_t, U_{tx}, U_x)$ and $\mathbf{c}' = (c_1, c_2, c_3) = (w_1^2, w_1 w_2, w_2^2)$.

The goal of maximizing $Z_{\mathbf{c}}$ is to obtain the optimal test statistic—which we will derive from (1.12) by replacing $\mathbf{c}$ by $\mathbf{c}_{opt}$, the vector of optimal weights that maximize $Z_{\mathbf{c}}$— that encompasses the contributions of the effects of treatment on both mortality (via U_t) and the nonfatal outcome (via U_x) as well as the impact of the corresponding proportions of deaths and survivors in both treatment groups (via U_{tx}) and their relative importance and magnitude, where each component is weighted accordingly through $\mathbf{c}_{opt}$.

From the definition of $\mathbf{U}$, we show in Appendix 2 that

$$\begin{aligned} E(\mathbf{U}) &= (E(U_t), E(U_{tx}), E(U_x))' \\ &= (\pi_{t1} p_1 p_2, p_1 q_2, \pi_{x1} q_1 q_2)'. \end{aligned} \tag{1.13}$$

and $Var(\mathbf{U}) = \Sigma$, where $\Sigma = (N_1 N_2)^{-1} \left(\Sigma_{ij}\right)_{1\le i,j\le 3}$ is a 3×3 matrix.

Without loss of generality, we have restricted the variance estimations in this section to the case where there are no ties. Under the null hypothesis of no difference between the two groups, with respect to both survival and nonfatal outcome, we have $p_1 = p_2 = p$, $q_1 = q_2 = q = 1 - p$, $\pi_{t1} = \pi_{x1} = 1/2$, and $\pi_{t2} = \pi_{x2} = \pi_{t3} = \pi_{x3} = 1/3$. Thus,

$$E_0(\mathbf{U}) = \frac{1}{2}\left(p^2, 2pq, q^2\right)' \quad \text{and} \quad Var_0(\mathbf{U}) = \Sigma_0, \tag{1.14}$$

where $\Sigma_0 = (N_1N_2)^{-1}\left(\Sigma_{0ij}\right)_{1\le i,j\le 3}$ is a symmetric matrix with

$$\Sigma_{011} = \frac{p^2}{12}A(p),\ \Sigma_{012} = \frac{p^2q}{2}\left[(N_2-1)q-N_1p\right],\ \Sigma_{013} = -\frac{p^2q^2}{4}(N_2+N_1-1)$$

$$\Sigma_{022} = pq\left[1-pq+(N_2-1)q^2+(N_1-1)p^2\right],\ \Sigma_{023} = \frac{pq^2}{2}\left((N_1-1)p-N_2q\right),$$

$$\Sigma_{033} = \frac{q^2}{12}A(q),\ \text{ where } A(x) = 6+4(N_2+N_1-2)x-3(N_2+N_1-1)x^2.$$

Moreover, since $Var_0(\mathbf{U}_w) = Var_0(\mathbf{c}'\mathbf{U}) = \mathbf{c}'\Sigma_0\mathbf{c} \ge 0$ by definition, the matrix Σ_0 is semi-positive definite.

The power formula for the weighted WMW, similar to Eq. (1.7), is

$$\Phi\left(\frac{\sigma_{0w}}{\sigma_{1w}}z_{\frac{\alpha}{2}} + \frac{\mu_{1w}-\mu_{0w}}{\sigma_{1w}}\right) + \Phi\left(\frac{\sigma_{0w}}{\sigma_{1w}}z_{\frac{\alpha}{2}} - \frac{\mu_{1w}-\mu_{0w}}{\sigma_{1w}}\right) \approx \Phi\left[\frac{\sigma_{0w}}{\sigma_{1w}}\left(z_{\frac{\alpha}{2}} + \frac{|\mu_{1w}-\mu_{0w}|}{\sigma_{0w}}\right)\right], \tag{1.15}$$

where $\mu_{1w} = \mathbf{c}'E(\mathbf{U})$, $\mu_{0w} = \mathbf{c}'E_0(\mathbf{U})$, $\sigma^2_{1w} = \mathbf{c}'\Sigma\mathbf{c}$, and $\sigma^2_{0w} = \mathbf{c}'\Sigma_0\mathbf{c}$.

Theorem 1.3 *Under the assumptions that, when $N = N_1 + N_2 \to \infty$,*

(i) N_1/N_2 converges to a constant ρ $(0 < \rho < 1)$,
(ii) both $\sqrt{N}\{F_1(t) - F_2(t)\}$ and $\sqrt{N}\{G_1(x) - G_2(x)\}$ are bounded, i.e., $\frac{\sigma_{0w}}{\sigma_{1w}}$ converges to 1,

a weight-vector $\mathbf{c}$ maximizes the power of the test statistic $Z_\mathbf{c}$ if and only if it maximizes $|\mu_{1w} - \mu_{0w}|/\sigma_{0w}$. The optimal weight vector $\mathbf{c}_{opt}$ is given by:

$$\mathbf{c}_{opt} = \frac{\Sigma_0^{-1}\mu}{\mathbf{b}'\Sigma_0^{-1}\mu}, \tag{1.16}$$

for $\mathbf{b}' = (1, 2, 1)$, $\boldsymbol{\mu} = E(\mathbf{U}) - E_0(\mathbf{U}) = \left(\pi_{t1}p_1p_2 - \frac{1}{2}p^2, p_1q_2 - pq, \pi_{x1}q_1q_2 - \frac{1}{2}q^2\right)'$ and $p = (N_1p_1 + N_2p_2)/(N_1 + N_2)$.

From Theorem 1.3, we derive the weights w_1 and w_2 as:

$$w_1 = \frac{\mathbf{d_1}'\Sigma_0^{-1}\mu}{\mathbf{b}'\Sigma_0^{-1}\boldsymbol{\mu}} \text{ and } w_2 = \frac{\mathbf{d_2}'\Sigma_0^{-1}\mu}{\mathbf{b}'\Sigma_0^{-1}\boldsymbol{\mu}}, \text{ where } \mathbf{d_1}' = (1, 1, 0) \text{ and } \mathbf{d_2}' = (0, 1, 1).$$

Before we continue to explore the statistical properties of the optimal weighted WMW test statistic we just defined, we need to make the following observations:

1. The weight vector $\mathbf{c}_{opt}$ is proportional to $\Sigma_0^{-1}\mu$, which is indicative of both the magnitude and the precision of the treatment effects. Therefore, it assigns larger weights to the effects of treatment on the components of the composite endpoint that have been estimated with greater precision or are of larger magnitude (or both). Moreover, the corresponding optimal weights yield a test statistic that has minimum variance.
2. The optimal weight vector $\mathbf{c}_{opt} = \Sigma_0^{-1}\mu$ depends on unknown population parameters π_{t1}, π_{x1}, p_1, p_2, and p which must be estimated in practice. The naive approach is to plug-in the sample estimates $\pi_{t1}, \widehat{\pi}_{x1}, \widehat{p}_1, \widehat{p}_2$, and $\widehat{p}$ of these population parameters. However, by doing so we annihilate the independence assumption that underlies the existence of the corresponding weights and facilitates their derivation. Indeed, with $U_w = c_1 U_t + c_2 U_{tx} + c_2 U_x$, we derived the corresponding mean and variance using the formulas:

$$\mu_1 = E(U_w) = c_1 E(U_t) + c_2 E(U_{tx}) + c_2 E(U_x)$$

and

$$\sigma_1 = Var(U_w) = c_1^2 Var(U_t) + c_2^2 Var(U_{tx}) + c_2^2 Var(U_x) + \\ 2c_1 c_2 Cov(U_t, U_{tx}) + 2c_1 c_3 Cov(U_t, U_x) + 2c_2 c_3 Cov(U_x, U_{tx}),$$

which presupposed that the weights c_1, c_2, c_3 were constant and independent. However, when we empirically estimate $\mathbf{U}' = (U_t, U_{tx}, U_x)$, $E_0(\mathbf{U})$, and $\mathbf{c}_{opt}$ using the same data set at hand, we are introducing a dependence among these quantities and thus, our independence assumption on which hinges the derivation of $\mathbf{c}_{opt}$ does not hold anymore. Such a circular, naive approach to estimate the weights should be avoided since it is more likely to introduce bias in the assessment of the treatment effect. Therefore, we need a better method to estimate the weights in such a way that the independence is preserved in order to calculate the test statistic Z_{opt} given by Eq. (1.12):
3. Known underlying distributions

   (a) When the distributions of the primary endpoint, X, and the survival time, t, are known approximately, we can estimate analytically the probabilities π_{t1} and π_{x1}, p_1, p_2 (as we have done in Appendix 4 for our simulation studies) and calculate an estimate of the probability p under the null hypothesis (H_0) as $\hat{p} = (N_1\hat{p}_1 + N_2\hat{p}_2)/(N_1 + N_2)$ (pooled sample proportion).

4. Unknown underlying distributions

   In general, the distributions of the primary endpoint and the survival time are not known. Optimal weights are estimated using either data from a pilot study (or from previous studies, when available) or the data at hand.

   (a) If we have data from prior studies, we can leverage them to estimate these parameters. Using Bayesian methods, we can elicit expert opinions to define prior distributions associated with Σ_0 and μ that best reflect the characteristics of the disease under study and determine posterior distributions to provide a more accurate assessment of the optimal weights (Minas et al. 2012). Alternatively, if the data is structured such that we have multiple strata available (e.g., different enrollment periods or different clinical centers for patients), we can use an adaptive weighting scheme to estimate Σ_0 and μ (Fisher 1998; Ramchandani et al. 2016).

(b) In the absence of data from prior studies, it is recommended to use a bootstrap approach to estimate the weights. To do this, we generate B bootstrap samples (e.g., $B = 500, 1000$, or 2000) and, for each bootstrap sample, we estimate the corresponding optimal weight vector $\mathbf{c}_{opt}$. Then, we compute the average weights from the B estimates. Finally, using these average weights, we compute the test statistic Z_{opt} on the original sample with the average weights estimated in the first part and test the null hypothesis.

(c) With the data at hand, we can also use a K-fold cross-validation. In that regard, we divide the data into K subsets of roughly equal size and estimate the weights $\mathbf{c}_{opt,k}$ and the test statistic $Z_{opt,k}$ exactly K times. At the k-th time, $k = 1, \ldots, K$, we use the k-th subset as *validation data* to calculate the weights $\mathbf{c}_{opt,k}$ and combine the remaining $K - 1$ subsets as *training data* to estimate the test statistic $Z_{opt,k}$ using the weights defined at the validation stage. Then, we estimate the test statistic Z_{opt} by averaging over all the K test statistics Z_{opt_k}, $k = 1, \ldots, K$ and run the hypothesis test.

1.3 Simulation Studies

To assess the performance of the weighted test statistic, we conducted simulation studies generating data to follow the pattern seen in stroke trials, where the outcome of interest (patient's improvement on the NIH stroke scale score over a follow-up period of $T_{max} = 3$ months) may be missing for some patients due to death.

For $n = m = 50$, we simulated death times T_{ij} under a proportional hazards model and the nonfatal outcome X_{ij} from normal distributions, that is:

$$T_{1k} \sim Exp(\lambda_1),\ \ T_{2l} \sim Exp(\lambda_2),\ \ \text{where } q_2 = \exp(-\lambda_2 T_{max}) \text{ and } HR = \lambda_1/\lambda_2;$$

$$X_{1k} \sim N(0, 1),\ \ X_{2l} \sim N(\sqrt{2}\Delta_x, 1),\ \text{ with } \Delta_x = (\mu_{x_2} - \mu_{x_1})/(\sigma_{x_1}\sqrt{2}).$$

We considered wide range of hazard ratios, from no difference to a highly significant difference in mortality, i.e., $HR = (1.0, 1.2, 1.4, 1.6, 2.0, 2.4, 3.0)$ and chose two different mortality rates in the treatment group $p_2 = (0.6, 0.8)$. We also set the standardized difference Δ_x to 0.0, 0.1, 0.2, 0.3, 0.4, 0.5, and 0.6. The conditional probabilities, $\pi_{t\gamma}$ and $\pi_{x\gamma}$, $\gamma = 1, 2, 3$, are given in Appendix 4.

Knowing the true underlying distributions of T_{ij} and X_{ij}, we calculated the conditional probabilities, the optimal weights, and the power for the weighted WMW test using the analytical power formula (1.15) for a two-sided $\alpha = 0.05$. In addition, we simulated 10,000 data sets and run both the optimal weighted WMW test and the ordinary WMW. This allowed us to estimate empirically the powers of these two tests by counting how many times we rejected the null hypothesis of no treatment difference out of the 10,000 simulated data sets. We compared the results on the basis of the type I error, the similitude between the empirical and analytical powers, the improvement in using the weighted WMW test as opposed to the ordinary WMW test. Our objective in running the ordinary WMW was to illustrate their differences and similarities and highlight the importance to consider optimally weighted WMW test when dealing with prioritized outcomes.

The results, given in Table 1.1, illustrate the accuracy of the analytical power formula (1.15). As we expected, the power of the test depends on the rate of mortality as well as the difference of treatment effects on mortality and the nonfatal outcome. As the rate of mortality increases, the power also increases under either the weighted or ordinary WMW test. For instance, with the same standardized difference and hazard ratio, we have more power to detect the difference between the treatment and the control when mortality rate is higher. Furthermore, the results regarding the analytical power formula and the empirical power formula are similar throughout the different hazard ratios and the standard differences.

The results given in Table 1.1 also indicate that the weighted WMW test statistic is more powerful than the ordinary WMW test for the worst-rank score composite outcome. The difference between the tests is more remarkable in two following cases:

1. the standardized difference in the nonfatal outcome Δ_x is small ($\Delta_x < 0.3$) and the difference in mortality is moderate or high (HR ≥ 1.2)
2. the difference in mortality is small (HR< 1.2) and the standard difference in the nonfatal outcome Δ_x is moderate or high ($\Delta_x \geq 0.3$).

Overall, these results indicate that whenever the effect of treatment on the primary, non-mortality outcome is small, the larger difference in mortality that could have been captured by the weighted WMW test is somewhat attenuated when assessing the overall difference through the ordinary WMW instead, where mortality and the nonfatal outcome are weighted equally. Likewise, if the difference in mortality is small, but the difference in the nonfatal outcome is moderate or high, the ordinary WMW test on the composite outcome has less power compared to the weighted WMW.

1.4 Application to a Stroke Clinical Trial

Matsouaka et al. (2016) reanalyzed data from a clinical trial of normobaric oxygen therapy (NBO) for acute ischemic stroke patients Singhal (2006, 2007). In this trial, 85 patients were randomly assigned either to NBO therapy (43 patients) or to room

Table 1.1 Power comparisons for a continuous outcome under proportional hazards for time to death

	Mortality rate under treatment= 40%							Mortality rate under treatment= 20%						
	HR							HR						
Δ_x	1.0	1.2	1.4	1.6	2.0	2.4	3.0	1.0	1.2	1.4	1.6	2.0	2.4	3.0
(a) Analytical power for the weighted WMW test														
0.0	0.05	0.11	0.24	0.41	0.73	0.90	0.98	0.05	0.08	0.15	0.24	0.45	0.68	0.87
0.1	0.08	0.12	0.25	0.42	0.73	0.90	0.98	0.09	0.12	0.18	0.28	0.51	0.70	0.88
0.2	0.15	0.19	0.30	0.46	0.75	0.91	0.98	0.21	0.24	0.30	0.38	0.58	0.75	0.90
0.3	0.27	0.30	0.40	0.53	0.78	0.92	0.98	0.39	0.41	0.46	0.53	0.69	0.82	0.93
0.4	0.41	0.44	0.51	0.61	0.82	0.93	0.98	0.59	0.61	0.64	0.69	0.79	0.88	0.95
0.5	0.55	0.57	0.62	0.70	0.86	0.94	0.99	0.76	0.77	0.79	0.81	0.87	0.92	0.97
0.6	0.68	0.68	0.72	0.77	0.89	0.95	0.99	0.88	0.88	0.89	0.90	0.93	0.96	0.98
(b) Empirical power for the weighted WMW test														
0.0	0.05[a]	0.10	0.23	0.40	0.72	0.91	0.99	0.05[a]	0.08	0.15	0.24	0.45	0.67	0.87
0.1	0.08	0.12	0.24	0.41	0.73	0.90	0.99	0.09	0.12	0.18	0.28	0.51	0.70	0.89
0.2	0.15	0.19	0.29	0.47	0.75	0.91	0.99	0.21	0.24	0.30	0.38	0.58	0.76	0.91
0.3	0.26	0.30	0.40	0.53	0.78	0.92	0.99	0.39	0.41	0.46	0.54	0.69	0.83	0.94
0.4	0.39	0.43	0.51	0.63	0.81	0.93	0.99	0.59	0.61	0.65	0.71	0.81	0.89	0.96
0.5	0.54	0.56	0.63	0.71	0.87	0.94	0.99	0.76	0.78	0.81	0.83	0.90	0.94	0.98
0.6	0.67	0.68	0.73	0.79	0.89	0.96	0.99	0.89	0.89	0.91	0.92	0.95	0.97	0.99

(continued)

Table 1.1 (continued)

	Mortality rate under treatment= 40%							Mortality rate under treatment= 20%						
	HR							HR						
Δ_x	1.0	1.2	1.4	1.6	2.0	2.4	3.0	1.0	1.2	1.4	1.6	2.0	2.4	3.0
	(c) Empirical power for the ordinary WMW test													
0.0	0.05	0.09	0.17	0.31	0.62	0.84	0.98	0.05	0.06	0.09	0.13	0.29	0.48	0.74
0.1	0.06	0.12	0.22	0.38	0.67	0.87	0.98	0.08	0.11	0.17	0.24	0.42	0.61	0.82
0.2	0.07	0.16	0.30	0.44	0.74	0.90	0.99	0.14	0.21	0.29	0.37	0.56	0.72	0.89
0.3	0.12	0.22	0.37	0.53	0.78	0.93	0.99	0.26	0.33	0.43	0.53	0.70	0.83	0.94
0.4	0.16	0.29	0.44	0.59	0.82	0.94	0.99	0.40	0.50	0.59	0.66	0.81	0.89	0.96
0.5	0.22	0.36	0.52	0.66	0.86	0.96	0.99	0.57	0.66	0.73	0.79	0.89	0.95	0.98
0.6	0.30	0.44	0.59	0.70	0.88	0.97	0.99	0.71	0.78	0.84	0.88	0.94	0.97	0.99

1. Power estimated based on: (a) the analytical formula (1.15), the proportion of simulated data sets for which (b) the weighted and (c) the ordinary WMW test statistics $|Z_{opt}| > 1.96$ and $|Z| > 1.96$

2. We assumed the treatment to be at least as good as the control on mortality or the nonfatal outcome; the survival times follow exponential distributions and the nonfatal outcome normal distributions. The number of patients is the same in each treatment group ($n_1 = n_2 = 50$)

3. Δ_x: standardized mean difference on the nonfatal outcome of interest; HR: hazard ratio; q_2: survival probability (proportion of patients alive) at 3 months in the treatment group

[a]The weights are equal and fixed to 1

air (control) for 8 h and assessed serially with clinical function scores. During the follow-up time of 3 months, 24 patients died (with seven from the control group) and 53 were discharged (among which 31 were in the control group). The primary efficacy and safety endpoints were, respectively, the mean change in NIHSS from baseline to 4 h (during therapy) and 24 h (after therapy) (Singhal 2006).

For illustration purposes, we focused on one of the secondary endpoints to examine the mean change in NIHSS scores from baseline to 3 months or at discharge. The log rank test of survival was significant ($\chi^2 = 6$ with 1 d.f., p-value $= 0.016$), indicating that the active treatment had an unfavorable effect on mortality. The ordinary WMW test applied to the survivors was not significant ($W = 572.5$, p-value $= 0.27$). However, using the same test on the worst-rank composite endpoint of death times and NIHSS scores, we found a significant difference between the two treatment groups ($W = 1112.5$, p-value $= 0.01$), mostly driven by the difference in mortality.

Finally, we applied our proposed method, the weighted WMW test, to estimate the weights and calculate the test statistic Z_w using B=2000 bootstrap samples. We obtained the estimated weight vector $\mathbf{c}'_{opt} = (0.45, 0.16, 0.24)$, the mean difference $\mu = -(0.016, 0.098, 0.073)$, the probability $p = 0.283$, and the variance–covariance matrix for U under the null probability

$$\Sigma_0 = \begin{pmatrix} 0.59 & 0.50 & -0.90 \\ 0.50 & 4.77 & -1.27 \\ -0.90 & -1.27 & 5.16 \end{pmatrix}.$$

The optimal weights $\mathbf{c}'_{opt} = (0.45, 0.16, 0.24)$ lead to the weights of the component outcomes of $w_1 = 0.61$ and $w_2 = 0.39$; meaning that mortality was weighted more heavily (61% of the weight) than NIHSS score, in addition to ranking death worse than any measure of the continuous outcome (NIHSS score).

The optimally weighted WMW test statistic Z_{opt} was equal to 3.42 with a corresponding p-value of 6.2×10^{-4}. The weighted WMW test statistic gave a clear result that was highly significant (than the result from the ordinary WMW test) as it optimally captured the significant difference in mortality between the two treatment groups and thus demonstrated its efficiency. The test provided strong evidence that in this specific trial, the performance of the NBO trial fell short of the room air and did not deliver on his promising results from the pilot study (Singhal et al. 2005).

1.5 Discussion

In this chapter, we have generalized the Wilcoxon–Mann–Whitney (WMW) test for a worst-rank composite outcome by deriving the optimally weighted WMW test. The weighted WMW test assigns weights to different components of the worse rank composite endpoint that maximize the power of the test. We have motivated the worst-rank composite outcome in the context of a randomized clinical trial of a non-mortality primary outcome, where the assessment of the primary outcome of interest at a prespecified time point may be precluded by death, any other debilitating event, or worsening of the disease condition. The corresponding composite outcome takes into account all patients enrolled in the trial, including those who had terminal events before the end of follow-up.

When there exists a hierarchy of the constituent outcomes of a composite endpoint, the method we have presented in this chapter enables these components to be weighted differentially. Using weights allows for an additional level of discrimination between the component outcomes beyond the prespecified outcome ranks alone, which incorporate their individual contributions to the overall treatment effect. While the worst-rank score mechanism pertains with how the different component outcomes of the composite endpoint are aggregated, assigning weights strengthens (or lessens) the influence these prioritized component outcomes exert on the overall composite. Although we have considered the possibility of using weights obtained or elicited from expert judgments (utility weights), this chapter focused on weights that are determined in a way that the corresponding WMW test statistic has a maximum power.

Therefore, based on a U-statistic method, we first provided the test statistic and the power of the weighted WMW test when utilities (or severity) weights, determined *a priori*, are available. In addition, we demonstrated that the ordinary (unweighted) WMW test on the worst-rank score outcome is a special case of the weighted WMW test, i.e., when the weights are all equal. Then, we derived the optimal weights such that the power of the corresponding weighted WMW test statistic is maximal. Finally, we conducted simulation studies to evaluate the accuracy of our power formula and confirmed, in the process, that the weighted WMW is more powerful than ordinary WMW test.

We applied the proposed method to the data from a clinical trial of normobaric oxygen therapy (NBO) for patients with acute ischemic stroke. Patients' improvement was assessed using the National Institutes of Health Stroke Scale (NIHSS) Scores. Against the null hypothesis of no difference on both mortality and continuous endpoint, we have focused on the alternative hypothesis that "the active treatment has a preponderance of positive effects on the multiple outcomes considered, while not being harmful for any" (Lachin and Bebu 2015). The results indicated a statistically significant difference between NBO therapy and room air—using either the proposed method or the ordinary WMW test on the worst-rank composite outcome of death and change in NIHSS—which we could not detect using the ordinary WMW on the survivors alone.

The difference between NBO therapy and room air was driven by the difference in mortality since there was a disproportionate number of NBO-treated patients who died. It is actually for this reason the trial was stopped by the Data and Safety Monitoring Board (DSMB) after 85 patients out of the projected 240 were enrolled. The stark imbalance between the two treatment group, although not attributed to the treatment, made it untenable to continue the trial (Singhal 2006; Samson 2013).

The end result of the NBO trial is one of the dreaded scenarios in the (traditional) analysis of composite endpoints. That the active treatment must be better than the control for one or both of the constituent outcomes (mortality and nonfatal outcome) and not worse for either of them as suggested by our alternative hypothesis, H_1, was clearly not the case for the NBO trial. While the active treatment was equivalent to the control treatment in change in NIHSS, the data showed also that NBO therapy increased mortality. Ideally, components of a composite endpoint should have similar clinical importance, frequency, and treatment effect. However, this is rarely the case as outcomes of different levels of severity are usually combined to facilitate the interpretation of such results, several authors have suggested running complementary analyses on components of the composite outcome (Freemantle et al. 2003; Cordoba et al. 2010; Tomlinson and Detsky 2010; Ferreira-Gonzalez et al. 2009; Ferreira-González et al. 2007a,b; Lubsen et al. 1996; Lubsen and Kirwan 2002).

When the impact of the active treatment on mortality is of greater clinical importance than its effect on the primary outcome of interest, the weighted WMW test statistic we have presented can be included into a set of testing procedures that ensure that the treatment is not inferior on both mortality and the outcome of interest and that it is superior on a least one of these endpoints. In the context of ischemic stroke, the clinical investigators desired a treatment that would have a positive impact on mortality while also improving survivors' functional outcomes. Testing procedures that incorporate contributions of each individual component of the composite while penalizing for any disadvantage in the active treatment when the treatment operates in opposite directions on the components of the composite outcome have been discussed (Huque et al. 2011; Mascha and Turan 2012; Dmitrienko et al. 2013; Sankoh et al. 2014).

For the analysis of NBO clinical trial, we propose two different stepwise procedures to analyze data using this weighted test: (1) two individual non-inferiority tests on mortality and nonfatal outcome, followed (if non-inferiority established) by a global test using the optimal weighted WMW test on the worst-rank composite endpoint; or (2) a global test using the optimal weighted WMW test on the worst-rank composite endpoint, then (if significant global test) two individual non-inferiority tests followed by individual superiority tests on mortality and nonfatal outcome. In either scenario, the overall type I error is preserved (Mascha and Turan 2012; Logan and Tamhane 2008; Röhmel et al. 2006; Huque et al. 2011).

The method presented in this chapter can be applied or extended to many other settings of composite endpoints beyond the realm of death-censored observations. The rationale, advantages (and limitations), and recommendations for using composite outcomes—based on clinical information, expert knowledge, or practical matters—abound in the literature (Moyé 2003; Gómez and Lagakos 2013; Ferreira-González et al. 2007a). One can also accommodate ties as well as non-informative censoring in the definition of the WMW U-statistic (see Matsouaka and Betensky 2015). In particular, when non-informative censoring is present (and, without loss of generality, assuming that there are no ties), survival times can be assessed using

Gehan's U-statistic, which is an extension of the WMW U-statistic to right censored data (Gehan 1965). In this case, $I(t_{1k} < t_{2l})$ will be equal to 1 if subject l in group 2 lived longer than subject k in group 1 and 0 if it is uncertain which subject lived longer.

Our proposed method can be applied in many disease areas in which different outcomes are clinically related and represent the manifestation of the same underlying condition. Clinical trials of unstable angina and non-ST segment elevation myocardial infarction are examples of such an application (Braunwald et al. 2002; Grech and Ramsdale 2003). The method can also be applied in clinical trials where the overall effect of treatment on a disease depends on hierarchy of meaningful—yet of different importance, magnitude, and impact—heterogenous outcomes. For instance, in clinical trials of asthma or of benign prostatic hyperplasia (BPH), several outcomes are necessary to capture the multifaceted manifestations of the disease. For patients with asthma, four outcomes (forced expiratory volume in 1 s (FEV_1), peak expiratory flow (PEF) rate, symptom score, and additional rescue medication use) are necessary to measure the different manifestations of the disease (National Asthma Education and Prevention Program (National Heart, Lung, and Blood Institute) 2007). Due to subjective nature of benign prostatic hyperplasia (BPH) symptoms, in addition to BPH symptom score index, measures to assess disease progression include: prostate-specific antigen (PSA), urinary cytology, post-void residual volume (PVR), urine flow rate, cystoscopy, urodynamic pressure-flow study, and ultrasound of the kidney or the prostate.

Our method does not immediately apply to the case where the treatment effect is assessed by stratifying for a confounding variable (baseline scores, baseline disease severity, age, ...) prespecified in the study design (Van Elteren 1960; Zhao 2006; Kawaguchi et al. 2011). For the NBO trial, had the investigators anticipated the imbalance between subjects on some baseline variables (e.g., large infarcts, advanced age, comorbidities, and most importantly, withdrawal of care based on pre-expressed wishes or family preference), they could have stratified the study population with respect to these variables (Singhal 2006; Samson 2013). The test statistic we have proposed does not adjust for such baseline covariates as the appropriate weighted WMW test for this case must take into account the stratum-specific characteristics in addition to the specificities of the worst-ranking procedure; this is a topic for future investigations.

A strong case may be made on why one should prefer analysis of covariance to the analysis of change from baseline score as we have done in this chapter Senn (2006). In reality, however, issues are more nuanced and the approach to use depends closely on the nature of the data as well as the clinical question of interest Fitzmaurice (2001); van Breukelen (2013); Shahar and Shahar (2012); Pearl (2014); Oakes and Feldman (2001); Willett (1988). For the difference in NIHSS scores (from baseline to 3 months) used in this chapter as outcome of interest, the fundamental question of interest was "on average, how much did the NBO-treated patients change over 3-month period compared to patients assigned to room air?" The change-from-baseline-score paradigm assumes that the same measure is used before and after the treatment and that these two measures are highly correlated Bonate (2000); Campbell and Kenny (1999).

In the stroke literature, it is proven that change from baseline in NIHSS satisfies this assumption since baseline NIHSS is a strong predictor of outcome after stroke (Young et al. 2005; Adams Jr et al. 1999). Moreover, it has been shown that change in the NIHSS score is a useful tool to measure treatment effect in acute stroke trials (see, for instance, the chapters by Bruno et al. (2006) and by Parsons et al. 2012). Hence, this justified the choice of improvement (or change) in NIHSS score as outcome of interest in this chapter.

We have assumed throughout this chapter that mortality is worse than any impact ischemic stroke may have on patients. Our assumption stems from the common view that ranks death as inferior to any quality-of-life measure; such a view is advocated in several medical fields (Follmann et al. 1992; Brittain et al. 1997; Felker et al. 2008; Felker and Maisel 2010; Allen et al. 2009; Sun et al. 2012; Subherwal et al. 2012; Berry et al. 2013). However, some people (patients, their family members, or caregivers) may argue otherwise and affirm that there are levels of stroke that are worse than death. For instance, in a study of the effects of thrombolytic therapy in reducing damage from a myocardial infarction, the hierarchy of the quality of component outcomes was "stroke resulting in a vegetative state, death, serious morbidity requiring major assistance, serious morbidity but capable of self-care, excess spontaneous hemorrhage ($\geq$ 3 blood transfusions), and 1–2 transfusions" (Hallstrom et al. 1992). There are a number of chapters in the causal inference literature that offer an alternative approach based on Rosenbaum's proposal of using different "placements of death" (Rosenbaum 2006). However, as Rubin pointed out, this elegant idea "maybe difficult to convey to consumers" (Rubin 2006) and we have not pursued this avenue here.

Finally, the null hypothesis H_0 for WMW test stipulates that the treatment does not change the outcome distribution, which means that the treatment has no effect on any patient. However, some studies may require a weaker version of the null hypothesis, i.e., the treatment does not affect the average group response (Fay and Proschan 2010; Gail et al. 1996). In such a case, the WMW is not an asymptotically valid test for the weaker null hypothesis (Pratt 1964; Chung and Romano 2016). As an alternative, one can use the Brunner–Munzel test (Brunner and Munzel 2000) where the marginal distribution functions of the two treatment groups are not assumed to be equal and may have different shapes, even under the null hypothesis. The use of a weighted Brunner–Munzel test for analysis of the worst-rank composite outcome of death and a quality-of-life (such as the NIHSS score) warrants further investigations and is beyond the scope of this chapter. In this chapter, we have chosen the WMW test because it is simple, widely used, efficient, and robust against parametric distributional assumptions. It allows to focus on how we leverage the contribution and the variation of each component of the composite outcome in the data at hand to better capture the overall treatment effect by assigning optimal weights accordingly.

Acknowledgements This work was supported by grants P50-NS051343, R01-CA075971, T32 NS048005, 1RO1HL118336-01, and UL1TR001117 awarded by the National Institutes of Health. The content of this paper is solely the responsibility of the authors and does not necessarily represent the official view of the National Institutes of Health.
Conflict of Interest: None declared.

Appendix 1: Proof of Theorem 1.1

We consider $\widetilde{X}_{ij} = \delta_{ij}(\eta + T_{ij}) + (1 - \delta_{ij})X_{ij}$, $i = 1, 2$ and $j = 1, \ldots, N_i$

$$\begin{aligned}
I(\widetilde{X}_{1k} < \widetilde{X}_{2l}) &= \delta_{1k}\delta_{2l}I(T_{1k} < T_{2l}|T_{1k} \leq T_{max},\ T_{2l} \leq T_{max}) + \delta_{1k}(1 - \delta_{2l}) \\
&\quad + (1 - \delta_{1k})(1 - \delta_{2l})I(X_{1k} < X_{2l}), \\
I(\widetilde{X}_{1k} = \widetilde{X}_{2l}) &= \delta_{1k}\delta_{2l}I(T_{1k} = T_{2l}|T_{1k} \leq T_{max},\ T_{2l} \leq T_{max}) \\
&\quad + (1 - \delta_{1k})(1 - \delta_{2l})I(X_{1k} = X_{2l}).
\end{aligned}$$

For $q_i = 1 - p_i$, we have

$$\begin{aligned}
\mu &= E(U) \\
&= E(U_{kl}) = E\left[I(\widetilde{X}_{1k} < \widetilde{X}_{2l}) + \frac{1}{2}I(\widetilde{X}_{1k} = \widetilde{X}_{2l})\right] \\
&= p_1p_2\left[P(T_{1k} < T_{2l}|T_{1k} \leq T_{max},\ T_{2l} \leq T_{max})\right. \\
&\qquad\qquad \left. + \frac{1}{2}P(T_{1k} = T_{2l}|T_{1k} \leq T_{max},\ T_{2l} \leq T_{max})\right] \\
&\quad + p_1q_2 + q_1q_2\left[P(X_{1k} < X_{2l}) + \frac{1}{2}P(X_{1k} = X_{2l})\right] \\
&= p_1p_2\pi_{t1} + p_1q_2 + q_1q_2\pi_{x1} \equiv \pi_{U1},
\end{aligned}$$

where

$$\begin{aligned}
\pi_{t1} &= P(T_{1k} < T_{2l}|T_{1k} \leq T_{max},\ T_{2l} \leq T_{max}) \\
&\quad + \frac{1}{2}P(T_{1k} = T_{2l}|T_{1k} \leq T_{max},\ T_{2l} \leq T_{max}) \\
\pi_{x1} &= P(X_{1k} < X_{2l}) + \frac{1}{2}P(X_{1k} = X_{2l}).
\end{aligned}$$

Define $U_{kl} = I(\widetilde{X}_{1k} < \widetilde{X}_{2l}) + \frac{1}{2}I(\widetilde{X}_{1k} = \widetilde{X}_{2l})$, for $k = 1, \ldots, N_1$ and $l = 1, \ldots, N_2$. The binary variable $U_{kl} = I(\widetilde{X}_{1k} < \widetilde{X}_{2l})$ follows Bernoulli distribution with probability π_{U1}. Its mean and variance, respectively, $E(U_{kl}) = \pi_{U1}$ and $Var(U_{kl}) = E(U_{kl})[1 - E(U_{kl})] = \pi_{U1}(1-\pi_{U1})$. Thus, we can use these results to derive the variance of U using the following formula:

$$Var(U) = (N_1N_2)^{-2}\left[\sum_{k=1}^{N_1}\sum_{l=1}^{N_2} Var(U_{kl}) + \sum_{k=1}^{N_1}\sum_{l=1}^{N_2}\sum_{k'=1}^{N_1}\sum_{l'=1}^{N_2} Cov(U_{kl}, U_{k'l'})\right]$$

$$= (N_1N_2)^{-1}\left[Var(U_{kl}) + (N_1 - 1)Cov(U_{kl}, U_{k'l}) + (N_2 - 1)Cov(U_{kl}, U_{kl'})\right].$$

Note that when $k \neq k'$ and $l \neq l'$, the covariance

$$Cov(U_{kl}, U_{k'l'}) = E(U_{kl}U_{k'l'}) - E(U_{kl})E(U_{k'l'}) = 0.$$

When $k \neq k'$ or $l \neq l'$, we have

$$Cov(U_{kl}, U_{k'l}) = E(U_{kl}U_{k'l}) - E(U_{kl})E(U_{k'l}) = \pi_{U2} - \pi_{U1}^2;$$

$$Cov(U_{kl}, U_{kl'}) = E(U_{kl}U_{kl'}) - E(U_{kl})E(U_{kl'}) = \pi_{U3} - \pi_{U1}^2.$$

where $\pi_{U2} = E(U_{kl}U_{k'l})$ and $\pi_{U3} = E(U_{kl}U_{kl'})$.

Therefore,

$$Var(U) = (N_1N_2)^{-1}\Big[\pi_{U1}(1 - \pi_{U1}) + (N_1 - 1)(\pi_{U2} - \pi_{U1}^2) + (N_2 - 1)(\pi_{U3} - \pi_{U1}^2)\Big]$$

1. **No ties:**
   When there are no ties, $I(\widetilde{X}_{1k} = \widetilde{X}_{2l}) = 0$. In which case, $U_{kl} = I(\widetilde{X}_{1k} < \widetilde{X}_{2l}) = \delta_{1k}\delta_{2l}I(T_{1k} < T_{2l}|T_{1k} \leq T_{max},\ T_{2l} \leq T_{max}) + \delta_{1k}(1 - \delta_{2l}) + (1 - \delta_{1k})(1 - \delta_{2l})I(X_{1k} < X_{2l})$, for $k = 1, \ldots, N_1$ and $l = 1, \ldots, N_2$,. We have

$$\begin{aligned}E(U_{kl}U_{k'l}) &= P(T_{1k} < T_{2l}, T_{1k'} < T_{2l}|\delta_{1k}\delta_{1k'}\delta_{2l} = 1)E(\delta_{1k}\delta_{1k'}\delta_{2l} = 1)\\ &+ P(X_{1k'} < X_{2l})P(\delta_{1k} = 1,\ \delta_{1k'} = \delta_{2l} = 0)\\ &+ P(X_{1k} < X_{2l})E(\delta_{1k} = \delta_{2l} = 0)E(\delta_{1k'} = 1)\\ &+ P(X_{1k} < X_{2l}, X_{1k'} < X_{2l})E(\delta_{1k} = \delta_{1k'} = \delta_{2l} = 0)\\ &+ E(\delta_{1k}\delta_{1k'} = 1)E(\delta_{2l} = 0)\\ &= p_1^2p_2\pi_{t2} + 2p_1q_1q_2\pi_{x1} + q_1^2q_2\pi_{x2} + p_1^2q_2 \equiv \pi_{U2}\end{aligned}$$

$$E(U_{kl}U_{kl'}) = P(T_{1k} < T_{2l},\ t_{1k} < t_{2l'}|\delta_{1k}\delta_{2l}\delta_{2l'} = 1)E(\delta_{1k}\delta_{2l}\delta_{2l'} = 1)$$

$$
\begin{aligned}
&+ P(T_{1k} < T_{2l}|\delta_{1k}\delta_{2l} = 1,\ \delta_{2l'} = 0)E(\delta_{1k}\delta_{2l} = 1)E(\delta_{2l'} = 0)\\
&+ P(t_{1k} < t_{2l'}|\delta_{1k} = 1,\ \delta_{2l} = 0,\ \delta_{2l'} = 1)E(\delta_{1k}\delta_{2l'} = 1)E(\delta_{2l} = 0)\\
&+ P(X_{1k} < X_{2l}, X_{1k} < X_{2l'})E(\delta_{1k} = \delta_{2l} = \delta_{2l'} = 0)\\
&+ E(\delta_{1k} = 1)E(\delta_{2l} = \delta_{2l'} = 0)\\
&= p_1p_2^2\pi_{t3} + 2p_1p_2q_2\pi_{t1} + q_1q_2^2\pi_{x3} + p_1q_2^2 \equiv \pi_{U3}
\end{aligned}
$$

with
$$
\begin{aligned}
\pi_{t2} &= P(T_{1k} < T_{2l}, T_{1k'} < T_{2l}|T_{1k} \le T_{max},\ T_{1k'} \le T_{max},\ T_{2l} \le T_{max}),\\
\pi_{x2} &= P(X_{1k} < X_{2l}, X_{1k'} < X_{2l}),\\
\pi_{t3} &= P(T_{1k} < T_{2l}, t_{1k} < t_{2l'}|T_{1k} \le T_{max},\ T_{2l} \le T_{max},\ T_{2l'} \le T_{max}),\\
\pi_{x3} &= P(X_{1k} < X_{2l}, X_{1k} < X_{2l'}).
\end{aligned}
$$

Under the null hypothesis of no difference between the two groups, with respect to survival and nonfatal outcome, we have $F_1 = F_2 = F$, $G_1 = G_2 = G$ and $p_1 = p_2 = p$, $q_1 = q_2 = q$. This implies

$$
\begin{aligned}
\pi_{t1} &= P(T_{1k} < T_{2l}|T_{1k} \le T_{max}, T_{2l} \le T_{max})\\
&= \frac{1}{2p^2}\left[F(T_{max})^2 - F(0)^2\right] = \frac{1}{2}\\
\pi_{t2} &= P(T_{1k} < T_{2l}, T_{1k'} < T_{2l}|T_{1k} \le T_{max}, T_{1k'} \le T_{max}, T_{2l} \le T_{max})\\
&= \frac{1}{p^3}\int_0^{T_{max}} F(t)^2 dF(t)\\
&= \frac{1}{3p^3}\left[F(T_{max})^3 - F(0)^3\right] = \frac{1}{3}\\
\pi_{t3} &= P(T_{1k} < T_{2l}, T_{1k} < T_{2l'}|T_{1k} \le T, T_{2l} \le T, T_{2l'} \le T))\\
&= \frac{1}{p^3}\int_0^{T_{max}} [1 - F(t)]^2\, dF(t)\\
&= \frac{1}{3p^3}\left\{[1 - F(T_{max})]^3 - [1 - F(0)]^3\right\} = \frac{1}{3}\\
\pi_{x1} &= P(X_{1k} < X_{2l}) = \int_{-\infty}^{\infty} G(x)dG(x) = \frac{1}{2}\left[G(x)^2\right]_{-\infty}^{\infty} = \frac{1}{2}\\
\pi_{x2} &= P(X_{1k} < X_{2l}, X_{1k'} < X_{2l}) = \int_{-\infty}^{\infty} G(t)^2 dG(t) = \frac{1}{3}\left[G(x)^3\right]_{-\infty}^{\infty} = \frac{1}{3}\\
\pi_{x3} &= P(X_{1k} < X_{2l}, X_{1k} < X_{2l'}) \int_{-\infty}^{\infty} [1 - G(t)]^2 dG(t)\\
&= -\frac{1}{3}\left\{[1 - G(x)]^3\right\}_{-\infty}^{\infty} = \frac{1}{3}.
\end{aligned}
$$

Therefore,

$$
\begin{aligned}
\pi_{U1} &= p_1 p_2 \pi_{t1} + p_1 q_2 + q_1 q_2 \pi_{x1} \\
&= \frac{1}{2}p^2 + pq + \frac{1}{2}q^2 = \frac{1}{2}(p+q)^2 = \frac{1}{2} \\
\pi_{U2} &= p_1^2 q_2 + p_1^2 p_2 \pi_{t2} + 2 p_1 q_1 q_2 \pi_{x1} + q_1^2 q_2 \pi_{x2} \\
&= p^2 q + \frac{1}{3}p^3 + pq^2 + \frac{1}{3}q^3 = \frac{1}{3}(p+q)^3 = \frac{1}{3} \\
\pi_{U3} &= p_1 q_2^2 + p_1 p_2^2 \pi_{t3} + 2 p_1 p_2 q_2 \pi_{x1} + q_1 q_2^2 \pi_{x3} \\
&= pq^2 + \frac{1}{3}p^3 + p^2 q + \frac{1}{3}q^3 = \frac{1}{3}(p+q)^3 = \frac{1}{3}.
\end{aligned}
$$

The mean and variance become

$$
\begin{aligned}
\mu_0 &= E_0(U) = \pi_{U1} = \frac{1}{2}; \\
\sigma_0^2 &= Var_0(U) \\
&= (N_1 N_2)^{-1} \Big[\pi_{U1} \left(1 - \pi_{U1}\right) + (N_1 - 1)\left(\pi_{U2} - \pi_{U1}^2\right) \\
&\qquad + (N_2 - 1)\left(\pi_{U3} - \pi_{U1}^2\right) \Big] \\
&= (N_1 N_2)^{-1} \left[\frac{1}{2}\left(1 - \frac{1}{2}\right) + (N_1 - 1)\left(\frac{1}{3} - \left(\frac{1}{2}\right)^2\right) \right. \\
&\qquad \left. + (N_2 - 1)\left(\frac{1}{3} - \left(\frac{1}{2}\right)^2\right) \right] \\
&= (N_1 N_2)^{-1} \left[\frac{1}{4} + \frac{1}{12}(N_1 - 1) + \frac{1}{12}(N_2 - 1) \right] = \frac{N_1 + N_2 + 1}{12 N_1 N_2}.
\end{aligned}
$$

2. **Ties are present:** More generally, we can approximate the probabilities $\pi_{U2} = E(U_{kl} U_{k'l})$ and $\pi_{U3} = E(U_{kl} U_{kl'})$ using their unbiased estimators.

   Following Hanley and McNeil (1982), we can show that the variance $Var(U)$ can be estimated by:

$$
(N_1 N_2)^{-1} \left[\widehat{\pi}_{U1} \left(1 - \widehat{\pi}_{U1}\right) + (N_1 - 1)(\widehat{\pi}_{U2} - \widehat{\pi}_{U1}^2) + (N_2 - 1)(\widehat{\pi}_{U3} - \widehat{\pi}_{U1}^2) \right]
$$

where $\widehat{\pi}_{U1} = (N_1N_2)^{-1}\sum_{k=1}^{N_1}\sum_{l=1}^{N_2} U_{kl}$, $\widehat{\pi}_{U2} = (N_1N_2^2)^{-1}\sum_{k=1}^{N_1} U_{k\bullet}^2$, and $\widehat{\pi}_{U3} = (N_1^2N_2)^{-1}\sum_{l=1}^{N_2} U_{\bullet l}^2$. In absence of ties, $\widehat{\pi}_{U2}$ and $\widehat{\pi}_{U3}$ are, respectively, estimates of π_{U3} and π_{U3}.

One can also consider other possible approximations of the variance of U using the exposition provided by Newcombe (2006).

As we know,

$$P(\widetilde{X}_{1k} < \widetilde{X}_{2l}) + P(\widetilde{X}_{1k} > \widetilde{X}_{2l}) + P(\widetilde{X}_{1k} = \widetilde{X}_{2l}) = 1.$$

Under the null hypothesis, i.e., $\widetilde{X}_{1k}$ and $\widetilde{X}_{2l}$ are identically distributed, we have $P(\widetilde{X}_{1k} < \widetilde{X}_{2l}) = P(\widetilde{X}_{1k} > \widetilde{X}_{2l})$ which implies $P(\widetilde{X}_{1k} < \widetilde{X}_{2l}) + \frac{1}{2}P(\widetilde{X}_{1k} = \widetilde{X}_{2l}) = \frac{1}{2}$. Therefore,

$$E(U) = E(U_{kl}) = P(\widetilde{X}_{1k} < \widetilde{X}_{2l}) + \frac{1}{2}P(\widetilde{X}_{1k} = \widetilde{X}_{2l}) = \frac{1}{2}.$$

The variance reduces to:

$$\sigma_0^2 = Var_0(U) = \frac{1}{12N_1N_2}\left(N_1 + N_2 + 1 - \frac{\sum_{\nu=1}^{g} t_\nu(t_\nu^2 - 1)}{(N_1 + N_2)(N_1 + N_2 - 1)}\right)$$

where t_ν is the number of observations with the same value in the ν-th block of tied observations sharing the same value and g is the number of such blocks (see, for instance, Rosner 2015).

Appendix 2: Mean and Variance of the Weighted U-Statistic

Consider the weights $\mathbf{w} = (w_1, w_2)$, we define the vector $\mathbf{c}' = (c_1, c_2, c_3) = (w_1^2, w_1w_2, w_2^2)$. Let $\widetilde{X}_{1k} = w_1\delta_{1k}(\eta + t_{1k}) + w_2(1 - \delta_{1k})X_{1k}$, for $k = 1, \ldots, N_1$ and $\widetilde{X}_{2l} = w_1\delta_{2l}(\eta + t_{2l}) + w_2(1 - \delta_{2l})X_{2l}$, for $l = 1, \ldots, N_2$.

We define the weighted WMW U-statistic by: $\mathbf{c}'\mathbf{U} = (U_t, U_{tx}, U_x)$ where $\mathbf{U}' = (U_t, U_{tx}, U_x)$ and

$$U_t = (N_1N_2)^{-1}\sum_{k=1}^{N_1}\sum_{l=1}^{N_2}\delta_{1k}\delta_{2l}\left[I(T_{1k} < T_{2l}) + \frac{1}{2}I(T_{1k} = T_{2l})\right], \text{ with}$$
$$T_{1k} \leq T_{max},\ T_{2l} \leq T_{max}$$
$$U_{tx} = (N_1N_2)^{-1}\sum_{k=1}^{N_1}\sum_{l=1}^{N_2}\delta_{1k}(1-\delta_{2l})$$
$$U_x = (N_1N_2)^{-1}\sum_{k=1}^{N_1}\sum_{l=1}^{N_2}(1-\delta_{1k})(1-\delta_{2l})\left[I(X_{1k} < X_{2l}) + I(X_{1k} = X_{2l})\right]$$

$$\begin{aligned}E(\mathbf{U}) &= (E(U_t), E(U_{tx}), E(U_x))' \\ &= \Big(E(\delta_{1k}=1)E(\delta_{2l}=1)\Big[P(T_{1k} < T_{2l}|T_{1k} \leq T_{max},\ T_{2l} \leq T_{max}) \\ &+\frac{1}{2}P(T_{1k} = T_{2l}|T_{1k} \leq T_{max},\ T_{2l} \leq T_{max})\Big],\ E(\delta_{1k}=1)E(\delta_{2l}=0), \\ &E(\delta_{1k}=0)E(\delta_{2l}=0)\left[P(X_{1k} < X_{2l}) + \frac{1}{2}P(X_{1k} = X_{2l})\right]\Big)' \\ &= (p_1p_2\pi_{t1}, p_1q_2, q_1q_2\pi_{x1})'\end{aligned}$$

In absence of ties, the variance $Var(\mathbf{U}) = \Sigma = (N_1N_2)^{-1}\left(\Sigma_{ij}\right)_{1\leq i,j\leq 3}$ is a 3×3 matrix such that

$$\begin{aligned}\Sigma_{11} &= E[(U_t - p_1p_2\pi_{t1})(U_t - p_1p_2\pi_{t1})] \\ &= p_1p_2\left[\pi_{t1}(1-p_1p_2\pi_{t1}) + p_1(N_1-1)(\pi_{t2} - p_2\pi_{t1}^2) + p_2(N_2-1)(\pi_{t3} - p_1\pi_{t1}^2)\right], \\ \Sigma_{12} &= \Sigma_{21} = E[(U_t - p_1p_2\pi_{t1})(U_{tx} - p_1q_2)] = \pi_{t1}p_1p_2q_2\left[(N_2-1)q_1 - N_1p_1\right], \\ \Sigma_{13} &= \Sigma_{31} = E[(U_t - p_1p_2\pi_{t1})(U_x - q_1q_2\pi_{x1})] = -\pi_{t1}\pi_{x1}(N_1+N_2-1)p_1q_1p_2q_2, \\ \Sigma_{22} &= E[(U_{tx} - p_1q_2)(U_{tx} - p_1q_2)] = p_1q_2\left[(1-p_1q_2) + (N_1-1)p_1p_2 + (N_2-1)q_1q_2\right] \\ \Sigma_{23} &= \Sigma_{32} = E[(U_{tx} - p_1q_2)(U_x - q_1q_2\pi_{x1})] = \pi_{x1}p_1q_1q_2\left[(N_1-1)p_2 - N_2q_2\right], \\ \Sigma_{33} &= E[(U_x - q_1q_2)(U_x - q_1q_2\pi_{x1})] \\ &= q_1q_2\left[\pi_{x1}(1-q_1q_2\pi_{x1}) + q_1(N_1-1)(\pi_{x2} - q_2\pi_{x1}^2) + q_2(N_2-1)(\pi_{x3} - q_1\pi_{x1}^2)\right].\end{aligned}$$

Therefore,

$$Var(\mathbf{c}'\mathbf{U}) = \mathbf{c}'\Sigma\mathbf{c}.$$

Under the null hypothesis of no difference between the two groups, with respect to both survival and nonfatal outcome, we have $p_1 = p_2 = p$, $q_1 = q_2 = q = 1 - p$, $\pi_{t1} = \pi_{x1} = 1/2$, and $\pi_{t2} = \pi_{x2} = \pi_{t3} = \pi_{x3} = 1/3$. Thus,

$$E_0(\mathbf{U}) = \frac{1}{2}\left(p^2, 2pq, q^2\right)' \quad \text{and} \quad Var_0(\mathbf{U}) = \Sigma_0, \tag{1.17}$$

where $\Sigma_0 = (N_1N_2)^{-1}\left(\Sigma_{0ij}\right)_{1\le i,j\le 3}$ is a symmetric matrix with

$$\Sigma_{011} = \frac{p^2}{12}A(p),\ \Sigma_{012} = \frac{p^2q}{2}\left[(N_2-1)q - N_1p\right],\ \Sigma_{013} = -\frac{p^2q^2}{4}(N_2+N_1-1)$$

$$\Sigma_{022} = pq\left[1-pq+(N_2-1)q^2+(N_1-1)p^2\right],\ \Sigma_{023} = \frac{pq^2}{2}\left((N_1-1)p-N_2q\right),$$

$$\Sigma_{033} = \frac{q^2}{12}A(q),\ \text{ where } A(x) = 6+4(N_2+N_1-2)x-3(N_2+N_1-1)x^2.$$

Moreover, since $Var_0(\mathbf{c}'\mathbf{U}) = \mathbf{c}'\Sigma_0\mathbf{c} \ge 0$ by definition, the matrix Σ_0 is positive semi-definite. In practice, p is estimated by the pooled sample proportion $\hat{p} = (N_1\widehat{p}_1+N_2\widehat{p}_2)/(N_1+N_2)$, and both $E_0(\mathbf{U})$ and $Var_0(\mathbf{U})$ are calculated accordingly. Finally, when ties are present, the foregoing formulas can be modified easily as we did in the non-weighted case to account for the ties in the variance estimations.

Appendix 3: Optimal Weights

From Eq. (1.15), we have

$$\mu_{1w}-\mu_{0w} = c_1\left(\pi_{t1}p_1p_2 - \frac{1}{2}p^2\right)+c_2\left(p_1q_2-pq\right)+c_3\left(\pi_{x1}q_1q_2-\frac{1}{2}q^2\right) = \mathbf{c}'\boldsymbol{\mu}$$

where $\boldsymbol{\mu}' = \left(\pi_{t1}p_1p_2 - \frac{1}{2}p^2, p_1q_2 - pq, \pi_{x1}q_1q_2 - \frac{1}{2}q^2\right)$, $\mathbf{c}' = (c_1, c_2, c_3)$ with $c_1 + 2c_2 + c_3 = 1$.

We assume that $det(\Sigma_0) > 0$, i.e., Σ_0 is positive definite. Maximizing $\dfrac{|\mu_{1w}-\mu_{0w}|}{\sigma_{0w}}$, subject to $c_1 + 2c_2 + c_3 = 1$, with respect to $\mathbf{c}$ corresponds to maximizing the Lagrange function:

$$O(\mathbf{c}, \lambda) = \left|\mathbf{c}'\mu\right|\left(\mathbf{c}'\Sigma_0\mathbf{c}\right)^{-\frac{1}{2}} - \lambda(\mathbf{c}'\mathbf{b}-1)$$

with respect to the vector $\mathbf{c}$ and λ, where λ is the Lagrange multiplier and $\mathbf{b}' = (1, 2, 1)$. Let $K(\mathbf{c}) = sign(\mathbf{c}'\mu)[(\mathbf{c}'\Sigma_0\mathbf{c})^{-\frac{3}{2}}]$, we have

$$\frac{\partial}{\partial \mathbf{c}} O(\mathbf{c}, \lambda) = K(\mathbf{c}) \left[(\mathbf{c}'\Sigma_0 \mathbf{c})\mu - (\Sigma_0 \mathbf{c})(\mathbf{c}'\mu) \right] - \lambda \mathbf{b} = 0 \tag{1.18}$$

$$\frac{\partial}{\partial \lambda} O(\mathbf{c}, \lambda) = \mathbf{c}'\mathbf{b} - 1 = 0 \tag{1.19}$$

From (1.18) and (1.19), we have

$$\begin{aligned} 0 &= \mathbf{c}' \left\{ K(\mathbf{c}) \left[(\mathbf{c}'\Sigma_0 \mathbf{c})\mu - (\Sigma_0 \mathbf{c})(\mathbf{c}'\mu) \right] - \lambda \mathbf{b} \right\} \\ &= K(\mathbf{c}) \left[(\mathbf{c}'\Sigma_0 \mathbf{c})\mathbf{c}'\mu - (\mathbf{c}'\Sigma_0 \mathbf{c})(\mathbf{c}'\mu) \right] - \lambda \mathbf{c}'\mathbf{b} = \lambda, \end{aligned}$$

because both $(\mathbf{c}'\Sigma_0\mathbf{c})$ and $(\mathbf{c}'\mu)$ are scalars and $\mathbf{c}'\mathbf{b} = c_1 + 2c_2 + c_3 = 1$.

Then, Eq. (1.18) implies $(\mathbf{c}'\Sigma_0\mathbf{c})\mu = (\Sigma_0\mathbf{c})(\mathbf{c}'\mu)$, i.e., $\mu = (\Sigma_0\mathbf{c})\frac{(\mathbf{c}'\mu)}{(\mathbf{c}'\Sigma_0\mathbf{c})} = \Sigma_0\frac{(\mathbf{c}'\mu)}{(\mathbf{c}'\Sigma_0\mathbf{c})}\mathbf{c}$. Since we assume that the matrix Σ_0^{-1} exists, this implies

$$\Sigma_0^{-1}\mu = \frac{(\mathbf{c}'\mu)}{(\mathbf{c}'\Sigma_0\mathbf{c})}\mathbf{c} \tag{1.20}$$

and thus, $\mathbf{b}'\Sigma_0^{-1}\mu = \frac{(\mathbf{c}'\mu)}{(\mathbf{c}'\Sigma_0\mathbf{c})}\mathbf{b}'\mathbf{c} = \frac{(\mathbf{c}'\mu)}{(\mathbf{c}'\Sigma_0\mathbf{c})}$.

Replacing $\frac{(\mathbf{c}'\mu)}{(\mathbf{c}'\Sigma_0\mathbf{c})}$ by $\mathbf{b}'\Sigma_0^{-1}\mu$ in Eq. (1.20) yields $\Sigma_0^{-1}\mu = (\mathbf{b}'\Sigma_0^{-1}\mu)\mathbf{c}$. Therefore, the optimal weight-vector is

$$\mathbf{c}_{opt} = \frac{\Sigma_0^{-1}\mu}{\mathbf{b}'\Sigma_0^{-1}\boldsymbol{\mu}}, \tag{1.21}$$

as long as $\mathbf{b}'\Sigma_0^{-1}\boldsymbol{\mu} \neq 0$. In addition,

$$\begin{aligned} \frac{\partial^2}{\partial \mathbf{c}^2}[O(\mathbf{c})]_{\mathbf{c}=\mathbf{c}_{opt}} &= sign(\mathbf{c}'\mu)(\mathbf{c}'\Sigma_0^{-1}\mathbf{c})^{-\frac{3}{2}}\left[2(\mathbf{c}'\Sigma_0)\mu - \mu'(\Sigma_0\mathbf{c}) - \Sigma_0(\mathbf{c}'\mu)\right]_{\mathbf{c}=\mathbf{c}_{opt}} \\ &\quad -3sign(\mathbf{c}'\mu)(\Sigma_0\mathbf{c})(\mu'\Sigma_0^{-1}\mu)^{-\frac{5}{2}}\left[(\mathbf{c}'\Sigma_0\mathbf{c})\mu - (\Sigma_0\mathbf{c})(\mathbf{c}'\mu)\right]_{\mathbf{c}=\mathbf{c}_{opt}} \\ &= 2sign(\mathbf{c}'\mu)(\mu'\Sigma_0^{-1}\mu)^{-\frac{3}{2}}(\mathbf{b}'\Sigma_0^{-1}\mu)^2\left[\mu\mu' - (\mu'\Sigma_0^{-1}\mu)\Sigma_0\right] \\ &= 2sign(\mathbf{b}'\Sigma_0^{-1}\mu)(\mu'\Sigma_0^{-1}\mu)^{-\frac{3}{2}}(\mathbf{b}'\Sigma_0^{-1}\mu)^2\left[\mu\mu' - (\mu'\Sigma_0^{-1}\mu)\Sigma_0\right]. \end{aligned}$$

Since Σ_0 is positive definite, we can show that the border-preserving principal minors of order $k > 2$ have sign $(-1)^k$. Therefore, $\mathbf{c}_{opt} = \dfrac{\Sigma_0^{-1}\mu}{\mathbf{b}'\Sigma_0^{-1}\boldsymbol{\mu}}$ maximizes $O(\mathbf{c})$.

Let us define two vectors $\mathbf{d_1}' = (1, 1, 0)$ and $\mathbf{d_2}' = \mathbf{b}' - \mathbf{d_1}' = (0, 1, 1)$. To calculate w_1 and w_2, we just need to consider the relationships $\mathbf{c} = (w_1^2, w_1 w_2, w_2^2)$ and $w_1 + w_2 = 1$. We have $\mathbf{d_1}'\mathbf{c} = w_1^2 + w_1(1 - w_1) = w_1$. Therefore, using the result given in Eq. (1.21), we can deduce $w_1 = \mathbf{d_1}'\mathbf{c} = \dfrac{\mathbf{d_1}'\Sigma_0^{-1}\mu}{\mathbf{b}'\Sigma_0^{-1}\boldsymbol{\mu}}$ and $w_2 = 1 - \mathbf{d_1}'\mathbf{c} = \dfrac{(\mathbf{b}' - \mathbf{d_1}')\Sigma_0^{-1}\mu}{\mathbf{b}'\Sigma_0^{-1}\boldsymbol{\mu}} = \dfrac{\mathbf{d_2}'\Sigma_0^{-1}\mu}{\mathbf{b}'\Sigma_0^{-1}\boldsymbol{\mu}}$.

Appendix 4: Conditional Probabilities

Exponential Distribution

Suppose that the death times t_1, t_2 follow exponential distributions with hazards λ_1, λ_2, respectively, and denote $\theta = \dfrac{\lambda_1}{\lambda_2}$, $q_1 = q_2^\theta$, and $q_2 = e^{-T\lambda_2}$. Given that $P(\delta_{1k} = 1) = p_1$, $P(\delta_{2l} = 1) = p_2$, we have

$$
\begin{aligned}
\pi_{t1} &= P(T_{1k} < T_{2l} | \delta_{1k} = \delta_{2l} = 1) = (p_1 p_2)^{-1} \int_0^{T_{max}} (1 - e^{-\lambda_1 u})\lambda_2 e^{-\lambda_2 u} du \\
&= \frac{1}{(1 - q_2^\theta)} \left[1 - \frac{1 - q_2^{(1+\theta)}}{(1+\theta)(1 - q_2)} \right]; \\
\pi_{t2} &= P(T_{1k} < T_{2l}, T_{1k'} < T_{2l} | \delta_{1k} = \delta_{1k'} = \delta_{2l} = 1) \\
&= p_1^{-2} p_2^{-1} \int_0^{T_{max}} (1 - e^{-\lambda_1 u})^2 \lambda_2 e^{-\lambda_2 u} du \\
&= (1 - q_2^\theta)^{-2} \left\{ 1 + \frac{1}{(1 - q_2)} \left[\frac{1 - q_2^{(1+2\theta)}}{1 + 2\theta} - \frac{2(1 - q_2^{(1+\theta)})}{1+\theta} \right] \right\} \\
\pi_{t3} &= P(T_{1k} < T_{2l}, t_{1k} < t_{2l'} | \delta_{1k} = \delta_{2l} = \delta_{2l'} = 1) \\
&= p_1^{-1} p_2^{-2} \int_0^{T} (e^{-\lambda_2 T} - e^{-\lambda_2 u})^2 \lambda_1 e^{-\lambda_1 u} du \\
&= \left(\frac{q_2}{1 - q_2} \right)^2 \left[1 + \frac{\theta(1 - q_2^{(2+\theta)})}{(2+\theta)(1 - q_2^\theta) q_2^2} - \frac{2\theta(1 - q_2^{(1+\theta)})}{(1+\theta)(1 - q_2^\theta) q_2} \right]
\end{aligned}
$$

Normal Distribution

Suppose that the nonfatal outcomes X_1, X_2 follow normal distributions $N(\mu_{x_1}, \sigma_{x_1})$ and $N(\mu_{x_2}, \sigma_{x_2})$, respectively.

Consider $\Delta_x = \frac{\mu_{x_2} - \mu_{x_1}}{\sqrt{\sigma_{x_1}^2 + \sigma_{x_2}^2}}$, $\rho_{x_j} = \frac{\sigma_{x_j}^2}{\sigma_{x_1}^2 + \sigma_{x_2}^2}$, and $Z_{kl} = \frac{X_{1k} - X_{2l} - (\mu_{x_1} - \mu_{x_2})}{\sqrt{\sigma_{x_1}^2 + \sigma_{x_2}^2}}$.

We can show that

$$\pi_{x1} = P(X_{1k} < X_{2l}) = \Phi(\Delta_x),$$
$$\pi_{x2} = P(X_{1k} < X_{2l}, X_{1k'} < X_{2l}) = P(Z_{kl} < \Delta_x,\ Z_{k'l} < \Delta_x),$$
$$\pi_{x3} = P(X_{1k} < X_{2l}, X_{1k} < X_{2l'}) = P(Z_{kl} < \Delta_x,\ Z_{kl'} < \Delta_x),$$

$$(Z_{kl}, Z_{k'l}) \sim N\left(\begin{pmatrix}0\\0\end{pmatrix}, \begin{pmatrix}1 & \rho_{x_2}\\ \rho_{x_2} & 1\end{pmatrix}\right) \text{ and } (Z_{kl}, Z_{kl'}) \sim N\left(\begin{pmatrix}0\\0\end{pmatrix}, \begin{pmatrix}1 & \rho_{x_1}\\ \rho_{x_1} & 1\end{pmatrix}\right).$$

References

Adams H., Jr., Davis, P., Leira, E., Chang, K., Bendixen, B., Clarke, W., et al. (1999). Baseline NIH stroke scale score strongly predicts outcome after stroke: A report of the Trial of Org 10172 in Acute Stroke Treatment (TOAST). *Neurology, 53*(1), 126.

Ahmad, Y., Nijjer, S., Cook, C. M., El-Harasis, M., Graby, J., Petraco, R., et al. (2015). A new method of applying randomised control study data to the individual patient: A novel quantitative patient-centred approach to interpreting composite end points. *International Journal of Cardiology, 195*, 216–224.

Allen, L. A., Hernandez, A. F., O'Connor, C. M., & Felker, G. M. (2009). End points for clinical trials in acute heart failure syndromes. *Journal of the American College of Cardiology, 53*(24), 2248–2258.

Anker, S. D., & Mcmurray, J. J. (2012). Time to move on from'time-to-first': Should all events be included in the analysis of clinical trials? *European Heart Journal, 33*(22), 2764–2765.

Anker, S. D., Schroeder, S., Atar, D., Bax, J. J., Ceconi, C., Cowie, M. R., et al. (2016). Traditional and new composite endpoints in heart failure clinical trials: Facilitating comprehensive efficacy assessments and improving trial efficiency. *European Journal of Heart Failure, 18*(5):482–489.

Anstrom, K. J., & Eisenstein, E. L. From batting average to wins above replacement to composite end points-refining clinical research using baseball statistical methods. *American Heart Journal, 161*(5), 805–806.

Armstrong, P. W., & Westerhout, C. M. (2013). The power of more than one. *Circulation 127*, 665–667.

Armstrong, P. W., & Westerhout, C. M. (2017). Composite end points in clinical research. *Circulation, 135*(23), 2299–2307.

Armstrong, P. W., Westerhout, C. M., Van de Werf, F., Califf, R. M., Welsh, R. C., Wilcox, R. G., et al. (2011). Refining clinical trial composite outcomes: An application to the assessment of the safety and efficacy of a new thrombolytic–3 (assent-3) trial. *American Heart Journal, 161*(5), 848–854.

Bakal, J. A., Roe, M. T., Ohman, E. M., Goodman, S. G., Fox, K. A., Zheng, Y., et al. (2015). Applying novel methods to assess clinical outcomes: Insights from the trilogy ACS trial. *European Heart Journal, 36*(6), 385–392.

Bakal, J. A., Westerhout, C. M., & Armstrong, P. W. (2012). Impact of weighted composite compared to traditional composite endpoints for the design of randomized controlled trials. *Statistical Methods in Medical Research, 24*(6), 980–988. https://doi.org/10.1177/0962280211436004

Bakal, J. A., Westerhout, C. M., Cantor, W. J., Fernández-Avilés, F., Welsh, R. C., Fitchett, D., et al. (2012). Evaluation of early percutaneous coronary intervention vs. standard therapy after fibrinolysis for st-segment elevation myocardial infarction: Contribution of weighting the composite endpoint. *European Heart Journal, 34*(12), 903–908.

Bebu, I., & Lachin, J. M. (2015). Large sample inference for a win ratio analysis of a composite outcome based on prioritized components. *Biostatistics, 17*(1), 178–187.

Berry, J. D., Miller, R., Moore, D. H., Cudkowicz, M. E., Van Den Berg, L. H., Kerr, D. A., et al. (2013). The combined assessment of function and survival (CAFS): A new endpoint for ALS clinical trials. *Amyotrophic Lateral Sclerosis and Frontotemporal Degeneration, 14*(3), 162–168.

Bonate, P. L. (2000). *Analysis of pretest-posttest designs*. Boca Raton: CRC Press.

Braunwald, E., Antman, E. M., Beasley, J. W., Califf, R. M., Cheitlin, M. D., Hochman, J. S., et al. (2002). ACC/AHA 2002 guideline update for the management of patients with unstable angina and non–st-segment elevation myocardial infarction–summary article: A report of the American college of cardiology/American heart association task force on practice guidelines (committee on the management of patients with unstable angina). *Journal of the American College of Cardiology, 40*(7), 1366–1374.

Brittain, E., Palensky, J., Blood, J., & Wittes, J. (1997). Blinded subjective rankings as a method of assessing treatment effect: A large sample example from the systolic hypertension in the elderly program (SHEP). *Statistics in Medicine, 16*(6), 681–693.

Brown, P. M., Anstrom, K. J., Felker, G. M., & Ezekowitz, J. A. (2016). Composite end points in acute heart failure research: Data simulations illustrate the limitations. *Canadian Journal of Cardiology, 32*(11), 1356.e21–1356.e28.

Brunner, E., & Munzel, U. (2000). The nonparametric Behrens-Fisher problem: Asymptotic theory and a small-sample approximation. *Biometrical Journal, 42*(1), 17–25.

Bruno, A., Saha, C., & Williams, L.S. (2006). Using change in the national institutes of health stroke scale to measure treatment effect in acute stroke trials. *Stroke, 37*(3), 920–921.

Buyse, M. (2010). Generalized pairwise comparisons of prioritized outcomes in the two-sample problem. *Statistics in Medicine, 29*(30), 3245–3257

Campbell, D. T., & Kenny, D. A. (1999). *A primer on regression artifacts*. New York: Guilford Publications.

Chung, E., & Romano, J. P. (2016). Asymptotically valid and exact permutation tests based on two-sample U-statistics. *Journal of Statistical Planning and Inference, 168*, 97–105.

Claggett, B., Wei, L.-J., & Pfeffer, M. A. (2013). Moving beyond our comfort zone. *European Heart Journal, 34*(12), 869–871.

Cordoba, G., Schwartz, L., Woloshin, S., Bae, H., & Gotzsche, P. (2010). Definition, reporting, and interpretation of composite outcomes in clinical trials: Systematic review. *British Medical Journal, 341*, c3920.

Davis, S. M., Koch, G. G., Davis, C., & LaVange, L. M. (2003). Statistical approaches to effectiveness measurement and outcome-driven re-randomizations in the clinical antipsychotic trials of intervention effectiveness (CATIE) studies. *Schizophrenia Bulletin, 29*(1), 73.

DeCoster, T., Willis, M., Marsh, J., Williams, T., Nepola, J., Dirschl, D., & Hurwitz, S. (1999). Rank order analysis of tibial plafond fractures: Does injury or reduction predict outcome? *Foot & Ankle International, 20*(1), 44–49.

Dmitrienko, A., D'Agostino, R. B., & Huque, M. F. (2013). Key multiplicity issues in clinical drug development. *Statistics in Medicine, 32*(7), 1079–1111.

Fay, M. P., & Proschan, M. A. (2010). Wilcoxon-Mann-Whitney or t-test? On assumptions for hypothesis tests and multiple interpretations of decision rules. *Statistics Surveys, 4*, 1–39.

Feldman, A., Baughman, K., Lee, W., Gottlieb, S., Weiss, J., Becker, L., & Strobeck, J. (1991). Usefulness of OPC-8212, a quinolinone derivative, for chronic congestive heart failure in patients with ischemic heart disease or idiopathic dilated cardiomyopathy. *The American Journal of Cardiology, 68*(11), 1203–1210.

Felker, G., Anstrom, K., & Rogers, J. (2008). A global ranking approach to end points in trials of mechanical circulatory support devices. *Journal of Cardiac Failure, 14*(5), 368–372.

Felker, G. M., & Maisel, A. S. (2010). A global rank end point for clinical trials in acute heart failure. *Circulation: Heart Failure, 3*(5), 643–646.

Ferreira-Gonzalez, I., Permanyer-Miralda, G., Busse, J., Devereaux, P., Guyatt, G., Alonso-Coello, P., et al. (2009). Composite outcomes can distort the nature and magnitude of treatment benefits in clinical trials. *Annals of Internal Medicine, 150*(8), 566.

Ferreira-González, I., Permanyer-Miralda, G., Busse, J. W., Bryant, D. M., Montori, V. M., Alonso-Coello, P., et al. (2007a). Methodologic discussions for using and interpreting composite endpoints are limited, but still identify major concerns. *Journal of Clinical Epidemiology, 60*(7), 651–657.

Ferreira-González, I., Permanyer-Miralda, G., Domingo-Salvany, A., Busse, J., Heels-Ansdell, D., Montori, V., et al. (2007b). Problems with use of composite end points in cardiovascular trials: Systematic review of randomised controlled trials. *The BMJ, 334*(7597), 786.

Finkelstein, D., & Schoenfeld, D. (1999). Combining mortality and longitudinal measures in clinical trials. *Statistics in Medicine, 18*(11), 1341–1354.

Fisher, L. D. (1998). Self-designing clinical trials. *Statistics in Medicine, 17*(14), 1551–1562.

Fitzmaurice, G. (2001). A conundrum in the analysis of change. *Nutrition, 17*(4), 360–361.

Follmann, D., Duerr, A., Tabet, S., Gilbert, P., Moodie, Z., Fast, P., et al. (2007). Endpoints and regulatory issues in HIV vaccine clinical trials: Lessons from a workshop. *Journal of Acquired Immune Deficiency Syndromes (1999), 44*(1), 49.

Follmann, D., Wittes, J., & Cutler, J. A. (1992). The use of subjective rankings in clinical trials with an application to cardiovascular disease. *Statistics in Medicine, 11*(4), 427–437.

Freemantle, N., Calvert, M., Wood, J., Eastaugh, J., & Griffin, C. (2003). Composite outcomes in randomized trials: Greater precision but with greater uncertainty? *JAMA, 289*(19), 2554.

Gail, M. H., Mark, S. D., Carroll, R. J., Green, S. B., & Pee, D. (1996). On design considerations and randomization-based inference for community intervention trials. *Statistics in Medicine, 15*(11), 1069–1092.

Gehan, E. A. (1965). A generalized Wilcoxon test for comparing arbitrarily singly-censored samples. *Biometrika, 52*(1–2), 203–223.

Gómez, G., & Lagakos, S. W. (2013). Statistical considerations when using a composite endpoint for comparing treatment groups. *Statistics in Medicine, 32*(5), 719–738.

Gould, A. (1980). A new approach to the analysis of clinical drug trials with withdrawals. *Biometrics, 36*(4), 721–727.

Grech, E., & Ramsdale, D. (2003). Acute coronary syndrome: Unstable angina and non-st segment elevation myocardial infarction. *The BMJ, 326*(7401), 1259.

Hallstrom, A., Litwin, P., & Douglas Weaver, W. (1992). A method of assigning scores to the components of a composite outcome: An example from the MITI trial. *Controlled Clinical Trials, 13*(2), 148–155.

Hanley, J. A., & McNeil, B. J. (1992). The meaning and use of the area under a receiver operating characteristic (roc) curve. *Radiology, 143*(1), 29–36.

Hariharan, S., McBride, M. A., & Cohen, E. P. (2003). Evolution of endpoints for renal transplant outcome. *American Journal of Transplantation, 3*(8), 933–941.

Heddle, N. M., & Cook, R. J. (2011). Composite outcomes in clinical trials: What are they and when should they be used? *Transfusion, 51*(1), 11–13.

Huang, P., Woolson, R. F., & O'Brien, P. C. (2008). A rank-based sample size method for multiple outcomes in clinical trials. *Statistics in Medicine, 27*(16), 3084–3104.

Huque, M. F., Alosh, M., & Bhore, R. (2011). Addressing multiplicity issues of a composite endpoint and its components in clinical trials. *Journal of Biopharmaceutical Statistics, 21*(4), 610–634.

Kaufman, K. D., Olsen, E. A., Whiting, D., Savin, R., DeVillez, R., Bergfeld, W., et al. (1998). Finasteride in the treatment of men with androgenetic alopecia. *Journal of the American Academy of Dermatology, 39*(4), 578–589.

Kawaguchi, A., Koch, G. G., & Wang, X. (2011). Stratified multivariate Mann–Whitney estimators for the comparison of two treatments with randomization based covariance adjustment. *Statistics in Biopharmaceutical Research, 3*(2), 217–231.

Lachin, J. (1999). Worst-rank score analysis with informatively missing observations in clinical trials. *Controlled Clinical Trials, 20*(5), 408–422.

Lachin, J. M., & Bebu, I. (2015). Application of the Wei–Lachin multivariate one-directional test to multiple event-time outcomes. *Clinical Trials, 12*(6), 627–633. https://doi.org/10.1177/1740774515601027.

Li, D., Zhao, G., Paty, D., University of British Columbia MS/MRI Analysis Research Group, The SPECTRIMS Study Group. (2001). Randomized controlled trial of interferon-beta-1a in secondary progressive MS MRI results. *Neurology, 56*(11), 1505–1513.

Lisa, A. B., & James, S. H. (1997). Rule-based ranking schemes for antiretroviral trials. *Statistics in Medicine, 16*, 1175–1191.

Logan, B., & Tamhane, A. (2008). Superiority inferences on individual endpoints following noninferiority testing in clinical trials. *Biometrical Journal, 50*(5), 693–703.

Lubsen, J., Just, H., Hjalmarsson, A., La Framboise, D., Remme, W., Heinrich-Nols, J., et al. (1996). Effect of pimobendan on exercise capacity in patients with heart failure: Main results from the Pimobendan in Congestive Heart Failure (PICO) trial. *Heart, 76*(3), 223.

Lubsen, J., & Kirwan, B.-A. (2002). Combined endpoints: Can we use them? *Statistics in Medicine, 21*(19), 2959–2970.

Luo, X., Qiu, J., Bai, S., & Tian, H. (2017). Weighted win loss approach for analyzing prioritized outcomes. *Statistics in Medicine, 36*(15), 2452–2465.

Manja, V., AlBashir, S., & Guyatt, G. (2017). Criteria for use of composite end points for competing risks–a systematic survey of the literature with recommendations. *Journal of Clinical Epidemiology, 82*, 4–11.

Mascha, E. J., & Turan, A. (2012). Joint hypothesis testing and gatekeeping procedures for studies with multiple endpoints. *Anesthesia & Analgesia, 114*(6), 1304–1317.

Matsouaka, R. A., & Betensky, R. A. (2015). Power and sample size calculations for the Wilcoxon–Mann–Whitney test in the presence of death-censored observations. *Statistics in Medicine, 34*(3), 406–431.

Matsouaka, R. A., Singhal, A. B., & Betensky, R. A. (2016). An optimal Wilcoxon–Mann–Whitney test of mortality and a continuous outcome. *Statistical Methods in Medical Research, 27*(8), 2384–2400. https://doi.org/10.1177/0962280216680524

Minas, G., Rigat, F., Nichols, T. E., Aston, J. A., & Stallard, N. (2012). A hybrid procedure for detecting global treatment effects in multivariate clinical trials: Theory and applications to fMRI studies. *Statistics in Medicine, 31*(3), 253–268.

Moyé, L. (2013). *Multiple analyses in clinical trials: Fundamentals for investigators*. Berlin: Springer.

Moyé, L., Davis, B., & Hawkins, C. (1992). Analysis of a clinical trial involving a combined mortality and adherence dependent interval censored endpoint. *Statistics in Medicine, 11*(13), 1705–1717.

National Asthma Education and Prevention Program (National Heart, Lung, and Blood Institute). (2007). Third expert panel on the management of asthma. *Expert panel report 3: Guidelines for the diagnosis and management of asthma*. US Department of Health and Human Services, National Institutes of Health, National Heart, Lung, and Blood Institute.

Neaton, J., Gray, G., Zuckerman, B., & Konstam, M. (2005). Key issues in end point selection for heart failure trials: Composite end points. *Journal of Cardiac Failure, 11*(8), 567–575.

Neaton, J. D., Wentworth, D. N., Rhame, F., Hogan, C., Abrams, D. I., & Deyton, L. (1994). Considerations in choice of a clinical endpoint for aids clinical trials. *Statistics in Medicine, 13*(19–20), 2107–2125.

Newcombe, R. G. (2006). Confidence intervals for an effect size measure based on the Mann–Whitney statistic. part 2: Asymptotic methods and evaluation. *Statistics in Medicine, 25*(4), 559–573.

Oakes, J. M., & Feldman, H. A. (2001). Statistical power for nonequivalent pretest-posttest designs the impact of change-score versus ANCOVA models. *Evaluation Review, 25*(1), 3–28.

O'Brien, P. C. (1984). Procedures for comparing samples with multiple endpoints. *Biometrics, 40*, 1079–1087.

Parsons, M., Spratt, N., Bivard, A., Campbell, B., Chung, K., Miteff, F., et al. (2012). A randomized trial of tenecteplase versus alteplase for acute ischemic stroke. *New England Journal of Medicine, 366*(12), 1099–1107.

Pearl, J. (2014). Lord's paradox revisited–(oh lord! kumbaya!). Tech. rep., Citeseer.

Pocock, S. J., Ariti, C. A., Collier, T. J., & Wang, D. (2011). The win ratio: A new approach to the analysis of composite endpoints in clinical trials based on clinical priorities. *European Heart Journal, 33*(2), 176–182.

Pratt, J. W. (1964). Robustness of some procedures for the two-sample location problem. *Journal of the American Statistical Association, 59*, 665–680.

Prieto-Merino, D., Smeeth, L., van Staa, T. P., & Roberts, I. (2013). Dangers of non-specific composite outcome measures in clinical trials. *The BMJ, 347*, f6782.

Ramchandani, R., Schoenfeld, D. A., & Finkelstein, D. M. (2016). Global rank tests for multiple, possibly censored, outcomes. *Biometrics, 72*, 926–935.

Röhmel, J., Gerlinger, C., Benda, N., & Läuter, J. (2006). On testing simultaneously non-inferiority in two multiple primary endpoints and superiority in at least one of them. *Biometrical Journal, 48*(6), 916–933.

Rosenbaum, P. R. (2006). Comment: The place of death in the quality of life. *Statistical Science, 21*(3), 313–316.

Rosner, B. (2015). *Fundamentals of biostatistics*. Toronto: Nelson Education.

Ross, S. (2007). Composite outcomes in randomized clinical trials: Arguments for and against. *American Journal of Obstetrics and Gynecology, 196*(2), 119–e1.

Rowan, J. A., Hague, W. M., Gao, W., Battin, M. R., & Moore, M. P. (2008). Metformin versus insulin for the treatment of gestational diabetes. *New England Journal of Medicine, 358*(19), 2003–2015.

Rubin, D. B. (2006). Rejoinder: Causal inference through potential outcomes and principal stratification: Application to studies with" censoring" due to death. *Statistical Science, 21*(3), 319–321.

Sampson, U. K., Metcalfe, C., Pfeffer, M. A., Solomon, S. D., & Zou, K. H. (2010). Composite outcomes: Weighting component events according to severity assisted interpretation but reduced statistical power. *Journal of Clinical Epidemiology, 63*(10), 1156–1158

Samson, K. (2013). News from the AAN annual meeting: Why a trial of normobaric oxygen in acute ischemic stroke was halted early. *Neurology Today, 13*(10), 34–35.

Sankoh, A. J., Li, H., & D'Agostino, R. B. (2014). Use of composite endpoints in clinical trials. *Statistics in Medicine, 33*(27), 4709–4714.

Senn, S. (2006). Change from baseline and analysis of covariance revisited. *Statistics in Medicine, 25*(24), 4334–4344.

Shahar, E., & Shahar, D. J. (2012). Causal diagrams and change variables. *Journal of Evaluation in Clinical Practice, 18*(1), 143–148.

Singhal, A., Benner, T., Roccatagliata, L., Koroshetz, W., Schaefer, P., Lo, E., et al. (2005). A pilot study of normobaric oxygen therapy in acute ischemic stroke. *Stroke, 36*(4), 797.

Singhal, A. B. (2006). Normobaric oxygen therapy in acute ischemic stroke trial. ClinicalTrials.gov Database. http://clinicaltrials.gov/ct2/show/NCT00414726

Singhal, A. B. (2007). A review of oxygen therapy in ischemic stroke. *Neurological Research, 29*(2), 173–183.

Spencer, S., Mayer, B., Bendall, K. L., & Bateman, E. D. (2007). Validation of a guideline-based composite outcome assessment tool for asthma control. *Respiratory Research, 8*(1), 26.

Subherwal, S., Anstrom, K. J., Jones, W. S., Felker, M. G., Misra, S., Conte, M. S., et al. (2012). Use of alternative methodologies for evaluation of composite end points in trials of therapies for critical limb ischemia. *American Heart Journal, 164*(3), 277.

Sun, H., Davison, B. A., Cotter, G., Pencina, M. J., & Koch, G. G. (2012). Evaluating treatment efficacy by multiple end points in phase ii acute heart failure clinical trials analyzing data using a global method. *Circulation: Heart Failure, 5*(6), 742–749.

Tomlinson, G., & Detsky, A. S. (2010). Composite end points in randomized trials: There is no free lunch. *JAMA, 303*(3), 267–268.

Tyler, K. M., Normand, S.-L. T., & Horton, N. J. (2011). The use and abuse of multiple outcomes in randomized controlled depression trials. *Contemporary Clinical Trials, 32*(2), 299–304.

van Breukelen, G. J. (2013). ANCOVA versus CHANGE from baseline in nonrandomized studies: The difference. *Multivariate Behavioral Research, 48*(6), 895–922.

Van Elteren, P. (1960). On the combination of independent two-sample tests of Wilcoxon. *Bulletin of the International Statistical Institute, 37*, 351–361.

Wen, X., Hartzema, A., Delaney, J. A., Brumback, B., Liu, X., Egerman, R., et al. (2017). Combining adverse pregnancy and perinatal outcomes for women exposed to antiepileptic drugs during pregnancy, using a latent trait model. *BMC Pregnancy and Childbirth, 17*(1), 10.

Willett, J. B. (1988). Questions and answers in the measurement of change. *Review of Research in Education, 15*, 345–422.

Wilson, R. F., & Berger, A. K. (2011). Are all end points created equal? The case for weighting. *Journal of the American College of Cardiology, 57*(5), 546–548.

Young, F. B., Weir, C. J., Lees, K. R., & GAIN International Trial Steering Committee and Investigators. (2005). Comparison of the national institutes of health stroke scale with disability outcome measures in acute stroke trials. *Stroke, 36*(10), 2187–2192.

Zhang, J., Quan, H., Ng, J., & Stepanavage, M. E. (1997). Some statistical methods for multiple endpoints in clinical trials. *Controlled Clinical Trials, 18*(3), 204–221.

Zhao, Y. (2006). Sample size estimation for the van Elteren test—a stratified Wilcoxon–Mann–Whitney test. *Statistics in Medicine, 25*(15), 2675–2687.

Chapter 2
A Selective Overview of Semiparametric Mixture of Regression Models

Sijia Xiang and Weixin Yao

2.1 Introduction

Finite mixture of regression models have been widely used in scenarios when a single regression fails to adequately explain the relationship between the variables at hand. This type of application is commonly seen in econometrics, where it is also known as switching regression models, and has been widely applied in various other fields, see, for example, in econometrics (Frühwirth-Schnatter 2001, 2006; Wedel and DeSarbo 1933) and in epidemiology (Green and Richardson 2002). Another wide application of finite mixture of regressions is in outlier detection or robust regression estimation (Young and Hunter 2010). Viele and Tong (2002) described masked outliers, which appeared in clusters and "cannot be detected individually by standard techniques." Pena et al. (2003) used a split and recombine (SAR) procedure to identify possible clusters in a sample, which can be extended to identify masked outliers.

In a typical finite mixture of regression models, assume $\{(\mathbf{x}_i, y_i), i = 1, \cdots, n\}$ is a random sample from the population $(\mathbf{x}, Y)$, where $\mathbf{x}_i = (x_{i,1}, \cdots, x_{i,p})^T$ for $p < n$ is a vector of predictors. The goal is to describe the conditional distribution of $Y_i|\mathbf{x}_i$ using a mixture of linear regressions with assumed Gaussian errors. That is, let $\mathscr{C}$ be a latent class variable with $P(\mathscr{C} = c|\mathbf{x}) = \pi_c$ for $c = 1, \cdots, C$. Given

S. Xiang (✉)
School of Data Sciences, Zhejiang University of Finance & Economics, Hangzhou, China
e-mail: sjxiang@zufe.edu.cn

W. Yao
Department of Statistics, University of California, Riverside, CA, USA
e-mail: weixin.yao@ucr.edu

Y. Zhao, D.-G. Chen (eds.), *New Frontiers of Biostatistics and Bioinformatics*,
ICSA Book Series in Statistics, https://doi.org/10.1007/978-3-319-99389-8_2

$\mathscr{C} = c$, suppose that the response y depends on $\mathbf{x}$ in a linear way $y = \mathbf{x}^T\boldsymbol{\beta}_c + \epsilon_c$, where $\epsilon_c \sim N(0, \sigma_c^2)$. Then, the conditional distribution of Y given $\mathbf{x}$ is

$$Y|\mathbf{x} \sim \sum_{c=1}^{C} \pi_c \phi(Y|\mathbf{x}^T\boldsymbol{\beta}_c, \sigma_c^2), \tag{2.1}$$

and the log-likelihood function for observations $\{(\mathbf{x}_1, y_1), \ldots, (\mathbf{x}_n, y_n)\}$ is

$$\ell_n(\boldsymbol{\theta}) = \sum_{i=1}^{n} \log \left[\sum_{c=1}^{C} \pi_c \phi(y_i; \mathbf{x}_i^T\boldsymbol{\beta}_c, \sigma_c^2) \right], \tag{2.2}$$

where $\boldsymbol{\theta} = (\pi_1, \cdots, \pi_C, \boldsymbol{\beta}_1, \cdots, \boldsymbol{\beta}_C, \sigma_1^2, \cdots, \sigma_C^2)^T$, $\phi(y|\mu, \sigma^2)$ is the normal density with mean μ and variance σ^2, $0 \le \pi_c \le 1$, and $\sum_{c=1}^{C} \pi_c = 1$.

It is well known that the mixture likelihood function (2.2) is unbounded, which can be seen if we let $y_i = \mathbf{x}_i^T\boldsymbol{\beta}_c$ and σ_c go to 0. Many efforts have been devoted to solve the unboundedness issue of mixture likelihood. For example, Hathaway (1985, 1986) proposed to find the maximum likelihood estimate (MLE) over a constrained parameter space. Chen and Tan (2009) and Chen et al. (2008) proposed to use maximum penalized likelihood estimator that adds a penalty term to the unequal variance. Yao (2010) proposed a profile log-likelihood method and a graphical way to find the local maximum points.

Since the advent of the Expectation-Maximization (EM) algorithm, maximum likelihood (ML) has been most commonly used to fit mixture models. Define a component label indicator

$$z_{ic} = \begin{cases} 1, & \text{ifobservation} i \text{isfromcomponent} c, \\ 0, & \text{otherwise.} \end{cases} \tag{2.3}$$

The EM algorithm to fit model (2.1) proceeds iteratively between the following two steps.

Algorithm 2.1.1

E-step:

Calculate the expectations of component labels based on estimates from lth iteration:

$$p_{ic}^{(l+1)} = E[z_{ic}|\mathbf{x}_i, \boldsymbol{\theta}^{(l)}] = \frac{\pi_c^{(l)} \phi(Y_i|\mathbf{x}_i^T\boldsymbol{\beta}_c^{(l)}, \sigma_c^{2(l)})}{\sum_{c'=1}^{C} \pi_{c'}^{(l)} \phi(Y_i|\mathbf{x}_i^T\boldsymbol{\beta}_{c'}^{(l)}, \sigma_{c'}^{2(l)})},$$

for $i = 1, \cdots, n$ and $c = 1, \cdots, C$.

M-step:
Update the estimates

$$\pi_c^{(l+1)} = \frac{\sum_{i=1}^n p_{ic}^{(l+1)}}{n},$$

$$\begin{aligned}\boldsymbol{\beta}_c^{(l+1)} &= \arg\max_{\beta_c} \left[\sum_{i=1}^n p_{ic}^{(l+1)} \log \phi(Y_i|\mathbf{x}_i^T \boldsymbol{\beta}_c, \sigma_c^{2(l)})\right] \\ &= (X^T W_c X)^{-1} X^T W_c \mathbf{y},\end{aligned}$$

$$\sigma_c^{2(l+1)} = \frac{\sum_{i=1}^n (y_i - \mathbf{x}_i^T \boldsymbol{\beta}_c^{(l+1)})^2 p_{ic}^{(l+1)}}{\sum_{i=1}^n p_{ic}^{(l+1)}},$$

for $c = 1, \cdots, C$, where $W_c = \text{diag}\{p_{1c}^{(l+1)}, \cdots, p_{nc}^{(l+1)}\}$, X is the design matrix and $\mathbf{y}$ is the vector of response variables. Iterate between the two steps until convergence.

Since Goldfeld and Quandt (1973) first introduced the mixture regression model, many efforts have been made to extend the traditional parametric mixture of linear regression models. In the following sections, we are going to give a selective overview of recently developed semiparametric mixture models, and their estimation methods. In order to be consistent throughout, we tried our best to use the same notation system, which might not be the same as the original articles.

2.2 Mixture of Regression Models with Varying Proportions

2.2.1 Continuous Response, $p = 1$

As assumed by model (2.1), the probability for each regression model to occur is a fixed value π_c, $c = 1, \cdots, C$. But if the covariates $\mathbf{x}$ contains some information about the relative weights, model (2.1) might not be accurate. Therefore, Young and Hunter (2010) replaced model (2.1) by

$$Y|\mathbf{x} \sim \sum_{c=1}^C \pi_c(\mathbf{x})\phi(Y|\mathbf{x}^T \boldsymbol{\beta}_c, \sigma_c^2). \tag{2.4}$$

If $\pi_c(\mathbf{x})$ is modeled as a logistic function, then model (2.4) becomes the hierarchical mixtures of experts (HME, Jacobs et al. 1991) in neural network. Young and Hunter (2010), on the other hand, modeled $\pi_c(\mathbf{x})$ nonparametrically for the purpose of a more flexible model assumption. To be more specific, $\pi_c(\mathbf{x})$ is modeled as

$$\pi_c(\mathbf{x}_i) = E[z_{ic}|\mathbf{x}_i],$$

where z_{ic} is defined in (2.3). Since z_{ic} is not known in reality, p_{ic}^{∞}, where ∞ denotes the converged value, is used as a response value. Applying the idea of local linear regression (Fan and Gijbels 1996), at each grid point $\mathbf{x}_i$, Young and Hunter (2010) proposed to estimate $\pi_c(\mathbf{x}_i)$ by

$$\arg\min_{\boldsymbol{\alpha}} \sum_{l=1}^{n} K_{\mathbf{h}}(\mathbf{x}_i - \mathbf{x}_l)\left[p_{l,c}^{(\infty)} - \left(\alpha_0 + \sum_{t=1}^{p}\alpha_t(x_{i,t} - x_{l,t})\right)\right]^2, \tag{2.5}$$

where $\boldsymbol{\alpha} = (\alpha_0, \alpha_1, \cdots, \alpha_p)^T$, p is the length of the predictor vector, and

$$K_{\mathbf{h}}(\mathbf{x}_i - \mathbf{x}_l) = \frac{1}{h_1 \cdots h_p} K\left(\frac{x_{i,1} - x_{l,1}}{h_1}, \cdots, \frac{x_{i,p} - x_{l,p}}{h_p}\right) \times I\left\{\left|\frac{x_{i,t} - x_{l,t}}{h_t}\right| \le 1 \forall t = 1, \cdots, p\right\}$$

is a multivariate kernel density function. The algorithm they used is a global/local EM-like algorithm, since it iterates between a global step to update $\boldsymbol{\beta}_c$ and σ_c and a local step to update $\pi_c(\mathbf{x}_i)$.

Algorithm 2.2.1
Global step: Update $\boldsymbol{\beta}_c$ and σ_c using standard EM algorithm updates:

$$\boldsymbol{\beta}_c^{(l+1)} = (X^T W_c^{(l)} X)^{-1} X^T W_c^{(l)} \mathbf{y},$$

$$\sigma_c^{2(l+1)} = \frac{\left\| W_c^{1/2(l)}(\mathbf{y} - X^T \boldsymbol{\beta}_c^{(l+1)})\right\|^2}{\operatorname{tr}(W_c^{(l)})},$$

for $c = 1, \cdots, C$, where X is the $n \times p$ design matrix and $W_c^{(l)} = \operatorname{diag}(p_{1c}^{(l)}, \cdots, p_{nc}^{(l)})$. Different from other classical EM algorithm, it is then to update p_{ic}, the classification probability, in the middle of the iteration to reflect the most recent updates of parameters. Let $r_{ic} = (y_i - \mathbf{x}_i^T \boldsymbol{\beta}_c^{(l+1)})/\sigma_c^{(l+1)}$, then

$$p_{ic}^{(l+0.5)} = \frac{\pi_c^{(l)}(\mathbf{x}_i)\phi(r_{ic}|0,1)/\sigma_c^{(l+1)}}{\sum_{c'=1}^{C} \pi_{c'}^{(l)}(\mathbf{x}_i)\phi(r_{ic'}|0,1)/\sigma_{c'}^{(l+1)}}, \tag{2.6}$$

for $i = 1, \cdots, n$ and $c = 1, \cdots, C$.
Local step: Update $\pi_c(\mathbf{x}_i)$ by solving (2.5). Re-update the estimates for p_{ic} as p_{ic}^{l+1} using (2.6).

However, due to the "curse of dimensionality," they only did simulation study for $p = 1$ case, and argued that extra caution should be given for high-dimensional predictor cases.

2.2.2 Continuous Response, $p > 1$

Huang and Yao (2012) also studied model (2.4), but different from Young and Hunter (2010), they allowed the predictors $\mathbf{x}$ to be of dimension $p > 1$, and model the mixing proportion π_c as $\pi_c(u)$, where u is of dimension one and could be part of $\mathbf{X}$. In other words, Huang and Yao (2012) proposed

$$Y|_{\mathbf{x},u} \sim \sum_{c=1}^{C} \pi_c(u)\phi(Y|\mathbf{x}^T\boldsymbol{\beta}_c, \sigma_c^2). \tag{2.7}$$

Note also that Huang and Yao (2012) provided the asymptotic properties of their estimators, where Young and Hunter (2010) only gave the computation algorithm without any theoretical results. The identifiability of model (2.7) was shown under mild conditions. They also proposed a new one-step backfitting estimation procedure for the model. Specifically,

1. Estimate $\boldsymbol{\pi}(\cdot)$ locally by maximizing the following local likelihood function

$$\ell_1(\boldsymbol{\pi}, \boldsymbol{\beta}, \boldsymbol{\sigma}^2) = \sum_{i=1}^{n} \log\left\{\sum_{c=1}^{C} \pi_c\phi(Y_i|\mathbf{x}_i^T\boldsymbol{\beta}_c, \sigma_c^2)\right\} K_h(U_i - u),$$

   and let $\tilde{\boldsymbol{\pi}}(u)$, $\tilde{\boldsymbol{\beta}}(u)$ and $\tilde{\boldsymbol{\sigma}}^2(u)$ be the solution.
2. Update the estimates of global parameters $\boldsymbol{\beta}$ and $\boldsymbol{\sigma}^2$ by maximizing

$$\ell_2(\boldsymbol{\beta}, \boldsymbol{\sigma}^2) = \sum_{i=1}^{n} \log\left\{\sum_{c=1}^{C} \tilde{\pi}_c(U_i)\phi(Y_i|\mathbf{x}_i^T\boldsymbol{\beta}_c, \sigma_c^2)\right\},$$

   and let $\hat{\boldsymbol{\beta}}(u)$, and $\hat{\boldsymbol{\sigma}}^2(u)$ be the solution.
3. Further improve the estimate of $\boldsymbol{\pi}(z)$ by maximizing

$$\ell_3(\boldsymbol{\pi}) = \sum_{i=1}^{n} \log\left\{\sum_{c=1}^{C} \pi_c\phi(Y_i|\mathbf{x}_i^T\hat{\boldsymbol{\beta}}_c, \hat{\sigma}_c^2)\right\} K_h(U_i - u),$$

   and let $\hat{\boldsymbol{\pi}}(u)$ be the solution.

$\hat{\boldsymbol{\pi}}(u)$, $\hat{\boldsymbol{\beta}}$ and $\hat{\boldsymbol{\sigma}}^2$ are called the one-step backfitting estimates. Huang and Yao (2012) proved that the one-step estimators of $\boldsymbol{\beta}$ and $\boldsymbol{\sigma}^2$ were $\sqrt{n}$-consistent, and followed an asymptotic normal distribution, and $\boldsymbol{\pi}(\cdot)$ based upon $\hat{\boldsymbol{\beta}}$ and $\hat{\boldsymbol{\sigma}}^2$ had the same first-order asymptotic bias and variance as the kernel estimates with true values of $\boldsymbol{\beta}$ and $\boldsymbol{\sigma}^2$.

2.2.3 Discrete Response

Model (2.4) can be extended to mixture of GLM with varying proportions (Wang et al. 2014) to account for discrete responses. That is,

$$Y|_{\mathbf{x}} \sim \sum_{c=1}^{C} \pi_c(\mathbf{x}) f_c(Y|\mathbf{x}, \boldsymbol{\theta}_c), \tag{2.8}$$

where f_c is the component specific density function coming from the exponential family of distributions, and $\boldsymbol{\theta}_c = (\boldsymbol{\beta}_c^T, \sigma_c^2)^T$. The mean of each component is then given by

$$\mu_c(\mathbf{x}) = g_c^{-1}(\mathbf{x}^T\boldsymbol{\beta}_c),$$

where $g_c(\cdot)$ is a component specific link function. Wang et al. (2014) established the identifiability result of the model (e2.15) and gave the corresponding conditions.

Motivated by rain data from a global climate model, Cao and Yao (2012) studied a special case of (2.8), a semiparametric mixture of binomial regression, where both the component proportions and the success probabilities depend on the predictors nonparametrically. That is,

$$Y|_{X=x} \sim \pi_1(x)\text{Bin}(Y; N, 0) + \pi_2(x)\text{Bin}(Y; N, p(x)), \tag{2.9}$$

where $\text{Bin}(Y; N, p)$ denotes the probability mass function of Y, which is binomially distributed with number of trials N and success probability p, $\pi_1(x)$ and $\pi_2(x)$ are two nonparametric functions with $\pi_1(x)+\pi_2(x) = 1$. Note that the first component is degenerating with mass 1 on 0, and therefore, model (2.9) has wide application in data with extra number of zeros. The local log-likelihood at any point x_0 is

$$\begin{aligned}\ell(\boldsymbol{\theta}(x_0)) =& \frac{1}{n}\sum_{i=1}^{n} K_h(x_i - x_0)\log[\pi_1(x_0)I(y_i = 0) \\ &+ \{1-\pi_1(x_0)\}\binom{N}{y_i} p(x_0)^{t_i}\{1-p(x_0)\}^{N-y_i}],\end{aligned} \tag{2.10}$$

where $\boldsymbol{\theta} = \{\pi_1, p\}^T$, x_i is the observation for Y at x_i, $i = 1, \cdots, n$, and $K_h(\cdot) = h^{-1}K(\cdot/h)$ is a scaled kernel function with bandwidth h. Since there is no explicit solution for maximizing (2.10), the authors proposed the following EM algorithm, which increased the local log-likelihood (2.10) monotonically.

Algorithm 2.2.2

E step: Calculate the classification probabilities

$$p_{i1}^{(l+1)} = \frac{\pi_1^{(l)}(x_0)\text{Bin}(y_i; N, 0)}{\pi_1^{(l)}(x_0)\text{Bin}(y_i; N, 0) + \{1 - \pi_1^{(l)}(x_0)\}\text{Bin}(y_i; N, p^{(l)}(x_0))},$$

$$p_{i2}^{(l+1)} = 1 - p_{i1}^{(l+1)}, i = 1, \cdots, n.$$

M step: Update the estimates

$$\pi_c^{(l+1)}(x_0) = \frac{\sum_{i=1}^n K_h(x_i - x_0)p_{ic}^{(l+1)}}{\sum_{i=1}^n \sum_{c'=1}^2 K_h(x_i - x_0)p_{ic'}^{(l+1)}}, c = 1, 2,$$

$$p^{(l+1)}(x_0) = \frac{\sum_{i=1}^n K_h(x_i - x_0)p_{i2}^{(l+1)} y_i}{N\sum_{i=1}^n K_h(x_i - x_0)p_{i2}^{(l+1)}}.$$

In Cao and Yao (2012), the researchers also considered a semiparametric mixture model with constant proportions, that is

$$Y|_{X=x} \sim \pi_1\text{Bin}(Y; N, 0) + \pi_2\text{Bin}(Y; N, p(x)),$$

as a special case of model (2.9). A one-step backfitting procedure was proposed to estimate the model, and the estimates were shown to achieve the optimal convergence rates.

1. Estimate $p(\cdot)$ and π_1 locally by maximizing the local log-likelihood function (2.10) and let $\tilde{p}(x)$ and $\tilde{\pi}_1(x)$ be the solution.
2. Update the estimates of global parameters π_1 by maximizing

$$\ell_1(\pi_1) = \frac{1}{n}\sum_{i=1}^n \log[\pi_1 I(y_i = 0) + \{1 - \pi_1\}\binom{N}{y_i}\tilde{p}(x_0)^{y_i}\{1 - \tilde{p}(x_0)\}^{N-y_i}],$$

   and let $\hat{\pi}_1$ be the solution.
3. Further improve the estimate of $p(t)$ by maximizing

$$\ell_2(p(x_0)) = \frac{1}{n}\sum_{i=1}^n K_h(x_i - x_0)\log[\hat{\pi}_1 I(y_i = 0) + \{1 - \hat{\pi}_1\}\binom{N}{y_i}p(x_0)^{y_i}\{1 - p(x_0)\}^{N-y_i}],$$

   and let $\hat{p}(x)$ be the solution.

$\hat{p}(x)$ and $\hat{\pi}_1$ are called the one-step backfitting estimates. The theoretical properties of both model estimates were discussed.

2.3 Nonparametric Errors

One main drawback of classic mixture of regression models (2.1) is the strong parametric assumption about the normal error density. The estimation results might be biased if the error density is misspecified. As a result, several new methods were proposed to relax the parametric assumption of the error densities.

2.3.1 Semiparametric EM Algorithm with Kernel Density Error

For each component, Hunter and Young (2012) also assumed the basic parametric linear mixture of regressions

$$Y_i = \mathbf{x}_i^T \boldsymbol{\beta}_c + \epsilon_i,$$

but instead of normality, the error term ϵ_i was modeled fully nonparametrically as $\epsilon_i \sim g$, where g was completely unspecified. Therefore, in this semiparametric model, the conditional distribution might be written as

$$Y|\mathbf{x} \sim \sum_{c=1}^{C} \pi_c g(Y - \mathbf{x}^T \boldsymbol{\beta}_c), \tag{2.11}$$

where, without loss of generality, g was assumed to have median zero. The authors showed that when regression planes were not parallel, the parameters were identifiable without any assumption on g, on the other hand, when the planes were parallel, identifiability could still be obtained given some additional assumptions on g. To estimate the parameters and the nonparametric functions, first define a nonlinear smoothing operator

$$\mathcal{N}_h g(x) = \exp \int \frac{1}{h} K\left(\frac{x-u}{h}\right) \log g(u) du,$$

then a smoothed version of the log-likelihood function of the parameters is defined as:

$$\ell_s(\boldsymbol{\pi}, \boldsymbol{\beta}, g) = \sum_{i=1}^{n} \log \left\{ \sum_{c=1}^{C} \pi_c \mathcal{N}_h g(y_i - \mathbf{x}_i^T \boldsymbol{\beta}_c) \right\}.$$

Hunter and Young (2012) proposed a minorization-conditional-maximization (MCM) algorithm.

Algorithm 2.3.1
Minorization step: Finding the "posterior" probabilities:

$$p_{ic}^{(l)} = \frac{\pi_c^{(l)} \mathcal{N}_h g^{(l)}(y_i - \mathbf{x}_i^T \boldsymbol{\beta}_c^{(l)})}{\sum_{c'=1}^{C} \pi_{c'}^{(l)} \mathcal{N}_h g^{(l)}(y_i - \mathbf{x}^T \boldsymbol{\beta}_{c'}^{(l)})},$$

for $i = 1, \cdots, n$ and $c = 1, \cdots, C$.
Maximization step: Update estimates for parameters

$$\pi_c^{(l+1)} = \frac{\sum_{i=1}^{n} p_{ic}^{(l)}}{n},$$

$$g^{(l+1)}(u) = \frac{1}{nh} \sum_{i=1}^{n} \sum_{c=1}^{C} p_{ic}^{(l)} K\left(\frac{u - y_i + \mathbf{x}_i^T \boldsymbol{\beta}_c^{(l)}}{h}\right),$$

$$\boldsymbol{\beta}_c^{(l+1)}(x) = \arg\max_{\beta} \sum_{i=1}^{n} \pi_c^{(l+1)} \log \mathcal{N}_h g^{(l+1)}(y_i - \mathbf{x}_i^T \boldsymbol{\beta}_c),$$

for $c = 1, \cdots, C$.

Hunter and Young (2012) showed that the above algorithm possessed the desirable ascent property enjoyed by all true EM algorithm. That is:

$$\ell_s(\boldsymbol{\pi}^{(l+1)}, \boldsymbol{\beta}^{(l+1)}, g^{(l+1)}) \geq \ell_s(\boldsymbol{\pi}^{(l)}, \boldsymbol{\beta}^{(l)}, g^{(l)}).$$

Simulation studies and real data applications showed the effectiveness of the new methods. One main drawback of this method is that by allowing for a completely flexible error distribution, the algorithm might not be able to identify all the components as they are asked to. In addition, those fully nonparametric methods also face difficulties, such as bandwidth selection for kernel smoothing.

2.3.2 *Log-Concave Density Error*

Hu et al. (2017) also tried to relax the normality assumption on the error terms. But instead of completely nonparametric, Hu et al. (2017) proposed to estimate the mixture regression parameters by only assuming the components to have log-concave error densities. That is, the model can still be written as (2.11), but in this case, g_c is assumed to be log-concave. That is, $g_c(x) = \exp\{\phi_c(x)\}$ for some unknown concave function $\phi_c(x)$. Examples of log-concave densities are normal, Laplace, chi-square, logistic, gamma with shape parameter greater than one, beta with both parameters greater than one, and so on.

When error terms for different components are assumed to have different distributions, an EM-type algorithm is proposed.

Algorithm 2.3.2

E step: Compute the classification probabilities

$$p_{ic}^{(l+1)} = \frac{\pi_c^{(l)} g_c^{(l)}(y_i - \mathbf{x}_i^T \boldsymbol{\beta}_c^{(l)})}{\sum_{c'=1}^{C} \pi_{c'}^{(l)} g_{c'}^{(l)}(y_i - \mathbf{x}_i^T \boldsymbol{\beta}_{c'}^{(l)})},$$

for $i = 1, \cdots, n$ and $c = 1, \cdots, C$.

M step:

1. Calculate the log-likelihood for each observation

$$\ell_i^{(l)} = \log \sum_{c=1}^{C} \pi_c^{(l)} g_c^{(l)}(y_i - \mathbf{x}_i^T \boldsymbol{\beta}_c^{(l)}),$$

   for $i = 1, \cdots, n$, and update the trimmed subset of size $n - s$, denoted by $I^{(l+1)}$, which has the $n - s$ largest log-likelihoods.
2. Update π as

$$\pi_c^{(l+1)} = \frac{1}{n-s} \sum_{i \in I^{(l+1)}} p_{ic}^{(l+1)}, c = 1, \cdots, C.$$

3. Update $\boldsymbol{\beta}$ as

$$\tilde{\boldsymbol{\beta}}_c^{(l+1)} = \arg\max_{\beta_j} \sum_{i \in I^{(l+1)}} p_{ic}^{(l+1)} \log g_c^{(l)}(y_i - \mathbf{x}_i^T \boldsymbol{\beta}_j), c = 1, \cdots, C.$$

4. Shift the intercept of $\tilde{\boldsymbol{\beta}}_c^{(l+1)}$ for residuals to have mean zero

$$\hat{\boldsymbol{\beta}}_c^{(l+1)} = (\hat{\beta}_{c,0}^{(l+1)}), \tilde{\beta}_{c,1}^{(l+1)}, \cdots, \tilde{\beta}_{c,p-1}^{(l+1)}),$$

   for $c = 1, \cdots, C$, where

$$\hat{\beta}_{c,0}^{(l+1)} = \tilde{\beta}_{c,0}^{(l+1)} + d_c^{(l+1)} \text{with} d_c^{(l+1)} = \frac{1}{n-s} \sum_{i \in I^{(l+1)}} p_{ic}^{(l+1)}(y_i - \mathbf{x}_i^T \tilde{\boldsymbol{\beta}}_c^{(l+1)}).$$

5. Update g_c by

$$g_c^{(l+1)} = \arg\max_{g_c \in \mathscr{G}} \sum_{i=1}^{n} p_{ic}^{(l+1)} \log g_c(y_i - \mathbf{x}_i^T \hat{\boldsymbol{\beta}}_c^{(l+1)}), c = 1, \cdots, C,$$

   where $\mathscr{G}$ is the family of all log-concave densities.

The updates of $\boldsymbol{\beta}_c$ and g_c in the above algorithm are through existing R package *optim* and *mlelcd*. Similar algorithm can be obtained for the special case when all components have the same error density g. The main difference lies in the estimate of g_c, where entire residuals, instead of residuals only from the corresponding component, are used for the estimation of g. Judging from numerical studies, the new method worked comparable to standard normal mixture EM algorithm when the underlying component error densities were normal, and much better otherwise.

2.3.3 Mixtures of Quantile Regressions

When the error pdfs are symmetric about zero, the methods studied in the previous two sections would seem to be reasonable. However, in cases of asymmetric error distributions, median or other quantile regressions should be considered.

Since quantiles are more robust than the mean, Wu and Yao (2016) studied a semiparametric mixture of quantile regressions model allowing regressions of the conditional quantiles on the covariates without any parametric assumption on the error densities. In practice, the median regression is a regular choice, but the discussion of the article is not restricted to it. Given $\mathscr{C} = c$, Wu and Yao (2016) assumed the response depending on the covariates through

$$Y = \mathbf{x}^T \boldsymbol{\beta}_c(\tau) + \epsilon_c(\tau), \tag{2.12}$$

where $\boldsymbol{\beta}_c(\tau) = (\beta_{0c}(\tau), \cdots, \beta_{pc}(\tau))^T$ are the τth quantile regression coefficient for the cth component. Independent of the covariates $\mathbf{x}$, $\epsilon_c(\tau)$ is assumed to have pdf's $g_c(\cdot)$, whose τth quantiles are zero. There are no more restrictions on the errors, since the distributions will be estimated nonparametrically. By fitting the data with varying conditional quantile functions, model (2.12) is more robust to non-normal component distributions and capable of revealing more detailed structure of the data. Since there is no parametric assumption made on the error densities, a kernel based EM-type algorithm is proposed to estimate the parameters $\boldsymbol{\theta} = (\pi_1, \boldsymbol{\beta}_1, \cdots, \pi_C, \boldsymbol{\beta}_C)$ and the error pdfs $\mathbf{g} = (g_1, \cdots, g_C)$.

Algorithm 2.3.3
E step: Compute the classification probabilities

$$p_{ic}^{(l+1)} = \frac{\pi_c^{(l)} g_c^{(l)}(r_{ic}^{(l)})}{\sum_{c'=1}^{C} \pi_{c'}^{(l)} g_{c'}^{(l)}(r_{ic'}^{(l)})},$$

for $i = 1, \cdots, n$ and $c = 1, \cdots, C$, where $r_{ic}^{(l)} = y_i - \mathbf{x}_i^T \boldsymbol{\beta}_c^{(l)}$.
M step: Update the estimates

$$\pi_c^{(l+1)} = \frac{1}{n}\sum_{i=1}^{n} p_{ic}^{(l+1)},$$

$$\boldsymbol{\beta}_c^{(l+1)} = \arg\min_{\beta_c} \sum_{i=1}^{n} p_{ic}^{(l+1)} \rho_\tau(y_i - \mathbf{x}_i^T \boldsymbol{\beta}_c),$$

$$g_c^{(l+1)}(t) = \sum_{i=1}^{n}\sum_{s=1}^{2} w_{sc}^{(l+1)} p_{ic}^{(l+1)} K_h(t - r_{ic}^{(l+1)}) I_s(r_{ic}^{(l+1)}),$$

for $c = 1, \cdots, C$, where $\rho_\tau(u) = u(\tau - I(u < 0))$, $I_1(u) = I(u \le 0)$, and $I_2(u) = I(u > 0)$. $w_{sc}^{(l+1)}$s are calculated by solving a system of linear equations

$$\sum_{i=1}^{n}\sum_{s=1}^{2} w_{sc}^{(l+1)} p_{ic}^{(l+1)} I_s(r_{ic}^{(l+1)}) = 1,$$

$$\sum_{i=1}^{n}\sum_{s=1}^{2} w_{sc}^{(l+1)} p_{ic}^{(l+1)} v_{ic}^{(l+1)} I_s(r_{ic}^{(l+1)}) = \tau,$$

where $v_{ic}^{(l+1)} = \int_{-\infty}^{0} K_h(t - r_{ic}^{(l+1)})dt$.

Through numerical studies, the mixture of quantile regressions were shown to be robust to non-normal error distributions and capable of revealing more data information.

2.4 Semiparametric Mixture of Nonparametric Regressions

In traditional finite mixture of regression models (2.1) and previously discussed semiparametric mixture models, the regressions functions are always assumed to be linear. In the following, different models were proposed to relax the linearity assumption to allow for more possibilities.

2.4.1 Nonparametric Mixture of Regressions

Motivated by a US house price index data, Huang et al. (2013) proposed a nonparametric mixture of regression models, where the mixing proportions, the mean functions, and the variance functions are all unknown but smooth functions. However, since the error term is still assumed to be normally distributed, we still treat it as a semiparametric mixture model. The conditional distribution of Y given X is

$$Y|_{X=x} \sim \sum_{c=1}^{C} \pi_c(x) N\{m_c(x), \sigma_c^2(x)\}, \tag{2.13}$$

where $\pi_c(\cdot), m_c(\cdot), \sigma_c^2(\cdot)$ are nonparametric functions, and $\sum_{c=1}^{C} \pi_c(\cdot) = 1$. The identifiability issue was studied for model (2.13), and kernel regression was applied for estimation. To be more specific, $\pi_c(x), m_c(x)$, and $\sigma_c^2(x)$ are estimated by maximizing the following local log-likelihood

$$\sum_{i=1}^{n} \log \left\{ \sum_{c=1}^{C} \pi_c \phi(Y_i | m_c, \sigma_c^2) \right\} K_h(X_i - x).$$

The EM algorithm corresponds to this method is:

Algorithm 2.4.1
E step: calculate the classification probability

$$p_{ic}^{(l+1)} = \frac{\pi_c^{(l)}(X_i)\phi\{Y_i | m_c^{(l)}(X_i), \sigma_c^{2(l)}(X_i)\}}{\sum_{c'=1}^{C} \pi_{c'}^{(l)}(X_i)\phi\{Y_i | m_{c'}^{(l)}(X_i), \sigma_{c'}^{2(l)}(X_i)\}}.$$

M step: update estimates

$$\pi_c^{(l+1)}(x) = \frac{\sum_{i=1}^{n} p_{ic}^{(l+1)} K_h(X_i - x)}{\sum_{i=1}^{n} K_h(X_i - x)},$$

$$m_c^{(l+1)}(x) = \frac{\sum_{i=1}^{n} w_{ic}^{(l+1)}(x) Y_i}{\sum_{i=1}^{n} w_{ic}^{(l+1)}(x)},$$

$$\sigma_c^{2(l+1)}(x) = \frac{\sum_{i=1}^{n} w_{ic}^{(l+1)}(x)\{Y_i - m_c^{(l+1)}(x)\}^2}{\sum_{i=1}^{n} w_{ic}^{(l+1)}(x)},$$

where $w_{ic}^{(l+1)}(x) = p_{ic}^{(l+1)} K_h(X_i - x)$.

The identifiability of the model was discussed, and the estimates were shown to have $\sqrt{nh}$ convergence rate.

2.4.2 *Nonparametric Component Regression Functions*

Although flexible, model (2.13) assumed by Huang et al. (2013) sacrificed the efficiency of the estimates in such a setting. Xiang and Yao (2016), on the other hand, discussed a new class of semiparametric mixture of regression models, where the mixing proportions and variances were constants, but the component regression functions were smooth functions of a covariate. The conditional distribution of Y given $X = x$ can be written as

$$Y|_{X=x} \sim \sum_{c=1}^{C} \pi_c \phi(Y|m_c(x), \sigma_c^2), \tag{2.14}$$

where $m_c(\cdot)$ are unknown smooth functions. Compared to the fully nonparametric mixture of regression models (2.13), this model improves the efficiency of the estimates of the mean functions by assuming the mixing proportions and variances to be constants, which are also presumed by the traditional mixture of linear regressions. Identifiability of model (2.14) is discussed under mild conditions, and two estimation methods are discussed.

Due to the existence of both global and local parameters, model (2.14) is more difficult to estimate. First, a regression spline based estimator is applied to transfer the semiparametric mixture model to a parametric mixture model. Since for any $m_c(x)$, it can be approximated by a cubic spline as

$$m_c(x) \approx \sum_{q=1}^{Q+4} \beta_{cq} B_q(x), c = 1, ..., C,$$

where $B_1(x)$, ..., $B_{Q+4}(x)$ is a cubic spline basis and Q is the number of internal knots. Then, model (2.14) is then turned into

$$Y|_{X=x} \sim \sum_{c=1}^{C} \pi_c \phi(Y| \sum_{q=1}^{Q+4} \beta_{cq} B_q(x), \sigma_c^2),$$

where the parameters can then be estimated by a traditional EM algorithm for mixture of linear regressions. As Xiang and Yao (2016) pointed out, this method was easy to carry, but more work remained to be down about its the theoretical properties. In addition, a more efficient one-step backfitting estimation procedure was proposed. Specifically,

1. Estimate $\boldsymbol{\pi} = \{\pi_1, ..., \pi_{C-1}\}^T$, $\mathbf{m} = \{m_1, ..., m_C\}^T$ and $\boldsymbol{\sigma}^2 = \{\sigma_1^2, ..., \sigma_C^2\}^T$ locally by maximizing the following local log-likelihood function:

$$\ell_1(\boldsymbol{\pi}(x), \mathbf{m}(x), \boldsymbol{\sigma}^2(x)) = \sum_{i=1}^{n} \log\{\sum_{j=1}^{k} \pi_j \phi(Y_i|m_j, \sigma_j^2)\} K_h(X_i - x).$$

   Let $\tilde{\boldsymbol{\pi}}(x)$, $\tilde{\mathbf{m}}(x)$, and $\tilde{\boldsymbol{\sigma}}^2(x)$ be the maximizer.
2. To improve the efficiency, update the estimates of $\boldsymbol{\pi}$ and $\boldsymbol{\sigma}^2$ by maximizing the following log-likelihood function:

$$\ell_2(\boldsymbol{\pi}, \boldsymbol{\sigma}^2) = \sum_{i=1}^{n} \log\{\sum_{j=1}^{k} \pi_j \phi(Y_i | \tilde{m}_j(X_i), \sigma_j^2)\}.$$

Denote by $\hat{\boldsymbol{\pi}}$ and $\hat{\boldsymbol{\sigma}}^2$ the solution of this step.

3. Further improve the estimate of $\mathbf{m}(\cdot)$ by maximizing the following local log-likelihood function:

$$\ell_3(\mathbf{m}(x)) = \sum_{i=1}^{n} \log\{\sum_{j=1}^{k} \hat{\pi}_j \phi(Y_i | m_j, \hat{\sigma}_j^2)\} K_h(X_i - x).$$

Let $\hat{\mathbf{m}}(x)$ be the solution. Then, $\hat{\boldsymbol{\pi}}$, $\hat{\mathbf{m}}(x)$, and $\hat{\boldsymbol{\sigma}}^2$ as the one-step backfitting estimates.

A generalized likelihood ratio test was proposed to compare between model (2.13) and model (2.14), which was shown to have the Wilks types of results. Numerical studies showed that the finite sample null distribution was quite close to a χ^2-distribution with the suggested degrees of freedom, especially for larger sample sizes.

2.4.3 Mixture of Regressions with Single-Index

Model (2.13) and model (2.14) can be used in many applications, but due to the "curse of dimensionality," they might not be suitable for predictors with high dimensions. Applying the ideas of single-index model (SIM) into finite mixture of regressions, Xiang and Yao (2017) proposed a mixture of single-index models (MSIM) and a mixture of regression models with varying single-index proportions (MRSIP), as extensions of some existing models. Many of recently proposed semiparametric/nonparametric mixture regression models can be considered as special cases of the proposed models. Efficient estimation methods were proposed to achieve the optimal convergence rate for both the parameters and the nonparametric functions, and theoretical results showed that nonparametric functions can be estimated with the same asymptotic accuracy as if the parameters were known and the index parameters can be estimated with the traditional parametric $\sqrt{n}$ convergence rate.

Mixture of Single-Index Models (MSIM) The conditional distribution of Y given $\mathbf{x}$ is assumed as

$$Y|_{\mathbf{x}} \sim \sum_{c=1}^{C} \pi_c(\boldsymbol{\alpha}^T\mathbf{x}) \phi(Y_i | m_c(\boldsymbol{\alpha}^T\mathbf{x}), \sigma_c^2(\boldsymbol{\alpha}^T\mathbf{x})), \tag{2.15}$$

where $\pi_c(\cdot)$, $m_c(\cdot)$, and $\sigma_c^2(\cdot)$ are unknown but smooth functions. Compared to model (2.13), model (2.15) focuses on index $\boldsymbol{\alpha}^T\mathbf{x}$, and so the so-called "curse of dimensionality" in fitting multivariate nonparametric regression functions is avoided. When $C = 1$, model (2.15) reduces to a single index model (Ichimura 1993; Härdle et al. 1993). If $\mathbf{x}$ is a scalar, then model (2.15) reduces to model (2.13). To achieve the optimal convergence rate for both the index parameter and nonparametric functions, Xiang and Yao (2017) proposed the following algorithm:

Algorithm 2.4.2

1. Given $\hat{\boldsymbol{\alpha}}$, apply a modified EM-type algorithm to update the estimates of nonparametric functions.
   E step: Calculate the classification probability

$$p_{ic}^{(l+1)} = \frac{\pi_c^{(l)}(\hat{\boldsymbol{\alpha}}^T\mathbf{x}_i)\phi(Y_i|m_c^{(l)}(\hat{\boldsymbol{\alpha}}^T\mathbf{x}_i), \sigma_c^{2(l)}(\hat{\boldsymbol{\alpha}}^T\mathbf{x}_i))}{\sum_{c'=1}^C \pi_{c'}^{(l)}(\hat{\boldsymbol{\alpha}}^T\mathbf{x}_i)\phi(Y_i|m_{c'}^{(l)}(\hat{\boldsymbol{\alpha}}^T\mathbf{x}_i), \sigma_{c'}^{2(l)}(\hat{\boldsymbol{\alpha}}^T\mathbf{x}_i))}.$$

   M-step:
   Update the estimates

$$\pi_c^{(l+1)}(z) = \frac{\sum_{i=1}^n p_{ic}^{(l+1)} K_h(\hat{\boldsymbol{\alpha}}^T\mathbf{x}_i - z)}{\sum_{i=1}^n K_h(\hat{\boldsymbol{\alpha}}^T\mathbf{x}_i - z)},$$

$$m_c^{(l+1)}(z) = \frac{\sum_{i=1}^n p_{ic}^{(l+1)} Y_i K_h(\hat{\boldsymbol{\alpha}}^T\mathbf{x}_i - z)}{\sum_{i=1}^n p_{ic}^{(l+1)} K_h(\hat{\boldsymbol{\alpha}}^T\mathbf{x}_i - z)},$$

$$\sigma_c^{2(l+1)}(z) = \frac{\sum_{i=1}^n p_{ic}^{(l+1)}(Y_i - m_c^{(l+1)}(z))^2 K_h(\hat{\boldsymbol{\alpha}}^T\mathbf{x}_i - z)}{\sum_{i=1}^n p_{ic}^{(l+1)} K_h(\hat{\boldsymbol{\alpha}}^T\mathbf{x}_i - z)}.$$

2. Update the estimate of $\boldsymbol{\alpha}$ by maximizing

$$\ell(\boldsymbol{\alpha}) = \sum_{i=1}^n \log\{\sum_{c=1}^C \hat{\pi}_c(\boldsymbol{\alpha}^T\mathbf{x}_i)\phi(Y_i|\hat{m}_c(\boldsymbol{\alpha}^T\mathbf{x}_i), \hat{\sigma}_c^2(\boldsymbol{\alpha}^T\mathbf{x}_i))\}.$$

One can further iterate the steps to improve the efficiency of the estimates.

Mixture of Regression Models with Varying Single-Index Proportions (MRSIP)
The MRSIP assumes that

$$Y|_{\mathbf{x}} \sim \sum_{c=1}^C \pi_c(\boldsymbol{\alpha}^T\mathbf{x})N(\mathbf{x}^T\boldsymbol{\beta}_c, \sigma_c^2). \tag{2.16}$$

That is, they still assumed linearity in the mean functions, and so kept the easy interpretation of the linear component regression functions while assuming that the mixing proportions are smooth functions of an index $\boldsymbol{\alpha}^T\mathbf{x}$. When $C = 1$, model (2.16) reduces to the traditional linear regression model. If $\mathbf{x}$ is a scalar, then model (2.16) reduces to (2.4) and (2.7), which were considered by Young and Hunter (2010) and Huang and Yao (2012). Similar to the estimation of MSIM, Xiang and Yao (2017) proposed to iteratively estimate the global parameters and nonparametric functions. Detailed algorithm is listed below:

1. Given $(\hat{\boldsymbol{\alpha}}, \hat{\boldsymbol{\beta}}, \hat{\boldsymbol{\sigma}}^2)$, update the estimates of $\pi_c(z)$ by the following modified EM-type algorithm.

Algorithm 2.4.3
E-step:
Calculate the expectations of component labels based on estimates from l^{th} iteration:

$$p_{ic}^{(l+1)} = \frac{\pi_c^{(l)}(\hat{\boldsymbol{\alpha}}^T\mathbf{x}_i)\phi(Y_i|\mathbf{x}_i^T\hat{\boldsymbol{\beta}}_c, \hat{\sigma}_c^2)}{\sum_{c'=1}^C \pi_{c'}^{(l)}(\hat{\boldsymbol{\alpha}}^T\mathbf{x}_i)\phi(Y_i|\mathbf{x}_i^T\hat{\boldsymbol{\beta}}_{c'}, \hat{\sigma}_{c'}^2)}.$$

M-step:
Update the estimate

$$\pi_c^{(l+1)}(z) = \frac{\sum_{i=1}^n p_{ic}^{(l+1)} K_h(\hat{\boldsymbol{\alpha}}^T\mathbf{x}_i - z)}{\sum_{i=1}^n K_h(\hat{\boldsymbol{\alpha}}^T\mathbf{x}_i - z)}.$$

2. Update $(\hat{\boldsymbol{\alpha}}, \hat{\boldsymbol{\beta}}, \hat{\boldsymbol{\sigma}}^2)$ by maximizing

$$\ell(\boldsymbol{\alpha}, \boldsymbol{\beta}, \boldsymbol{\sigma}^2) = \sum_{i=1}^n \log\{\sum_{c=1}^C \hat{\pi}_c(\boldsymbol{\alpha}^T\mathbf{x}_i)\phi(Y_i|\mathbf{x}_i^T\boldsymbol{\beta}_c, \sigma_c^2)\}.$$

Since there is no explicit solution for this, iterate between the following steps.

a. Given $\hat{\boldsymbol{\alpha}}$, update $(\boldsymbol{\beta}, \boldsymbol{\sigma}^2)$.

Algorithm 2.4.4
E-step:
Calculate the expectations of component identities:

$$p_{ic}^{(l+1)} = \frac{\hat{\pi}_c(\hat{\boldsymbol{\alpha}}^T\mathbf{x}_i)\phi(Y_i|\mathbf{x}_i^T\boldsymbol{\beta}_c^{(l)}, \sigma_c^{2(l)})}{\sum_{c'=1}^C \hat{\pi}_{c'}(\hat{\boldsymbol{\alpha}}^T\mathbf{x}_i)\phi(Y_i|\mathbf{x}_i^T\boldsymbol{\beta}_{c'}^{(l)}, \sigma_{c'}^{2(l)})}.$$

M-step:
Update $\boldsymbol{\beta}$ and $\boldsymbol{\sigma}^2$:

$$\boldsymbol{\beta}_c^{(l+1)} = (S^T R_c^{(l+1)} S)^{-1} S^T R_c^{(l+1)} \mathbf{y},$$

$$\sigma_c^{2(l+1)} = \frac{\sum_{i=1}^n p_{ic}^{(l+1)} (y_i - \mathbf{x}_i^T \boldsymbol{\beta}_c^{(l+1)})^2}{\sum_{i=1}^n p_{ic}^{(l+1)}},$$

where $R_c^{(l+1)} = \text{diag}\{p_{ic}^{(l+1)}, ..., p_{nc}^{(l+1)}\}$, $S = (\mathbf{x}_1, ..., \mathbf{x}_n)^T$, $\mathbf{y} = (y_1, ..., y_n)^T$.

b. Given $(\hat{\boldsymbol{\beta}}, \hat{\boldsymbol{\sigma}}^2)$, maximize $\ell(\boldsymbol{\alpha}) = \sum_{i=1}^n \log\{\sum_{c=1}^C \hat{\pi}_c(\boldsymbol{\alpha}^T \mathbf{x}_i)\phi(Y_i|\mathbf{x}_i^T \hat{\boldsymbol{\beta}}_c, \hat{\sigma}_c^2)\}$ to update the estimate of $\boldsymbol{\alpha}$, using some numerical methods.

2.5 Semiparametric Regression Models for Longitudinal/Functional Data

2.5.1 *Mixture of Time-Varying Effects for Intensive Longitudinal Data*

Intensive longitudinal data (ILD) are becoming increasingly popular in behavioral sciences, due to its richness in information. On the other hand, however, heterogeneous and nonlinear in nature, ILD is facing numerous analytical challenges. Dziak et al. (2015) proposed a mixture of time-varying effect models (MixTVCM), which incorporate time-varying effect model (TVEM) in a finite mixture model framework. To define the model, Dziak et al. (2015) modeled the latent class membership as a multiple-category logistic regression model. That is, conditional on time-invariant subject-level covariates $s_1, \cdots, s_Q$, the probability that individual i comes from class c is

$$\pi_{ic} = P(\mathscr{C}_i = c) = \frac{\exp\left(\gamma_{0,c} + \sum_{q=1}^Q \gamma_{1qc} s_q\right)}{\sum_{c'=1}^C \exp\left(\gamma_{0,c'} + \sum_{q=1}^Q \gamma_{1qc'} s_q\right)}.$$

Within each class, the response y_{ij} is modeled as

$$\mu_{ij} = E(y_{ij}|\mathscr{C}_i = c) = \beta_{0c}(t_{ij}) + \beta_{10}(t_{ij})x_{ij1} + \cdots + \beta_{pc}(t_{ij})x_{ijP}, \qquad (2.17)$$

where $x_1, \cdots, x_P$ is the observation-level covariates. Model (2.17) is essentially the same as the TVEM in Tan et al. (2012). The variance of y_{ij} is assumed as

$$\text{cov}(y_{ij}, y_{ij'}) = \sigma_a^2 \rho^{|t_{ij} - t_{ij'}|} + \sigma_e^2,$$

where σ_a^2 is the variance of a normally distributed subject-level error, and σ_e^2 is the variance of a normally distributed observation-level error. Although the mean of y is modeled nonparametrically, a normal probability distribution for the errors is assumed for computation feasibility, and therefore, MixTVEM is a semiparametric mixture model. EM algorithm is used to accommodate the mixture structure. For identifiability reasons, individuals should be assigned to one and only one latent class, and therefore, the EM algorithm assigns posterior probabilities to individuals as a whole rather than to particular observations. Penalized B-spline is used to approximate $\beta(\cdot)$, where the penalization is considered to ensure for a smooth and parsimonious shape.

2.5.2 *Mixtures of Gaussian Processes*

To incorporate both functional and inhomogeneous data, Huang et al. (2014) proposed a new estimation procedure for the mixture of Gaussian processes, which can be used to deal with data collected at irregular, possibly subject-depending time points. The model studied in Huang et al. (2014) is the following. Suppose that $P(\mathscr{C} = c) = \pi_c, c = 1, \cdots, C$. Conditional on $\mathscr{C} = c$,

$$y_{ij} = \mu_c(t_{ij}) + \sum_{q=1}^{\infty} \xi_{iqc}\nu_{qc}(t_{ij}) + \epsilon_{ij}, i = 1, \cdots, n; j = 1, \cdots, N_i, \tag{2.18}$$

where ϵ_{ij}'s are iid of $N(0, \sigma^2)$. $\mu_c(t)$ is the mean of the Gaussian process, whose corresponding covariance function is $G_c(s, t)$. ξ_{iqc} and $\nu_{qc}(t)$ are the functional principal component FPC score and eigenfunctions of $G_c(s, t)$ (Karhunen-Loève theorem; Sapatnekar 2011). Huang et al. (2014) also considered a reduced version of model (2.18). Given $\mathscr{C} = c$,

$$y_{ij} = \mu_c(t_{ij}) + \epsilon_{il}^*, \tag{2.19}$$

where ϵ_{ij}^*'s are independent with $E(\epsilon_{ij}^*) = 0$ and $\text{var}(\epsilon_{ij}^*) = \sigma_c^{*2}(t_{ij})$ with $\sigma_c^{*2}(t) = G_c(t, t) + \sigma^2$. That is, in this model, the data within subjects are independent ($G_c(s, t) = 0$ if $s \neq t$). The estimation procedure combines the idea of EM algorithm, kernel regression, and functional principal component (FPC) analysis. To estimate for model (2.19), Huang et al. (2014) proposed an EM-type algorithm:

Algorithm 2.5.1
E step: calculate the classification probability

$$p_{ic}^{(l+1)} = \frac{\pi_c^{(l)}[\prod_{j=1}^{N_i} \phi\{y_{ij}|\mu_c^{(l)}(t_{ij}), \sigma_c^{*2(l)}(t_{ij})\}]}{\sum_{c'=1}^{C} \pi_{c'}^{(l)}[\prod_{j=1}^{N_i} \phi\{y_{ij}|\mu_{c'}^{(l)}(t_{ij}), \sigma_{c'}^{*2(l)}(t_{ij})\}]}.$$

M step: To estimate the nonparametric functions $\mu_c(\cdot)$'s and $\sigma_c^{*2}(\cdot)$'s, maximize the following local log-likelihood

$$\sum_{i=1}^{n}\sum_{c=1}^{C} p_{ic}^{(l+1)} \sum_{j=1}^{N_i} [\log \phi\{y_{ij}|\mu_c(t_0), \sigma_c^{*2}(t_0)\}] K_h(t_{ij}-t_0),$$

where t_0 is in the neighborhood of t_{ij}. This yields

$$\pi_c^{(l+1)} = \frac{1}{n}\sum_{i=1}^{n} p_{ic}^{(l+1)},$$

$$\mu_c^{(l+1)}(t_0) = \frac{\sum_{i=1}^{n}\sum_{j=1}^{N_i} w_{ijc}^{(l+1)} y_{ij}}{\sum_{i=1}^{n}\sum_{j=1}^{N_i} w_{ijc}^{(l+1)}},$$

$$\sigma_c^{*2(l+1)}(t_0) = \frac{\sum_{i=1}^{n}\sum_{j=1}^{N_i} w_{ijc}^{(l+1)}\{y_{ij}-\mu_c^{(l+1)}(t_0)\}^2}{\sum_{i=1}^{n}\sum_{j=1}^{N_i} w_{ijc}^{(l+1)}},$$

where $w_{ijc}^{l+1)} = p_{ic}^{(l+1)} K_h(t_{ij}-t_0)$. Denote $\tilde{\pi}_c$, $\tilde{\mu}_c$, and $\tilde{\sigma}_c^{*2}$ as the resulting estimates.

To estimate for model (2.19), Huang et al. (2014) suggested to use $\tilde{\pi}_c$, $\tilde{\mu}_c$ as the initial estimates for the estimation of the covariances. Let $\bar{G}_{ic}(t_{ij}, t_{il}) = \{y_{ij} - \tilde{\mu}_c(t_{ij})\}\{y_{il} - \tilde{\mu}_c(t_{il})\}$, then $G_c(s,t)$ is estimated by minimizing

$$\sum_{i=1}^{n} p_{ic} \sum_{1\le j\neq l\le N} \{\bar{G}_{ic}(t_{ij}, t_{il}) - \beta_0\}^2 K_{h^*}(t_{ij}-s)(t_{il}-t)$$

with respect to β_0. Denote $\hat{G}_c(s,t) = \hat{\beta}_0$ as the estimates. The estimates of λ_{qc} and v_{qc} are calculated by

$$\int_T \hat{G}_c(s,t)\hat{v}_{qc}(s)ds = \hat{\lambda}_{qc}\hat{v}_{qc}(t),$$

where $\int_T \hat{v}_{qc}^2(t)dt = 1$ and $\int_T \hat{v}_{sc}(t)\hat{v}_{qc}(t)dt = 0$ if $s \neq q$.

2.5.3 *Mixture of Functional Linear Models*

Wang et al. (2016) proposed a mixture of functional linear models to analyze heterogeneous functional data. Conditional on $\mathscr{C} = c$, $\{y(t), t \in T\}$ follows a functional linear model

$$y(t)|_{\mathscr{C}=c} = \mathbf{X}(t)^T \boldsymbol{\beta}_c(t) + \epsilon_c(t), \tag{2.20}$$

where $\mathbf{X}(t)$ is a random covariate process of dimension p, and $\boldsymbol{\beta}_c(t)$ is a smooth regression coefficient function of component c. $\epsilon_c(t)$ is a Gaussian process with mean zero, independent of $\mathbf{X}(t)$, and is assumed as

$$\epsilon_c(t) = \zeta_c(t) + e(t),$$

where $\zeta(t)$ denotes a trajectory process with covariance $\Gamma_c(s,t) = \text{cov}\{\xi_c(s), \xi_c(t)\}$, and $e(t)$ is the measurement error with constant variance σ^2. Define $y_{ij} = y_i(t_{ij})$, $j = 1, \cdots, N_i$, and similarly, define ϵ_{cij}, e_{ij}, etc. Similar to Huang et al. (2014), by Karhunen-Loève theorem, model (2.20) is represented as

$$y_{ij} = \mathbf{X}_i(t_{ij})^T \boldsymbol{\beta}_c(t_{ij}) + \sum_{q=1}^{\infty} \xi_{iqc} v_{qc}(t_{ij}) + e_{ij},$$

where $v_{qc}(\cdot)$'s are eigenfunctions of $\Gamma_c(s,t)$ and λ_{qc}'s are corresponding eigenvalues, and ξ_{iqc}'s are uncorrelated FPC of $\zeta_c(t)$ satisfying $E(\xi_{iqc}) = 0$, $\text{var}(\xi_{iqc}) = \lambda_{qc}$. If ignoring the correlation structure, y_{ij} can be thought to be coming from the following mixture of Gaussian process

$$y(t) \sim \sum_{c=1}^{C} \pi_c N\{\mathbf{X}(t)^T \boldsymbol{\beta}_c(t), \sigma_c^{*2}(t)\},$$

where $\sigma_c^{*2}(t) = \Gamma_c(t.t) + \sigma^2$. Then, the parameters π_c, $\boldsymbol{\beta}_c(\cdot)$, and $\sigma_c^{*2}(\cdot)$ can be estimated by an EM-type algorithm.

Algorithm 2.5.2
E step: calculate the classification probability

$$p_{ic}^{(l+1)} = \frac{\pi_c^{(l)} \left[\prod_{j=1}^{N_i} \phi\{y_{ij} | \mathbf{X}_i(t_{ij})^T \boldsymbol{\beta}_c^{(l)}(t_{ij}), \sigma_c^{*2(l)}(t_{ij})\}\right]}{\sum_{c'=1}^{C} \pi_{c'}^{(l)} \left[\prod_{j=1}^{N_i} \phi\{y_{ij} | \mathbf{X}_i(t_{ij})^T \boldsymbol{\beta}_{c'}^{(l)}(t_{ij}), \sigma_{c'}^{*2(l)}(t_{ij})\}\right]}.$$

M step: Update the estimates

$$\pi_c^{(l+1)} = \frac{1}{n} \sum_{i=1}^{n} p_{ic}^{(l+1)},$$

$$\boldsymbol{\beta}_c^{(l+1)}(t_0) = \left\{\sum_{i=1}^{n} \mathbf{X}_i^T W_{ic}^{(l+1)}(t_0) \mathbf{X}_i\right\}^{-1} \left\{\sum_{i=1}^{n} \mathbf{X}_i^T W_{ic}^{(l+1)}(t_0) \mathbf{y}_i\right\},$$

$$\sigma_c^{*2(l+1)}(t_0) = \frac{\sum_{i=1}^{n} \sum_{j=1}^{N_i} w_{cij}^{(l+1)} \{y_{ij} - \mathbf{X}_i(t_{ij})^T \boldsymbol{\beta}_c^{(l+1)}(t_0)\}^2}{\sum_{i=1}^{n} \sum_{j=1}^{N_i} w_{cij}^{(l+1)}},$$

where $\mathbf{y}_i = (y_{i1}, \cdots, y_{iN_i})^T, \mathbf{X}_i = (\mathbf{X}_i(t_{i1}), \cdots, \mathbf{X}_i(t_{iN_i}))^T, w_{icj}^{(l+1)} = p_{ic}^{(l+1)} K_{h_\beta}(t_{ij} - t_0), W_{ic}^{(l+1)}(t_0) = \text{diag}\{p_{ic}^{(l+1)} K_{h_\beta}(t_{i1} - t_0), \cdots, p_{ic}^{(l+1)} K_{h_\beta}(t_{iN_i} - t_0)\}$, and h_β is a bandwidth. The estimates based on this procedure is denoted as $\hat{\pi}_c$, $\hat{\boldsymbol{\beta}}_c(\cdot)$, and $\hat{\sigma}_c^{*2}(\cdot)$.

To model for general covariance structure, the parameters can be estimated iteratively as follows.

1. Estimate the covariance function $\Gamma_c(s,t)$

$$\Gamma_c(s,t)^{(l+1)} = \frac{\sum_{i=1}^n p_{ic}^{(l)} \sum_{1 \le j \ne l \le N_i} \gamma_{ic}^{(l)}(t_{ij}, t_{il}) K_{h_\Gamma}(t_{ij} - s) K_{h_\Gamma}(t_{il} - t)}{\sum_{i=1}^n p_{ic}^{(l)} \sum_{1 \le j \ne l \le N_i} K_{h_\Gamma}(t_{ij} - s) K_{h_\Gamma}(t_{il} - t)},$$

   where $\gamma_{ic}^{(l)}(t_{ij}, t_{il}) = \{y_{ij} - \mathbf{X}_i(t_{ij})^T \boldsymbol{\beta}_c^{(l)}(t_{ij})\}\{y_{il} - \mathbf{X}_i(t_{il})^T \boldsymbol{\beta}_c^{(l)}(t_{il})\}$. Denote $\lambda_{qc}^{(l+1)}$ and $v_{qc}^{(l+1)}(\cdot)$ as the corresponding eigenvalues and eigenfunctions, respectively.
2. Calculate

$$\xi_{iqc}^{(l+1)} = \int_T \{y_i(t) - \mathbf{X}_i(t)^T \boldsymbol{\beta}_c^{(l)}(t)\} v_{qc}^{(l+1)}(t) dt,$$

$$y_c^{(l+1)}(t_{ij}) = y_{ij} - \sum_q \xi_{iqc}^{(l+1)} I\{\lambda_{qc}^{(l+1)} > 0\} v_{qc}^{(l+1)}(t_{ij}).$$

3. One cycle of E-step:

$$p_{ic}^{(l+1)} = \frac{\pi_c^{(l)} \left[\prod_{j=1}^{N_i} \phi\{y_c^{(l+1)}(t_{ij}) | \mathbf{X}_i(t_{ij})^T \boldsymbol{\beta}_c^{(l)}(t_{ij}), \sigma^{2(l)}\}\right]}{\sum_{c'=1}^C \pi_{c'}^{(l)} \left[\prod_{j=1}^{N_i} \phi\{y_c^{(l+1)}(t_{ij}) | \mathbf{X}_i(t_{ij})^T \boldsymbol{\beta}_{c'}^{(l)}(t_{ij}), \sigma^{2(l)}\}\right]}.$$

4. One cycle of M-step:

$$\pi_c^{(l+1)} = \frac{1}{n} \sum_{i=1}^n p_{ic}^{(l+1)},$$

$$\boldsymbol{\beta}_c^{(l+1)}(t_0) = \left\{\sum_{i=1}^n \mathbf{X}_i^T W_{ic}^{(l+1)}(t_0) \mathbf{X}_i\right\}^{-1} \left\{\sum_{i=1}^n \mathbf{X}_i^T W_{ic}^{(l+1)}(t_0) \mathbf{y}_{ic}^{(l+1)}\right\},$$

$$\sigma_c^{2(l+1)} = \frac{1}{\sum_{i=1}^n N_i} \sum_{i=1}^n \sum_{c=1}^C \sum_{j=1}^{N_i} p_{ic}^{(l+1)} \{y_c^{(l+1)}(t_{ij}) - \mathbf{X}_i(t_{ij})^T \boldsymbol{\beta}_c^{(l+1)}(t_{ij})\}^2,$$

   where $\mathbf{y}_{ic}^{(l+1)} = \{y_c^{(l+1)}(t_{i1}), \cdots, y_c^{(l+1)}(t_{iN_i})\}^T$.

2.6 Some Additional Topics

In addition to what we discussed above, there are some other interesting topics. For example, Vandekerkhove (2013) studied a two-component mixture of regressions model in which one component is entirely known while the mixing proportion, the slope, the intercept, and the error distribution of the other component are unknown. The method proposed by Vandekerkhove (2013) performs well for data sets of reasonable size, but since it is based on the optimization of a contrast function of size $O(n^2)$, the performance is not desirable as the sample size increases. Bordes et al. (2013) also studied the same model as Vandekerkhove (2013), and proposed a new method-of-moments estimator, whose order is of $O(n)$. Young (2014) extended the mixture of linear regression models to incorporate changepoints, by assuming one or more of the components are piecewise linear. Such model is a great combination of traditional mixture of linear regression models and standard changepoint regression model. Faicel (2016) proposed a new fully unsupervised algorithm to learn regression mixture models with unknown number of components. Unlike the standard EM for mixture of regressions, this method did not require accurate initialization. Yao et al. (2011) extended the classical functional linear models to a mixture of non-concurrent functional linear models, so that the regression structure could be different for different groups of subjects. Montuelle and Pennec (2014) studied a mixture of Gaussian regressions model with logistic weights, and proposed to estimate the number of components and other parameters through a penalized maximum likelihood approach. Huang et al. (2018) proposed a semiparametric hidden Markov model with non-parametric regression, in which the mean and variance of emission model are unknown smooth functions. Huang et al. (2017) established the identifiability and investigated the statistical inference for mixture of varying coefficient models, in which each mixture component follows a varying coefficient model.

2.7 Discussion

This article summarizes several semiparametric extensions to the standard parametric mixture of regressions model. Detailed model settings and corresponding estimation methods and algorithms are presented. As we have seen here, this field has received a lot of interest, but there are still a great number of questions and issues remained to be addressed. For example, most of the models discussed above assume the order of the mixture to be known and fixed, and so more work remains to be done regarding selection of the number of components. In addition, since lots of the models we discussed above are closely connected or even nested, then in addition to data driven methods, it is natural to develop some testing procedure to formally select the model. We hope that this review article could inspire more researchers to shine more lights on this topic.

Acknowledgements Xiang's research is supported by Zhejiang Provincial NSF of China [grant no. LQ16A010002] and NSF of China [grant no. 11601477]. Yao's research is supported by NSF [grant no. DMS-1461677] and Department of Energy with the award DE-EE0007328.

References

Bordes, L., Kojadinovic, I., & Vandekerkhove, P. (2013). Semiparametric estimation of a two-component mixture of linear regressions in which one component known. *Electronic Journal of Statistics, 7*, 2603–2644.

Cao, J., & Yao, W. (2012). Semiparametric mixture of binomial regression with a degenerate component. *Statistica Sinica, 22*, 27–46.

Chen, J., & Tan, X. (2009). Inference for multivariate normal mixtures. *Journal of Multivariate Analysis, 100*, 1367–1383.

Chen, J., Tan, X., & Zhang, R. (2008). Inference for normal mixture in mean and variance. *Statistica Sincia, 18*, 443–465.

Dziak, J. J., Li, R., Tan, X., Shiffman, S., & Shiyko, M. P. (2015). Modeling intensive longitudinal data with mixtures of nonparametric trajectories and time-varying effects. *Psychological Methods, 20*(4), 444–469.

Faicel, C. (2016). Unsupervised learning of regression mixture models with unknown number of components. *Journal of Statistical Computation and Simulation, 86*(12), 2308–2334.

Fan, J., & Gijbels, I. (1996). *Local polynomial modelling and its applications*. London: Chapman & Hall.

Frühwirth-Schnatter, S. (2001). Markov chain monte carlo estimation of classical and dynamic switching and mixture models. *Journal of American and Statistical Association, 96*, 194–209.

Frühwirth-Schnatter, S. (2006). *Finite mixture and Markov switching models*. New York: Springer.

Goldfeld, S. M., & Quandt, R. E. (1973). A Markov model for switching regression. *Journal of Econometrics, 1*, 3–15.

Green, P. J., & Richardson, S. (2002). Hidden Markov models and disease mapping. *Journal of American and Statistical Association, 97*, 1055–1070.

Härdle, W., Hall, P., & Ichimura, H. (1993). Optimal smoothing in single-index models. *Annals of Statistics, 21*, 157–178.

Hathaway, R. J. (1985). A constrained formulation of maximum-likelihood estimation for normal mixture distributions. *Annals of Statistics, 13*, 795–800.

Hathaway, R. J. (1986). A constrained EM algorithm for univariate mixtures. *Journal of Statistical Computation and Simulation, 23*, 211–230.

Hu, H., Yao, W., & Wu, Y. (2017). The robust EM-type algorithms for log-concave mixtures of regression models. *Computational Statistics & Data Analysis, 111*, 14–26.

Huang, M., & Yao, W. (2012). Mixture of regression models with varying mixing proportions: a semiparametric approach. *Journal of the American Statistical Association, 107*(498), 711–724.

Huang, M., Li, R., & Wang, S. (2013). Nonparametric mixture of regression models. *Journal of the American Statistical Association, 108*(503), 929–941.

Huang, M., Li, R., Wang, H., & Yao, W. (2014). Estimating mixture of Gaussian processes by kernel smoothing. *Journal of Business & Economic Statistics, 32*(2), 259–270.

Huang, M., Ji, Q., & Yao, W. (2017). Semiparametric hidden Markov model with nonparametric regression. *Communications in Statistics-Theory and Methods*. https://doi.org/10.1080/03610926.2017.1388398.

Huang, M., Yao, W., Wang, S., & Chen, Y. (2018). Statistical inference and application of mixture of varying coefficient models. *Scandinavian Journal of Statistical models, 45*(3), 618–643.

Hunter, D. R., & Young, D. S. (2012). Semiparametric mixtures of regressions. *Journal of Nonparametric Statistics, 24*(1), 19–38.

Ichimura, H. (1993). Semiparametric least squares (SLS) and weighted SLS estimation of single-index models. *Journal of Econometrics, 58*, 71–120.

Jacobs, R. A., Jordan, M. I., Nowlan, S. J., & Hinton, G. E. (1991). Adaptive mixtures of local experts. *Neural Computation, 3*(1), 79–87.

Montuelle, L., & Le Pennec, E. (2014). Mixture of Gaussian regressions model with logistic weights, a penalized maximum likelihood approach. *Electronic Journal of Statistics, 8*, 1661–1695.

Pena, D., Rodríguez, J., & Tiao, G. C. (2003). Identifying mixtures of regression equations by the SAR procedure. In J. M. Bernardo, M. J. Bayarri, J. O. Berger, A. P. Dawid, D. Heckerman, A. F. M. Smith, & M. West (Eds.), *Bayesian statistics* (Vol. 7, pp. 327–348). Oxford: Clarendon Press.

Sapatnekar, S. S. (2011). Overcoming variations in nanometer-scale technologies. *IEEE Journal on Emerging and Selected Topics in Circuits and Systems, 1*(1), 5–18.

Tan, X., Shiyko, M. P., Li, R., Li, Y., & Dierker, L. (2012). A time-varying effect model for intensive longitudinal data. *Psychological Methods, 17*(1), 61–77.

Vandekerkhove, P. (2013). Estimation of a semiparametric mixture of regressions model. *Journal of Nonparametric Statistics, 25*, 181–208.

Viele, K., & Tong, B. (2002). Modeling with mixtures of linear regressions. *Statistics and Computing, 12*, 315–330.

Wang, S., Yao, W., & Huang, M. (2014). A note on the identifiability of nonparametric and semiparametric mixtures of GLMs. *Statistics and Probability Letters, 93*, 41–45.

Wang, S., Huang, M., Wu, X., & Yao, W. (2016). Mixture of functional linear models and its application to CO2-GDP functional data. *Computational Statistics & Data Analysis, 97*, 1–15.

Wedel, M., & DeSarbo, W. S. (1993). A latent class binomial logit methodology for the analysis of paired comparison data. *Decision Sciences, 24*, 1157–1170.

Wu, Q., & Yao, W. (2016). Mixtures of quantile regressions. *Computational Statistics & Data Analysis, 93*, 162–176.

Xiang, S., & Yao, W. (2016). Semiparametric mixtures of nonparametric regressions. *Annals of the Institute of Statistical Mathematics*. https://doi.org/10.1007/s10463-016-0584-7.

Xiang, S., & Yao, W. (2017). Semiparametric mixtures of regressions with single-index for model based clustering. arXiv:1708.04142v1.

Yao, W. (2010). A profile likelihood method for normal mixture with unequal variance. *Journal of Statistical Planning and Inference, 140*, 2089–2098.

Yao, F., Fu, Y., & Lee, T. C. M. (2011). Functional mixture regression. *Biostatistics, 12*, 341–353.

Young, D. S. (2014). Mixtures of regressions with changepoints. *Statistical Computations, 24*, 265–281.

Young, D. S., & Hunter, D. R. (2010). Mixtures of regressions with predictor-dependent mixing proportions. *Computational Statistics & Data Analysis, 54*, 2253–2266.

Chapter 3
Rank-Based Empirical Likelihood for Regression Models with Responses Missing at Random

Huybrechts F. Bindele and Yichuan Zhao

3.1 Introduction

Missing data problems have captured a lot of attention within the few last decades, and have become a hot topic of research in the statistical community. The occurrence of missing data is subject to a number of common reasons. Among them, we have equipment malfunction, contamination of samples, manufacturing defects, drop out in clinical trials, weather conditions, incorrect data entry to name a few. As an example, one may be interested in a survey of families in a city that includes many socioeconomic variables and a follow-up survey a few months or years later for the recording of new observations. By the time of the new recording, some families may have left the city, died, or cannot be located thereby resulting in missing observations. Also in a clinical trial study, some patients may decide to drop out during the course of study, which leads to some missing information.

In statistical analysis, the first step to take when dealing with missing data is to understand the mechanism that causes such missingness. There are many missing data mechanisms but the most commonly encountered in the literature are *missing at random* (MAR), *missing completely at random* (MCAR), and *missing not at random* (MNAR). An elaborate discussion is given in Rubin (1976). These missing mechanisms exist and are encountered in many fields of study such as social sciences, survey analysis, biomedical studies, survival analysis, agriculture economics, phychology among others. Our interest is on a robust and efficient inference about regression parameters in a general regression model, where some

H. F. Bindele (✉)
Department of Mathematics and Statistics, University of South Alabama, Mobile, AL, USA
e-mail: hbindele@southalabama.edu

Y. Zhao
Department of Mathematics and Statistics, Georgia State University, Atlanta, GA, USA
e-mail: yichuan@gsu.edu

Y. Zhao, D.-G. Chen (eds.), *New Frontiers of Biostatistics and Bioinformatics*,
ICSA Book Series in Statistics, https://doi.org/10.1007/978-3-319-99389-8_3

responses are MAR. The MAR assumption asserts that the probability that a response variable is observed can only depend on the values of those other variables that have been observed. Although it is not trivial to verify the MAR assumption in practice, it has been proven to be the most plausible assumption to make for missing data problems in many scenarios (Little and Rubin 2002).

Consider the general regression model

$$y_i = g(\mathbf{x}_i, \boldsymbol{\beta}_0) + \varepsilon_i, \qquad 1 \le i \le n, \tag{3.1}$$

where $g : \mathbb{R}^p \times \mathscr{B} \to \mathbb{R}$ is fully specified, and $\boldsymbol{\beta} \in \mathscr{B}$ is a vector of parameters with $\mathscr{B}$ a compact vector space, $\mathbf{x}_i$'s are independent and identically distributed (i.i.d.) p-variable random covariate vectors, and the model errors ε_i are i.i.d. with conditional mean zero given the covariates and positive variance. We are interested in a robust and efficient inference about $\boldsymbol{\beta}_0$, when there are some responses missing at random in model (3.1). There is a rich literature on how to handle model (3.1) for the complete case analysis, that is, when ignoring observations with missing responses.

Missing data at random in the context of model (3.1) present a lot of challenges, as both the response probability and the regression parameters need to be estimated. Several approaches have been proposed in the literature on how to handle them. The main approach relies on replacing missing values by some plausible values and then performs the statistical analysis as if the data were complete. This approach known as *imputation* was originated in the early 1970s in applications to social surveys and has gained popularity over the years. Since then, several imputation methods have been proposed in the literature. In regression modeling, seminal papers include Healy and Westmacott (1956), Cheng (1994), Zhao et al. (1996), Wang et al. (1997), Wang and Rao (2002), Rubin (2004), Wang et al. (2004), Wang and Sun (2007) among others.

When it comes to estimating the regression parameters in model (3.1) with responses missing at random, several methods have also been proposed in the literature. Such methods include the least squares (LS) and the maximum likelihood (ML), among others. Statistical inference based on the LS approach is efficient when the model assumptions such as normality of the model error distribution and homogeneity (constant variance of the model errors) are satisfied. Under the violation of these assumptions, the LS could lead to a misleading inference. The ML approach, on the other hand, is a very powerful alternative to the LS, but requires the model distribution specification. One of the main disadvantages of this approach is that in real life situations, mainly with missing data, it is very unrealistic to specify the model error distribution. Under the MAR assumption, as pointed out in Little and Rubin (2002), the complete case analysis (ignoring observations with missing response) may lead to an efficient ML estimator. It is worth pointing out that even for the complete case analysis, the rank-based (R) approach introduced by Jaeckel (1972) outperforms the aforementioned approaches in terms of robustness and efficiency when dealing with heavy-tailed model errors and/or in the presence of outliers; see Hettmansperger and McKean (2011) for linear models and Bindele and Abebe (2012) for nonlinear models. Recently, for missing responses under the

MAR assumption, a rank-based approach has been proposed by Bindele (2015) for model (3.1), and by Bindele and Abebe (2015) for the semiparametric linear model. Most of the approaches listed above are based on the normal approximation as a way to handle statistical inference. Unfortunately, normal approximation approaches require estimating the estimator's covariance matrix, which has been shown to be complicated in many situations, mainly considering the rank-based objective function (Brunner and Denker 1994).

Empirical likelihood (EL) approach, on the other hand, is a way of avoiding estimating such a covariance matrix, conducting a direct inference about the true parameters and overcoming the drawback of normal approximation method (Owen 1988, 1990). Qin and Lawless (1994) developed the EL inference procedure for general estimating equations for complete data, and Owen (2001) makes an excellent summary about the theory and applications of the EL methods. Recent progress in EL method includes linear transformation models with censoring data (Yu et al. 2011), the jackknife EL procedure (Jing et al. 2009; Gong et al. 2010; Zhang and Zhao 2013), high dimensional EL methods (Chen et al. 2009; Hjort et al. 2009; Tang and Leng 2010; Lahiri and Mukhopadhyay 2012), the signed-rank regression using EL (Bindele and Zhao 2016), and rank-based EL methods with non-ignorable missing data (Bindele and Zhao in press), etc.

In this paper, we propose an empirical likelihood approach based on the general rank dispersion function in an effort to construct robust confidence regions for $\boldsymbol{\beta}_0$ in model (3.1), where some responses are MAR. We also investigate the adverse effects of contaminated, heavy-tailed model error distributions and gross outliers on the least squares estimator of the regression parameter. The motivation beyond the use of Jaeckel (1972) objective function is that it results in a robust and efficient estimator compared with many of the existing estimation methods such as the least-squares (LS), the maximum likelihood (ML), and many other methods of moments including the least absolute deviation (LAD). Moreover, it has a simple geometric interpretability. For all these facts, see Hettmansperger and McKean (2011).

The rest of the paper is organized as follows: Imputation of missing responses under MAR, the normal approximation, and the proposed empirical likelihood method are discussed in Sect. 3.2. While in Sect. 3.2.1 we give a brief discussion about the normal approximation approach as well as the estimation of the rank-based estimator's covariance matrix, the empirical likelihood approaches are developed in Sect. 3.2.2. Section 3.3 provides a simulation study and a real data example. Section 3.4 gives a conclusion of our findings. The proofs of the main results are provided in the Appendix.

3.2 Imputation

Recall that the missing at random (MAR) asserts that the missingness of y depends only on the observed inputs $\mathbf{x}$. Parameter estimation under this missing mechanism has been shown to be a very challenging task, as the response mechanism is generally unknown and the parameters in the regressing settings need to be estimated. To fix some ideas, let $\delta_i = 1$, if y_i is observed, and $\delta_i = 0$, if y_i is missing. Clearly δ_i is binary and can be assumed to follow a Bernoulli distribution with parameter $P(\delta_i = 1|\mathbf{x}_i, y_i)$. The MAR assumption is obtained when δ and y are conditionally independent given $\mathbf{x}$, that is, $P(\delta = 1|\mathbf{x}, y) = P(\delta = 1|\mathbf{x})$. For more discussion about this assumption, see Little and Rubin (1987). From model (3.1), if we consider the complete case analysis, one can obtain a preliminary rank estimator of the true regression parameter $\boldsymbol{\beta}_0$. Indeed, premultiplying Eq. (3.1) by δ_i, we get $\delta_i y_i = \delta_i g(\mathbf{x}_i, \boldsymbol{\beta}) + \delta_i \varepsilon_i$ and setting $e_i = \delta_i \varepsilon_i$, a preliminary rank estimator $\widehat{\boldsymbol{\beta}}_n^c$ of $\boldsymbol{\beta}_0$ is the minimizer of

$$D_n^c(\boldsymbol{\beta}) = \frac{1}{n}\sum_{i=1}^{n} \varphi\Big(\frac{R(e_i(\boldsymbol{\beta}))}{n+1}\Big) e_i(\boldsymbol{\beta}),$$

where $\varphi : (0, 1) \to \mathbb{R}$ is some bounded, nondecreasing, and squared integrable function, and $R(t) = \sum_{j=1}^{n} I\{e_j(\boldsymbol{\beta}) \le t\}$. Asymptotic properties of $\widehat{\boldsymbol{\beta}}_n^c$ are obtained in a similar manner as discussed in Hettmansperger and McKean (2011) for linear models, and Bindele and Abebe (2012) and Bindele (2017) for general regression models. Note that the above rank objective function does not take into account the missing responses. As pointed out in Little (1992) and Little and Rubin (2002), the statistical analysis that ignores the missing data reduces the estimation efficiency and can lead to biased estimates of the regression parameters.

3.2.1 Imputation Under MAR

As in Bindele (2015), denoting $\pi(\mathbf{x}) = P(\delta = 1|\mathbf{x} = \mathbf{x})$, one can perform data augmentation in two ways: simple random imputation ($j = 1$) and the inverse marginal probability weighting imputation ($j = 2$) as follows:

$$y_{ij} = \begin{cases} \delta_i y_i + (1 - \delta_i)\, g(\mathbf{x}_i, \boldsymbol{\beta}), & \text{if} j = 1; \\ \dfrac{\delta_i}{\pi(\mathbf{x}_i)} y_i + \left(1 - \dfrac{\delta_i}{\pi(\mathbf{x}_i)}\right) g(\mathbf{x}_i, \boldsymbol{\beta}), & \text{if} j = 2. \end{cases} \tag{3.2}$$

While the simple imputation is relatively easy to deal with, it does not account for the response probability $\pi(\mathbf{x})$. The MAR assumption implies that $P(\delta = 1 \mid \mathbf{x}, y) = \pi(\mathbf{x})$ and from the fact that $E(\varepsilon|\mathbf{x}_i) = 0$, we have $E(Y_{ij}|\mathbf{x}_i) = g(\mathbf{x}_i, \boldsymbol{\beta}) = E(Y_i \mid \mathbf{x}_i)$. The missing mechanism being usually unknown, $\pi(\mathbf{x})$ needs to be estimated. From the assumption $\inf_{\mathbf{x}} \pi(\mathbf{x}) > 0$ in (I_4), one can carry out a nonparametric estimation of $\pi(\mathbf{x})$ by $\widehat{\pi}(\mathbf{x}) = \sum_{j=1}^{n} \omega_{nj}(\mathbf{x})\delta_j$, where

$$\omega_{nj}(\mathbf{x}) = \frac{K\left((\mathbf{x}-\mathbf{x}_j)/h_n^p\right)}{\sum_{i=1}^n K\left((\mathbf{x}-\mathbf{x}_i)/h_n^p\right)},$$

with K being a kernel function and h_n a bandwidth satisfying $h_n \to 0$ as $n \to \infty$. Following Einmahl and Mason (2005), by assumptions $(I_2)-(I_4)$, it is obtained that $\widehat{\pi}(\mathbf{x}) \to \pi(\mathbf{x})$ $a.s.$, which further gives $\widehat{\pi}^{-1}(\mathbf{x}) \to \pi^{-1}(\mathbf{x})$ $a.s.$ Now the missing responses can be imputed as follows:

$$\tilde{y}_{ijn} = \begin{cases} \delta_i y_i + (1-\delta_i) g(\mathbf{x}_i, \widehat{\boldsymbol{\beta}}_n^c) & j=1 \\ \dfrac{\delta_i}{\widehat{\pi}(\mathbf{x}_i)} y_i + \left(1 - \dfrac{\delta_i}{\widehat{\pi}(\mathbf{x}_i)}\right) g(\mathbf{x}_i, \widehat{\boldsymbol{\beta}}_n^c) & j=2. \end{cases}$$

In practice, the inverse marginal probability weighting imputation approach may be very computationally expensive, as it usually depends on high dimensional kernel smoothing for estimating the completely unknown probability function $\pi(\mathbf{x})$, especially for large p. This issue, known as the *curse of dimensionality*, may restrict the use of the resulting estimator. For $\mathbf{x}_i = (x_{1i}, \ldots, x_{pi})^\tau$ and under the assumption that the x_{ki}, $k = 1, \ldots, p$ are i.i.d., one way to avoid the curse of dimensionality issue is to assume that $K_h(\mathbf{x}-\mathbf{x}_i) = \prod_{k=1}^p \Omega_h(x - x_{ki})$, where $\Omega_h(\cdot) = \Omega(\cdot/h)$ with Ω being a univariate kernel function and h the corresponding bandwidth. The other issue that comes with kernel smoothing estimation is bandwidth selection. In general, the m-fold cross-validation (m-CV) procedure can be used to select the bandwidth h.

The continuity of g together with the MAR assumption, and the fact that $E(\varepsilon|\mathbf{x}_i) = 0$, we have $E\left[\tilde{Y}_{ijn}|\mathbf{x}_i\right] - g(\mathbf{x}_i, \boldsymbol{\beta}) = o(1)$ with probability 1 and for n large enough. Setting the residuals as $\nu_{ij}(\boldsymbol{\beta}) = \tilde{y}_{ijn} - g(\mathbf{x}_i, \boldsymbol{\beta})$, the imputed rank-based objective function is defined as

$$D_n^j(\boldsymbol{\beta}) = \frac{1}{n} \sum_{i=1}^n \varphi\big(R(\nu_{ij}(\boldsymbol{\beta}))/(n+1)\big) \nu_{ij}(\boldsymbol{\beta}),$$

where $R(\nu_{ij}(\boldsymbol{\beta})) = \sum_{k=1}^n I\{\nu_{kj}(\boldsymbol{\beta}) \le \nu_{ij}(\boldsymbol{\beta})\}$ is the rank of $\nu_{ij}(\boldsymbol{\beta})$ among $\nu_{1j}(\boldsymbol{\beta}), \ldots, \nu_{nj}(\boldsymbol{\beta})$, $j = 1, 2$. The rank-based estimator is obtained as

$$\widehat{\boldsymbol{\beta}}_n^j = \operatorname*{Argmin}_{\boldsymbol{\beta} \in \mathscr{B}} D_n^j(\boldsymbol{\beta}).$$

By the continuity of g and consistency of $\widehat{\boldsymbol{\beta}}_n^c$, it is not hard to see that $\nu_{ij}(\boldsymbol{\beta}_0) = \xi_{ij}\varepsilon_i + o_p(1)$. This together with the continuity of φ' gives $\varphi'(\nu_{ij}(\boldsymbol{\beta}_0)) = \varphi'(\xi_{ij}\varepsilon_i) + o_p(1)$, where $\xi_{i1} = \delta_i$ and $\xi_{i2} = \delta_i/\pi(\mathbf{x}_i)$.

Consider the following notations: $\boldsymbol{\lambda}_i = \nabla_{\boldsymbol{\beta}} g(\mathbf{x}_i, \boldsymbol{\beta}_0)$ and $F_{ij\boldsymbol{\beta}}(s) = P(\nu_{ij}(\boldsymbol{\beta}) < s)$ with $F_{ij}(s) = P\big(\nu_{ij}(\boldsymbol{\beta}_0) \le s\big)$. The following theorem establishes the strong consistency and the asymptotic distribution of the proposed estimator.

Theorem 3.1 *Under assumptions* $(I_1)-(I_5)$, $\widehat{\boldsymbol{\beta}}_n^j \to \boldsymbol{\beta}_0$ *a.s., as* $n \to \infty$. *Moreover, set* $S_n^j(\boldsymbol{\beta}) = -\nabla_{\boldsymbol{\beta}} D_n^j(\boldsymbol{\beta})$ *and* $Q_n^j(\boldsymbol{\beta}) = \frac{1}{n}\sum_{i=1}^n \lambda_i \varphi\left(F_{ij\boldsymbol{\beta}}(\nu_{ij}(\boldsymbol{\beta}))\right)$. *Under* $(I_1)-(I_7)$, $\forall\, \mu > 0$, *we have*

(*i*) $\lim_{n\to\infty} \sup_{\boldsymbol{\beta}\in\mathscr{B}} P\left(\sqrt{n}\|S_n^j(\boldsymbol{\beta}) - Q_n^j(\boldsymbol{\beta})\| > \mu\right) = 0$ *a.s. and* $\lim_{n\to\infty} \nabla_{\boldsymbol{\beta}} Q_n^j(\boldsymbol{\beta}_0) = \mathbf{V}_j$ *a.s., with*

$$\nabla_{\boldsymbol{\beta}} Q_n^j(\boldsymbol{\beta}_0) = \frac{1}{n}\sum_{i=1}^n \nabla_{\boldsymbol{\beta}} g(\mathbf{x}_i, \boldsymbol{\beta}_0)\nabla_{\boldsymbol{\beta}}^T g(\mathbf{x}_i, \boldsymbol{\beta}_0) f_{ij}(\nu_{ij}(\boldsymbol{\beta}_0))\varphi'(U_{ij})$$
$$+\frac{1}{n}\sum_{i=1}^n \nabla_{\boldsymbol{\beta}}^2 g(\mathbf{x}_i, \boldsymbol{\beta}_0)\varphi(U_{ij})$$

and

$$\mathbf{V}_j = E[\boldsymbol{\Lambda}\boldsymbol{\Lambda}^T f_j(F_j^{-1}(U_j))\varphi'(U_j)] + E\left[\nabla_{\boldsymbol{\beta}}^2 g(\mathbf{X}, \boldsymbol{\beta}_0)\varphi(U_j)\right],$$

where $U_{ij} = F_{ij}(\nu_{ij}(\boldsymbol{\beta}_0))$, $i = 1, \ldots, n$, *and* $j = 1, 2$ *are i.i.d. uniformly distributed on* $(0, 1)$, $\boldsymbol{\Lambda} = \nabla_{\boldsymbol{\beta}_0} g(\mathbf{X}, \boldsymbol{\beta}_0)$, *and* $f_j(t) = dF_j(t)/dt$.

(*ii*) $\sqrt{n}S_n^j(\boldsymbol{\beta}_0) \xrightarrow{\mathscr{D}} N_p(\mathbf{0}, \Sigma_{\boldsymbol{\beta}_0}^j)$, $j = 1, 2$, *and*

(*iii*) $\sqrt{n}(\widehat{\boldsymbol{\beta}}_n^j - \boldsymbol{\beta}_0) \xrightarrow{\mathscr{D}} N_p(\mathbf{0}, \mathbf{M}_j)$, *where* $\mathbf{M}_j = \mathbf{V}_j^{-1}\Sigma_{\boldsymbol{\beta}_0}^j \mathbf{V}_j^{-1}$.

The proof of the strong consistency can be constructed along the lines as given in Bindele (2017). For the sake of brevity, it is omitted. The proof of (i) is obtained in a straightforward manner, while that of (ii) relies on Lemma 3.1 given in the Appendix. Finally the proof of (iii) follows from (ii).

The normal approximation-based inference relies on the estimation of $\mathbf{M}_j$, which turns out to be a difficult task in the rank-based estimation framework. A sandwich-type estimator of $\mathbf{M}_j$ can be obtained following Brunner and Denker (1994). Such an estimator under the MAR assumption, say $\widehat{\mathbf{M}}_j$, was derived in Bindele and Abebe (2015), who considered a semi-parametric linear regression model. Therefore, readers seeking for details on how to estimate $\mathbf{M}_j$ are referred to the aforementioned paper.

3.2.2 Empirical Likelihood Method

In this subsection, we adopt the empirical likelihood approach to make inference about the true regression parameters. The motivation comes from the well-known fact that the empirical likelihood provides more accurate confidence intervals/regions compared to its normal approximation competitor for the regression parameters. We consider $S_n^j(\boldsymbol{\beta})$ defined as

$$S_n^j(\boldsymbol{\beta}) = -\nabla D_n^j(\boldsymbol{\beta}) = \frac{1}{n}\sum_{i=1}^{n}\varphi\Big(\frac{R(v_{ij}(\boldsymbol{\beta}))}{n+1}\Big)\nabla_{\boldsymbol{\beta}} g(\mathbf{x}_i, \boldsymbol{\beta}), \quad j = 1, 2.$$

Set $\eta_{ij}(\boldsymbol{\beta}) = \varphi\big(R(v_{ij}(\boldsymbol{\beta}))/(n+1)\big)\nabla_{\boldsymbol{\beta}} g(\mathbf{x}_i, \boldsymbol{\beta})$, and recall that the rank-based estimator is obtained by solving the estimating equation $S_n^j(\boldsymbol{\beta}) = \mathbf{0}$. Under assumption (I_5) and with probability 1, we have $E[S_n^j(\boldsymbol{\beta}_0)] \to \mathbf{0}$ as $n \to \infty$. Therefore such an estimating equation, $S_n^j(\boldsymbol{\beta}) = \mathbf{0}$, is asymptotically unbiased. Letting $(p_{1j}, \cdots, p_{nj})^\tau$ be a vector of probabilities satisfying $\sum_{i=1}^n p_{ij} = 1$, with $p_{ij} \geq 0$, $j = 1, 2$ the empirical likelihood function of $\boldsymbol{\beta}_0$ is defined as follows:

$$L_n^j(\boldsymbol{\beta}_0) = \sup_{(p_{1j},\ldots,p_{nj})\in(0,1)^n}\left\{\prod_{i=1}^{n} p_{ij} : \sum_{i=1}^{n} p_{ij} = 1,\ p_{ij} \geq 0,\ \sum_{i=1}^{n} p_{ij}\eta_{ij}(\boldsymbol{\beta}_0) = 0,\right\}$$
$$j = 1, 2.$$

The empirical likelihood ratio at $\boldsymbol{\beta}_0$ is given by

$$R_n^j(\boldsymbol{\beta}_0) = \sup_{(p_1,\ldots,p_n)\in(0,1)^n}\left\{\prod_{i=1}^{n} np_{ij} : \sum_{i=1}^{n} p_{ij}=1,\ p_{ij} \geq 0,\ \sum_{i=1}^{n} p_{ij}\eta_{ij}(\boldsymbol{\beta}_0) = 0\right\},$$
$$j = 1, 2. \tag{3.3}$$

From the Lagrange multiplier method, it is obtained that $R_n^j(\boldsymbol{\beta}_0)$ attains its maximum when

$$p_{ij} = \frac{1}{n(1 + \boldsymbol{\xi}^\tau \eta_{ij}(\boldsymbol{\beta}_0))},$$

where $\boldsymbol{\xi} \in \mathscr{B}$ is some vector of the same dimension as $\boldsymbol{\beta}$, satisfying the nonlinear equation:

$$\frac{1}{n}\sum_{i=1}^{n}\frac{\eta_{ij}(\boldsymbol{\beta}_0)}{1 + \boldsymbol{\xi}^\tau \eta_{ij}(\boldsymbol{\beta}_0)} = \mathbf{0}. \tag{3.4}$$

In practice, since $\boldsymbol{\beta}_0$ is usually unknown, the solution of Eq. (3.4), say $\widehat{\boldsymbol{\xi}}$, can be obtained using the Newton-Raphson method, where $\boldsymbol{\beta}_0$ could be replaced by $\widehat{\boldsymbol{\beta}}_n^j$. Now equating (3.3) and (3.4) together gives

$$-2\log R_n^j(\boldsymbol{\beta}_0) = -2\log\prod_{i=1}^{n}\left(1+\boldsymbol{\xi}^\tau\eta_{ij}(\boldsymbol{\beta}_0)\right)^{-1} = 2\sum_{i=1}^{n}\log\left(1+\boldsymbol{\xi}^\tau\eta_{ij}(\boldsymbol{\beta}_0)\right), \quad (3.5)$$

which leads to the following Wilks theorem.

Theorem 3.2 *Under assumptions $(I_1)-(I_4)$ and (I_7), one has*

$$-2\log R_n^j(\boldsymbol{\beta}_0) \xrightarrow{\mathscr{D}} \chi_p^2 \quad as \quad n\to\infty.$$

The empirical likelihood (EL) confidence region for $\boldsymbol{\beta}_0$ with a confidence level $1-\alpha$ is given by

$$\mathscr{R}_1^j = \{\boldsymbol{\beta} : -2\log R_n^j(\boldsymbol{\beta}) \le \chi_p^2(\alpha)\},$$

where $\chi_p^2(\alpha)$ is the $(1-\alpha)$th percentile of the χ_p^2-distribution with p degrees of freedom. This confidence region enables us to perform statistical inference about $\boldsymbol{\beta}_0$.

In practice, when the dimension of $\boldsymbol{\beta}_0$ is more than 1, confidence regions are no longer easily interpretable. In this case, it is more preferable to derive confidence intervals of the components of $\boldsymbol{\beta}_0$. To this end, we partition $\boldsymbol{\beta}_0$ as $\boldsymbol{\beta}_0 = (\boldsymbol{\beta}_{01}^\tau, \boldsymbol{\beta}_{02}^\tau)^\tau$, where $\boldsymbol{\beta}_{01}$ is the true value of q-dimensional parameter and $\boldsymbol{\beta}_{02}$ is the true value of $(p-q)$-dimensional parameter. Similarly, for any $\boldsymbol{\beta}\in\mathscr{B}$, write $\boldsymbol{\beta} = (\boldsymbol{\beta}_1^\tau, \boldsymbol{\beta}_2^\tau)^\tau$, where $\boldsymbol{\beta}_1$ is a q-dimensional parameter and $\boldsymbol{\beta}_2$ is a $(p-q)$-dimensional parameter. Setting $\Gamma_n^j(\boldsymbol{\beta}) = -2\log R_n^j(\boldsymbol{\beta})$ and following Qin and Lawless (1994) and Yang and Zhao (2012), the profile empirical likelihood is obtained as

$$\omega_n^j(\boldsymbol{\beta}_1) = \inf_{\boldsymbol{\beta}_2}\Gamma_n^j(\boldsymbol{\beta}).$$

Following the similar arguments in Yang and Zhao (2012), we establish the following Wilks theorem.

Theorem 3.3 *Under the assumptions of Theorem 3.2, we have that as $n\to\infty$, $\omega_n^j(\boldsymbol{\beta}_{01})\to\chi_q^2$, $j=1,2$.*

The $100(1-\alpha)\%$ profile EL confidence region for $\boldsymbol{\beta}_{01}$ is obtained as

$$\mathscr{R}_2^j = \{\boldsymbol{\beta}_1 : \omega_n^j(\boldsymbol{\beta}_1) \le \chi_q^2(\alpha)\},$$

where $\chi_q^2(\alpha)$ is the $(1-\alpha)$th percentile of the χ^2-distribution with q degrees of freedom. Our simulation studies and real data example are based on this profile empirical likelihood confidence region.

3.3 Simulation Study

3.3.1 Simulation Settings

To compare the performance of the proposed rank-based empirical likelihood approach with the normal approximation approach with MAR responses, an extensive simulation under different settings is conducted, from which coverage probabilities (CP) and average lengths of confidence intervals/regions of the true regression coefficients are reported.

First, in model (3.1), we consider the simple regression function g defined as $g(x, \boldsymbol{\beta}) = \beta_1 + \beta_2 x$ with $\boldsymbol{\beta} = (\beta_1, \beta_2) = (1.7, 0.7)$. The random errors ($\varepsilon$) are generated from the contaminated normal distribution $\mathscr{CN}(\kappa, \sigma) = (1 - \kappa)N(0, 1) + \kappa N(1, \sigma^2)$ for different rates of contamination ($\kappa = 0, 0.1, 0.25$) with $\sigma = 2$, the t-distribution with various degrees of freedom ($df = 5, 15, 25$), and the standard Laplace distribution (the location parameter set at 0 and the dispersion parameter set at 1) with different sample sizes ($n = 35, 50, 100$). These distributions allow us to study the effect of contamination, tail thickness, and sample sizes, respectively. The standard normal distribution is recovered by setting $\kappa = 0$ in $\mathscr{CN}(\kappa, \sigma)$. The covariate x is generated from $N(1, 1)$.

Second, to accommodate a nonlinear case, as in Bindele (2015), we also consider the sinusoidal model, where in (3.1), $g(x, \boldsymbol{\beta}) = C + A \sin(\pi\sqrt{12}/1.645)\omega x + \phi$, with $\boldsymbol{\beta} = (C, A, \omega, \phi)^{\tau}$. C is a constant defining the mean level, A is an amplitude for the sine wave, ω is the frequency, x is a time variable generated from the uniform distribution in the interval $[0, 2\pi]$, and ϕ is the scale parameter known as the phase. For simplicity, we set $C = 0$, $A = 1$, and $\phi = (1.645/\sqrt{12} - 3/2)/(3.29/\sqrt{12}) \approx -0.412$. We are interested in the estimation of the true frequency set at $\omega_0 = 1/\sqrt{3}$, and for the sake of brevity, the random errors are generated just from $\mathscr{CN}(0.9, 2)$ and t_3.

In these two scenarios, δ is generated from a Bernoulli distribution with probability $\pi(x)$, where we investigate three different response probabilities:

Case 1: $\pi(x) = 0.8 - 0.7|x - 1|$ if $|x - 1| \leq 1$, and 0.65 elsewhere.
Case 2: $\pi(x) = \exp\{-0.5x + 0.4x^2\}/\big(1 + \exp\{-0.5x + 0.4x^2\}\big)$.
Case 3: $\pi(x) = \exp\{-0.8 \sin x\}/\big(1 + \exp\{-0.8 \sin x\}\big)$.

These three cases give about 47%, 46%, and 42% of missing responses. As in practice the functional form of the response probability is more likely to be unknown, one can consider $\widehat{\pi}(\mathbf{x})$ defined in Sect. 3.2. It is well understood in the framework of kernel smoothing estimation that the choice of the kernel function has less importance as long as it satisfies the required assumptions (Einmahl and Mason 2005). To that end, we consider the Gaussian kernel function, that is, $K(t) = (2\pi)^{-1/2}\exp(-t^2/2)$. For the bandwidth selection, which is also important in kernel smoothing estimation, our optimal bandwidth is chosen to be proportional to

$n^{-1/(2r+1)}$, where r is the order of smoothness of the kernel function K (Delecroix et al. 2006). For this simulation study, we have found that the rule of thumb for bandwidth selection suggested by Silverman (1986) works reasonably well. Such rule of thumb suggests to choose h as $h = 1.06sn^{-1/5}$, where s be the median of the standard deviations of the predictor variables from the complete case analysis and n is the sample size. In our objective function, we considered the Wilcoxon score function $\varphi(u) = \sqrt{12}(u - 0.5)$.

From 5000 replications, coverage probabilities (CP) and average lengths (AL) of β_2 based on the EL approach are reported and are compared with those based on the normal approximation (NA) approach; these under both simple imputation (SI) and inverse marginal probability imputation (IP). The CP and AL based on the EL approaches for both the LS and the R are obtained with respect to their corresponding estimating equations, while those for the NA approaches are based on the estimators of covariance matrices. The results of the simulation study are given in Tables 3.1, 3.2, 3.3, 3.4.

Considering the t-distribution with a sample size of $n = 150$ for the three cases, based on simple imputation (SI) or the inverse marginal probability weighting

Table 3.1 95% Coverage probabilities (average lengths) of β_2 under the t-distribution with different degrees of freedom $df = 5, 15, 25$ and $n = 150$ under SI and IP

Imputation	Cases	df	NA_{LS}	EL_{LS}	NA_R	EL_R
		5	96.76% (1.03)	95.66% (0.79)	95.93% (0.84)	95.09% (0.53)
	Case 1	15	96.25% (0.87)	95.27% (0.54)	95.72% (0.63)	95.05% (0.37)
		25	95.89% (0.78)	95.08% (0.35)	95.17% (0.47)	95.00% (0.22)
		5	96.87% (1.12)	95.79% (0.89)	95.97% (0.94)	95.14% (0.65)
SI	Case 2	15	96.43% (0.95)	95.43% (0.68)	95.53% (0.75)	95.04% (0.42)
		25	96.02% (0.84)	95.12% (0.46)	95.18% (0.53)	94.98% (0.27)
		5	96.96% (1.22)	95.83% (0.94)	96.01% (0.97)	95.13% (0.71)
	Case 3	15	96.31% (0.98)	95.62% (0.74)	95.79% (0.83)	95.06% (0.53)
		25	96.12% (0.87)	95.17% (0.57)	95.23% (0.62)	95.01% (0.33)
		5	96.71% (0.92)	95.57% (0.68)	95.78% (0.75)	95.03% (0.47)
	Case 1	15	96.55% (0.81)	95.38% (0.48)	95.54% (0.54)	95.01% (0.26)
		25	96.19% (0.69)	95.07% (0.31)	95.13% (0.43)	94.99% (0.14)
		5	96.81% (0.95)	95.65% (0.71)	95.91% (0.87)	95.07% (0.52)
IP	Case 2	15	96.24% (0.87)	95.25% (0.49)	95.58% (0.61)	95.06% (0.31)
		25	95.87% (0.79)	95.07% (0.32)	95.14% (0.46)	94.94% (0.18)
		5	96.98% (1.02)	95.79% (0.77)	95.97% (0.89)	95.11% (0.54)
	Case 3	15	96.37% (0.91)	95.32% (0.54)	95.63% (0.66)	95.02% (0.35)
		25	96.07% (0.83)	95.12% (0.36)	95.25% (0.49)	94.96% (0.24)

Table 3.2 95% Coverage probabilities (average lengths) of β_2 under the contaminated normal distribution with different rates of contamination $\kappa = 0, 0.10, 0.25$ and $n = 150$ under SI and IP

Imputation	Cases	κ	NA_{LS}	EL_{LS}	NA_R	EL_R
		0.00	95.96% (0.264)	94.98% (0.223)	96.01% (0.345)	95.07% (0.251)
	Case 1	0.10	96.52% (0.653)	95.88% (0.289)	96.26% (0.401)	95.03% (0.268)
		0.25	96.76% (0.778)	95.96% (0.295)	96.47% (0.457)	95.05% (0.271)
		0.00	95.77% (0.260)	95.01% (0.221)	96.05% (0.336)	95.11% (0.243)
SI	Case 2	0.10	96.83% (0.761)	95.76% (0.378)	96.41% (0.445)	94.99% (0.279)
		0.25	96.97% (0.953)	95.79% (0.499)	96.59% (0.631)	95.03% (0.301)
		0.00	95.56% (0.293)	95.10% (0.225)	96.24% (0.432)	95.03% (0.234)
	Case 3	0.10	96.73% (0.698)	95.55% (0.441)	96.34% (0.513)	94.93% (0.257)
		0.25	96.99% (0.889)	95.61% (0.567)	96.49% (0.613)	95.02% (0.297)
		0.00	95.37% (0.233)	95.02% (0.219)	96.27% (0.331)	95.04% (0.243)
	Case 1	0.10	96.15% (0.473)	95.59% (0.275)	96.38% (0.434)	94.97% (0.257)
		0.25	96.39% (0.664)	95.73% (0.287)	96.42% (0.437)	95.05% (0.263)
		0.00	95.53% (0.242)	95.07% (0.216)	96.12% (0.322)	95.01% (0.237)
IP	Case 2	0.10	96.47% (0.565)	95.53% (0.357)	96.33% (0.397)	95.00% (0.261)
		0.25	96.68% (0.872)	95.64% (0.473)	96.47% (0.539)	95.03% (0.297)
		0.00	95.34% (0.271)	95.06% (0.220)	96.13% (0.343)	95.05% (0.223)
	Case 3	0.10	96.64% (0.467)	95.34% (0.419)	96.25% (0.471)	95.03% (0.248)
		0.25	96.77% (0.674)	95.56% (0.517)	96.38% (0.527)	94.99% (0.288)

imputation (IP) (see Table 3.1), the EL approach based on LS (EL_{LS}) provides better coverage probabilities compared to the normal approximation methods (NA_{LS} and NA_R). The NA method gives coverage probabilities that are larger than nominal levels for small degrees of freedom, and the EL approach based on R (EL_R) gives consistent coverage probabilities that are closer to the nominal confidence level. While EL_{LS} dominates both NA_R and NA_{LS} in terms of average lengths of the confidence interval, EL_R remains superior to all. Moreover, NA_R dominates NA_{LS} in terms of both CP and AL. As the degrees of freedom increase, all the considered methods have coverage probabilities that converge to the nominal confidence level and their respective average lengths decrease. It is worth pointing that average lengths obtained based on IP tend to be smaller compared with those obtained based on SI.

For the contaminated normal distribution model error with a sample size of $n = 150$ under the three cases, based on simple imputation (SI) or the inverse marginal probability imputation (IP) (see Table 3.2), at $\kappa = 0$, EL_R dominates both NA_{LS} and NA_{LS} in terms of CP and AL, and EL_{LS} is superior to all. However, as κ increases,

Table 3.3 95% Coverage probabilities (average lengths) of β_2 under the Laplace distribution with different sample sizes $n = 35, 50, 100$ under SI and IP

Imputation	Cases	n	NA_{LS}	EL_{LS}	NA_R	EL_R
		35	97.17% (1.345)	95.98% (0.978)	96.29% (1.079)	95.16% (0.843)
	Case 1	50	96.73% (0.916)	95.45% (0.727)	95.85% (0.932)	95.08% (0.634)
		100	95.69% (0.891)	95.18% (0.635)	95.46% (0.713)	95.01% (0.478)
		35	97.04% (1.281)	95.57% (0.842)	96.63% (0.965)	95.03% (0.739)
SI	Case 2	50	96.48% (0.857)	95.34% (0.678)	95.67% (0.783)	94.99% (0.573)
		100	95.43% (0.768)	95.09% (0.553)	95.21% (0.652)	95.05% (0.368)
		35	97.09% (1.333)	95.86% (0.961)	96.17% (0.997)	95.12% (0.812)
	Case 3	50	96.63% (0.909)	95.39% (0.713)	95.77% (0.858)	95.04% (0.619)
		100	95.58% (0.879)	95.15% (0.623)	95.42% (0.706)	94.97% (0.459)
		35	96.35% (0.941)	95.65% (0.756)	95.93% (0.853)	95.05% (0.625)
	Case 1	50	96.04% (0.832)	95.24% (0.654)	95.47% (0.741)	95.02% (0.421)
		100	95.34% (0.725)	95.11% (0.489)	95.22% (0.657)	95.01% (0.325)
		35	96.28% (0.918)	95.41% (0.721)	95.52% (0.837)	95.01% (0.597)
IP	Case 2	50	95.96% (0.817)	95.18% (0.636)	95.27% (0.711)	95.07% (0.415)
		100	95.24% (0.714)	95.05% (0.474)	95.12% (0.613)	95.02% (0.316)
		35	96.31% (0.924)	95.52% (0.735)	95.81% (0.844)	95.09% (0.613)
	Case 3	50	95.99% (0.821)	95.19% (0.647)	95.38% (0.733)	95.02% (0.402)
		100	95.31% (0.719)	95.06% (0.483)	95.18% (0.638)	94.99% (0.321)

EL_R performs better than NA_{LS}, NA_{LS}, and EL_R by providing consistent CP that are closer to the nominal level and shorter AL. As observed for the $t-$distribution, NA_R outperforms the NA_{LS} as the rate of contamination increases.

When it comes to the Laplace distribution model error for the three cases and based on either simple imputation (SI) or the inverse marginal probability imputation (IP) as can be seen in Table 3.3, similar observations are made as in the previous two error distributions. It is observed that as the sample size increases, CP converge to the nominal confidence level and AL decrease, as expected, with EL_R showing its superiority over the other three approaches.

When we consider the nonlinear model with the considered model error distributions, similar observations are made as for the linear model, see Table 3.4. The contaminated normal distribution provides shorter AL compared to the other two error distributions considered.

Table 3.4 95% coverage probabilities (average lengths of 95% confidence intervals) of ω_0 for the nonlinear Micheaelis-Menten model under t_3 and $\mathscr{CN}(0.9)$ with $n = 150$ and regression simple imputation (SI)

Imputation	Cases	Distribution	NA_{LS}	EL_{LS}	NA_R	EL_R
	Case 1	$\mathscr{CN}(0.9, 2)$	96.53% (2.76)	95.27% (1.19)	95.77% (1.47)	95.07% (0.98)
		t_3	96.75% (2.93)	95.43% (1.32)	95.92% (1.64)	95.10% (1.05)
SI	Case 2	$\mathscr{CN}(0.9, 2)$	96.18% (2.39)	95.21% (1.13)	95.76% (1.32)	95.03% (0.87)
		t_3	96.26% (2.76)	95.33% (1.27)	95.51% (1.52)	94.98% (0.93)
	Case 3	$\mathscr{CN}(0.9, 2)$	96.64% (2.82)	95.31% (1.25)	95.82% (1.58)	95.12% (1.08)
		t_3	96.87% (3.07)	95.54% (1.36)	96.02% (1.75)	95.17% (1.18)
	Case 1	$\mathscr{CN}(0.9, 2)$	96.36% (1.97)	95.14% (0.91)	95.28% (1.13)	95.02% (0.84)
		t_3	96.52% (2.08)	95.25% (1.07)	95.49% (1.23)	94.95% (0.91)
IP	Case 2	$\mathscr{CN}(0.9, 2)$	96.23% (1.86)	95.07% (0.83)	95.18% (1.06)	95.00% (0.78)
		t_3	96.37% (1.99)	95.13% (0.94)	95.31% (1.15)	95.04% (0.83)
	Case 3	$\mathscr{CN}(0.9, 2)$	96.47% (2.08)	95.26% (1.08)	95.43% (1.17)	95.15% (0.92)
		t_3	96.62% (2.33)	95.37% (1.23)	95.79% (1.41)	95.21% (1.03)

3.3.2 Real Data

To illustrate our methodology, we consider the Air Quality data in R that were obtained from the New York State Department of Conservation (ozone data) and the National Weather Service (meteorological data). The data consist of 153 observations on 6 variables. The response of interest is the mean ozone (y) in parts per billion from 1300 to 1500 h at Roosevelt Island and contains about 24% of missing information. The covariates are solar radiation (x_1) in Langleys in the frequency band 4000–7700 Angstroms from 800 to 1200 h at Central Park, the average wind speed (x_2) in miles per hour at 700 and 1000 h at LaGuardia Airport, and the maximum daily temperature (x_3) in degrees Fahrenheit at La Guardia Airport. The data contain other two covariates, which are month and day but are not being considered in the analysis to avoid the temporal dependence effect. We fit a linear model, that is, in model (3.1), $g(\mathbf{x}, \boldsymbol{\beta}) = \mathbf{x}^\tau \boldsymbol{\beta}$, where $\mathbf{x} = (x_1, x_2, x_3)^\tau$ and $\boldsymbol{\beta} = (\beta_1, \beta_2, \beta_3)^\tau$. As our interest is in modeling the missing responses under the MAR mechanism, we assume that the missingness of the mean ozone can only depend on the observed covariates. The same data was considered by Van Buuren (2012) for the implementation of the `mice` package. In the inverse marginal probability weighting imputation process, $\pi(\mathbf{x}) = P(\delta = 1|\mathbf{x})$, where $\mathbf{x} = (x_1, x_2, x_3)$ is estimated following the discussion below Eq. (3.2). As $\mathbf{x}$ is multivariate, to avoid the high-dimensional kernel smoothing issue in the estimation

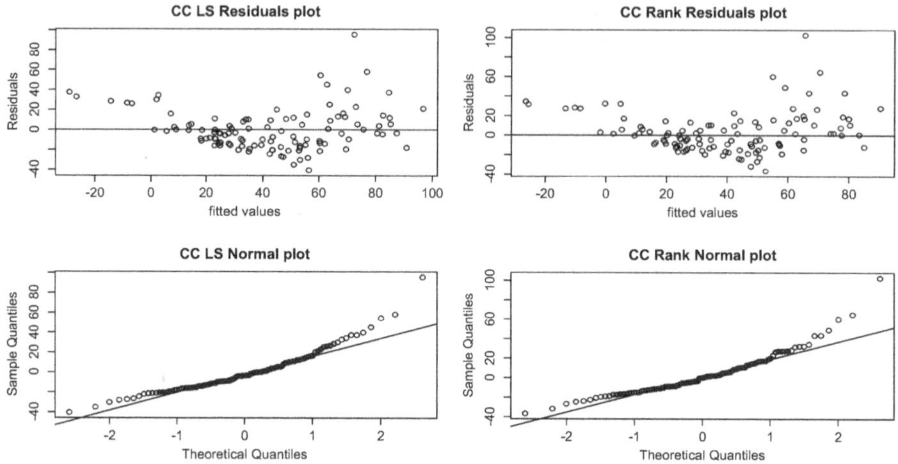

Fig. 3.1 Residuals and normal plots of the LS and rank for the complete case (CC)

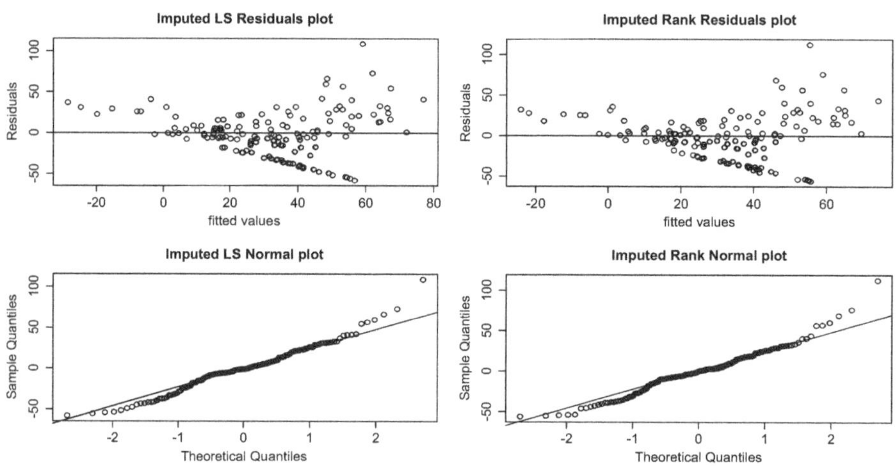

Fig. 3.2 Studentized residuals and normal plots of the LS and rank for imputed responses

of $\pi(\mathbf{x})$, we set $K_h(\mathbf{x}) = \prod_{i=1}^{3} \Omega_h(\mathrm{x}_i)$, where $\Omega_h(t) = \Omega(t/h)$ with $\Omega(t)$ taken as the univariate standard normal probability density function. Figures 3.1, 3.2 and Tables 3.4, 3.5 display the results of the statistical analysis.

Figures 3.1 and 3.2 reveal that whether we consider the CC analysis or the imputed analysis, there are issues of heteroscedasticity with few outliers in the response space. While residuals seem to deviate from the normality for the CC analysis, imputed residuals can fairly be approximated by a normal distribution, which is contaminated by few outliers. Considering the normal approximation approaches, we observe that although both methods provide similar estimates, the R

Table 3.5 Estimates (SEs) of the regression parameters for the imputed responses based on SI and IP

Method	Variable	CC	SI	IP
	Solar	0.060 (0.023)	0.060 (0.022)	0.060 (0.018)
LS	Wind	−3.334 (0.654)	−3.274 (0.636)	−3.334 (0.490)
	Temperature	1.652 (0.254)	1.672 (0.251)	1.652 (0.194)
	Solar	0.053 (0.021)	0.053 (0.013)	0.053 (0.010)
Rank	Wind	−2.754 (0.590)	−2.754 (0.352)	−2.754 (0.265)
	Temperature	1.724 (0.229)	1.724 (0.139)	1.724 (0.105)

Table 3.6 Lengths of 95% confidence intervals for the regression parameters for the Air Quality data with about 24% of missing responses

		CC		SI		IP	
Method	Variable	NA	EL	NA	EL	NA	EL
	Solar	0.090	0.078	0.086	0.069	0.071	0.063
LS	Wind	2.564	2.045	2.493	1.885	1.921	1.239
	Temperature	0.996	0.872	0.984	0.747	0.760	0.712
	Solar	0.082	0.069	0.051	0.036	0.039	0.027
Rank	Wind	2.313	1.978	1.380	0.993	1.039	0.945
	Temperature	0.898	0.786	0.545	0.397	0.412	0.383

estimator is more efficient by providing smaller SEs and lengths of 95% confidence intervals compared to the LS estimator; see Tables 3.4 and 3.5. While the estimators are consistent, there is significant reduction of biases, mainly for the R estimator. As observed in the simulation study, the analysis based on responses imputed from IP gives smaller SEs than that based on responses imputed from SI.

Considering EL methods, similar observations are made with the rank-based empirical likelihood showing its dominance. Also, EL approaches based on imputed responses show their superiority compared to the EL methods based on the CC analysis; see Table 3.5. Finally, whether we consider the NA approaches or the EL approaches, R is more efficient than LS, as R provides shorter lengths compared to the LS (Table 3.6).

3.4 Conclusion

This paper provides an empirical likelihood approach derived based on the rank-based objective function. The proposed approach provides better coverage probabilities and shorter lengths of confidence intervals/regions of parameters in regression models with responses missing at random compared to its normal approximation counterpart. Moreover, the proposed method is robust and more efficient compared to the empirical likelihood derived based on the least squares objective function for contaminated, skewed, and heavy-tailed model error distributions and/or when data contain gross outliers.

Acknowledgements The authors would like to thank the two reviewers for their helpful comments. The research of Yichuan Zhao is supported by the National Security Agency grant.

Appendix

This Appendix contains assumptions used in the development of theoretical results as well as the proof of the main results.

Assumptions

(I_1) φ is a nondecreasing, bounded, and twice continuously differentiable score function with bounded derivatives, defined on (0, 1), and, satisfying:

$$\int_0^1 \varphi(u)du = 0 \quad and \quad \int_0^1 \varphi^2(u)du = 1.$$

(I_2) $g(\cdot)$ being a function of two variables $\mathbf{x}$ and $\boldsymbol{\beta}$, it is required that g has continuous derivatives with respect to $\boldsymbol{\beta}$ that are bounded up to order 3 by p-integrable functions of $\mathbf{x}$, independent of $\boldsymbol{\beta}$, $p \geq 1$.

(I_3) $K(\cdot)$ is a regular kernel of order $r > 2$, with window b_n satisfying $nb_n^{4r} \to 0$, $C(\log n/n)^\gamma < b_n < h_n$, for any $C > 0$, $\gamma = 1 - 2/p$, $p > 2$ and h_n is a bandwidth such that $C(\log n/n)^\gamma < h_n < 1$ with $h_n \to 0$ as $n \to \infty$.

(I_4) $\sup_{\mathbf{x}} E[|Y|^p|\mathbf{x} = \mathbf{x}] < \infty$, for $p \geq 1$ and $\inf_{\mathbf{x}} \Delta(\mathbf{x}) > 0$.

(I_5) For fixed n, $\boldsymbol{\beta}_{0,n} \in Int(\mathscr{B})$ is the unique minimizer of $E[D_n(\boldsymbol{\beta})]$ such that $\lim_{n\to\infty} \boldsymbol{\beta}_{0,n} = \boldsymbol{\beta}_0$

(I_6) The model error has a distribution with a finite Fisher information.

(I_6) $Var(\sqrt{n}S_n^j(\boldsymbol{\beta}_0)) \to \Sigma_{\boldsymbol{\beta}_0}^j$, where $\Sigma_{\boldsymbol{\beta}_0}^j$ is positive definite.

(I_7) Set $\mathbf{H} = \mathbf{B}(\mathbf{B}^\tau\mathbf{B})^{-1}\mathbf{B}^\tau$, where $\mathbf{B} = \nabla_{\boldsymbol{\beta}} g(\mathbf{x}, \boldsymbol{\beta}_0)$. $\mathbf{H}$ is the projection matrix onto the column space of $\mathbf{B}$, which in this case represents the tangent space generated by $\mathbf{B}$ and let h_{iin}, $i = 1, \cdots, n$, be the leverage values that stand for the diagonal entries of $\mathbf{H}$. We assume that $\lim_{n\to\infty} \max_{1\leq i\leq n} h_{iin} = 0$

Assumptions $(I_2)-(I_4)$ are necessary and sufficient to ensure the strong consistency of $\widehat{\pi}(\mathbf{x})$ used in the imputation process. On the other hand, assumptions (I_1), $(I_5) - (I_7)$ together with the previous assumptions are necessary to establish the asymptotic properties (consistency and asymptotic normality distribution) of the rank-based estimators of $\boldsymbol{\beta}_0$. An elaborate discussion about these assumptions can be found in Hettmansperger and McKean (2011), Bindele and Abebe (2012), and Bindele (2017).

By definition of $S_n^j(\boldsymbol{\beta})$, $\widehat{\boldsymbol{\beta}}_n^j$ is solution to the equation $S_n^j(\boldsymbol{\beta}) = \mathbf{0}$. As in Brunner and Denker (1994), assume without loss of generality that $\|\boldsymbol{\lambda}_i\| = 1$ and define

$$J_{jn}(s) = \frac{1}{n}\sum_{i=1}^{n} F_{ij}(s),\quad \hat{J}_{jn}(s) = \frac{1}{n}\sum_{i=1}^{n} I(\nu_{ij}(\boldsymbol{\beta}_0) \leq s),\quad F_{jn}(s) = \frac{1}{n}\sum_{i=1}^{n}\boldsymbol{\lambda}_i F_{ij}(s)$$

$$\hat{F}_{jn}(s) = \frac{1}{n}\sum_{i=1}^{n}\boldsymbol{\lambda}_i I(\nu_{ij}(\boldsymbol{\beta}_0) \leq s),\qquad T_n^j(\boldsymbol{\beta}_0) = S_n^j(\boldsymbol{\beta}_0) - E\big[S_n^j(\boldsymbol{\beta}_0)\big].$$

The following lemma due to Brunner and Denker (1994) is a key for establishing asymptotic normality of the rank gradient function for dependent data.

Lemma 3.1 *Let ς_{jn} be the minimum eigenvalue of $\mathbf{W}_{jn} = Var(U_{jn})$ with U_{jn} given by*

$$U_{jn} = \int \varphi(J_{jn}(s))(\hat{F}_{jn} - F_{jn})(ds) + \int \varphi'(J_{jn}(s))(\hat{J}_{jn}(s) - J_{jn}(s))F_{jn}(ds)\,.$$

Suppose that $\varsigma_{jn} \geq Cn^a$ for some constants $C, a \in \mathbb{R}$ and $m(n)$ is such that $M_0 n^\gamma \leq m(n) \leq M_1 n^\gamma$ for some constants $0 < M_0 \leq M_1 < \infty$ and $0 < \gamma < (a+1)/2$. Then $m(n)\mathbf{W}_{jn}^{-1}T_n^j(\boldsymbol{\beta}_0)$ is asymptotically standard multivariate normal, provided φ is twice continuously differentiable with bounded second derivative.

We provide a sketch of the proof of this lemma. A detailed proof can be found in Brunner and Denker (1994).

Proof Set

$$B_{jn} = -\int (\hat{F}_{jn} - F_{jn})d\varphi(J_{jn}) + \int (\hat{J}_{jn} - J_{jn})\frac{dF_{jn}}{dJ_{jn}}d\varphi(J_{jn}).$$

Brunner and Denker (1994) showed that $\mathbf{W}_{jn} = n^2 Var(B_{jn})$, as $U_n = nB_n$. From its definition, $S_n^j(\boldsymbol{\beta})$ can be rewritten as

$$S_n^j(\boldsymbol{\beta}_0) = \frac{1}{n}\sum_{i=1}^{n}\boldsymbol{\lambda}_i\varphi\Big(\frac{R(\nu_{ij}(\beta_0))}{n+1}\Big) = \int \varphi\Big(\frac{n}{n+1}\hat{J}_{jn}\Big)dF_{jn}.$$

By (I_5), since $\boldsymbol{\beta}_0 = \lim_{n\to\infty} \operatorname{Argmin}_{\boldsymbol{\beta}\in\mathscr{B}} E\{D_n^j(\boldsymbol{\beta})\}$, we have $E\{S_n^j(\boldsymbol{\beta}_0)\} \to \mathbf{0}$ as $n \to \infty$. From the fact that $\mathrm{Var}(\varepsilon|\mathbf{x}) > 0$, there exists a positive constant C such that $\varsigma_{jn} \geq Cn^2$ which satisfies the assumptions of Lemma 3.1, as φ is twice continuously differentiable with bounded derivatives, $\gamma < (a+1)/2$ with $a = 2$, $M_0 = M_1 = 1$, $\gamma = 1$, and $m(n) = n$. Thus, for n large enough, $n\mathbf{W}_{jn}^{-1}T_n^j(\boldsymbol{\beta}_0) \approx n\mathbf{W}_{jn}^{-1}S_n^j(\boldsymbol{\beta}_0)$, which converges to a multivariate standard normal, by Lemma 3.1. A direct application of Slutsky's Lemma and putting $\Sigma_{jn} = n^{-1/2}\mathbf{W}_{jn}$ give $\sqrt{n}S_n^j(\boldsymbol{\beta}_0) \xrightarrow{\mathscr{D}} N_p(\mathbf{0}, \Sigma^j_{\boldsymbol{\beta}_0})$, $j = 1, 2$, where $\Sigma^j_{\boldsymbol{\beta}_0} = \lim_{n\to\infty} \Sigma_{jn}\Sigma^\tau_{jn}$.

Proof of Theorem 3.2. Let C be an arbitrary positive constant. Recall from Eq. (3.5) that the log likelihood ratio of $\boldsymbol{\beta}_0$ is given by

$$-2\log R_n^j(\boldsymbol{\beta}_0) = -2\log\prod_{i=1}^n \left(1 + \boldsymbol{\xi}^\tau\eta_{ij}(\boldsymbol{\beta}_0)\right)^{-1} = 2\sum_{i=1}^n \log\left(1 + \boldsymbol{\xi}^\tau\eta_{ij}(\boldsymbol{\beta}_0)\right).$$

Under (I_1) and (I_2), there exist a positive constant M and a function $h \in L^p$, $p \geq 1$ such that $|\varphi(t)| \leq M$ for all $t \in (0, 1)$, and $\|\nabla_{\boldsymbol{\beta}} g(\mathbf{x}_i, \boldsymbol{\beta}_0)\| \leq h(\mathbf{x}_i)$, where $\|\cdot\|$ stands for the L^2- norm. From this, $\max_{1\leq i\leq n} \|\nabla_{\boldsymbol{\beta}} g(\mathbf{x}_i, \boldsymbol{\beta}_0)\| = o_p(n^{1/2})$ since $E(|h(\mathbf{x}_i)|^P) < \infty$, $p \geq 1$. Also, since $\Sigma_{jn}\Sigma^\tau_{jn} \to \Sigma^j_{\boldsymbol{\beta}_0}$ *a.s.*, Σ_{jn} is almost surely bounded. Thus, $\|\eta_{ij}(\boldsymbol{\beta}_0)\| \leq M \times \max_{1\leq i\leq n} h(\mathbf{x}_i)$, which implies that

$$\max_{1\leq i\leq n} \|\eta_{ij}(\boldsymbol{\beta}_0)\| = o_p(n^{1/2}) \quad and \quad \frac{1}{n}\sum_{i=1}^n \|\eta_{ij}(\boldsymbol{\beta}_0)\|^3 = o_p(n^{1/2}). \tag{3.6}$$

Moreover, $\boldsymbol{\Lambda}_{nj} = Var(\sqrt{n}S_n^j(\boldsymbol{\beta}_0)) = n^{-1}\sum_{i=1}^n \eta_{ij}(\boldsymbol{\beta}_0)\eta^\tau_{ij}(\boldsymbol{\beta}_0) = \Sigma^j_{\boldsymbol{\beta}_0} + o_p(1)$ by assumption (I_6), from which $\Sigma^j_{\boldsymbol{\beta}_0}$ is assumed to be positive definite. Hence, following the proof of Lemma 3.1, we have $\Sigma_{jn}\Sigma^\tau_{jn} - \boldsymbol{\Lambda}_{nj} \to 0$ *a.s.* Since $\sqrt{n}S_n^j(\boldsymbol{\beta}_0) \xrightarrow{\mathscr{D}} N(0, \Sigma^j_{\boldsymbol{\beta}_0})$, we have $\|S_n^j(\boldsymbol{\beta}_0)\| = O_p(n^{-1/2})$. Now from Eq. (3.4), using similar arguments as those in Owen (1990), it is obtained that $\|\boldsymbol{\xi}\| = O_p(n^{-1/2})$. On the other hand, performing a Taylor expansion to the right-hand side of Eq. (3.5) results in

$$-2\log R_n^j(\boldsymbol{\beta}_0) = 2\sum_{i=1}^n \left[\boldsymbol{\xi}^\tau\eta_{ij}(\boldsymbol{\beta}_0) - \frac{1}{2}\left(\boldsymbol{\xi}^\tau\eta_{ij}(\boldsymbol{\beta}_0)\right)^2\right] + \boldsymbol{\gamma}_n,$$

where $\boldsymbol{\gamma}_n = O_P(1)\sum_{i=1}^n |\boldsymbol{\xi}^\tau\eta_{ij}(\boldsymbol{\beta}_0)|^3$. Now, using similar arguments as in Owen (2001), we have

$$\begin{aligned}
-2\log R_n^j(\boldsymbol{\beta}_0) &= \sum_{i=1}^n \boldsymbol{\xi}^\tau\eta_{ij}(\boldsymbol{\beta}_0) + o_P(1)\\
&= \Big(\frac{1}{n}\sum_{i=1}^n \eta_{ij}(\boldsymbol{\beta}_0)\Big)^\tau (n\boldsymbol{\Lambda}_{nj})^{-1}\Big(\frac{1}{n}\sum_{i=1}^n \eta_{ij}(\boldsymbol{\beta}_0)\Big) + o_p(1)\\
&= \left(\sqrt{n}\boldsymbol{\Lambda}_{nj}^{-1/2}S_n^j(\boldsymbol{\beta}_0)\right)^\tau\left(\sqrt{n}\boldsymbol{\Lambda}_{nj}^{-1/2}S_n^j(\boldsymbol{\beta}_0)\right) + o_p(1).
\end{aligned}$$

Using Slutsky's lemma, we have $\sqrt{n}\boldsymbol{\Lambda}_{nj}^{-1/2} S_n^j(\boldsymbol{\beta}_0) \xrightarrow{\mathscr{D}} N_p(0, I_p)$ as $n \to \infty$, and therefore,

$$-2 \log R_n^j(\boldsymbol{\beta}_0) \xrightarrow{\mathscr{D}} \chi_p^2.$$

References

Bindele, H. F. (2015). The signed-rank estimator for nonlinear regression with responses missing at random. *Electronic Journal of Statistics, 9*(1), 1424–1448.

Bindele, H. F., & Abebe, A. (2012). Bounded influence nonlinear signed-rank regression. *Canadian Journal of Statistics, 40*(1), 172–189.

Bindele, H. F., & Abebe, A. (2015). Semi-parametric rank regression with missing responses. *Journal of Multivariate Analysis, 142*, 117–132.

Bindele, H. F., & Zhao, Y. (2016). Signed-rank regression inference via empirical likelihood. *Journal of Statistical Computation and Simulation, 86*(4), 729–739.

Bindele, H. F., & Zhao, Y. (in press). Rank-based estimating equation with non-ignorable missing responses via empirical likelihood. *Statistica Sinica*.

Bindele, H. F. A. (2017). Strong consistency of the general rank estimator. *Communications in Statistics - Theory and Methods, 46*(2), 532–539.

Brunner, E., & Denker, M. (1994). Rank statistics under dependent observations and applications to factorial designs. *Journal of Statistical Planning and Inference, 42*(3), 353–378.

Chen, S. X., Peng, L., & Qin, Y.-L. (2009). Effects of data dimension on empirical likelihood. *Biometrika, 96*(3), 711–722.

Cheng, P. E. (1994). Nonparametric estimation of mean functionals with data missing at random. *Journal of the American Statistical Association, 89*(425), 81–87.

Delecroix, M., Hristache, M., & Patilea, V. (2006). On semiparametric estimation in single-index regression. *Journal of Statistical Planning and Inference, 136*(3), 730–769.

Einmahl, U., & Mason, D. M. (2005). Uniform in bandwidth consistency of kernel-type function estimators. *The Annals of Statistics, 33*(3), 1380–1403.

Gong, Y., Peng, L., & Qi, Y. (2010). Smoothed jackknife empirical likelihood method for ROC curve. *Journal of Multivariate Analysis, 101*(6), 1520–1531.

Healy, M., & Westmacott, M. (1956). Missing values in experiments analysed on automatic computers. *Journal of the Royal Statistical Society. Series C (Applied Statistics), 5*(3), 203–206.

Hettmansperger, T. P., & McKean, J. W. (2011). *Robust Nonparametric Statistical Methods*. Monographs on Statistics and Applied Probability (Vol. 119, 2nd ed.). Boca Raton, FL: CRC Press.

Hjort, N. L., McKeague, I. W., & Van Keilegom, I. (2009). Extending the scope of empirical likelihood. *The Annals of Statistics, 37*(3), 1079–1111.

Jaeckel, L. A. (1972). Estimating regression coefficients by minimizing the dispersion of the residuals. *Annals of Mathematical Statistics, 43*, 1449–1458.

Jing, B.-Y., Yuan, J., & Zhou, W. (2009). Jackknife empirical likelihood. *Journal of the American Statistical Association, 104*(487), 1224–1232.

Lahiri, S. N., & Mukhopadhyay, S. (2012). A penalized empirical likelihood method in high dimensions. *The Annals of Statistics, 40*(5), 2511–2540.

Little, R. J. (1992). Regression with missing x's: A review. *Journal of the American Statistical Association, 87*(420), 1227–1237.

Little, R. J. A., & Rubin, D. B. (1987). *Statistical analysis with missing data*. Wiley Series in Probability and Mathematical Statistics: Applied Probability and Statistics. New York: Wiley.

Little, R. J. A., & Rubin, D. B. (2002). *Statistical analysis with missing data*. Wiley Series in Probability and Statistics (2nd ed.). Hoboken, NJ: Wiley-Interscience [John Wiley & Sons].

Owen, A. B. (1988). Empirical likelihood ratio confidence intervals for a single functional. *Biometrika, 75*(2), 237–249.

Owen, A. B. (1990). Empirical likelihood ratio confidence regions. *The Annals of Statistics, 18*(1), 90–120.

Owen, A. B. (2001). *Empirical Likelihood*. Chapman & Hall/CRC Monographs on Statistics & Applied Probability. Boca Raton, FL: CRC Press.

Qin, J., & Lawless, J. (1994). Empirical likelihood and general estimating equations. *The Annals of Statistics, 22*(1), 300–325.

Rubin, D. B. (1976). Inference and missing data. *Biometrika, 63*(3), 581–592.

Rubin, D. B. (2004). *Multiple imputation for nonresponse in surveys*. Wiley Classics Library (Vol. 81). Hoboken, NJ: Wiley.

Silverman, B. W. (1986). *Density estimation for statistics and data analysis* (Vol. 26). Boca Raton, FL: CRC Press.

Tang, C. Y., & Leng, C. (2010). Penalized high-dimensional empirical likelihood. *Biometrika, 97*(4), 905–920.

Van Buuren, S. (2012). *Flexible imputation of missing data*. Chapman & Hall/CRC Interdisciplinary Statistics. Boca Raton, FL: Taylor & Francis.

Wang, C. Y., Wang, S., Zhao, L.-P., & Ou, S.-T. (1997). Weighted semiparametric estimation in regression analysis with missing covariate data. *Journal of the American Statistical Association, 92*(438), 512–525.

Wang, Q., Linton, O., & Härdle, W. (2004). Semiparametric regression analysis with missing response at random. *Journal of the American Statistical Association, 99*(466), 334–345.

Wang, Q., & Rao, J. N. K. (2002). Empirical likelihood-based inference under imputation for missing response data. *The Annals of Statistics, 30*(3), 896–924.

Wang, Q., & Sun, Z. (2007). Estimation in partially linear models with missing responses at random. *Journal of Multivariate Analysis, 98*(7), 1470–1493.

Yang, H., & Zhao, Y. (2012). New empirical likelihood inference for linear transformation models. *Journal of Statistical Planning and Inference, 142*(7), 1659–1668.

Yu, W., Sun, Y., & Zheng, M. (2011). Empirical likelihood method for linear transformation models. *Annals of the Institute of Statistical Mathematics, 63*(2), 331–346.

Zhang, Z., & Zhao, Y. (2013). Empirical likelihood for linear transformation models with interval-censored failure time data. *Journal of Multivariate Analysis, 116*, 398–409.

Zhao, L. P., Lipsitz, S., & Lew, D. (1996). Regression analysis with missing covariate data using estimating equations. *Biometrics, 52*(4), 1165–1182.

Chapter 4
Bayesian Nonparametric Spatially Smoothed Density Estimation

Timothy Hanson, Haiming Zhou, and Vanda Inácio de Carvalho

4.1 Introduction

Geographic information systems (GIS) technology has exploded over the last several decades due to impactful advances in data storage, computing power, sophisticated processing techniques, and visualization software. Accordingly, there has been an increasing need for the development of the state-of-the-art statistical models for spatial data as well (for an overview of developed methods, see Gelfand et al. 2010). Much recent literature has focused on spatially varying trends in the form of random fields. Although fundamental, spatially varying density estimation has received much less attention, perhaps due to challenges inherent in adapting existing methods to the spatial setting. This paper is then motivated by the need to fill a particular gap in the literature: provide a conceptually simple and computationally feasible, yet competitive, approach to modeling spatially dependent distributions.

The field of Bayesian nonparametrics has offered several viable spatially varying density estimators over the last decade, the majority of which are based on convolutions of a continuous kernel with a spatially varying discrete measure. More specifically, all proposed methods have been extensions of Dirichlet process (DP) mixture models (Escobar and West 1995) towards dependent Dirichlet process (DDP) mixtures (MacEachern 2001). The first such contribution was proposed by Gelfand et al. (2005). These authors developed a DDP for point-referenced spatial

T. Hanson (✉)
Medtronic Inc., Minneapolis, MN, USA
e-mail: tim.hanson2@medtronic.com

H. Zhou
Northern Illinois University, DeKalb, IL, USA

V. I. de Carvalho
University of Edinburgh, Edinburgh, Scotland, UK

Y. Zhao, D.-G. Chen (eds.), *New Frontiers of Biostatistics and Bioinformatics*, ICSA Book Series in Statistics, https://doi.org/10.1007/978-3-319-99389-8_4

data where the underlying DP base measure was taken to be a Gaussian process over Euclidean space; as a result, the density estimate is a discrete mixture of normal distributions where the components' means are Gaussian process realizations observed at a spatial location. An extension of this model allowing different surface selection at different sites was proposed by Duan et al. (2007). In turn, Griffin and Steel (2006) proposed a spatial DP model that permutes the random variables building the weights in the stick-breaking representation, allowing the occurrence of the stick-breaking atoms to be more or less likely in different regions of the spatial domain. Further, Reich and Fuentes (2007), motivated by the need to analyze hurricane surface wind fields, developed a spatial stick-breaking prior where the weights are spatially correlated. Related proposals include Petrone et al. (2009) and Rodríguez et al. (2010), both developing spatial DP's where the stick-breaking weights follow a copula representation. Very recently, Zhou et al. (2015) considered a spatial model where the marginal distributions follow the linear DDP of De Iorio et al. (2009), but a copula induces dependence for georeferenced data. The local DP (Chung and Dunson 2011), developed to accommodate predictor-dependent weights in a DDP with identical margins, offers an approach to the localized spatial "sharing" of atoms within neighborhoods of fixed size that could be extended to the spatial setting. Additionally, Fuentes and Reich (2013) generalize the models of Reich and Fuentes (2007) and Dunson and Park (2008) to the multivariate spatial setting with nonseparable and nonstationary covariance functions. A related approach, although not relying on the stick-breaking representation, was developed by Jo et al. (2017), who considered spatial conditional autoregressive (CAR) species sampling models. In contrast, the frequentist literature on spatial density estimation is very scarce, with an exception being the spatially weighted kernel density estimator proposed by Fotheringham et al. (2002, Section 8.4, pp. 202–203).

All of the above methods, as already mentioned, rely on discrete mixtures of smooth kernels; in fact, each is a particular mixture of normal distributions with some subset of model parameters changing smoothly in space. To the best of our knowledge, the only approach to spatial density estimation that does not rely on mixtures is the very recent Polya tree approach of Tansey et al. (2017). This approach follows Zhao and Hanson (2011) by taking the conditional probabilities that define the Polya tree to have a spatial structure. Whereas Zhao and Hanson (2011) consider multiple logistic-transformed independent CAR priors for the Polya tree conditional probabilities over a lattice (i.e., areal data), in Tansey et al. (2017) the logistic-transformed Polya tree conditional probabilities from adjacent spatial locations are shrunk towards each other via a graph-fused LASSO prior. This latter approach is especially fast and easy to compute when spatial locations lie on a rectangular grid. However, since the approach is not marginalized, density estimates at every spatial location need to be computed, and therefore the method is not immediately amenable to multivariate outcomes. Furthermore, an important drawback of Tansey et al. (2017) is that the fitting algorithm requires spatial locations to fall on a rectangular grid, something that rarely happens with typical observational data, data arising from irregularly placed monitoring stations, et cetera.

Our proposed estimator is built upon a modification of the predictive density from a marginalized Polya tree to accommodate localized behavior. The modification is readily implemented in existing Markov chain Monte Carlo (MCMC) schemes for models using marginalized Polya trees (e.g., Hanson 2006). Because we rely on marginalization, the method is fast, even for multivariate outcomes. Additionally, and unlike DP priors-based methods, our method does not rely on mixtures; in fact, a nice feature of Polya trees is that they can be centered at a parametric family (e.g., a normal distribution). In a similar fashion to the work of Dunson (2007), Dunson et al. (2007), and Dunson and Park (2008), observations are weighted according to a distance measure. However, unlike these approaches, a key property of our model is that for extreme covariate values, the estimator essentially follows the parametric family centering the Polya tree. That is, in spatial regions where data are sparse, the estimate is smoothed towards the parametric family, whereas areas that are data-heavy provide a more data-driven estimate. Also, our estimator can handle spatial locations (e.g., longitude and latitude), or covariates, or a mixture of spatial location and covariates. Additional contributions of our work include a refinement to accommodate arbitrarily censored data and a test for whether the density changes across space (and/or covariates). It is worth mentioning that although we are mainly interested in density regression, the methods developed here can be used to model the error in a general regression setup.

The remainder of the paper is organized as follows. In Sect. 4.2, we present our model, associated MCMC scheme, a generalization to censored data, and a permutation test. Section 4.3 provides several applications of the methods to real data. Concluding remarks are provided in Sect. 4.4.

4.2 The Predictive Model

The Polya tree and other partition models lead to a beautifully simple updating rule. A family of densities $\{G_{\boldsymbol{\theta}} : \boldsymbol{\theta} \in \boldsymbol{\Theta}\}$ is assumed to approximately hold, and $\mathbb{R}$ is broken up into regions of equal probability $\frac{1}{2^j}$ under $G_{\boldsymbol{\theta}}$ at level j. As data are collected, the proportion falling into a region is compared to what is expected under $G_{\boldsymbol{\theta}}$; if this proportion is higher than expected under $G_{\boldsymbol{\theta}}$, then the region is assigned higher predictive probability, and vice versa. However, the amount of the increase or decrease relative to $G_{\boldsymbol{\theta}}$ is attenuated through a smoothing parameter c that signifies how much confidence one has in the family $G_{\boldsymbol{\theta}}$ to begin with. This simple updating rule requires only counting the numbers of observations falling into the regions. This is now developed formally.

Initially, assume that data follow a univariate Polya tree centered at a normal distribution:

$$y_1, \ldots, y_n | G \overset{\text{iid}}{\sim} G, \quad G \sim PT_J(c, N(\mu, \sigma^2)),$$

truncated to level J (Hanson 2006). The predictive density for an observation y_i given the previous values $\mathbf{y}_{1:i-1} = (y_1, y_2, \ldots, y_{i-1})$ is given by:

$$p(y_i | \mathbf{y}_{1:i-1}, c, \boldsymbol{\theta}) = \phi(y_i|\boldsymbol{\theta}) \prod_{j=1}^{J} \frac{cj^2 + \sum_{k=1}^{i-1} I\{\lceil 2^j \Phi\{\frac{y_i-\mu}{\sigma}\}\rceil = \lceil 2^j \Phi\{\frac{y_k-\mu}{\sigma}\}\rceil\}}{cj^2 + \frac{1}{2}\sum_{k=1}^{i-1} I\{\lceil 2^{j-1} \Phi\{\frac{y_i-\mu}{\sigma}\}\rceil = \lceil 2^{j-1} \Phi\{\frac{y_k-\mu}{\sigma}\}\rceil\}}, \tag{4.1}$$

where $\phi(y|\boldsymbol{\theta})$ is the density function for an $N(\mu, \sigma^2)$ random variable, $\boldsymbol{\theta} = (\mu, \log \sigma)$, $I\{A\}$ is the usual indicator function for event A, $\Phi(\cdot)$ is the cumulative distribution function of $N(0, 1)$, $\lceil \cdot \rceil$ is the ceiling function, and c is a precision parameter controlling how closely G follows the parametric centering distribution $N(\mu, \sigma^2)$ in terms of L_1 distance (Hanson 2006). Large values of c (e.g., 100 or 1000) lead to a strong belief that y_is are closely iid from $N(\mu, \sigma^2)$. On the other hand, smaller values of c (e.g., 0.01 or 0.1) allow more pronounced deviations of G from $N(\mu, \sigma^2)$. Therefore, the predictive density in (4.1) can change dramatically with c. To mitigate the effect of c on the posterior inference, Hanson (2006) suggests a gamma prior on c which will also be used for our proposal in Eq. (4.2) below.

Now, consider spatial data where the observation y_i is observed at spatial location $\mathbf{x}_i$. The full data are then $\{(\mathbf{x}_i, y_i)\}_{i=1}^n$. Note that $\mathbf{x}_i$ can simply be spatial location, e.g., longitude and latitude, or covariates, or a mixture of spatial location and covariates. In the Polya tree, an observation y_k ($k < i$) contributes the same weight to the predictive density (4.1) of y_i, regardless of how close corresponding spatial locations $\mathbf{x}_i$ and $\mathbf{x}_k$ are. If we replace the indicator functions in (4.1) by a distance measure $d_\psi(\mathbf{x}_i, \mathbf{x}_k)$ only giving a "whole observation" when $\mathbf{x}_i = \mathbf{x}_k$ and some fraction of unity that is a function of the distance between $\mathbf{x}_i$ and $\mathbf{x}_k$ otherwise, we obtain a predictive process with a tailfree flavor, but the additional flexibility to be able to adapt locally:

$$p(y_i | \mathbf{y}_{1:i-1}, c, \boldsymbol{\theta}, \psi) = \phi(y_i|\boldsymbol{\theta}) \prod_{j=1}^{J} \frac{cj^2 + \sum_{k=1}^{i-1} I\{\lceil 2^j \Phi\{\frac{y_i-\mu}{\sigma}\}\rceil = \lceil 2^j \Phi\{\frac{y_k-\mu}{\sigma}\}\rceil\} d_\psi(\mathbf{x}_i, \mathbf{x}_k)}{cj^2 + \frac{1}{2}\sum_{k=1}^{i-1} I\{\lceil 2^{j-1} \Phi\{\frac{y_i-\mu}{\sigma}\}\rceil = \lceil 2^{j-1} \Phi\{\frac{y_k-\mu}{\sigma}\}\rceil\} d_\psi(\mathbf{x}_i, \mathbf{x}_k)} \tag{4.2}$$

The distance measure used herein is a function of the sample Mahalanobis distance $d_\psi(\mathbf{x}_i, \mathbf{x}_k) = \exp\{-\psi(\mathbf{x}_i - \mathbf{x}_k)'\mathbf{S}^{-1}(\mathbf{x}_i - \mathbf{x}_k)\}$, defining an effective window in which data can affect the predictive density. When $\psi = 0$, the prediction rule from a Polya tree with exchangeable observations is obtained as $d_\psi(\mathbf{x}_i, \mathbf{x}_k) = 1$ for all i and k. When $d_\psi(\mathbf{x}_i, \mathbf{x}_k) \approx 0$, it is essentially as if y_k is not in the sample. Note for $\mathbf{x}_i$ very far from the sample mean $\bar{\mathbf{x}}$, the distances will all essentially be zero and the parametric model is obtained. As a consequence, in regions that are data scarce the estimator will essentially follow the parametric $N(\mu, \sigma^2)$ centering the Polya tree.

The Mahalanobis distance gives the commonly used spatial Gaussian correlation function when Euclidean coordinates are independently observed. Furthermore, the Mahalanobis distance is anisotropic, allowing for quite different "x" and "y" scales if present in data, as well as correlated spatial coordinates. Distance measures such as this are also used in "local Dirichlet process" approaches, e.g., Chung and Dunson (2011) as well as earlier versions for densities that change smoothly with covariates, e.g., Dunson (2007), although not incorporating correlation among variables. The accommodation of correlation is important because it essentially obviates concerns about multicollinearity as well as mitigates a need for variable selection. Superfluous dimensions of $\mathbf{x}_i$ are handled naturally within the Mahalanobis distance wherein the distances between highly positively correlated variables are much less than the same distances between uncorrelated or negatively correlated variables. We reiterate that the "locations" $\mathbf{x}_i$ can be spatial locations, covariates, or mixtures of spatial locations and covariates.

The joint density for all observations is given by the Markov expansion:

$$p(y_1, \ldots, y_n|c, \boldsymbol{\theta}, \psi) = \prod_{i=1}^{n} p(y_i|\mathbf{y}_{1:i-1}, c, \boldsymbol{\theta}, \psi), \tag{4.3}$$

where $p(y_1|c, \boldsymbol{\theta}, \psi) = \phi(y_1|\boldsymbol{\theta})$. This defines a valid joint probability model; however, $p(y_1, \ldots, y_n|c, \boldsymbol{\theta}, \psi)$ is not invariant to the order in which pairs $(\mathbf{x}_i, y_i)$ enter into the model. That is, $p(y_1, y_2, y_3|c, \boldsymbol{\theta}, \psi)$ is not necessarily equal to $p(y_2, y_3, y_1|c, \boldsymbol{\theta}, \psi)$. This has ramifications in terms of drawing inferences for $(c, \boldsymbol{\theta}, \psi|\mathbf{y}_{1:n})$.

The posterior

$$p(c, \boldsymbol{\theta}, \psi|\mathbf{y}_{1:n}) \propto p(y_1, \ldots, y_n|c, \boldsymbol{\theta}, \psi)p(c, \boldsymbol{\theta}, \psi)$$

depends on the order $y_1, \ldots, y_n$. To make the posterior invariant to the ordering of the observed data, we propose to instead use the permutation density:

$$p(c, \boldsymbol{\theta}, \psi|\mathbf{y}_{1:n}) \propto p(c, \boldsymbol{\theta}, \psi)\frac{1}{n!}\sum_{(i_1,\ldots,i_n)\in\mathcal{P}} p(y_{i_1}, \ldots, y_{i_n}|c, \boldsymbol{\theta}, \psi),$$

where $\mathcal{P}$ are all permutations of $\{1, \ldots, n\}$. Dahl et al. (2017) consider a related scenario in defining a partition distribution indexed by a permutation. Computation of all permutations is not feasible so Dahl et al. (2017) place a uniform prior distribution over all permutations, allowing the MCMC to numerically perform the marginalization. In our setting, there is no reason to favor one permutation over another so equal weight is placed on all partitions by the *model* rather than through only the prior. Thus, the partition is not "updated" during MCMC according to a Metropolis–Hastings step; rather a different permutation is forced at each MCMC iteration.

Although the permutation density avoids order dependence, this dependence is empirically observed to be quite weak. Simply using the data order as observed changes posterior inference only slightly from that obtained through the permutation density. Similarly, Newton and Zhang (1999) noted in a Bayesian nonparametric setting involving the Dirichlet process with censored data, where exchangeability is lost, the order of updating affects the predictive distribution negligibly.

We proceed to develop a reasonable prior for ψ. With probability q, assume that the density is not spatially varying, i.e. $P(\psi = 0) = q$; setting $q = 0.5$ gives the posterior odds as the Bayes factor so we consider that here. For spatially varying densities, assume $\psi|\psi > 0 \sim \Gamma(a_\psi, b_\psi)$, where $\Gamma(a, b)$ denotes a gamma distribution with mean a/b. We set $a_\psi = 2$ and $b_\psi = (a_\psi - 1)/\psi_0$ so that the prior of $\psi|\psi > 0$ has mode at ψ_0, where ψ_0 satisfies $\exp\{-\psi_0 \max_{1\le i\le n}(\mathbf{x}_i - \bar{\mathbf{x}})'\mathbf{S}^{-1}(\mathbf{x}_i - \bar{\mathbf{x}})\} = 0.001$. Thus, the final prior on ψ is the mixture

$$\psi \sim q\,\delta_0 + (1-q)\,\Gamma(a_\psi, b_\psi),$$

where δ_x is Dirac measure at x and the default is $q = \frac{1}{2}$. A Bayes factor for comparing the spatially varying model to the exchangeable model is given by:

$$BF = \frac{P(\psi > 0|\mathbf{y}_{1:n})/P(\psi = 0|\mathbf{y}_{1:n})}{P(\psi > 0)/P(\psi = 0)} = \frac{q\ P(\psi > 0|\mathbf{y}_{1:n})}{(1-q)\ P(\psi = 0|\mathbf{y}_{1:n})}.$$

Two options are taken for $(\mu, \log\sigma)$. The default follows Hanson et al. (2008) and Chen and Hanson (2014) by simply fixing them at their maximum likelihood estimates (MLEs) under the parametric $N(\mu, \sigma^2)$ model as $c \to \infty$; for uncensored data, these are of course the sample moments $\hat{\mu} = \frac{1}{n}\sum_{i=1}^n y_i$, $\hat{\sigma}^2 = \frac{1}{n}\sum_{i=1}^n (y_i - \hat{\mu})^2$. When data are censored, the MLEs are not available in closed form but are readily computed by most statistical software packages, e.g., `survreg` in R using `family="gaussian"`. The second option is to take a bivariate normal prior $N_2(\boldsymbol{\theta}_0, \mathbf{V}_0)$ on $\boldsymbol{\theta} = (\mu, \log\sigma)$ with $\boldsymbol{\theta}_0 = (\hat{\mu}, \log\hat{\sigma})$ and $\mathbf{V}_0$ being the estimated asymptotic "plug-in" covariance for the MLE. The only remaining parameter is c; $c \sim \Gamma(a_c, b_c)$ is assumed with the default $c \sim \Gamma(5, 1)$.

The function `SpatDensReg` is developed within the R package `spBayesSurv` (Zhou and Hanson 2018) for the implementation. The usage syntax is

```
SpatDensReg(formula, data, na.action, prior=NULL,
            state=NULL,
           mcmc=list(nburn=3000, nsave=2000,
            nskip=0, ndisplay=500),
           permutation=TRUE, fix.theta=TRUE)
```

Here, `formula` is a formula expression with the response returned by the `Surv` function in the `survival` package. It supports right-censoring, left-censoring, interval-censoring, and mixtures of them. For survival data, the input response should be log survival times. The argument `prior` is a list giving the prior information. The list includes the following elements.

`prior` element	`maxL`	`a0`	`b0`	`theta0`	`V0`	`phia0`	`phib0`	`phiq0`
Corresponding symbol	L	a_c	b_c	$\boldsymbol{\theta}_0$	$\mathbf{V}_0$	a_ψ	b_ψ	q

The argument `permutation` is a logical flag to indicate whether a random data permutation will be implemented in the beginning of each iterate; the default is `TRUE`. The argument `fix.theta` is a logical flag to indicate whether the parameters $\boldsymbol{\theta}$ are fixed; the default is `TRUE` indicating that they are fixed at $(\hat{\mu}, \log\hat{\sigma})$.

4.2.1 Markov Chain Monte Carlo

The MCMC algorithm uses blockwise adaptive MCMC (Haario et al. 2001). The blocks are $(\mu, \log\sigma)$ when the option `fix.theta=FALSE` is chosen, c, and ψ; ψ requires some special care, detailed below. The permutation density can be used by fixing `permutation=TRUE` which is the default.

For the mixture prior $\psi \sim q\delta_0 + (1-q)\Gamma(a_\psi, b_\psi)$, we need to first conditionally sample whether $\psi = 0$ or not. Bayes rule gives

$$P(\psi = 0|\mathbf{y}_{1:n}, c, \boldsymbol{\theta})$$
$$= \frac{q\ p(\mathbf{y}_{1:n}|c, \boldsymbol{\theta}, \psi = 0)}{q\ p(\mathbf{y}_{1:n}|c, \boldsymbol{\theta}, \psi = 0) + (1-q)\int_0^\infty p(\mathbf{y}_{1:n}|c, \boldsymbol{\theta}, \psi)\gamma(\psi; a_\psi, b_\psi)d\psi},$$

where $\gamma(\cdot; a, b)$ refers to the density of $\Gamma(a, b)$. Set $\Gamma(\cdot; a, b)$ to be the cumulative distribution function of $\Gamma(a, b)$. The integral can be approximated by a Riemann sum:

$$\int_0^\infty p(\mathbf{y}_{1:n}|c, \boldsymbol{\theta}, \psi)\gamma(\psi; a_\psi, b_\psi)d\psi$$
$$\approx \sum_{k=1}^{K} \frac{1}{\psi_k - \psi_{k-1}} p(\mathbf{y}_{1:n}|c, \boldsymbol{\theta}, \psi_k)\gamma(\psi_k; a_\psi, b_\psi),$$

where $\psi_k = \Gamma^{-1}(\frac{k}{K+1}; a_\psi, b_\psi)$ for $k = 0, 1, \ldots, K$, e.g., $K = 20$. If $\psi = 0$ is sampled, we are done, otherwise we need to sample $\psi|\psi > 0$ using usual adaptive M-H. When computing the adaptive variance, only those sample values of ψ that are positive are included. An option forcing $\psi > 0$ only ($q = 0$) is also available which speeds up the MCMC considerably but disallows the computation of a Bayes factor to test whether spatial location and/or covariates affect the distribution of the response.

Given the posterior sample $\{\psi^{(j)}, j = 1, \ldots, M\}$, the Bayes factor for the spatial model vs. exchangeable model is simply $[\frac{1-\bar{q}}{\bar{q}}]/[\frac{1-q}{q}]$ where $\bar{q} = \frac{1}{M}\sum_{j=1}^{M} I\{\psi^{(j)} = 0\}$.

4.2.2 Censored Data

Censored observations are readily sampled from Metropolis–Hastings proposals based on the underlying centering distribution. Define

$$q(y_i|\mathbf{y}_{-i}, c, \boldsymbol{\theta}, \psi) \\ = \prod_{j=1}^{J} \frac{cj^2 + \sum_{k\neq i} I\{\lceil 2^j \Phi\{\frac{y_i-\mu}{\sigma}\}\rceil = \lceil 2^j \Phi\{\frac{y_k-\mu}{\sigma}\}\rceil\} d_\psi(\mathbf{x}_i, \mathbf{x}_k)}{cj^2 + \frac{1}{2}\sum_{k\neq i} I\{\lceil 2^{j-1} \Phi\{\frac{y_i-\mu}{\sigma}\}\rceil = \lceil 2^{j-1} \Phi\{\frac{y_k-\mu}{\sigma}\}\rceil\} d_\psi(\mathbf{x}_i, \mathbf{x}_k)},$$

where $\mathbf{y}_{-i}$ is $\mathbf{y}_{1:n}$ with the ith observation removed.

Let $\mathcal{C} = \{i : \delta_i = 0\}$ be the indices of censored observations where $\delta_i = 0$ if y_i is only known to lie in the interval $y_i \in (a_i, b_i)$, $a_i < b_i$, and $\delta_i = 1$ if y_i is observed exactly. The latent values of $\mathcal{Y}_c = \{y_i : i \in \mathcal{C}\}$ are updated via MCMC along with the model parameters $\boldsymbol{\theta}$, c, and ψ. If $i \in \mathcal{C}$ propose $y_i^* \sim N(\mu, \sigma^2)$ truncated to (a_i, b_i) and accept with probability:

$$1 \wedge \frac{q(y_i^*|\mathbf{y}_{-i}, c, \boldsymbol{\theta}, \psi)}{q(y_i|\mathbf{y}_{-i}, c, \boldsymbol{\theta}, \psi)},$$

otherwise leave y_i at its current value.

4.2.3 Direct Estimation and a Permutation Test p-Value

For uncensored data, estimation can be sped up substantially by avoiding MCMC entirely and simply using maximum a posteriori (MAP) estimates coupled with an empirical Bayes approach to fixing the centering distribution. Dunson (2007) considered a somewhat related approach in a covariate-weighted Dirichlet process mixture. Chen and Hanson (2014) consider a hybrid approach for uncensored data by setting $\boldsymbol{\theta} = (\hat{\mu}, \log\hat{\sigma})$, the MLE's under normality, and maximizing (4.3) with $\psi = 0$ over a grid of c values $c_i = \exp\{\frac{14}{19}(i-1) - 7\}$ for $i = 1, \ldots, 20$. The spatial version (4.3) allowing for $\psi > 0$ can similarly be maximized over a lattice of (c_i, ψ_j) values. From considerations in Sect. 4.2.1, quantiles of the prior $\psi_j = \Gamma^{-1}(\frac{j}{11}; a_\psi, b_\psi)$ for $i = 1, \ldots, 10$ are reasonable; call these values $\mathcal{S}$. Similarly, $c_i \in \mathcal{C} = \{0.001, 0.01, 0.1, 0.5, 1, 5, 10, 50, 100, 1000\}$ could be used giving 100 values of $\{(c_i, \psi_j)\}$ to compute and maximize (4.3) over.

The Bayes factor described in Sects. 4.2.1 and 4.2.2 tests the hypothesis H_0 : $\psi > 0$ relative to $H_0 : \psi = 0$ via MCMC using the priors described in these sections. A "maximized Bayes factor" from direct estimation is given by:

$$BF = \frac{\max_{(c,\psi)\in\mathcal{C}\times\mathcal{S}} p(c, \hat{\boldsymbol{\theta}}, \psi|\mathbf{y}_{1:n})}{\max_{c\in\mathcal{C}} p(c, \hat{\boldsymbol{\theta}}, 0|\mathbf{y}_{1:n})}. \tag{4.4}$$

This Bayes factor gives the "most evidence" in favor of the spatially varying model and akin to a likelihood ratio test, albeit with added prior information and a plug-in estimate for $\boldsymbol{\theta}$. Consider that the null $H_0 : \mathbf{x} \in \mathcal{X}$ is independent of $y \in \mathcal{Y}$. Under this null, we can repeatedly take random, uniformly distributed permutations $(i_1, \ldots, i_n) \in \mathcal{P}$, form "data" $\{(\mathbf{x}_j, y_{i_j})\}_{j=1}^n$, and compute Bayes factors from (4.4). The proportion of these larger than the one based on the original data is a permutation test p-value (Fisher 1935) for testing association between the response and spatial location (and/or covariates).

The function `BF.SpatDensReg` is developed within the R package `spBayesSurv` to obtain the BF in (4.4) and the permutation test p-value. The usage syntax is

```
BF.SpatDensReg(y, X, prior = NULL, nperm = 100,
               c_seq = NULL,  phi_seq = NULL)
```

Here, `y` is a vector of uncensored responses, rows of `X` are spatial locations and/or covariates, `prior` is the same as the one used in `SpatDensReg`, `nperm` is an integer giving the total number of permutations, `c_seq` is an vector giving grid values for c, and `phi_seq` is a vector giving grid values for ψ. To illustrate the use of this method, we generate the data as follows: $y_i \overset{ind}{\sim} N(\beta x_i, 0.2^2)$, $x_i \overset{iid}{\sim}$ Beta(0.3, 0.3), $i = 1, \ldots, 300$, where $\beta = 0.01, 0.05, 0.1, 0.5, 1$. The following R code is used to obtain these BFs and p-values. As expected, we see that as β increases, so does the Bayes factor while the p-value is approaching zero.

```
library(spBayesSurv)
set.seed(2017)
beta = c(0.01, 0.05, 0.1, 0.5, 1);
BFs = rep(NA, length(beta));
Pvalues = rep(NA, length(beta));
for(sim in 1:length(beta)){
print(sim);
## Generate data
n = 300;
x = rbeta(n, 0.3, .3)
y = rep(0, n);
uu = runif(n);
for(i in 1:n){
y[i] = rnorm(1, beta[sim]*x[i], .2);
}
```

```
prior = list(maxL=6);
res1 = BF.SpatDensReg(y, x, prior=prior, nperm=500);
BFs[sim] = res1$BF;
Pvalues[sim] = res1$pvalue;
}
###### Outputs:
> BFs
   beta=0.01     beta=0.05      beta=0.1      beta=0.5
        beta=1
4.624762e-01 3.283394e+00 8.828178e+03 1.290706e+30
        4.403196e+83
> Pvalues
beta=0.01 beta=0.05  beta=0.1  beta=0.5    beta=1
    0.974     0.922     0.000     0.000     0.000
```

4.3 Examples

4.3.1 IgG Distribution Evolving with Age

Jara and Hanson (2011) and Schörgendorfer and Branscum (2013) considered serum immunoglobulin G (IgG) concentrations from $n = 298$ children aged 6 months to 6 years old. Like these authors, we consider the log-transformation of the data y_i; the log-IgG values are plotted versus age in Fig. 4.1f. We consider the spatially smoothed Polya tree for estimating the log-IgG density as smoothly varying function of age. Unlike previous authors, we rely on only the Polya tree and do not explicitly model an IgG trend via fractional polynomials.

The following R code is used to fit the proposed model with $J = 4$ and the default prior settings in Sect. 4.2 except for the option `fix.theta=FALSE` which provides much smoother posterior density estimates.

```
#needed packages
library(survival)
library(spBayesSurv)
library(coda)
library(DPpackage)

#data management
data(igg); d = igg; n = nrow(d);
d$logIgG = log(d$igg)

#fitting the model
nburn=20000; nsave=5000; nskip=9;
mcmc=list(nburn=nburn, nsave=nsave, nskip=nskip,
```

```
     ndisplay=500);
prior = list(maxL=4, phiq0=0.5);
res1 = SpatDensReg(formula = Surv(logIgG)~age, data=d,
     prior=prior,
   mcmc=mcmc, permutation = TRUE, fix.theta=FALSE);

#output from summary
summary(fit) # most output removed to save space
Posterior inference of centering distribution
parameters
(Adaptive M-H acceptance rate: 0.1054):
             Mean       Median     Std. Dev.   95%CI-Low
              95%CI-Upp
location     1.47804    1.49332    0.05165     1.33244
              1.53099
log(scale)  -0.70716   -0.70207    0.04997    -0.81437
             -0.62347

Posterior inference of precision parameter
(Adaptive M-H acceptance rate: 0.35896):
       Mean     Median   Std. Dev.   95%CI-Low   95%CI-Upp
alpha  0.6967   0.6368   0.3025      0.2866      1.4248

Posterior inference of distance function range phi
(Adaptive M-H acceptance rate: 0.38898):
       Mean    Median   Std. Dev.   95%CI-Low   95%CI-Upp
range  3.527   3.326    1.212       1.805       6.484

Bayes Factor for the spatial model vs. the
exchangeable model: Inf Number of subjects: n=298
```

The traceplots for $\boldsymbol{\theta}$, ψ, and c mixed very well (not shown). The Bayes factor for testing association between age and log-IgG is ∞ indicating a decisive evidence of dependency. Figure 4.1 presents the posterior mean and 95% pointwise credible interval of the log-IgG density at five different ages. These fitted densities are similar to those obtained by Jara and Hanson (2011). The following R code can be used to provide these plots.

```
ygrid = seq(min(d$logIgG), max(d$logIgG), length.out
      = 200);
xpred = data.frame(age=c(11, 25, 38, 52, 65, 79)/12);
estimates=plot(res1, xnewdata=xpred, ygrid=ygrid);
for(i in 1:nrow(xpred)){
pdf(file =paste("IgG-densities-age", xpred[i,]*12,
      "-bt.pdf", sep=""),
paper="special", width=8, height=6)
par(cex=1.5,mar=c(4.1,4.1,2,1),cex.lab=1.4,
```

```
    cex.axis=1.1)
plot(estimates$ygrid, estimates$fhat[,i], "l",
main=paste("IgG data, age=", xpred[i,]*12, "months",
    sep=""),
xlab="log IgG", ylab="density",
ylim=c(0,1.5), xlim=c(0, 2.5), lty=1, lwd=3);
lines(estimates$ygrid, estimates$fhatup[,i], lty=2,
    lwd=2);
lines(estimates$ygrid, estimates$fhatlow[,i], lty=2,
    lwd=2);
dev.off()
}
pdf(file ="IgG-scatter.pdf", paper="special", width=8,
    height=6)
par(cex=1.5,mar=c(4.1,4.1,2,1),cex.lab=1.4,
    cex.axis=1.1)
plot(d$age*12, d$logIgG, main="IgG data",
xlab="age (months)", ylab="log IgG")
for(i in 1:5){
points(xpred[i,]*12, -0.2, pch = 16, cex=1.3,
col = "red", las = 1,xpd = TRUE)
text(xpred[i,]*12, 0, paste("", xpred[i,]*12, sep=""),
col = "red", adj = c(-0.1, .5))
}
dev.off()
```

4.3.2 Time to Infection in Amphibian Populations

Spatial data on the number of years from discovery to the time-to-arrival of the fungus *Batrachochytrium dendrobatidis* (Bd) in mountain yellow-legged frog populations throughout Sequoia-Kings Canyon National Park was considered by Zhou et al. (2015). Once infected, the Bd fungus can wipe out a frog population in a few weeks, and it is of interest to determine the distribution of time-to-infection and how it varies spatially. The data consist of $n = 309$ frog populations (Fig. 4.2f) initially discovered during park-wide surveys conducted from 1997 to 2002, and then resurveyed regularly through 2011. The observed event time is calculated as the number of years from the initial survey to either Bd arrival (time actually observed) or the last resurvey (right censored). By the end of the study, about 11% of the frog populations remained Bd-negative (right censored), and the rest of populations are interval censored.

We fit the spatially smoothed Polya tree with the same settings as Sect. 4.3.1. The Bayes factor for testing spatial variation of time-to-Bd is estimated to be ∞, strong evidence that the time-to-Bd distribution spatially varies. Figure 4.2 shows

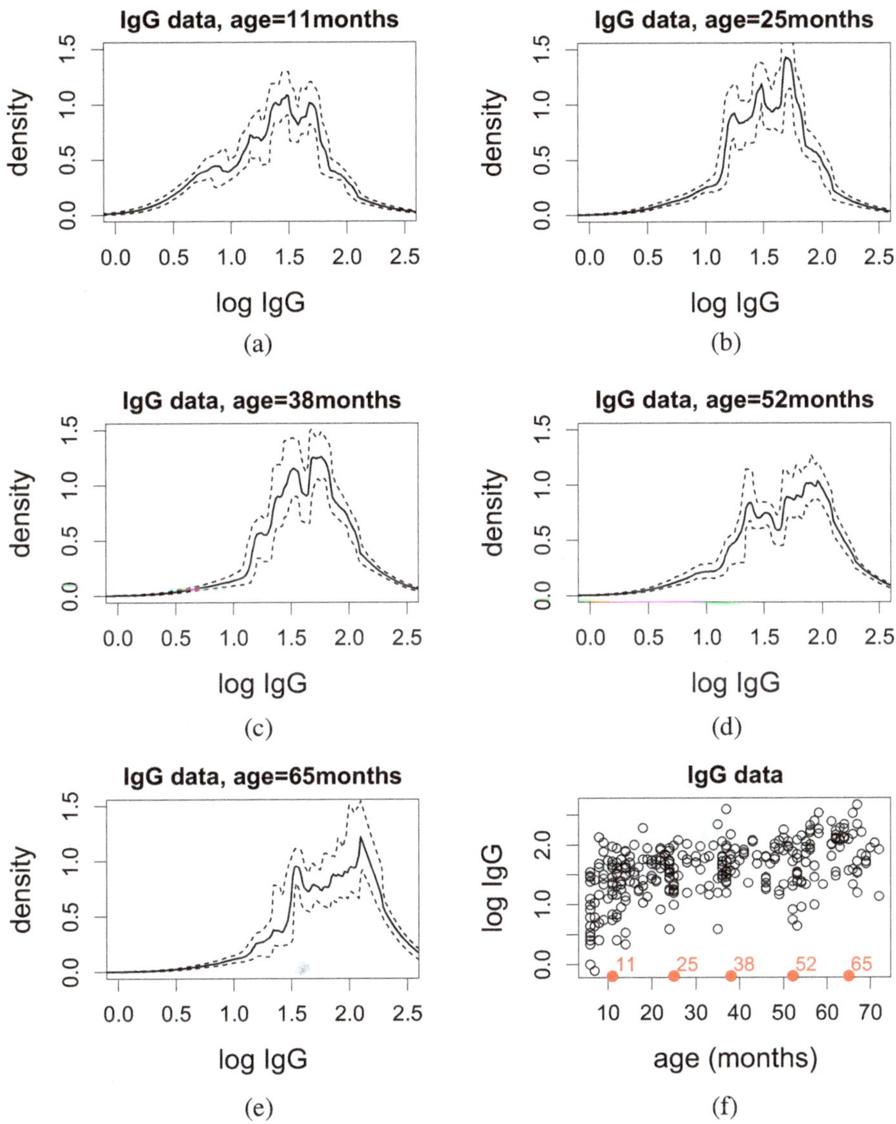

Fig. 4.1 IgG data. Panels (**a**)–(**e**) show the posterior mean (solid) and 95% pointwise credible interval (dashed) of the density of log IgG at five different ages. Panel (**f**) shows the five age points and the scatterplot of the data

the posterior mean and 95% pointwise credible intervals of the log time-to-Bd density at five different locations (marked in Fig. 4.2f). The distribution of log time-to-Bd at location 5 has two modes which can also be seen from the predictive log times-to-Bd around this location.

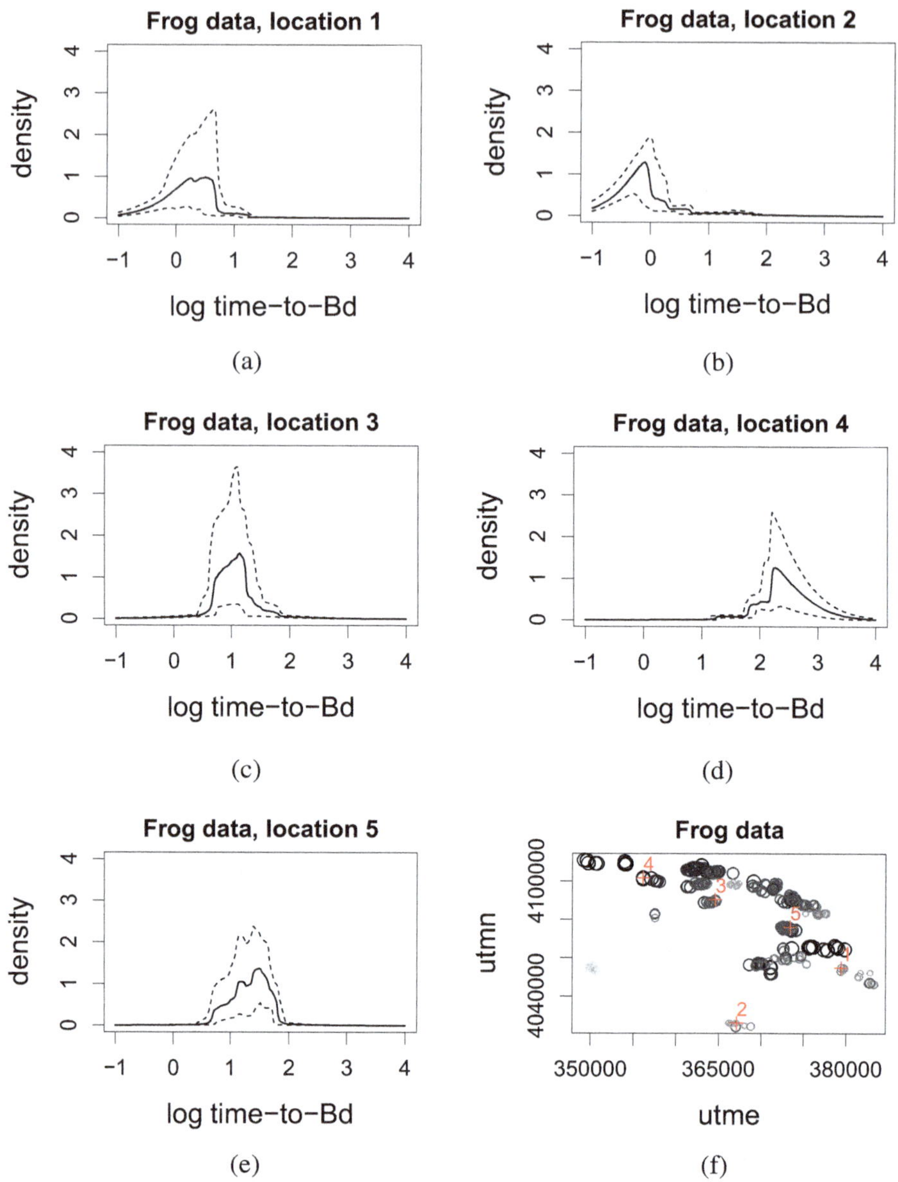

Fig. 4.2 Frog data. Panels (**a**)–(**e**) show the posterior mean (solid) and 95% pointwise credible interval (dashed) of the log time-to-Bd density at five different locations. Panel (**f**) shows the five considered locations and the data locations with circle size representing the posterior mean of log times-to-Bd

4.3.3 Simulated Data

We generate two datasets from the following two scenarios, respectively.

1. $y_i \overset{ind}{\sim} N(x_i, 0.2^2)$, $x_i \overset{iid}{\sim}$ Beta(0.3, 0.3), $i = 1, \ldots, 300$,
2. $y_i \overset{iid}{\sim} 0.5N(0.5, 0.2^2) + 0.5N(1, 0.3^2)$, $x_i \overset{iid}{\sim}$ Beta(0.3, 0.3), $i = 1, \ldots, 300$.

Here, we expect a large BF value for scenario 1 and a BF less than 1 for scenario 2. The censoring times are generated from Uniform(0.5, 2) so that the censoring rate is 0.13 under scenario 1 and 0.21 under scenario 2.

The spatially smoothed Polya tree is fit with $J = 6$ and the same prior settings as Sect. 4.3.1. We retain 5000 scans thinned from 50,000 after a burn-in period of 20,000 iterations. The BF factor for scenario 1 is ∞ as expected, while it is 0.01 under scenario 2. Figures 4.3 and 4.4 present the posterior mean and 95% pointwise credible interval of the conditional density of y at three different x values under

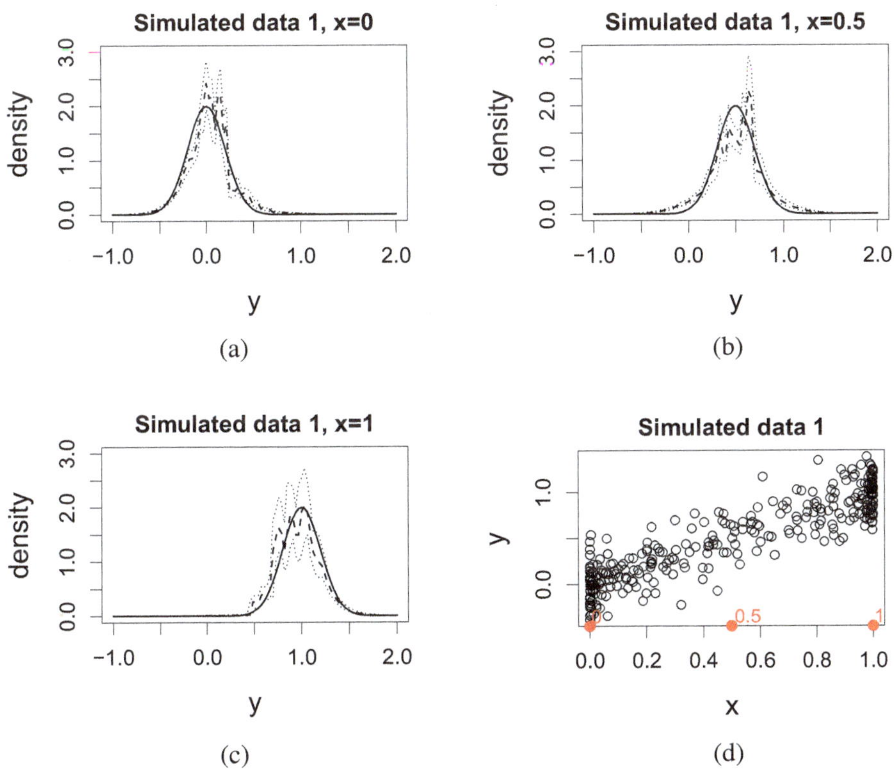

Fig. 4.3 Simulated data 1. Panels (**a**)–(**c**) show the posterior mean (dashed) and 95% pointwise credible interval (dotted) of the conditional density at three different x values; the solid curves are the corresponding true densities. Panel (**d**) shows the three x points and the scatterplot of the data

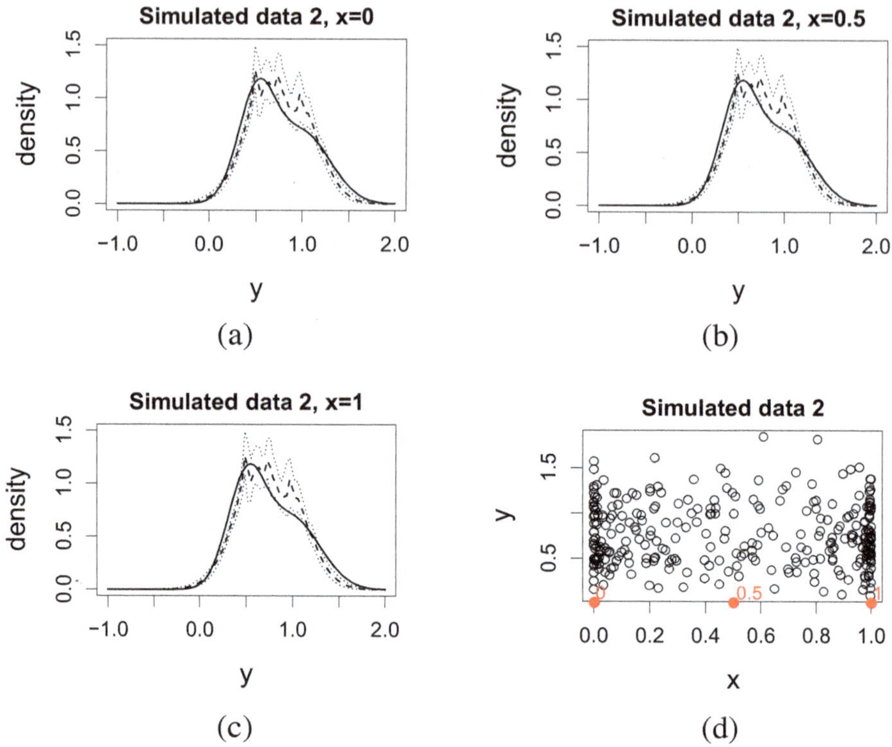

Fig. 4.4 Simulated data 2. Panels (**a**)–(**c**) show the posterior mean (dashed) and 95% pointwise credible interval (dotted) of the conditional density at three different x values; the solid curves are the corresponding true densities. Panel (**d**) shows the three x points and the scatterplot of the data

each scenario. The results demonstrate that the proposed model can capture the conditional densities quite well *without any spatial trend component*, although the estimates are a bit spiky.

To investigate the impact of censoring on our model performance, we use the simulated data 1 (Fig. 4.3d, uncensored version) again under the following two cases: (1) right-censoring with high censoring rate, and (2) interval-censoring. For case (1), the censoring times are generated from $N(x_i - 0.2, 0.5^2)$ yielding a 0.67 right-censoring rate. For case (2), we first generate right-censored times from Uniform(0.5, 2), then transfer uncensored times into internal-censored times using the endpoints $\{0, 0.2, 0.4, \ldots, 1.8, 2\}$, yielding a rate of 0.13 for right-censoring, 0.14 for left-censoring and 0.73 for interval-censoring. The BF factors for both cases are ∞ as expected. The posterior conditional density estimates (Fig. 4.5) are all close to the truth except for the $x = 0$ and $x = 1$ under case (1) for which increasing the sample size can be helpful. In addition, wider credible intervals are also observed as expected. Overall, our method still performs reasonably well for right-censored data with high censoring rate and interval-censored data.

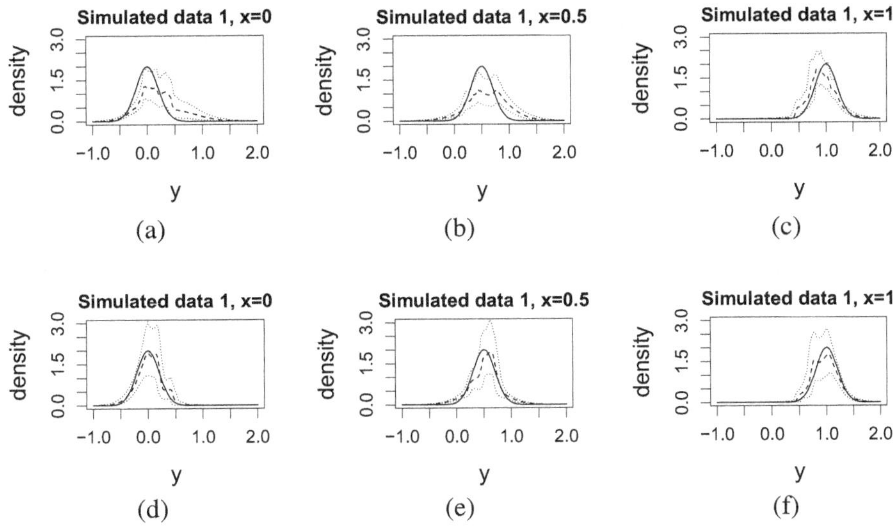

Fig. 4.5 Simulated data 1 with high right-censoring rate (panels (**a**)–(**c**)) and interval-censoring (panels (**c**)–(**e**)). Each panel provides the posterior mean (dashed) and 95% pointwise credible interval (dotted) of the conditional density at three different x values; the solid curves are the corresponding true densities

4.4 Conclusion

The prediction rule from a marginalized Polya tree is generalized to spatially smooth densities over spatial regions, weighing data from proximal locations more heavily than remote ones. Although ideas presented are quite simple and easy-to-implement, the approach has several advantages quite distinct from other approaches. First, it is the only method that we are aware of that smooths the density estimate towards a parametric estimate in data-lean portions of space, e.g., a normal density. The method is fairly fast and competitive with methods based on the Dirichlet process. Finally, a freely available `R` function `SpatDensReg` is available in the `spBayesSurv` package that makes use of compiled C++ to fit the Bayesian model and report the Bayes factor, for arbitrarily censored (or uncensored) data.

As with nonspatial Polya trees, the spatially smoothed version is easily constrained to be median-zero. Thus median regression with a spatially weighted error density is possible leading to heteroscedastic accelerated failure time models that retain the interpretability of acceleration factors in terms of the median (e.g., Jara and Hanson 2011; Zhou et al. 2017). For example, the Polya tree can be shifted and/or stretched via regressions on the centering distribution parameters such as $\mu_{\mathbf{x}} = \mathbf{x}'\boldsymbol{\beta}$ and $\log \sigma_{\mathbf{x}} = \mathbf{x}'\boldsymbol{\tau}$. Similarly, extension to multivariate outcomes is straightforward, including the computation of the Bayes factor for testing spatial dependence; however, obtaining marginal density estimates requires simulating from the Polya tree and using univariate smoothers, e.g., Hanson et al. (2008).

One modification of the model as developed that could potentially improve prediction is the use of a spatially varying centering distribution $\boldsymbol{\theta}_{\mathbf{x}}$ rather than static $\boldsymbol{\theta}$. Spatially weighted $\boldsymbol{\theta}_{\mathbf{x}}$

$$\mu_{\mathbf{x}} = \frac{\sum_{i=1}^{n} d_{\psi}(\mathbf{x}_i, \mathbf{x}) y_i}{\sum_{i=1}^{n} d_{\psi}(\mathbf{x}_i, \mathbf{x})}, \quad \sigma_{\mathbf{x}}^2 = \frac{\sum_{i=1}^{n} d_{\psi}(\mathbf{x}_i, \mathbf{x})(y_i - \mu_{\mathbf{x}})^2}{\sum_{i=1}^{n} d_{\psi}(\mathbf{x}_i, \mathbf{x})},$$

are used in the predictive density $p(y|\mathbf{y}_{1:n}, c, \boldsymbol{\theta}_{\mathbf{x}}, \psi)$ so that the location and spread of the centering normal distribution now change with spatial location. It is unclear, however, how to create a valid likelihood for the remaining parameters (c, ψ) in (4.2) with spatial $\boldsymbol{\theta}_{\mathbf{x}}$. A possible approach is to simply estimate (c, ψ) via cross-validation methods. This, an exploration of the permutation test for spatial association in Sect. 4.2.3, and the median-regression version of the model are topics for the future research.

References

Chen, Y., & Hanson, T. (2014). Bayesian nonparametric k-sample tests for censored and uncensored data. *Computational Statistics and Data Analysis, 71*, 335–346.

Chung, Y., & Dunson, D. B. (2011). The local Dirichlet process. *Annals of the Institute of Statistical Mathematics, 63*, 59–80.

Dahl, D. B., Day, R., & Tsai, J. W. (2017). Random partition distribution indexed by pairwise information. *Journal of the American Statistical Association, 112*, 721–732.

De Iorio, M., Johnson, W. O., Müller, P., & Rosner, G. L. (2009). Bayesian Nonparametric nonproportional hazards survival modeling. *Biometrics, 65*, 762–771.

Duan, J., Guindani, M., & Gelfand, A. E. (2007). Generalized spatial Dirichlet process models. *Biometrika, 94*, 809–825.

Dunson, D. B. (2007). Empirical Bayes density regression. *Statistica Sinica, 17*, 481–504.

Dunson, D. B., & Park, J.-H. (2008). Kernel stick-breaking processes. *Biometrika, 95*, 307–323.

Dunson, D. B., Pillai, N. S., & Park, J-H. (2007). Bayesian density regression. *Journal of the Royal Statistical Society, Series B, 69*, 163–183.

Escobar, M. D., & West, M. (1995). Bayesian density estimation and inference using mixtures. *Journal of the American Statistical Association, 90*, 577–588.

Fisher, R. (1935). *The design of experiments*. Edinburgh: Oliver & Boyd.

Fotheringham, A. S., Brunsdon, C., & Charlton, M. (2002). *Geographically weighted regression: The analysis of spatially varying relationships*, Hoboken: Wiley.

Fuentes, M., & Reich, B. (2013). Multivariate spatial nonparametric modeling via kernel process mixing. *Statistica Sinica, 23*, 75–97.

Gelfand A. E., Diggle, P. J., Fuentes, M., & Guttorp, P. (Eds.) (2010). *Handbook of spatial statistics. Chapman&Hall/CRC handbooks of modern statistical methods*. Boca Raton: CRC Press.

Gelfand, A. E., Kottas, A., & Maceachern, S. N. (2005). Bayesian nonparametric spatial modelling with Dirichlet process mixing. *Journal of the American Statistical Association, 100*, 1021–1035.

Griffin, J. E., & Steel, M. F. J. (2006). Order based dependent Dirichlet processes. *Journal of the American Statistical Association, 101*, 179–194.

Haario, H., Saksman, E., & Tamminen, J. (2001). An adaptive Metropolis algorithm. *Bernoulli, 7*, 223–242.

Hanson, T. (2006). Inference for mixtures of finite Polya tree models. *Journal of the American Statistical Association, 101*, 1548–1565.

Hanson, T., Branscum, A., & Gardner, I. (2008). Multivariate mixtures of Polya trees for modelling ROC data. *Statistical Modelling, 8*, 81–96.

Jara, A., & Hanson, T. (2011). A class of mixtures of dependent tailfree processes. *Biometrika, 98*, 553–566.

Jo, S., Lee, J., Müller, P., Quintana, F., & Trippa, L. (2017). Dependent species sampling model for spatial density estimation. *Bayesian Analysis, 12*, 379–406.

MacEachern, S. N. (2001). Decision theoretic aspects of dependent nonparametric processes. In E. George (Ed.), *Bayesian methods with applications to science, policy and official statistics* (pp. 551–560). Luxembourg City: Eurostat.

Newton, M. A., & Zhang, Y. (1999). A recursive algorithm for nonparametric analysis with missing data. *Biometrika, 86*, 15–26.

Petrone, S., Guindani, M., & Gelfand, A. E. (2009). Hybrid Dirichlet mixture models for functional data. *Journal of the Royal Statistical Society, Series B, 71*, 755–782.

Reich, B., & Fuentes, M. (2007). A multivariate semiparametric Bayesian spatial modeling framework for hurricane surface wind fields. *Annals of Applied Statistics, 1*, 249–264.

Rodríguez, A., Dunson, D., & Gelfand, A. (2010). Latent stick-breaking processes. *Journal of the American Statistical Association, 105*, 647–659.

Schörgendorfer, A., & Branscum, A. J. (2013). Regression analysis using dependent Polya trees. *Statistics in Medicine, 32*, 4679–4695.

Tansey, W., Athey, A., Reinhart, A., & Scott, J. G. (2017). Multiscale spatial density smoothing: An application to large-scale radiological survey and anomaly detection. *Journal of the American Statistical Association, 112*, 1047–1063.

Zhao, L., & Hanson, T. (2011). Spatially dependent Polya tree modeling for survival data. *Biometrics, 67*, 391–403.

Zhou, H., & Hanson, T. (2018). ***spBayesSurv****: Bayesian Modeling and Analysis of Spatially Correlated Survival Data*. R package version 1.1.3 or higher. http://CRAN.R-project.org/package=spBayesSurv

Zhou, H., Hanson, T., & Knapp, R. (2015). Marginal Bayesian nonparametric model for time to disease arrival of threatened amphibian populations. *Biometrics, 71*, 1101–1110.

Zhou, H., Hanson, T., & Zhang, J. (2017). Generalized accelerated failure time spatial frailty model for arbitrarily censored data. *Lifetime Data Analysis, 23*, 495–515.

Part II
Wavelet-Based Approach for Complex Data

Chapter 5
Mammogram Diagnostics Using Robust Wavelet-Based Estimator of Hurst Exponent

Chen Feng, Yajun Mei, and Brani Vidakovic

5.1 Introduction

Breast cancer is one of the major health concerns among women. It has been estimated by the National Cancer Institute that 1 in 8 women will be diagnosed with breast cancer during their lifetime. Early detection is proven to be the best strategy for improving prognosis. Most of the references dealing with automated breast cancer detection are based on microcalcifications (El-Naqa et al. 2002; Kestener et al. 2011; Bala and Audithan 2014; Netsch and Peitgen 1999; Wang and Karayiannis 1998). Recently, predicting disease using image data becomes an active research area in statistics and machine learning (Reiss and Ogden 2010; Zhou et al. 2013; Zipunnikov et al. 2011; Reiss et al. 2005). For example, Reiss and Ogden proposed a functional generalized linear regression model with images as predictors (Reiss and Ogden 2010). However, predicting breast cancer based on the tissue images directly is like a black-box. Physicians will have a hard time to summarize the common features from the cancerous images, and the prediction results are not easily interpreted. In this paper, we study the scaling information from the tissue image and then predict breast cancer based on the estimated scaling parameter. It has been found in literatures that the scaling information is efficient and accurate in early detection of breast cancer (Hamilton et al. 2011; Nicolis et al. 2011; Ramírez-Cobo and Vidakovic 2013; Jeon et al. 2014). In fact, regular scaling is a common

C. Feng (✉) · Y. Mei
H. Milton Stewart School of Industrial and Systems Engineering, Georgia Institute of Technology, Atlanta, GA, USA
e-mail: cfeng@gatech.edu; ymei@isye.gatech.edu

B. Vidakovic
H. Milton Stewart School of Industrial and Systems Engineering and Wallace H. Coulter Department of Biomedical Engineering, Georgia Institute of Technology, Atlanta, GA, USA
e-mail: brani@gatech.edu

Y. Zhao, D.-G. Chen (eds.), *New Frontiers of Biostatistics and Bioinformatics*, ICSA Book Series in Statistics, https://doi.org/10.1007/978-3-319-99389-8_5

phenomenon in high-frequency signals and high-resolution digital images collected in real life. Examples can be found in a variety of fields including economics, telecommunications, physics, geosciences, as well as in biology and medicine (Feng and Vidakovic 2017; Engel Jr et al. 2009; Gregoriou et al. 2009; Katul et al. 2001; Park and Willinger 2000; Woods et al. 2016; Zhou 1996).

The standard measure of regular scaling is the Hurst exponent, denoted by H in the sequel. Recall that a stochastic process $\left\{X(\boldsymbol{t}), \boldsymbol{t} \in \mathbb{R}^d\right\}$ is self-similar with Hurst exponent H if, for any $\lambda \in \mathbb{R}^+$, $X(\boldsymbol{t}) \stackrel{\mathrm{d}}{=} \lambda^{-H} X(\lambda \boldsymbol{t})$. Here the notation $\stackrel{\mathrm{d}}{=}$ means the equality in all finite-dimensional distributions. The Hurst exponent quantifies the self-similarity and describes the rate at which autocorrelations decrease as the lag between two realizations in a time series increases. A value H in the range 0–0.5 indicates a zig-zagging intermittent time series with long-term switching between high and low values in adjacent pairs. A value H in the range 0.5 to 1 indicates a time series with long-term positive autocorrelations, which preserves trends on a longer time horizon and gives a time series more regular appearance.

Multiresolution analysis is one of the many methods to estimate the Hurst exponent. An overview can be found in Abry et al. (2000, 1995, 2013). In particular, the non-decimated wavelet transforms (NDWT) (Nason and Silverman 1995; Vidakovic 2009; Percival and Walden 2006) has several potential advantages when employed for Hurst exponent estimation. Input signals and images of arbitrary size can be transformed in a straightforward manner due to the absence of decimation. As a redundant transform, the NDWT can decrease variance in the scaling estimation (Kang and Vidakovic 2017). Least square regression can be fitted to estimate H instead of weighted least square regression since the variances of the level-wise derived distributions based on log NDWT coefficients do not depend on level. Local scaling can be assessed due to the time-invariance property. Of course, the dependence of coefficients in NDWT is much more pronounced. Similar to Soltani et al. (2004), we will control this dependence by systematic sampling of coefficients on which the estimator is based.

Different wavelet-based methods for estimation of H have been proposed in the literature for the one-dimensional case. Abry et al. (2000) suggested the estimation of H by weighted least square regression using the level-wise $\log_2\left(\overline{d_j^2}\right)$, In addition, the authors corrected for the bias caused by the order of taking the logarithm and the average in $\log_2\left(\overline{d_j^2}\right)$, where d_j indicates any detail coefficient at level j. We use $d_{j,k}$ to denote the kth coefficient at level j in the sequel. Soltani et al. (2004) defined a mid-energy as $D_{j,k} = \left(d_{j,k}^2 + d_{j,k+N_j/2}^2\right)/2$, and showed that the level-wise averages of $\log_2 D_{j,k}$ are asymptotically normal and more stable, which is used to estimate H by regression. The estimators in Soltani et al. (2004) consistently outperform the estimators in Abry et al. (2000). Shen et al. (2007) showed that the method of Soltani et al. (2004) yields more accurate estimators since it takes the logarithm of the mid-energy first and then averages.

The robust estimation of H has recently become a topic of interest due to the presence of outlier coefficients and outlier multiresolution levels, inter and within level dependences, and distributional contaminations (Franzke et al. 2012; Park and Park 2009; Shen et al. 2007; Sheng et al. 2011). Hamilton et al. (2011) came up with a robust approach based on Theil-type weighted regression (Theil 1992), a method for robust linear regression that selects the weighted average of all slopes defined by different pairs of regression points. Like the VA method, they regress the level-wise $\log_2\left(\overline{d_j^2}\right)$ against the level indices, but instead of weighted least square regression, they use the Theil-type weighted regression to make it less sensitive to outlier levels. Kang and Vidakovic (2017) proposed MEDL and MEDLA methods based on non-decimated wavelets to estimate H. MEDL estimates H by regressing the medians of $\log d_j^2$ on level j, while MEDLA uses the level-wise medians of $\log\left(\left(d_{j,k_1}^2 + d_{j,k_2}^2\right)/2\right)$ to estimate H, where k_1 and k_2 are properly selected locations at level j to approximate the independence.

Both MEDL and MEDLA use the median of the derived distribution instead of the mean, because the medians are more robust to potential outliers that can occur when logarithmic transform of a squared wavelet coefficient is taken and the magnitude of coefficient is close to zero. Although median is outlier-resistant, it can behave unexpectedly as a result of its non-smooth character. The fact that the median is not "universally the best outlier-resistant estimator" motivates us to develop the general trimean estimators of the level-wise derived distributions to estimate H, where the general trimean estimator was derived as a weighted average of the distribution's median and two quantiles symmetric about the median, combining the median's emphasis on center values with the quantiles' attention to the tails. Tukey's trimean estimator (Tukey 1977; Andrews and Hampel 2015) and Gastwirth estimator (Gastwirth 1966; Gastwirth and Cohen 1970; Gastwirth and Rubin 1969) are two special cases under such general framework.

In this paper, we are concerned with the robust estimation of Hurst exponent in self-similar signals. Here, the focus is on images, but the methodology applies to multiscale context of arbitrary dimension. The properties of the proposed Hurst exponent estimators are studied both theoretically and numerically. The performance of the robust approach is compared with other standard wavelet-based methods (Veitch and Abry (VA) method, Soltani, Simard, and Boichu (SSB) method, median based estimators MEDL and MEDLA, and Theil-type (TT) weighted regression method).

The rest of the paper consists of six additional sections and an Appendix. Section 5.2 discusses background of non-decimated wavelet transforms and wavelet-based spectrum in the context of estimating the Hurst exponent for fractional Brownian motion (fBm). Section 5.3 introduces the general trimean estimators and discusses two special estimators following that general framework; Sect. 5.4 describes estimation of Hurst exponent using the general trimean estimators, presents distributional results on which the proposed methods are based, and derives optimal weights that minimize the variances of the estimators. Section 5.5 provides the simulation results and compares the performance of the proposed methods to other standardly used, wavelet-based methods. The proposed methods are applied to classify the digitized mammogram images as cancerous or non-cancerous in Sect. 5.6. The paper is concluded with a summary and discussion in Sect. 5.7.

5.2 Background

5.2.1 Non-decimated Wavelet Transforms

The non-decimated wavelet transforms (NDWT) (Nason and Silverman 1995; Vidakovic 2009; Percival and Walden 2006) are redundant transforms because they are performed by repeated filtering with a minimal shift, or a maximal sampling rate, at all dyadic scales. Subsequently, the transformed signal contains the same number of coefficients as the original signal at each multiresolution level. We start by describing algorithmic procedure of 1-D NDWT and then expand to 2-D NDWT. Traditionally, we perform a wavelet transformation as a convolution of an input data with wavelet and scaling filters. A principal difference between NDWT and DWT is the sampling rate.

Any square integrable function $f(x) \in \boldsymbol{L}_2(\mathbb{R})$ can be expressed in the wavelet domain as

$$f(x) = \sum_k c_{J_0,k}\phi_{J_0,k}(x) + \sum_{j \geq J_0}^{\infty} \sum_k d_{j,k}\psi_{j,k}(x),$$

where $c_{J_0,k}$ denote coarse coefficients, $d_{j,k}$ indicate detail coefficients, $\phi_{J_0,k}(x)$ represent scaling functions, and $\psi_{j,k}(x)$ signify wavelet functions. For specific choices of scaling and wavelet functions, the basis for NDWT can be formed from the atoms

$$\phi_{J_0,k}(x) = 2^{J_0/2}\phi\left(2^{J_0}(x-k)\right) \text{ and}$$

$$\psi_{j,k}(x) = 2^{j/2}\psi\left(2^{j}(x-k)\right),$$

where $x \in \mathbb{R}$, j is a resolution level, J_0 is the coarsest level, and k is the location of an atom. Notice that atoms for NDWT have the constant location shift k at all levels, yielding the finest sampling rate on any level. The coarse coefficients $c_{J_0,k}$ and detail coefficients $d_{j,k}$ can be obtained via

$$c_{J_0,k} = \int f(x)\,\phi_{J_0,k}(x)dx \quad \text{and} \quad d_{j,k} = \int f(x)\,\psi_{j,k}(x)dx. \tag{5.1}$$

In a J-level decomposition of an 1-D input signal of size N, an NDWT will yield $N \times (J+1)$ wavelet coefficients, including $N \times 1$ coarse coefficients and $N \times J$ detail coefficients.

Expanding on the 1-D definitions, we could easily describe 2-D NDWT of $f(x, y)$ with $(x, y) \in \mathbb{R}^2$. Several versions of 2-D NDWT exist, but we only focus on the scale-mixing version based on which our methods are proposed. For the scale-mixing 2-D NDWT, the wavelet atoms are

$$\phi_{J_{01},J_{02};\boldsymbol{k}}(x, y) = 2^{(J_{01}+J_{02})/2}\phi(2^{J_{01}}(x - k_1))\phi(2^{J_{02}}(y - k_2)),$$

$$\psi_{J_{01},j_2;\boldsymbol{k}}(x, y) = 2^{(J_{01}+j_2)/2}\phi(2^{J_{01}}(x - k_1))\psi(2^{j_2}(y - k_2)),$$

$$\psi_{j_1,J_{02};\boldsymbol{k}}(x, y) = 2^{(j_1+J_{02})/2}\psi(2^{j_1}(x - k_1))\phi(2^{J_{02}}(y - k_2)),$$

$$\psi_{j_1,j_2;\boldsymbol{k}}(x, y) = 2^{(j_1+j_2)/2}\psi(2^{j_1}(x - k_1))\psi(2^{j_2}(y - k_2)),$$

where $\boldsymbol{k} = (k_1, k_2)$ is the location index, J_{01} and J_{02} are coarsest levels, $j_1 > J_{01}$, and $j_2 > J_{02}$. The wavelet coefficients for $f(x, y)$ after the scale-mixing NDWT can be obtained as

$$\begin{aligned}
c_{J_{01},J_{02};\boldsymbol{k}} &= \iint f(x, y)\, \phi_{J_{01},J_{02};\boldsymbol{k}}(x, y)dxdy,\\
h_{J_{01},j_2;\boldsymbol{k}} &= \iint f(x, y)\, \psi_{J_{01},j_2;\boldsymbol{k}}(x, y)dxdy,\\
v_{j_1,J_{02};\boldsymbol{k}} &= \iint f(x, y)\, \psi_{j_1,J_{02};\boldsymbol{k}}(x, y)dxdy,\\
d_{j_1,j_2;\boldsymbol{k}} &= \iint f(x, y)\, \psi_{j_1,j_2;\boldsymbol{k}}(x, y)dxdy.
\end{aligned} \tag{5.2}$$

Note that $c_{J_{01},J_{02};\boldsymbol{k}}$ are coarse coefficients and represent the coarsest approximation, $h_{J_{01},j_2;\boldsymbol{k}}$ and $v_{j_1,J_{02}}$ represent the mix of coarse and detail information, and $d_{j_1,j_2;\boldsymbol{k}}$ carry information about details only. In our methods, only detail coefficients $d_{j_1,j_2;\boldsymbol{k}}$ are used to estimate H.

5.2.2 The fBm: Wavelet Coefficients and Spectra

Among models having been proposed for analyzing the self-similar phenomena, arguably the most popular is the fractional Brownian motion (fBm) first described by Kolmogorov (1940) and formalized by Mandelbrot and Van Ness (1968).

In this section, an overview of 1-D fBm and its extension to 2-D fBm is provided. Consider a stochastic process $\{X(t), t \in \mathbb{R}\}$ is self-similar with Hurst exponent H, then the 1-D detail coefficients defined in (5.1) satisfy

$$d_{jk} \overset{d}{=} 2^{-j(H+1/2)} d_{0k},$$

for a fixed level j (Abry et al. 2003). If the process has stationary increments, i.e., $X(t+h) - X(t)$ is independent of t, then $\mathbb{E}(d_{0k}) = 0$ and $\mathbb{E}(d_{0k}^2) = \mathbb{E}(d_{00}^2)$. We obtain

$$\mathbb{E}\left(d_{jk}^2\right) \propto 2^{-j(2H+1)}. \tag{5.3}$$

The Hurst exponent can be estimated by taking logarithms on both sides of Eq. (5.3). The wavelet spectrum is defined by the sequence $\left\{S(j) = \log \mathbb{E}\left(d_{jk}^2\right), j \in \mathbb{Z}\right\}$. Fractional Brownian motion (fBm), denoted as $B_H(t)$ is the unique Gaussian process with stationary increments that is self-similar (Abry et al. 2003; Abry 2003). The definition of the one-dimensional fBm can be extended to the multivariate case. In particular, a two-dimensional fBm, $B_H(\boldsymbol{t})$, for $\boldsymbol{t} \in [0, 1] \times [0, 1]$ and $H \in (0, 1)$, is a Gaussian process with stationary zero-mean increments, satisfying

$$B_H(a\boldsymbol{t}) \stackrel{\mathrm{d}}{=} a^H B_H(\boldsymbol{t}).$$

It can be shown that the detail coefficients $d_{j_1,j_2;\boldsymbol{k}}$ defined in Eq. (5.2) satisfy

$$\log_2 \mathbb{E}\left(|d_{j_1,j_2;\boldsymbol{k}}|^2\right) = -(2H+2)j + C,$$

which defines the two-dimensional wavelet spectrum, from which the Hurst exponent can be estimated. Our proposed methods in next sections are based on but improve from this spectrum.

5.3 General Trimean Estimators

Let $X_1, X_2, \ldots, X_n$ be i.i.d. continuous random variables with pdf $f(x)$ and cdf $F(x)$. Let $0 < p < 1$, and let ξ_p denote the pth quantile of F, so that $\xi_p = \inf\{x | F(x) \geq p\}$. If F is monotone, the pth quantile is simply defined as $F(\xi_p) = p$.

Let $Y_p = X_{\lfloor np \rfloor : n}$ denote a sample pth quantile. Here $\lfloor np \rfloor$ denotes the greatest integer that is less than or equal to np. The general trimean estimator is defined as a weighted average of the distribution's median and its two quantiles Y_p and Y_{1-p}, for $p \in (0, 1/2)$:

$$\hat{\mu} = \frac{\alpha}{2}\, Y_p + (1-\alpha)\, Y_{1/2} + \frac{\alpha}{2}\, Y_{1-p}. \tag{5.4}$$

The weights for the two quantiles are the same for Y_p and Y_{1-p}, and $\alpha \in [0, 1]$. This is equivalent to the weighted sum of the median and the average of Y_p and Y_{1-p} with weights $1-\alpha$ and α:

$$\hat{\mu} = (1-\alpha)\ Y_{1/2} + \alpha\left(\frac{Y_p + Y_{1-p}}{2}\right).$$

This general trimean estimator turns out to be more robust than mean but smoother than the median. To derive its asymptotic distribution, the asymptotic joint distribution of sample quantiles is needed, as shown in Lemma 5.1; detailed proof can be found in DasGupta (2008).

Lemma 5.1 *Consider* r *sample quantiles,* $Y_{p_1}, Y_{p_2}, \ldots, Y_{p_r}$*, where* $1 \le p_1 < p_2 < \ldots < p_r \le n$*. If for any* $1 \le i \le r$*,* $\sqrt{n}\left(\lfloor np_i \rfloor / n - p_i\right) \to 0$ *is satisfied, then the asymptotic joint distribution of* $Y_{p_1}, Y_{p_2}, \ldots, Y_{p_r}$ *is:*

$$\sqrt{n}\left(\left(Y_{p_1}, Y_{p_2}, \ldots, Y_{p_r}\right) - \left(\xi_{p_1}, \xi_{p_2}, \ldots, \xi_{p_r}\right)\right) \overset{approx}{\sim} \mathcal{MVN}\left(0, \Sigma\right),$$

where

$$\Sigma = \left(\sigma_{ij}\right)_{r\times r},$$

and

$$\sigma_{ij} = \frac{p_i\left(1-p_j\right)}{f\left(x_{p_i}\right) f\left(x_{p_j}\right)},\ i \le j. \tag{5.5}$$

From Lemma 5.1, the asymptotic distribution of general trimean estimator will be normal as a linear combination of the components each with an asymptotic normal distribution. The general trimean estimator itself may be defined in terms of order statistics as

$$\hat{\mu} = A \cdot \boldsymbol{y},$$

where

$$A = \begin{bmatrix} \frac{\alpha}{2} & 1-\alpha & \frac{\alpha}{2} \end{bmatrix},\ \text{and}\ \boldsymbol{y} = \begin{bmatrix} Y_p & Y_{1/2} & Y_{1-p} \end{bmatrix}^T.$$

It can be easily verified that $\sqrt{n}\left(\lfloor pn \rfloor / n - p\right) \to 0$ for $p \in (0, 1/2]$. If we denote $\boldsymbol{\xi} = \begin{bmatrix} \xi_p & \xi_{1/2} & \xi_{1-p} \end{bmatrix}^T$ the population quantiles, the asymptotic distribution of $\boldsymbol{y}$ is

$$\sqrt{n}\left(\boldsymbol{y} - \boldsymbol{\xi}\right) \overset{\text{approx}}{\sim} \mathcal{MVN}\left(0, \Sigma\right),$$

where $\Sigma = \left(\sigma_{ij}\right)_{3\times 3}$, and σ_{ij} follows Eq. (5.5) for $p_1 = p,\ p_2 = 1/2,$ and $p_3 = 1-p$. Therefore

$$\hat{\mu} \overset{\text{approx}}{\sim} \mathcal{N}\left(\mathbb{E}\left(\hat{\mu}\right), \text{Var}\left(\hat{\mu}\right)\right),$$

with the theoretical expectation and variance being

$$\mathbb{E}\left(\hat{\mu}\right) = \mathbb{E}\left(A \cdot \boldsymbol{y}\right) = A \cdot \mathbb{E}\left(\boldsymbol{y}\right) = A \cdot \boldsymbol{\xi}, \tag{5.6}$$

and

$$\text{Var}\left(\hat{\mu}\right) = \text{Var}\left(A \cdot \boldsymbol{y}\right) = A\,\text{Var}\left(\boldsymbol{y}\right) A^T = \frac{1}{n} A \Sigma A^T. \tag{5.7}$$

5.3.1 Tukey's Trimean Estimator

Tukey's trimean estimator is a special case of the general trimean estimators, with $\alpha = 1/2$ and $p = 1/4$ in Eq. (5.4). To compute this estimator, we first sort the data in ascending order. Next, we take the values that are one-fourth of the way up this sequence (the first quartile), half way up the sequence (i.e., the median), and three-fourths of the way up the sequence (the third quartile). Given these three values, we then form the weighted average, giving the central (median) value a weight of $1/2$ and the two quartiles a weight of $1/4$ each.

If we denote Tukey's trimean estimator as $\hat{\mu}_T$, then

$$\hat{\mu}_T = \frac{1}{4}\,Y_{1/4} + \frac{1}{2}\,Y_{1/2} + \frac{1}{4}\,Y_{3/4}.$$

The asymptotic distribution is

$$\hat{\mu}_T \overset{\text{approx}}{\sim} \mathcal{N}\left(A_T \cdot \boldsymbol{\xi}_T, \frac{1}{n} A_T \Sigma_T A_T^T\right),$$

where $A_T = \left[\frac{1}{4}\ \frac{1}{2}\ \frac{1}{4}\right]$, $\boldsymbol{\xi}_T = \left[\xi_{1/4}\ \xi_{1/2}\ \xi_{3/4}\right]^T$, $\Sigma_T = \left(\sigma_{ij}\right)_{3\times 3}$ is the covariance matrix of the asymptotic multivariate normal distribution, and σ_{ij} follows Eq. (5.5) with $p_1 = 1/4,\ p_2 = 1/2,$ and $p_3 = 3/4$.

5.3.2 Gastwirth Estimator

As Tukey's estimator, the Gastwirth estimator is another special case of the general trimean estimators, with $\alpha = 0.6$ and $p = 1/3$ in Eq. (5.4).

If we denote this estimator as $\hat{\mu}_G$, then

$$\hat{\mu}_G = 0.3\,Y_{1/3} + 0.4\,Y_{1/2} + 0.3\,Y_{2/3}.$$

The asymptotic distribution can be derived as

$$\hat{\mu}_G \overset{\text{approx}}{\sim} \mathcal{N}\left(A_G \cdot \boldsymbol{\xi}_G, \frac{1}{n} A_G \Sigma_G A_G^T\right),$$

where $A_G = [0.3 \ \ 0.4 \ \ 0.3]$, $\boldsymbol{\xi}_G = \left[\xi_{1/3} \ \ \xi_{1/2} \ \ \xi_{2/3}\right]^T$, $\Sigma_G = \left(\sigma_{ij}\right)_{3\times 3}$, and σ_{ij} follows Eq. (5.5) with $p_1 = 1/3,\ p_2 = 1/2,$ and $p_3 = 2/3$.

5.4 Methods

Our proposal for robust estimation of Hurst exponent H is based on non-decimated wavelet transforms (NDWT). In a J-depth decomposition of a 2-D fBm of size $N \times N$, a scale-mixing 2-D NDWT generates $(J+1)\times(J+1)$ blocks of coefficients, with each block the same size as original image, i.e., $N \times N$. The tessellation of coefficients of scale-mixing 2-D NDWT is shown in Fig. 5.1a. From the 2-D NDWT wavelets coefficients, our methods use the diagonal blocks ($j_1 = j_2 = j$) of the detail coefficients $d_{j_1,j_2;\boldsymbol{k}}$ to predict H, as is shown in Fig. 5.1b.

At each detail level j, the corresponding level-j diagonal block is of size $N \times N$, the same size as original image. Note that those coefficients $d_{j,j;\boldsymbol{k}}$ in level-j diagonal block are not independent, however, their autocorrelations decay exponentially, that is, they possess only the short memory. We reduce such within block dependency by dividing the block into $M \times M$ equal grids and then random sampling one coefficient from each grid, therefore increasing the distance between two consecutive coefficients. To improve the efficiency, here we apply symmetric sampling. To be specific, we partition the level-j diagonal block into four equal parts (top left, top right, bottom left, and bottom right), only sample from the $M^2/4$

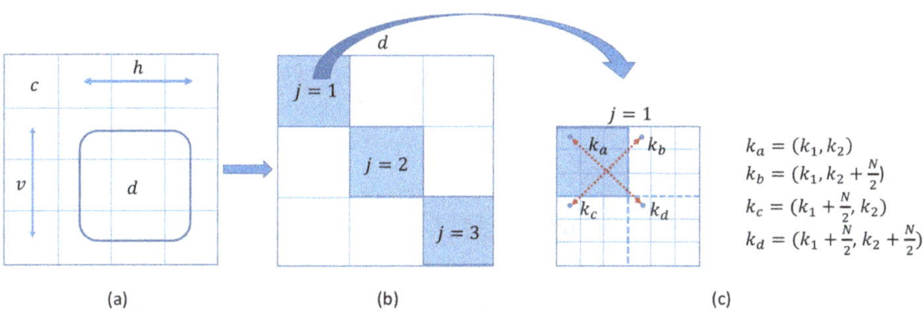

Fig. 5.1 (**a**) Four types of wavelet coefficients with their locations in the tessellation of a 2-D scale mixing NDWT of depth of 3 ($J = 3$), with each block the size of $N \times N$. Coefficients c represent the coarsest approximation, h and v are the mix of coarse and detail information, and d carry detail information only. (**b**) Detail coefficients d and its diagonal blocks corresponding to 3 ($J = 3$) levels. (**c**) Symmetric random sampling from level-1 ($j = 1$) diagonal block divided into 6×6 ($M = 6$) grids

grids at the top left, and then get the corresponding coefficients that have the same location in other parts, which is shown in Fig. 5.1c.

If assuming the coefficient $d_{j,j;(k_{i1},k_{i2})}$ is randomly sampled from grid $i \in \{1, \ldots, \frac{M^2}{4}\}$ at the top left part of level-j diagonal block, and $k_{i1}, k_{i2} \in \{1, 2, \ldots, \frac{N}{2}\}$ being the corresponding location indexes, then we can extract corresponding coefficients $d_{j,j;(k_{i1},k_{i2}+\frac{N}{2})}$, $d_{j,j;(k_{i1}+\frac{N}{2},k_{i2})}$, and $d_{j,j;(k_{i1}+\frac{N}{2},k_{i2}+\frac{N}{2})}$ from the top right, bottom left, and bottom right parts, respectively. From the set

$$\{d_{j,j;(k_{i1},k_{i2})}, d_{j,j;(k_{i1},k_{i2}+\frac{N}{2})}, d_{j,j;(k_{i1}+\frac{N}{2},k_{i2})}, d_{j,j;(k_{i1}+\frac{N}{2},k_{i2}+\frac{N}{2})}\},$$

we could generate two mid-energies as

$$\begin{aligned} D_{i,j} &= \frac{d^2_{j,j;(k_{i1},k_{i2})} + d^2_{j,j;(k_{i1}+\frac{N}{2},k_{i2}+\frac{N}{2})}}{2} \\ D'_{i,j} &= \frac{d^2_{j,j;(k_{i1},k_{i2}+\frac{N}{2})} + d^2_{j,j;(k_{i1}+\frac{N}{2},k_{i2})}}{2}, \quad i \in \{1, \ldots, \frac{M^2}{4}\}, \end{aligned} \tag{5.8}$$

where $D_{i,j}$ and $D'_{i,j}$ denote the two mid-energies corresponding to grid i at level j. If we denote D_j as the set of all mid-energies at level j, then

$$D_j = \{D_{1,j}, D'_{1,j}, D_{2,j}, D'_{2,j}, \ldots, D_{\frac{M^2}{4},j}, D'_{\frac{M^2}{4},j}\}. \tag{5.9}$$

The $M^2/2$ mid-energies at each level j are treated as if they are independent. Note that M must be divisible by 2.

Our methods have two different versions, one is based on mid-energies D_j, while the other is using logged mid-energies $\log D_j$ (in bracket). First, the distribution of D_j $(\log D_j)$ is derived under the independence approximation between $d_{j,j;(k_{i1},k_{i2})}$, $d_{j,j;(k_{i1},k_{i2}+\frac{N}{2})}$, $d_{j,j;(k_{i1}+\frac{N}{2},k_{i2})}$, and $d_{j,j;(k_{i1}+\frac{N}{2},k_{i2}+\frac{N}{2})}$. Next, we calculate the general trimean estimators from the level-wise derived distributions to estimate H.

5.4.1 General Trimean of the Mid-energy (GTME) Method

At each decomposition level j, the asymptotic distribution of the general trimean estimator on $M^2/2$ mid-energies in D_j is derived, from which we find the relationship between the general trimean estimators and H. The general trimean of the mid-energy (GTME) method is described in the following theorem:

Theorem 5.1 *Let $\hat{\mu}_j$ be the general trimean estimator based on the $M^2/2$ mid-energies in D_j defined by (5.9) at level j in a J-level NDWT of a 2-D fBm of size $N \times N$ with Hurst exponent H. Then, the asymptotic distribution of $\hat{\mu}_j$ is normal,*

$$\hat{\mu}_j \overset{approx}{\sim} \mathcal{N}\left(c(\alpha, p)\lambda_j, \frac{2}{M^2} f(\alpha, p)\lambda_j^2\right), \tag{5.10}$$

where

$$c(\alpha, p) = \frac{\alpha}{2}\log\left(\frac{1}{p(1-p)}\right) + (1-\alpha)\log 2,$$

$$f(\alpha, p) = \frac{\alpha(1-2p)(\alpha-4p)}{4p(1-p)} + 1,$$

$$\lambda_j = \sigma^2 \cdot 2^{-(2H+2)j},$$

and σ^2 *is the variance of wavelet coefficients from level 0, the Hurst exponent can be estimated as*

$$\hat{H} = -\frac{\hat{\beta}}{2} - 1, \tag{5.11}$$

where $\hat{\beta}$ *is the regression slope in the least square linear regression on pairs* $\left(j, \log_2\left(\hat{\mu}_j\right)\right)$ *from level* J_1 *to* J_2, $J_1 \leq j \leq J_2$. *The estimator* $\hat{H}$ *follows the asymptotic normal distribution*

$$\hat{H} \overset{approx}{\sim} \mathcal{N}(H, V_1), \tag{5.12}$$

where the asymptotic variance V_1 *is a constant number independent of simple size* N *and level* j,

$$V_1 = \frac{6f(\alpha, p)}{(\log 2)^2 M^2 c^2(\alpha, p)\, q(J_1, J_2)},$$

and

$$q(J_1, J_2) = (J_2 - J_1)(J_2 - J_1 + 1)(J_2 - J_1 + 2). \tag{5.13}$$

The proof of Theorem 5.1 is deferred to the Appendix.

To find the optimal α and p by minimizing the asymptotic variance of $\hat{\mu}_j$, we take partial derivatives of $f(\alpha, p)$ with respect to α and p and set them to 0. The optimal $\hat{\alpha}$ and $\hat{p}$ can be obtained by solving

$$\begin{aligned} \frac{\partial f(\alpha, p)}{\partial \alpha} &= -\frac{2p-1}{2p(1-p)}\alpha + \frac{1+p}{2(1-p)} - \frac{3}{2} = 0, \\ \frac{\partial f(\alpha, p)}{\partial p} &= \frac{\alpha(2-\alpha)}{2(1-p)^2} + \frac{\alpha^2(2p-1)}{4p^2(1-p)^2} = 0. \end{aligned} \tag{5.14}$$

Since $\alpha \in [0, 1]$ and $p \in (0, 1/2)$, we get the unique solution $\alpha = 2p \approx 0.6$ and $p = 1 - \sqrt{2}/2 \approx 0.3$. The Hessian matrix of $f(\alpha, p)$ is

$$\begin{bmatrix} \frac{\partial^2 f(\alpha,p)}{\partial \alpha^2} & \frac{\partial^2 f(\alpha,p)}{\partial \alpha \partial p} \\ \frac{\partial^2 f(\alpha,p)}{\partial \alpha \partial p} & \frac{\partial^2 f(\alpha,p)}{\partial p^2} \end{bmatrix} =$$

$$\begin{bmatrix} -\frac{2p-1}{2p(1-p)} & \frac{2p^2-2\alpha p^2+\alpha(2p-1)}{2p^2(1-p)^2} \\ \frac{2p^2-2\alpha p^2+\alpha(2p-1)}{2p^2(1-p)^2} & \frac{2p^3\alpha(2-\alpha)+\alpha^2 p(1-p)+\alpha^2(2p-1)^2}{2p^3(1-p)^3} \end{bmatrix}.$$

Since $-\frac{2p-1}{2p(1-p)} > 0$ and the determinant is $5.66 > 0$ when $\alpha = 2p \approx 0.6$ and $p = 1 - \sqrt{2}/2 \approx 0.3$, the above Hessian matrix is positive definite. Therefore, $\hat{\alpha} = 2 - \sqrt{2}$ and $\hat{p} = 1 - \sqrt{2}/2$ provide the global minima of $f(\alpha, p)$, minimizing also the asymptotic variance of $\hat{\mu}_{j,i}$. In comparing these optimal $\hat{\alpha} \approx 0.6$ and $\hat{p} \approx 0.3$ with $\alpha = 0.6$ and $p = 1/3$ from the Gastwirth estimator, curiously, we find that the optimal general trimean estimator is very close to the Gastwirth estimator.

5.4.2 General Trimean of the Logarithm of Mid-energy (GTLME) Method

Previously discussed the GTME method calculates the general trimean estimator of the mid-energy first and then takes the logarithm. In this section, we will calculate the general trimean estimator of the logged mid-energies at each level j. The following theorem describes the general trimean of the logarithm of mid-energy, the GTLME method.

Theorem 5.2 *Let $\hat{\mu}_j$ be the general trimean estimator based on $\log(D_j)$, which is the set of $M^2/2$ logged mid-energies at level j in a J-level NDWT of a 2-D fBm of size $N \times N$ with Hurst exponent H, and $1 \le j \le J$. Then, the asymptotic distribution of $\hat{\mu}_j$ is normal,*

$$\hat{\mu}_j \overset{approx}{\sim} \mathcal{N}\left(c(\alpha, p) + \log(\lambda_j), \frac{2}{M^2} f(\alpha, p)\right), \tag{5.15}$$

where

$$c(\alpha, p) = \frac{\alpha}{2} \log\left(\log \frac{1}{1-p} \cdot \log \frac{1}{p}\right) + (1-\alpha)\log(\log 2),$$

$$f(\alpha, p) = \frac{\alpha^2}{4g_1(p)} + \frac{\alpha(1-\alpha)}{2g_2(p)} + \frac{(1-\alpha)^2}{(\log 2)^2},$$

$g_1(p)$ and $g_2(p)$ are two functions of p given in the Appendix,

$$\lambda_j = \sigma^2 \cdot 2^{-(2H+2)j},$$

and σ^2 is the variance of wavelet coefficients from level 0. The Hurst exponent can be estimated as

$$\hat{H} = -\frac{1}{2\log 2}\hat{\beta} - 1, \tag{5.16}$$

where $\hat{\beta}$ is the regression slope in the least square linear regressions on pairs $\left(j, \hat{\mu}_j\right)$ from level J_1 to J_2, $J_1 \leq j \leq J_2$. The estimator $\hat{H}$ follows the asymptotic normal distribution

$$\hat{H} \overset{approx}{\sim} \mathcal{N}(H, V_2), \tag{5.17}$$

where the asymptotic variance V_2 is a constant number independent of simple size N and level j,

$$V_2 = \frac{6f(\alpha, p)}{(\log 2)^2 M^2 q(J_1, J_2)},$$

and $q(J_1, J_2)$ is given in Eq. (5.13).

The proof of Theorem 5.2 is provided in the Appendix. Similarly, as for the GTME, the optimal α and p which minimize the asymptotic variance of $\hat{\mu}_j$ can be obtained by solving

$$\frac{\partial f(\alpha, p)}{\partial \alpha} = 0, \text{ and } \frac{\partial f(\alpha, p)}{\partial p} = 0. \tag{5.18}$$

From the first equation in (5.18) it can be derived that

$$\alpha = \frac{\frac{2}{\log(2)^2} - \frac{1}{2}g_2(p)}{\frac{1}{2}g_1(p) - g_2(p) + \frac{2}{(\log 2)^2}}.$$

The second equation in (5.18) cannot be simplified to a finite form. As an illustration, we plot the $f(\alpha, p)$ with p ranging from 0 to 0.5 and α being a function of p. The plot of α against p is also shown in Fig. 5.2. Numerical computation gives $\hat{\alpha} = 0.5965$ and $\hat{p} = 0.24$. These optimal parameters are close to $\alpha = 0.5$ and $p = 0.25$ in the Tukey's trimean estimator, but put some more weight on the median.

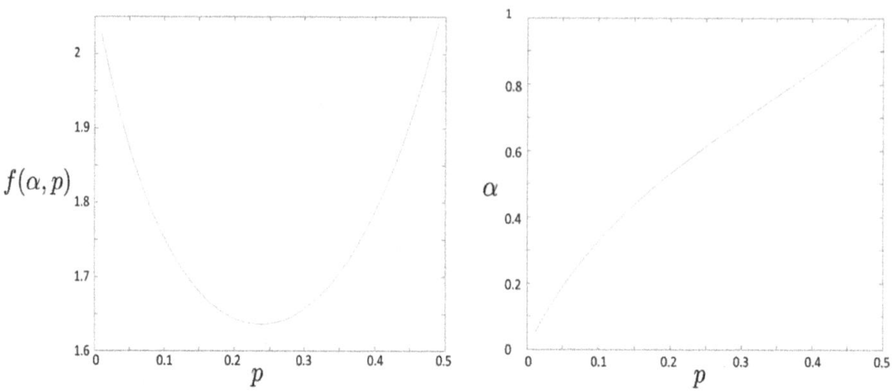

Fig. 5.2 Plot of $f(\alpha, p)$ against p on the left; plot of α against p on the right

5.4.3 Special Cases: Tukey's Trimean and Gastwirth Estimators

The Tukey's trimean of the mid-energy (TTME) method and Gastwirth of the mid-energy (GME) method are described in the following Lemma.

Lemma 5.2 *Let $\hat{\mu}_j^T$ and $\hat{\mu}_j^G$ be the Tukey's trimean and Gastwirth estimators based on D_j defined in (5.9). Then the asymptotic distributions of $\hat{\mu}_j^T$ and $\hat{\mu}_j^G$ are normal:*

$$\hat{\mu}_j^T \overset{approx}{\sim} \mathcal{N}\left(c_1\lambda_j, \frac{5}{3M^2}\lambda_j^2\right), \tag{5.19}$$

$$\hat{\mu}_j^G \overset{approx}{\sim} \mathcal{N}\left(c_2\lambda_j, \frac{1.67}{M^2}\lambda_j^2\right), \tag{5.20}$$

where c_1 and c_2 are constant numbers and can be found in the Appendix, $\lambda_j = \sigma^2 \cdot 2^{-(2H+2)j}$, and σ^2 is the variance of wavelet coefficients from level 0. The Hurst exponent can be estimated as

$$\hat{H}^T = -\frac{\hat{\beta}^T}{2} - 1, \text{ and } \hat{H}^G = -\frac{\hat{\beta}^G}{2} - 1, \tag{5.21}$$

where $\hat{\beta}^T$ and $\hat{\beta}^G$ are the regression slopes in the least square linear regression on pairs $\left(j, \log_2\left(\hat{\mu}_j^T\right)\right)$ and pairs $\left(j, \log_2\left(\hat{\mu}_j^G\right)\right)$ from level J_1 to J_2, $J_1 \leq j \leq J_2$. The estimators $\hat{H}^T$ and $\hat{H}^G$ follow the asymptotic normal distributions

$$\hat{H}^T \overset{approx}{\sim} \mathcal{N}\left(H, V_1^T\right), \text{ and } \hat{H}^G \overset{approx}{\sim} \mathcal{N}\left(H, V_1^G\right), \tag{5.22}$$

where the asymptotic variances V_1^T *and* V_1^G *are constant numbers,*

$$V_1^T = \frac{5}{(\log 2)^2 M^2 c_1^2 q(J_1, J_2)},$$

$$V_1^G = \frac{5.01}{(\log 2)^2 M^2 c_2^2 q(J_1, J_2)}.$$

The function $q(J_1, J_2)$ *is the same as Eq. (5.13) in Theorem 5.1.*

The following Lemma describes the Tukey's trimean (TTLME) and Gastwirth (GLME) of the logarithm of mid-energy method.

Lemma 5.3 *Let* $\hat{\mu}_j^T$ *and* $\hat{\mu}_j^G$ *be the Tukey's trimean estimator and Gastwirth estimator based on* $\log(D_j)$ *defined in the Theorem 5.2. The asymptotic distributions of* $\hat{\mu}_j^T$ *and* $\hat{\mu}_j^G$ *are normal,*

$$\hat{\mu}_j^T \overset{approx}{\sim} \mathcal{N}\left(-(2H+2)\log 2j + c_3, V_T\right), \tag{5.23}$$

$$\hat{\mu}_j^G \overset{approx}{\sim} \mathcal{N}\left(-(2H+2)\log 2j + c_4, V_G\right), \tag{5.24}$$

where c_3 *,*V_T*,* c_4*, and* V_G *are constant numbers and can be found in the Appendix. The Hurst exponent can be estimated as*

$$\hat{H}^T = -\frac{\hat{\beta}^T}{2\log 2} - 1, \text{ and } \hat{H}^G = -\frac{\hat{\beta}^G}{2\log 2} - 1, \tag{5.25}$$

where $\hat{\beta}^T$ *and* $\hat{\beta}^G$ *are the regression slopes in the least square linear regression on pairs* $\left(j, \hat{\mu}_j^t\right)$ *and pairs* $\left(j, \hat{\mu}_j^g\right)$ *from level* J_1 *to* J_2*,* $J_1 \leq j \leq J_2$*. The estimators* $\hat{H}^T$ *and* $\hat{H}^G$ *follow the asymptotic normal distributions*

$$\hat{H}^T \overset{approx}{\sim} \mathcal{N}\left(H, V_2^T\right), \text{ and } \hat{H}^G \overset{approx}{\sim} \mathcal{N}\left(H, V_2^G\right), \tag{5.26}$$

where the asymptotic variances V_2^T *and* V_2^G *are constant numbers,*

$$V_2^T = \frac{3V_T}{(\log 2)^2 q(J_1, J_2)},$$

$$V_2^G = \frac{3V_G}{(\log 2)^2 q(J_1, J_2)}.$$

The function $q(J_1, J_2)$ *is provided in Eq. (5.13).*

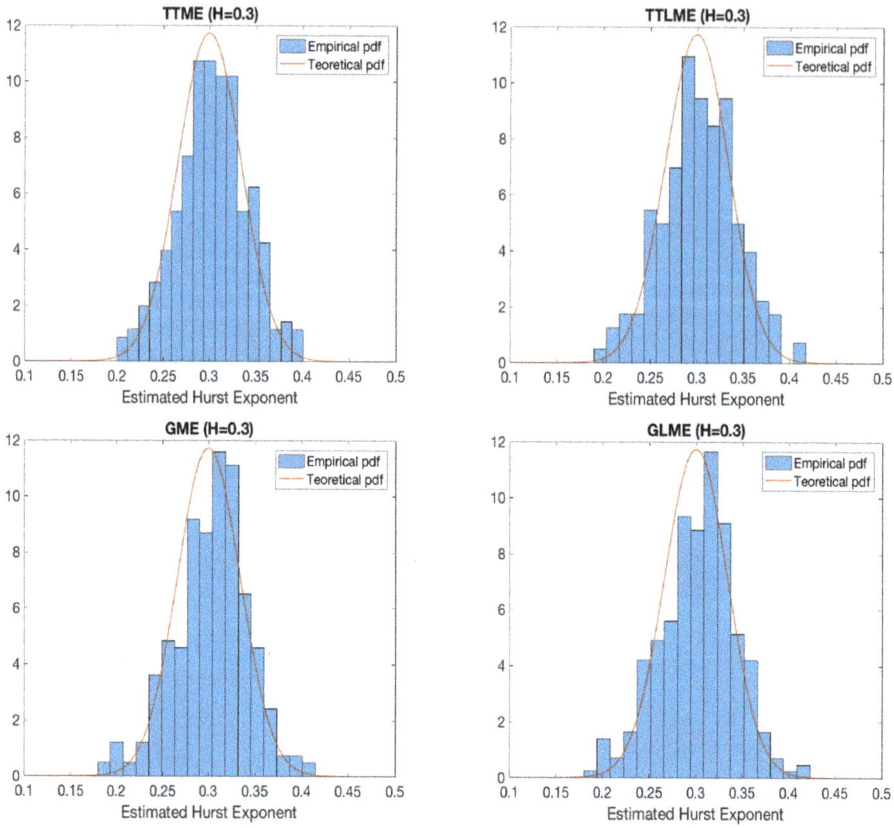

Fig. 5.3 Histograms and theoretical distributions of $\hat{H}$

The proofs of Lemmas 5.2 and 5.3 are provided in the Appendix. To verify the asymptotic normal distributions of predictors in Lemmas 5.2 and 5.3, we perform an NDWT of depth 10 on 300 simulated fBm's with $H = 0.3$. We use resulting wavelet coefficients from levels 4 to 10 inclusive to estimate H. Figure 5.3 shows the histograms and theoretical distributions of $\hat{H}$ using TTME, TTLME, GME, and GLME methods, respectively.

5.5 Simulation

We simulate 2-D fBm of sizes $2^{10} \times 2^{10}$ $(N = 2^{10})$ with Hurst exponent $H = 0.3, 0.5, 0.7, 0.8, 0.9$, respectively. NDWT of depth $J = 10$ using Haar wavelet is performed on the simulated signal to obtain wavelet coefficients. The two-

Table 5.1 Simulation results for $2^{10} \times 2^{10}$ fBm using Haar wavelet (300 replications)

	Existing methods					Proposed methods					
H	VA	SSB	MEDL	MEDLA	TT	TTME	TTLME	GME	GLME	GTME	GTLME
$\hat{H}$											
0.3	0.3103	0.3055	0.3018	0.3031	0.3054	0.3032	0.3028	0.3032	0.3034	0.3028	0.3030
0.5	0.5220	0.5132	0.5095	0.5102	0.5151	0.5126	0.5111	0.5108	0.5100	0.5118	0.5116
0.7	0.7382	0.7235	0.7175	0.7165	0.7326	0.7193	0.7179	0.7193	0.7184	0.7199	0.7181
0.8	0.8458	0.8261	0.8200	0.8204	0.8398	0.8222	0.8214	0.8208	0.8206	0.8212	0.8221
0.9	0.9593	0.9328	0.9241	0.9274	0.9641	0.9303	0.9282	0.9287	0.9278	0.9295	0.9287
Variances											
0.3	**0.0014**	0.0016	0.0026	0.0020	0.0017	0.0015	0.0016	0.0016	0.0016	0.0015	0.0016
0.5	0.0020	0.0017	0.0027	0.0018	0.0034	**0.0013**	0.0016	0.0014	0.0016	0.0014	0.0016
0.7	0.0037	0.0019	0.0030	0.0026	0.0086	**0.0018**	0.0021	0.0020	0.0021	0.0019	0.0020
0.8	0.0050	0.0021	0.0027	0.0023	0.0095	**0.0018**	0.0020	0.0020	0.0021	0.0019	0.0020
0.9	0.0073	0.0021	0.0028	0.0022	0.0168	**0.0018**	0.0019	0.0019	0.0020	**0.0018**	0.0019
MSEs											
0.3	**0.0015**	0.0016	0.0026	0.0020	0.0017	**0.0015**	0.0016	0.0016	0.0016	**0.0015**	0.0016
0.5	0.0025	0.0019	0.0027	0.0019	0.0037	**0.0015**	0.0017	0.0016	0.0017	**0.0015**	0.0017
0.7	0.0052	0.0025	0.0033	0.0028	0.0097	**0.0022**	0.0024	0.0024	0.0025	0.0023	0.0024
0.8	0.0070	0.0027	0.0031	0.0028	0.0110	**0.0023**	0.0024	0.0024	0.0025	**0.0023**	0.0025
0.9	0.0108	0.0032	0.0033	0.0030	0.0208	**0.0027**	**0.0027**	**0.0027**	0.0028	**0.0027**	**0.0027**

dimensional fBm signals were simulated based on the method of Wood and Chan (1994).

The proposed methods (with six variations) are applied on the NDWT detail coefficients to estimate Hurst exponent H. Each level diagonal block is divided into 16×16 grids ($M = 16$) for all proposed methods, and we use wavelet coefficients from levels 4 to 10 for the least square linear regression. The estimation performance of the proposed methods is compared to five other existing methods: Veitch and Abry (VA) method, Soltani, Simard, and Boichu (SSB) method, MEDL method, MEDLA method, and Theil-type regression (TT) method. The GTME and GTLME methods are based on the optimal parameters which minimize the variances. Estimation performance is reported in terms of mean, variance, and mean square error (MSE) based on 300 repetitions for each case.

The simulation results are shown in Table 5.1. For each H (corresponding to each row in the table), the smallest variances and MSEs are highlighted in bold. From simulations results, all our six variations outperform SSB, MEDL, MEDLA, and TT methods for all H's regarding variances and MSEs. Compared with VA method, our methods yield significantly smaller variances and MSEs when $H > 0.5$. When $H = 0.3$, our methods are still comparable to VA. Although the performances of our six variations are very similar regarding variances and MSEs, the TTME method based on Tukey's trimean estimator of the mid-energy has the best performance among all of them. The variances of GTME based on the optimal parameters are very close or equal to those of GME and TTME methods in most cases. Besides, in most cases the optimized GTLME method is superior to other logged mid-energy methods TTLME and GLME with respect to variances; however, such superiority is not significant, since the variances are close to each other.

5.6 Application

In this section, we apply the proposed methodology to classification of digitized mammogram images. The digitized mammograms were obtained from the University of South Florida's Digital Database for Screening Mammography (DDSM) (Heath et al. 2000). All cases examined had biopsy results which served as ground truth. Researchers used the HOWTEK scanner at the full 43.5-micron per pixel spatial resolution to scan 45 mammograms from patients with normal studies (control group) and 79 from patients with confirmed breast cancer (study group). Figure 5.4 shows an example of mammograms from study group, and it is almost impossible for physicians to distinguish a cancerous mammogram with a non-cancerous mammogram just by eyes. Each subject contains two mammograms from a screening exam, one craniocaudal projection for each side breast. We only keep one projection for each subject, either right side or left side breast image. A sub-image of size 1024 × 1024 was taken manually from each mammogram.

Our methods were then applied on each sub-image to estimate the Hurst exponent parameter for each subject. To be specific, the NDWT of depth $J = 10$ using Haar wavelet was performed on each sub-image to obtain wavelet coefficients. The proposed methods (with six variations) are applied on the NDWT detail coefficients

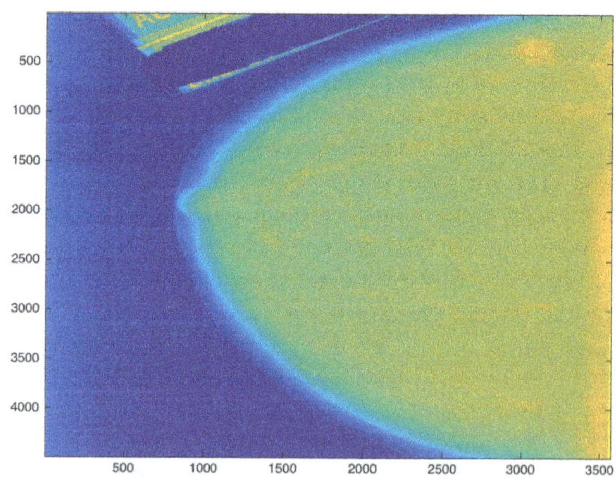

Fig. 5.4 An example of mammograms with breast cancer

Table 5.2 Descriptive statistics group summary

	Existing methods					Proposed methods					
Group	VA	SSB	MEDL	MEDLA	TT	TTME	TTLME	GME	GLME	GTME	GTLME
Mean of $\hat{H}$											
Control	0.3570	0.3457	0.3323	0.3403	0.3716	0.3454	0.3422	0.3444	0.3420	0.3450	0.3430
Study	0.4310	0.4038	0.3935	0.4023	0.4203	0.4061	0.4026	0.4031	0.4019	0.4053	0.4027
Median of $\hat{H}$											
Control	0.3368	0.3339	0.3326	0.3140	0.3871	0.3248	0.3188	0.3198	0.3240	0.3263	0.3278
Study	0.4286	0.4147	0.3865	0.4165	0.4204	0.4194	0.4211	0.4178	0.4150	0.4168	0.4209
Variance of $\hat{H}$											
Control	0.0267	0.0270	0.0268	0.0298	0.0305	0.0284	0.0279	0.0285	0.0279	0.0281	0.0277
Study	0.0159	0.0172	0.0198	0.0175	0.0128	0.0169	0.0173	0.0174	0.0175	0.0175	0.0174

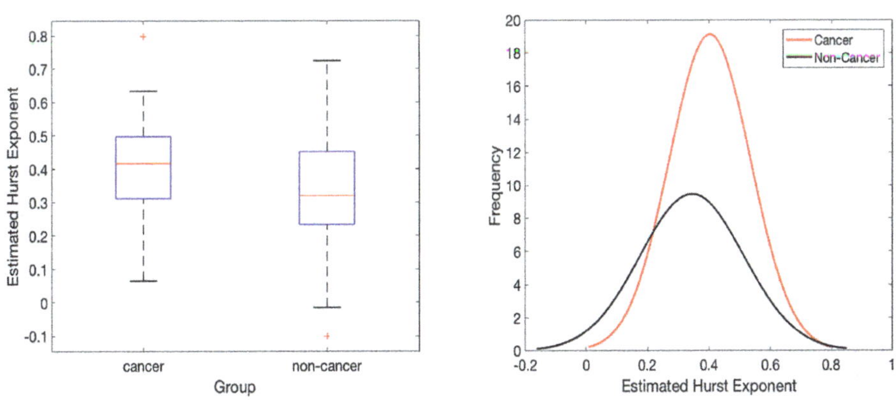

Fig. 5.5 Using GME method to estimate Hurst exponent, boxplots in cancer and non-cancer groups on the left; normal density curves fitted in cancer and non-cancer groups on the right

to estimate Hurst exponent H. Each level diagonal block is divided into 16×16 grids ($M = 16$) for all proposed methods, and we use levels 4 to 10 for the least square linear regression. Veitch and Abry (VA) method, Soltani, Simard, and Boichu (SSB) method, MEDL method, MEDLA method, and Theil-type regression (TT) method were applied, as well, to compare with our methods.

Table 5.2 provides descriptive statistics of the estimated Hurst exponent $\hat{H}$ in each group using our proposed methods and other standard methods to compare with. To visualize the difference in $\hat{H}$ across cancer and non-cancer groups, we present in Fig. 5.5 the boxplots of estimated H and fitted normal density curves in two groups based on proposed GME method. As can be seen, the non-cancer group exhibited a smaller value for $\hat{H}$ in both the mean and median, and the variance of

Table 5.3 Results of classification by logistic regression

	Existing methods					Proposed methods					
	VA	SSB	MEDL	MEDLA	TT	TTME	TTLME	GME	GLME	GTME	GTLME
Overall accuracy	0.629	0.597	0.645	0.589	0.547	0.622	0.613	**0.654**	0.605	0.628	0.645
Sensitivity	0.695	0.659	0.709	0.623	0.543	0.659	0.659	**0.722**	0.647	0.684	0.708
Specificity	0.511	0.491	0.532	0.534	**0.553**	**0.553**	0.534	0.536	0.532	0.528	0.534

$\hat{H}$ is slightly larger. In fact, images with smaller Hurst exponent tend to be more disordered and unsystematic, therefore healthy individuals tend to have more rough breast tissue images.

For subject i, we generated the data $\{Y_i, H_i\}$, where H_i represents the estimated Hurst exponent, and Y_i is the indicator of the disease status with 1 and 0 signifying cancer and non-cancer, respectively. The subjects were classified using a logistic regression model by treating H_i as the predictor and Y_i as the response. The overall classification accuracy, true positive rate (sensitivity), and true negative rate (specificity) were obtained by using a fourfold-cross validation. Instead of the constant 0.5 threshold, we used a training-data-determined adaptive threshold, i.e., each time the threshold of the logistic regression was first chosen to maximize Youden index on the training set and then applied to the testing set to classify.

Table 5.3 summarizes the results of the classification for each estimation method. The best classification rate (0.6538) and sensitivity (0.7217) were both achieved using GME estimator, and the best specificity (0.5530) was achieved using TT or TTME estimator (highlighted in bold). In general, the six variations of our robust method performed better as compared to other methods in classification of breast cancers using mammograms.

Real-world images like mammograms may be characterized by non-stationary conditions such as extreme values, causing outlier coefficients in multiresolution levels after NDWT. VA method estimates H by weighted least square regression using the level-wise $\log_2\left(\overline{d_{j,j}^2}\right)$, and SSB method uses $\overline{\log_2 D_j}$, with D_j defined in (5.9), they are easily affected by those within level outliers, in that they both use mean of derived distributions on level-wise detail coefficients to estimate H. Besides, potential outliers can also occur when logarithmic transform is taken and the magnitude of coefficient is close to zero. Like the VA method, TT method regress the level-wise $\log_2\left(\overline{d_{j,j}^2}\right)$ against the level indices, but instead of weighted least square regression, they use the Theil-type weighted regression, the weighted average of all slops between different pairs of regression points, to make it less sensitive to outlier levels. However, it is still not robust to within level outlier coefficients. MEDL and MEDLA use the median of the derived distribution instead of the mean. Although median is outlier-resistant, it can behave unexpectedly as a result of its non-smooth character. To improve, our methods (six derivations) use the general trimean estimator on non-decimated wavelet detail coefficients of the transformed data, combining the median's emphasis on central values with the quantiles' attention to the extremes. Besides, in the context of our scenario, Theil-type regression is equivalent to least square regression, since the variance of our pair-wise slop is independent of levels and sample size. Those explain why our robust methods performed the best in classification of mammograms.

5.7 Conclusions

In this paper, we proposed methodologies and derived six variations to improve the robustness of estimation of Hurst exponent H in two-dimensional setting. Non-decimated wavelet transforms (NDWT) are utilized for its redundancy and time-invariance. Instead of using mean or median of the derived distribution on level-wise wavelet coefficients, we defined the general trimean estimators that combine the median's emphasis on center values with the quantiles' attention to the extremes and used them on the level-wise derived distributions to estimate H.

The proposed variations were: (1) Tukey's trimean of the mid-energy (TTME) method; (2) Tukey's trimean of the logged mid-energy (TTLME) method; (3) Gastwirth of the mid-energy (GME) method; (4) Gastwirth of the logged mid-energy (GLME) method; (5) general trimean of the mid-energy (GTME) method; (6) general trimean of the logarithm of mid-energy (GTLME) method. The GTME and GTLME methods are based on the derived optimal parameters in general trimean estimators to minimize the asymptotic variances. Tukey's trimean and Gastwirth estimators are two special cases following the general trimean estimators' framework. These estimators are applied on both mid-energy (as defined by Soltani et al. 2004) and logarithm of the mid-energy at each NDWT level detail coefficient diagonal block. The estimation performance of the proposed methods is compared to five other existing methods: Veitch and Abry (VA) method, Soltani, Simard, and Boichu (SSB) method, MEDL method, MEDLA method, and Theil-type regression (TT) method.

Simulation results indicate all our six variations outperform SSB, MEDL , MEDLA, and TT methods for all H's regarding variances and MSEs. Compared with VA method, our methods yield significantly smaller variances and MSEs when $H > 0.5$. When $H = 0.3$, our methods are still comparable to VA. Although the performances of our six variations are very similar regarding variances and MSEs, the TTME method based on Tukey's trimean estimator of the mid-energy has the best performance among all of them.

The proposed methods have been applied to digitized mammograms to classify patients with and without breast cancer. Our methods helped to differentiate individuals based on the estimated Hurst parameters $\hat{H}$. Higher values for $\hat{H}$ have been found in cancer group, and individuals with breast cancer have smoother breast tissue images. This increase of regularity with increase of the degree of pathology is common for many other biometric signals: EEG, EKG, high frequency protein mass-spectra, high resolution medical images of tissue, to list a few.

Appendix

Proof of Theorem 5.1

Proof A single wavelet coefficient in a non-decimated wavelet transform of a 2-D fBm of size $N \times N$ with Hurst exponent H is normally distributed, with variance depending on its level j. The four coefficients in each set

$$\{d_{j,j;(k_{i1},k_{i2})}, d_{j,j;(k_{i1},k_{i2}+\frac{N}{2})}, d_{j,j;(k_{i1}+\frac{N}{2},k_{i2})}, d_{j,j;(k_{i1}+\frac{N}{2},k_{i2}+\frac{N}{2})}\}$$

are assumed to be independent and follow the same normal distribution.

$$\begin{aligned} & d_{j,j;(k_{i1},k_{i2})}, d_{j,j;(k_{i1},k_{i2}+\frac{N}{2})}, d_{j,j;(k_{i1}+\frac{N}{2},k_{i2})}, d_{j,j;(k_{i1}+\frac{N}{2},k_{i2}+\frac{N}{2})} \\ & \sim \mathscr{N}\left(0, 2^{-(2H+2)j}\sigma^2\right). \end{aligned}$$

Then the mid-energies in D_j defined in (5.9) and (5.8) can be readily shown to have exponential distribution with scale parameter $\lambda_j = \sigma^2 \cdot 2^{-(2H+2)j}$. Therefore at each detail level j, the mid-energies in D_j are i.i.d. $\mathscr{E}xp\left(\lambda_j^{-1}\right)$, and when applying general trimean estimator $\hat{\mu}_j$ on D_j, following the derivation in Sect. 5.3, we have

$$\boldsymbol{\xi} = \left[\log\left(\frac{1}{1-p}\right)\lambda_j \quad \log(2)\,\lambda_j \quad \log\left(\frac{1}{p}\right)\lambda_j\right]^T,$$

and

$$\Sigma = \begin{bmatrix} \frac{p}{(1-p)}\lambda_j^2 & \frac{p}{(1-p)}\lambda_j^2 & \frac{p}{(1-p)}\lambda_j^2 \\ \frac{p}{(1-p)}\lambda_j^2 & \lambda_j^2 & \lambda_j^2 \\ \frac{p}{(1-p)}\lambda_j^2 & \lambda_j^2 & \frac{1-p}{p}\lambda_j^2 \end{bmatrix}_{3\times 3},$$

therefore, the asymptotic distribution of $\hat{\mu}_{j,i}$ is normal with mean

$$\begin{aligned} \mathbb{E}\left(\hat{\mu}_{j,i}\right) &= A \cdot \boldsymbol{x} \\ &= \left(\frac{\alpha}{2}\log\left(\frac{1}{p(1-p)}\right) + (1-\alpha)\log 2\right)\lambda_j \\ &\triangleq c(\alpha, p)\,\lambda_j, \end{aligned}$$

and variance

$$\begin{aligned}
\mathrm{Var}\left(\hat{\mu}_{j,i}\right) &= \frac{2}{M^2} A \Sigma A^T \\
&= \frac{2}{M^2}\left(\frac{\alpha(1-2p)(\alpha-4p)}{4p(1-p)}+1\right)\lambda_j^2 \\
&\triangleq \frac{2}{M^2} f(\alpha, p)\, \lambda_j^2 .
\end{aligned}$$

Since the Hurst exponent can be estimated as

$$\hat{H} = -\frac{\hat{\beta}}{2} - 1, \tag{5.27}$$

where $\hat{\beta}$ is the regression slope in the least square linear regression on pairs $\left(j, \log_2\left(\hat{\mu}_j\right)\right)$ from level J_1 to J_2, $J_1 \le j \le J_2$. It can be easily derived that $\hat{\beta}$ is a linear combination of $\log_2\left(\hat{\mu}_j\right)$,

$$\hat{\beta} - \sum_{j=J_1}^{J_2} a_j \log_2\left(\hat{\mu}_j\right), \quad a_j = \frac{j-(J_1+J_2)/2}{\sum_{j=J_1}^{J_2}\left(j-(J_1+J_2)/2\right)^2}.$$

We can check that $\sum_{j=J_1}^{J_2} a_j = 0$ and $\sum_{j=J_1}^{J_2} a_j j = 1$. Also, if $X \sim \mathcal{N}(\mu, \sigma^2)$, the approximate expectation and variance of $g(X)$ are

$$\mathbb{E}\left(g(X)\right) = g(\mu) + \frac{g''(\mu)\sigma^2}{2}, \quad \text{and} \quad \mathrm{Var}\left(g(X)\right) = \left(g'(\mu)\right)^2 \sigma^2,$$

based on which we calculate

$$\mathbb{E}\left(\log_2\left(\hat{\mu}_j\right)\right) = -(2H+2)j + \text{Constant}, \text{ and } \mathrm{Var}\left(\log_2\left(\hat{\mu}_j\right)\right) = \frac{\frac{2}{M^2} f(\alpha, p)}{(\log 2)^2 c^2(\alpha, p)}.$$

Therefore

$$\begin{aligned}
\mathbb{E}\left(\hat{\beta}\right) &= \sum_{j=J_1}^{J_2} a_j \mathbb{E}\left(\log_2\left(\hat{\mu}_j\right)\right) = -(2H+2), \text{ and } \mathrm{Var}\left(\hat{\beta}\right) \\
&= \sum_{j=J_1}^{J_2} a_j^2 \,\mathrm{Var}\left(\log_2\left(\hat{\mu}_j\right)\right) := 4V1,
\end{aligned}$$

and

$$\mathbb{E}\left(\hat{H}\right) = H, \text{ and } \mathrm{Var}\left(\hat{H}\right) = V1, \tag{5.28}$$

where the asymptotic variance V_1 is a constant number independent of simple size N and level j,

$$V_1 = \frac{6f(\alpha, p)}{(\log 2)^2 M^2 c^2(\alpha, p)\, q(J_1, J_2)},$$

and

$$q(J_1, J_2) = (J_2 - J_1)(J_2 - J_1 + 1)(J_2 - J_1 + 2).$$

Proof of Theorem 5.2

Proof We have stated that each mid-energy in D_j follows $\mathcal{E}xp\left(\lambda_j^{-1}\right)$ with scale parameter $\lambda_j = \sigma^2 \cdot 2^{-(2H+2)j}$. If we denote the kth element in $\log\left(D_j\right)$ as $y_{j,k}$ for $k = 1, \ldots, \frac{M^2}{2}$ and $j = 1, \ldots, J$, the pdf and cdf of $y_{j,k}$ are

$$f\left(y_{j,k}\right) = \lambda_j^{-1} e^{-\lambda_j^{-1} e^{y_{j,k}}} e^{y_{j,k}},$$

and

$$F\left(y_{j,k}\right) = 1 - e^{-\lambda_j^{-1} e^{y_{j,k}}}.$$

The p-quantile can be obtained by solving $F\left(y_p\right) = 1 - e^{-\lambda_j^{-1} e^{y_p}} = p$, and $y_p = \log\left(-\lambda_j \log(1-p)\right)$. Then it can be shown that $f\left(y_p\right) = -(1-p)\log(1-p)$. When applying the general trimean estimator $\hat{\mu}_j$ on $\log\left(D_j\right)$, following the derivation in Sect. 5.3, we get

$$\boldsymbol{\xi} = \begin{bmatrix} \log\left(\log\left(\frac{1}{1-p}\right)\right) + \log\left(\lambda_j\right) \\ \log(\log 2) + \log\left(\lambda_j\right) \\ \log\left(\log\left(\frac{1}{p}\right)\right) + \log\left(\lambda_j\right) \end{bmatrix},$$

and

$$\Sigma = \begin{bmatrix} \frac{p}{(1-p)(\log(1-p))^2} & \frac{p}{(1-p)\log(1-p)\log\left(\frac{1}{2}\right)} & \frac{p}{(1-p)\log(1-p)\log p} \\ \frac{p}{(1-p)\log(1-p)\log\left(\frac{1}{2}\right)} & \frac{1}{(\log 2)^2} & \frac{1}{\log\left(\frac{1}{2}\right)\log p} \\ \frac{p}{(1-p)\log(1-p)\log p} & \frac{1}{\log\left(\frac{1}{2}\right)\log p} & \frac{1-p}{p(\log p)^2} \end{bmatrix},$$

thus, the asymptotic distribution of $\hat{\mu}_{j,i}$ is normal with mean

$$\begin{aligned}\mathbb{E}\left(\hat{\mu}_{j,i}\right) &= A \cdot \boldsymbol{\xi} \\ &= \frac{\alpha}{2}\log\left(\log\frac{1}{1-p}\cdot\log\frac{1}{p}\right) + (1-\alpha)\log(\log 2) + \log\left(\lambda_j\right) \\ &\triangleq c(\alpha, p) + \log\left(\lambda_j\right),\end{aligned}$$

and variance

$$\begin{aligned}\mathrm{Var}\left(\hat{\mu}_{j,i}\right) &= \frac{2}{M^2}A\Sigma A^T \\ &= \frac{2}{M^2}\left(\frac{\alpha^2}{4}g_1(p) + \frac{\alpha(1-\alpha)}{2}g_2(p) + \frac{(1-\alpha)^2}{(\log 2)^2}\right) \\ &\triangleq \frac{2}{M^2}f(\alpha, p),\end{aligned}$$

where

$$\begin{aligned}g_1(p) =& \frac{p}{(1-p)(\log(1-p))^2} + \\ & \frac{1-p}{p(\log p)^2} + \frac{2p}{(1-p)\log(1-p)\log p},\end{aligned}$$

and

$$g_2(p) = \frac{2p}{(1-p)\log(1-p)\log\frac{1}{2}} + \frac{2}{\log\frac{1}{2}\log p}.$$

Since the Hurst exponent can be estimated as

$$\hat{H} = -\frac{1}{2\log 2}\hat{\beta} - 1, \tag{5.29}$$

where $\hat{\beta}$ is the regression slope in the least square linear regressions on pairs $\left(j, \hat{\mu}_j\right)$ from level J_1 to J_2, $J_1 \leq j \leq J_2$. It can be easily derived that $\hat{\beta}$ is a linear combination of $\hat{\mu}_j$,

$$\hat{\beta} = \sum_{j=J_1}^{J_2} a_j\hat{\mu}_j, \quad a_j = \frac{j-(J_1+J_2)/2}{\sum_{j=J_1}^{J_2}(j-(J_1+J_2)/2)^2}.$$

Again, we can check that $\sum_{j=J_1}^{J_2} a_j = 0$ and $\sum_{j=J_1}^{J_2} a_j j = 1$. Therefore

$$\mathbb{E}\left(\hat{\beta}\right) = \sum_{j=J_1}^{J_2} a_j \mathbb{E}\left(\hat{\mu}_{j,i}\right) = -(2H+2)\log 2, \text{ and } \operatorname{Var}\left(\hat{\beta}\right)$$

$$= \sum_{j=J_1}^{J_2} a_j^2 \operatorname{Var}\left(\hat{\mu}_{j,i}\right) := 4(\log 2)^2 V_2,$$

and

$$\mathbb{E}\left(\hat{H}\right) = H, \text{ and } \operatorname{Var}\left(\hat{H}\right) = V_2, \tag{5.30}$$

where the asymptotic variance V_2 is a constant number independent of simple size N and level j,

$$V_2 = \frac{6f(\alpha, p)}{(\log 2)^2 M^2 q(J_1, J_2)},$$

and $q(J_1, J_2)$ is given in Eq. (5.13).

Proof of Lemma 5.2

Proof When applying Tukey's trimean estimator $\hat{\mu}_j^T$ on D_j, following the derivation in Sect. 5.3.1, we have

$$\boldsymbol{\xi}_T = \begin{bmatrix} \log\left(\frac{4}{3}\right)\lambda_j \\ \log(2)\lambda_j \\ \log(4)\lambda_j \end{bmatrix},$$

and

$$\Sigma_T = \begin{bmatrix} \frac{1}{3}\lambda_j^2 & \frac{1}{3}\lambda_j^2 & \frac{1}{3}\lambda_j^2 \\ \frac{1}{3}\lambda_j^2 & \lambda_j^2 & \lambda_j^2 \\ \frac{1}{3}\lambda_j^2 & \lambda_j^2 & \frac{1}{3}\lambda_j^2 \end{bmatrix}_{3\times 3},$$

therefore, the asymptotic distribution of $\hat{\mu}_j^T$ is normal with mean

$$\mathbb{E}\left(\hat{\mu}_j^T\right) = A_T \cdot \boldsymbol{\xi}_T = \frac{1}{4}\log\left(\frac{64}{3}\right)\lambda_j \triangleq c_1\lambda_j,$$

and variance

$$\mathrm{Var}\left(\hat{\mu}_{j,i}^{T}\right) = \frac{2}{M^2} A_T \Sigma_T A_T^T = \frac{5}{3M^2}\lambda_j^2.$$

When applying Gastwirth estimator $\hat{\mu}_j^G$ on D_j, following the derivation in Sect. 5.3.2, we have

$$\boldsymbol{\xi}_G = \begin{bmatrix} \log\left(\frac{3}{2}\right)\lambda_j \\ \log(2)\lambda_j \\ \log(3)\lambda_j \end{bmatrix},$$

and

$$\Sigma_G = \begin{bmatrix} \frac{1}{2}\lambda_j^2 & \frac{1}{2}\lambda_j^2 & \frac{1}{2}\lambda_j^2 \\ \frac{1}{2}\lambda_j^2 & \lambda_j^2 & \lambda_j^2 \\ \frac{1}{2}\lambda_j^2 & \lambda_j^2 & \frac{1}{2}\lambda_j^2 \end{bmatrix},$$

therefore, the asymptotic distribution of $\hat{\mu}_j^G$ is normal with mean

$$\begin{aligned} \mathbb{E}\left(\hat{\mu}_{j,i}^{G}\right) &= A_G \cdot \boldsymbol{\xi}_G \\ &= \left(0.3 \times \log\left(\frac{9}{2}\right) + 0.4 \times \log(2)\right)\lambda_j \\ &\triangleq c_2\lambda_j, \end{aligned}$$

and variance

$$\mathrm{Var}\left(\hat{\mu}_{j,i}^{G}\right) = \frac{2}{M^2} A_G \Sigma_G A_G^T = \frac{1.67}{M^2}\lambda_j^2.$$

Based on Eq. (5.28), we have

$$\hat{H}^T \overset{\text{approx}}{\sim} \mathscr{N}\left(H, V_1^T\right), \text{ and } \hat{H}^G \overset{\text{approx}}{\sim} \mathscr{N}\left(H, V_1^G\right), \qquad (5.31)$$

where the asymptotic variances V_1^T and V_1^G are constant numbers,

$$V_1^T = \frac{5}{(\log 2)^2 M^2 c_1^2 q(J_1, J_2)},$$

$$V_1^G = \frac{5.01}{(\log 2)^2 M^2 c_2^2 q(J_1, J_2)}.$$

The function $q(J_1, J_2)$ is the same as Eq. (5.13) in Theorem 5.1.

Proof of Lemma 5.3

Proof When applying Tukey's trimean estimator $\hat{\mu}_j^T$ on $\log\left(D_j\right)$, following the derivation in Sect. 5.3.1, we have

$$\boldsymbol{\xi}_T = \begin{bmatrix} \log\left(\log\left(\frac{4}{3}\right)\right) + \log\left(\lambda_j\right) \\ \log\left(\log 2\right) + \log\left(\lambda_j\right) \\ \log\left(\log 4\right) + \log\left(\lambda_j\right) \end{bmatrix},$$

and

$$\Sigma_T = \begin{bmatrix} \frac{1}{3\left(\log\left(\frac{3}{4}\right)\right)^2} & \frac{1}{3\log\left(\frac{3}{4}\right)\log\left(\frac{1}{2}\right)} & \frac{1}{3\log\left(\frac{3}{4}\right)\log\left(\frac{1}{4}\right)} \\ \frac{1}{3\log\left(\frac{3}{4}\right)\log\left(\frac{1}{2}\right)} & \frac{1}{(\log 2)^2} & \frac{1}{\log\left(\frac{1}{2}\right)\log\left(\frac{1}{4}\right)} \\ \frac{1}{3\log\left(\frac{3}{4}\right)\log\left(\frac{1}{4}\right)} & \frac{1}{\log\left(\frac{1}{2}\right)\log\left(\frac{1}{4}\right)} & \frac{3}{(\log 4)^2} \end{bmatrix},$$

therefore, the asymptotic distribution of $\hat{\mu}_j^T$ is normal with mean

$$\begin{aligned} \mathbb{E}\left(\hat{\mu}_{j,i}^T\right) &= A_T \cdot \boldsymbol{\xi}_T \\ &= -\left(2H+2\right)\log 2 \cdot j + \log\sigma^2 + \\ &\quad \frac{1}{4}\log\left(\log\left(\frac{4}{3}\right)\cdot\log 4\right) + \frac{1}{2}\log\left(\log 2\right) \\ &\triangleq -\left(2H+2\right)\log 2 \cdot j + c_3 \end{aligned}$$

and variance

$$\begin{aligned} \mathrm{Var}\left(\hat{\mu}_{j,i}^T\right) &= \frac{2}{M^2} A_T \Sigma_T A_T^T \\ &= \frac{2}{M^2}\left(\frac{1}{48\left(\log\frac{3}{4}\right)^2} + \frac{1}{12\log\frac{3}{4}\log\frac{1}{2}} + \frac{1}{24\log\frac{3}{4}\log\frac{1}{4}} + \right. \\ &\quad \left. \frac{1}{4\left(\log 2\right)^2} + \frac{1}{4\log\frac{1}{2}\log\frac{1}{4}} + \frac{3}{16\left(\log\frac{1}{4}\right)^2}\right) \\ &\triangleq V_T. \end{aligned}$$

When applying Gastwirth estimator $\hat{\mu}_j^G$ on $\log\left(D_{j,i}\right)$, following the derivation in Sect. 5.3.2, we have

$$\boldsymbol{\xi}_G = \begin{bmatrix} \log\left(\log\left(\frac{3}{2}\right)\right) + \log\left(\lambda_j\right) \\ \log\left(\log 2\right) + \log\left(\lambda_j\right) \\ \log\left(\log 3\right) + \log\left(\lambda_j\right) \end{bmatrix},$$

and

$$\Sigma_G = \begin{bmatrix} \frac{1}{2\left(\log\frac{2}{3}\right)^2} & \frac{1}{2\log\left(\frac{2}{3}\right)\log\left(\frac{1}{2}\right)} & \frac{1}{2\log\left(\frac{1}{3}\right)\log\left(\frac{2}{3}\right)} \\ \frac{1}{2\log\left(\frac{2}{3}\right)\log\left(\frac{1}{2}\right)} & \frac{1}{(\log 2)^2} & \frac{1}{\log\left(\frac{1}{2}\right)\log\left(\frac{1}{3}\right)} \\ \frac{1}{2\log\left(\frac{1}{3}\right)\log\left(\frac{2}{3}\right)} & \frac{1}{\log\left(\frac{1}{2}\right)\log\left(\frac{1}{3}\right)} & \frac{2}{(\log 3)^2} \end{bmatrix},$$

therefore, the asymptotic distribution of $\hat{\mu}_j^G$ is normal with mean

$$\begin{aligned} \mathbb{E}\left(\hat{\mu}_{j,i}^G\right) &= A_g \cdot \boldsymbol{\xi}_G \\ &= -(2H+2)\log 2 \cdot j + \log\sigma^2 + \\ &\quad 0.3 \times \log\left(\log\left(\frac{3}{2}\right) \cdot \log 3\right) + 0.4 \times \log\left(\log 2\right) \\ &\triangleq -(2H+2)\log 2 \cdot j + c_4 \end{aligned}$$

and variance

$$\begin{aligned} \mathrm{Var}\left(\hat{\mu}_{j,i}^G\right) &= \frac{2}{M^2} A_G \Sigma_G A_G^T \\ &= \frac{2}{M^2}\left(\frac{0.09}{2\left(\log\frac{2}{3}\right)^2} + \frac{0.12}{\log\frac{2}{3}\log\frac{1}{2}} + \frac{0.09}{\log\frac{1}{3}\log\frac{2}{3}} + \right. \\ &\quad \left. \frac{0.16}{\left(\log\frac{1}{2}\right)^2} + \frac{0.24}{\log\frac{1}{2}\log\frac{1}{3}} + \frac{0.18}{\left(\log\frac{1}{3}\right)^2}\right) \\ &\triangleq V_G. \end{aligned}$$

Based on Eq. (5.30), we can easily derive

$$\hat{H}^T \overset{\text{approx}}{\sim} \mathcal{N}\left(H, V_2^T\right), \text{ and } \hat{H}^G \overset{\text{approx}}{\sim} \mathcal{N}\left(H, V_2^G\right), \tag{5.32}$$

where the asymptotic variances V_2^T and V_2^G are constant numbers,

$$V_2^T = \frac{3V_T}{(\log 2)^2 q(J_1, J_2)},$$

$$V_2^G = \frac{3V_G}{(\log 2)^2 q(J_1, J_2)}.$$

The function $q(J_1, J_2)$ is provided in Eq. (5.13).

References

Abry, P. (2003). Scaling and wavelets: An introductory walk. In *Processes with long-range correlations* (pp. 34–60). Berlin: Springer.

Abry, P., Gonçalvés, P., & Flandrin, P. (1995). Wavelets, spectrum analysis and 1/f processes. In: *Wavelets and statistics* (pp. 15–29). Springer: New York.

Abry, P., Goncalves, P., & Véhel, J. L. (2013). *Scaling, fractals and wavelets*. New York: Wiley.

Abry, P., Flandrin, P., Taqqu, M. S., & Veitch, D. (2000). Wavelets for the analysis, estimation and synthesis of scaling data. *Self-similar network traffic and performance evaluation* (pp. 39–88). New York: Wiley.

Abry, P., Flandrin, P., Taqqu, M. S., & Veitch, D. (2003). Self-similarity and long-range dependence through the wavelet lens. In *Theory and applications of long-range dependence* (pp. 527–556). Boston, MA: Birkhauser.

Andrews, D. F., & Hampel, F. R. (2015). *Robust estimates of location: Survey and advances*. Princeton: Princeton University Press.

Bala, B. K., & Audithan, S. (2014). Wavelet and curvelet analysis for the classification of microcalcifiaction using mammogram images. In *Second International Conference on Current Trends in Engineering and Technology - ICCTET 2014* (pp. 517–521). https://doi.org/10.1109/ICCTET.2014.6966351

DasGupta, A. (2008). Edgeworth expansions and cumulants. In *Asymptotic theory of statistics and probability* (pp. 185–201). New York: Springer.

El-Naqa, I., Yang, Y., Wernick, M. N., Galatsanos, N. P., & Nishikawa, R. M. (2002). A support vector machine approach for detection of microcalcifications. *IEEE Transactions on Medical Imaging, 21*(12), 1552–1563.

Engel, Jr J., Bragin, A., Staba, R., & Mody, I. (2009). High-frequency oscillations: What is normal and what is not? *Epilepsia, 50*(4), 598–604.

Feng, C., & Vidakovic, B. (2017), Estimation of the hurst exponent using trimean estimators on nondecimated wavelet coefficients. arXiv preprint arXiv:170908775.

Franzke, C. L., Graves, T., Watkins, N. W., Gramacy, R. B., & Hughes, C. (2012). Robustness of estimators of long-range dependence and self-similarity under non-gaussianity. *Philosophical Transactions of the Royal Society A, 370*(1962), 1250–1267.

Gastwirth, J. L. (1966). On robust procedures. *Journal of the American Statistical Association, 61*(316), 929–948.

Gastwirth, J. L., & Cohen, M.L (1970) Small sample behavior of some robust linear estimators of location. *Journal of the American Statistical Association, 65*(330), 946–973

Gastwirth, J. L, & Rubin, H. (1969). On robust linear estimators. *The Annals of Mathematical Statistics, 40*(1), 24–39.

Gregoriou, G. G., Gotts, S. J., Zhou, H., & Desimone, R. (2009). High-frequency, long-range coupling between prefrontal and visual cortex during attention. *Science, 324*(5931), 1207–1210.

Hamilton, E. K., Jeon, S., Cobo, P. R., Lee, K. S., & Vidakovic, B. (2011). Diagnostic classification of digital mammograms by wavelet-based spectral tools: A comparative study. In *2011 IEEE International Conference on Bioinformatics and Biomedicine (BIBM), IEEE* (pp. 384–389).

Heath, M., Bowyer, K., Kopans, D., Moore, R., & Kegelmeyer, W. P. (2000). The digital database for screening mammography. In *Proceedings of the 5th International Workshop on Digital Mammography* (pp. 212–218). Medical Physics Publishing.

Jeon, S., Nicolis, O., & Vidakovic, B. (2014). Mammogram diagnostics via 2-d complex wavelet-based self-similarity measures. *The São Paulo Journal of Mathematical Sciences, 8*(2), 265–284.

Kang, M., & Vidakovic, B. (2017). Medl and medla: Methods for assessment of scaling by medians of log-squared nondecimated wavelet coefficients. ArXiv Preprint ArXiv:170304180.

Katul, G., Vidakovic, B., & Albertson, J. (2001). Estimating global and local scaling exponents in turbulent flows using discrete wavelet transformations. *Physics of Fluids, 13*(1), 241–250.

Kestener, P., Lina, J. M., Saint-Jean, P., & Arneodo, A. (2011). Wavelet-based multifractal formalism to assist in diagnosis in digitized mammograms. *Image Analysis & Stereology, 20*(3), 169–174.

Kolmogorov, A. N. (1940). Wienersche spiralen und einige andere interessante kurven in hilbertscen raum, cr (doklady). *Academy of Sciences URSS (NS), 26*, 115–118.

Mandelbrot, B. B., & Van Ness, J. W. (1968). Fractional Brownian motions, fractional noises and applications. *SIAM Review, 10*(4), 422–437.

Nason, G. P., & Silverman, B. W. (1995). The stationary wavelet transform and some statistical applications. In *Wavelets and statistics* (pp. 281–299). New York: Springer.

Netsch, T., & Peitgen, H. O. (1999). Scale-space signatures for the detection of clustered microcalcifications in digital mammograms. *IEEE Transactions on Medical Imaging, 18*(9), 774–786.

Nicolis, O., Ramírez-Cobo, P., & Vidakovic, B. (2011). 2d wavelet-based spectra with applications. *Computational Statistics & Data Analysis, 55*(1), 738–751.

Park, J., & Park, C. (2009). Robust estimation of the hurst parameter and selection of an onset scaling. *Statistica Sinica, 19*, 1531–1555.

Park, K., & Willinger, W. (2000). *Self-similar network traffic and performance evaluation*. Wiley Online Library. https://doi.org/10.1002/047120644X.

Percival, D. B., & Walden, A. T. (2006). *Wavelet methods for time series analysis* (vol. 4). New York: Cambridge University Press.

Ramírez-Cobo, P., & Vidakovic, B. (2013). A 2d wavelet-based multiscale approach with applications to the analysis of digital mammograms. *Computational Statistics & Data Analysis, 58*, 71–81.

Reiss, P. T., & Ogden, R. T. (2010). Functional generalized linear models with images as predictors. *Biometrics, 66*(1), 61–69.

Reiss, P. T., Ogden, R. T., Mann, J. J., & Parsey, R. V. (2005). Functional logistic regression with pet imaging data: A voxel-level clinical diagnostic tool. *Journal of Cerebral Blood Flow & Metabolism, 25*(1_suppl), S635–S635.

Shen, H., Zhu. Z., & Lee, T. C. (2007). Robust estimation of the self-similarity parameter in network traffic using wavelet transform. *Signal Processing, 87*(9), 2111–2124.

Sheng, H., Chen, Y., & Qiu, T. (2011). On the robustness of hurst estimators. *IET Signal Processing, 5*(2), 209–225.

Soltani, S., Simard, P., & Boichu, D. (2004). Estimation of the self-similarity parameter using the wavelet transform. *Signal Processing, 84*(1), 117–123.

Theil, H. (1992). A rank-invariant method of linear and polynomial regression analysis. In *Henri Theils contributions to economics and econometrics* (pp. 345–381). The Netherlands: Springer.

Tukey, J. W. (1977). *Exploratory data analysis* (Vol. 2). Reading, MA. Addison Wesley.

Vidakovic, B. (2009). *Statistical modeling by wavelets* (Vol. 503). New York: Wiley.

Wang, T. C., & Karayiannis, N. B. (1998). Detection of microcalcifications in digital mammograms using wavelets. *IEEE Transactions on Medical Imaging, 17*(4), 498–509.

Wood, A. T., & Chan, G. (1994). Simulation of stationary gaussian processes in [0, 1] d. *Journal of Computational and Graphical Statistics, 3*(4), 409–432.

Woods, T., Preeprem, T., Lee, K., Chang, W., & Vidakovic, B. (2016). Characterizing exons and introns by regularity of nucleotide strings. *Biology Direct, 11*(1), 6.

Zhou, B. (1996). High-frequency data and volatility in foreign-exchange rates. *Journal of Business and Economic Statistics, 14*(1), 45–52.

Zhou, H., Li, L., & Zhu, H. (2013). Tensor regression with applications in neuroimaging data analysis. *Journal of the American Statistical Association, 108*(502), 540–552.

Zipunnikov, V., Caffo, B., Yousem, D. M., Davatzikos, C., Schwartz, B. S., Crainiceanu. C. (2011). Functional principal component model for high-dimensional brain imaging. *NeuroImage, 58*(3), 772–784.

Chapter 6
Wavelet-Based Profile Monitoring Using Order-Thresholding Recursive CUSUM Schemes

Ruizhi Zhang, Yajun Mei, and Jianjun Shi

6.1 Introduction

With the rapid development of advanced sensing technologies, rich and complex real-time profile or curve data are available in many processes in biomedical sciences, manufacturing, and engineering. For instance, physiologic monitoring systems generated real-time profile conditions of a patient in intensive care units. In modern manufacturing, profile data are generated to provide valuable information about the quality or reliability performance of the process or product. In these applications, it is often desirable to utilize the observed profile data to develop efficient methodologies for process monitoring and fault diagnosing.

A concrete motivating example of profile data in this article is from a progressive forming process with five die stations including preforming, blanking, initial forming, forming, and trimming, see Fig. 6.1 for illustration. Ideally, when the process is in control, a work piece should pass through these five stations. However, a missing part problem, which means that the work piece is not settled in the right die station but is conveyed to the downstream stations, may occur in this process (Lei et al. 2010; Zhou et al. 2016). Such a fault often leads to unfinished or nonconforming products and/or severe die damage. The tonnage signal measured by the press tonnage sensor, which is the summation of all stamping forces, contains rich process information of forming operations and widely used for monitoring the forming process. Figure 6.2 shows the tonnage profiles collected under normal condition and five faulty conditions corresponding to missing operations occurring in each of the five die stations. It is clear from the figure that each profile is

R. Zhang (✉) · Y. Mei · J. Shi
H. Milton Stewart School of Industrial and Systems Engineering, Georgia Institute of Technology, Atlanta, GA, USA
e-mail: rzhang320@gatech.edu; yajun.mei@isye.gatech.edu; jianjun.shi@isye.gatech.edu

Y. Zhao, D.-G. Chen (eds.), *New Frontiers of Biostatistics and Bioinformatics*, ICSA Book Series in Statistics, https://doi.org/10.1007/978-3-319-99389-8_6

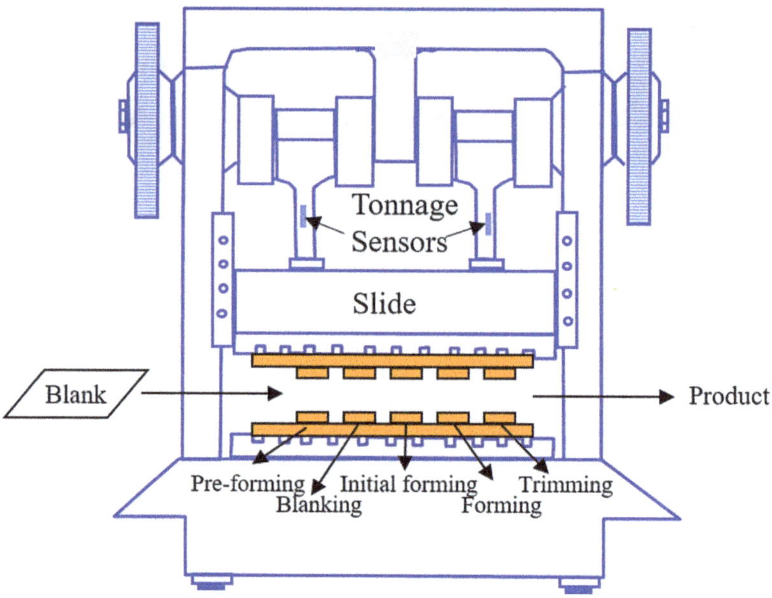

Fig. 6.1 Illustration of a progressive forming process

highly nonlinear, since the observed forces at different segments correspond to different stages of the operation within one production cycle. In addition, the difference between normal profiles and fault profiles is also nonlinear. For some particular faults, i.e., Fault 4, profiles are quite overlapping with the normal profiles. Under a high-production rate environment, it is highly desirable but challenging to effectively online monitor these profiles and detect those different types of unknown but subtle changes quickly.

In the profile monitoring literature, much research has been done for monitoring linear profiles, see, for example, Kang and Albin (2000), Chang and Gan (2006), Zou et al. (2007b,a), Kazemzadeh et al. (2008). However, in many real-world applications including those profiles in Fig. 6.2, the form of the profile data is too complicated to be expressed as a linear or parametric function. Several nonlinear profile monitoring procedures have been developed in the literature based on nonparametric regression techniques such as smoothing splines (Gardner et al. 1997; Chang and Yadama 2010), Fourier analysis (Chen and Nembhard 2011), local kernel regression (Qiu et al. 2010; Zou et al. 2009), and functional principal component analysis (FPCA) (Hall et al. 2001; Paynabar et al. 2016). However all these approaches tend to monitor a smooth, in-control profile, and thus may loss information about local structures such as jumps or cusps. Moreover, all these approaches are based on monitoring the changes of selected model coefficients, while it will be difficult to interpret their meanings back to the original profiles.

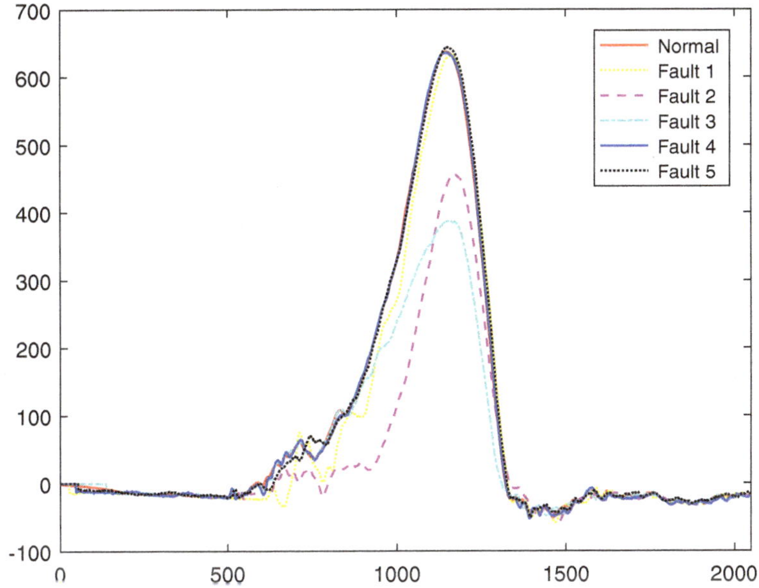

Fig. 6.2 Six profile samples from a forming process: one is in-control, normal sample and the other five are out-of-control, fault samples

In this article, we propose to monitor nonlinear profiles based on the discrete wavelet transform (DWT). Besides a useful dimensional reduction tool, wavelet-based approaches have other advantages: the multi-resolution decomposition of the wavelets could be useful to locate the anomaly of the profile, and fast computational algorithms of the DWT are available (Mallat 1989). Indeed, DWT has been applied to detect and diagnose process faults in the offline context, see Fan (1996) and Jin and Shi (1999). In the online monitoring context, many existing methods follow the suggestions of Donoho and Johnstone (1994) to first conduct wavelet shrinkage for dimension reduction under the in-control state, and then monitor the changes on the selected wavelet coefficients for the out-of-control state, see Hotelling T^2 control chart (Jeong et al. 2006; Zhou et al. 2006), and the CUSUM-type control chart (Lee et al. 2012). However, one will lose detection power if the change of the out-of-control state is on the wavelet coefficients that are not selected under the in-control state. To illustrate the importance of the out-of-control state on the wavelet coefficients selection, we provide a simple two-dimensional example in Fig. 6.3. As can be seen in this figure, the magnitude of wavelet coefficient 2 is very small compared with wavelet coefficient 1. However, if we just select wavelet coefficient 1 based on the in-control estimation, it would be difficult to detect the out-of-control samples since the changes occurred on the wavelet coefficient 2. To address this issue, it was proposed in Chicken et al. (2009) to use all wavelet coefficients to conduct a likelihood ratio test. However, as we will show later

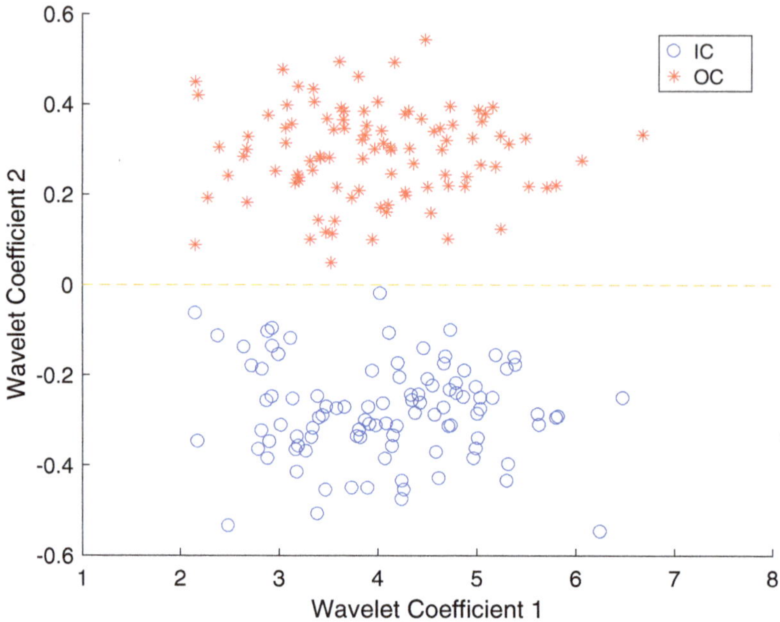

Fig. 6.3 A simulated dataset in the 2-dimensional wavelet domain, where blue circles indicate IC observations and red stars indicate OC observations. The mean shift is along the second wavelet coefficient, and the change is undetectable if using the first wavelet coefficient

in the simulation and case study, their methods are based on some asymptotic approximated likelihood ratio statistics, therefore may lose some detection power especially when the changed wavelet coefficients are sparse. Moreover, their method is not scalable and requires a lot of memory to store past observations.

In this paper, we propose to first construct the local adaptive CUSUM statistics as in Lorden and Pollak (2008) and Liu et al. (2017) for monitoring all wavelet coefficients by the hard-shrinkage estimation of the mean of in-control coefficients. Then we use the order-shrinkage to select those wavelet coefficients that are involved in the change significantly. Thus, from the methodology point of view, our proposed methodologies are analogous to those offline statistical methods such as (adaptive) truncation, soft-, hard-, and order-thresholding, see Neyman (1937), Donoho and Johnstone (1994), Fan and Lin (1998), and Kim et al. (2010). However, our motivation here is different and our application to profile monitoring is new.

The remainder of this article is as follows. In Sect. 6.2, we present problem formulation and background information of wavelet transform. In Sect. 6.3, we develop our proposed schemes for online nonlinear profile monitoring. In Sect. 6.4, a case study about monitoring tonnage signature is presented. In Sect. 6.5, a simulation study about monitoring the Mallet's piecewise smooth function is conducted.

6.2 Problem Formulation and Wavelet Background

In this section, we will first present the mathematical formulation of the profile monitoring problem based on an additive change point model. Then we give a brief review of wavelet transformation that will be used for our proposed profile monitoring procedure.

Assume we obverse p-dimensional profile data, $y_1, y_2, \cdots$, sequentially from a process. Each profile y_k consists of p coordinates $y_k(x_i)$, for $i = 1, 2, \ldots, p$, with x_i equispaced over the interval [0, 1], and can be thought of as the realization of a profile function $y_k(x)$. In the profile monitoring problem, we assume that the profile functions $y_k(x)$'s are from the additive change-point model:

$$y_k(x) = \begin{cases} f_0(x) + \epsilon_k(x), & \text{for } k = 1, 2, .., \nu \\ f_1(x) + \epsilon_k(x), & \text{for } k = \nu + 1, \ldots \end{cases} \tag{6.1}$$

where $f_0(\cdot)$ and $f_1(\cdot)$ are the mean functions that need be estimated from the data, and $\epsilon_k(x)$'s are the random noise, which are assumed to be normally distributed with mean 0 that are independent across different time k. The problem is to utilize the observed profile data $y_k(x_i)$'s to detect the unknown change-time ν as quickly as possible when it occurs.

Since our proposed methods are based on monitoring the coefficients of the wavelet transformations of $y_k(x)$'s, let us provide a brief review of wavelet transformation of profile data and discrete wavelet transform (DWT). For any square-integrable function $f(x)$ on $\mathbb{R}$, it can be written as an (infinite) linear combinations of wavelet basis functions:

$$f(x) = \sum_{k \in \mathbb{Z}} c_{j_0}^k \phi_{j_0 k}(x) + \sum_{j=j_0}^{\infty} \sum_{k \in \mathbb{Z}} d_j^k \psi_{jk}(x). \tag{6.2}$$

Here the sets of two bases, $\phi_{jk}(x)$'s and $\psi_{jk}(x)$'s, are known as scaling and wavelet basis functions, respectively, and are generated from two parent wavelets: one is the father wavelet $\phi(x)$ that characterizes basic wavelet scale, and the other is the mother wavelet $\psi(x)$ that characterizes basic wavelet shape. Mathematically, $\phi_{jk}(x) = 2^{j/2}\phi(2^j x - k)$ and $\psi_{jk}(x) = 2^{j/2}\psi(2^j x - k)$, and the decomposed coefficients $c_{j_0}^k$ and d_j^k are called the scaling and detail coefficients, which represent the low-frequency and high-frequency components of original function $f(x)$.

The discrete wavelet transform (DWT) is a numeric and fast algorithm to determine the wavelet coefficients $\mathbf{c}$ when the observed data are discrete and dyadic, i.e., $\mathbf{y} = (y(x_1), y(x_2), \ldots, y(x_p))^T$ with p a dyadic integer, $p = 2^J$. The matrix form of DWT is represented as $\mathbf{c} = W\mathbf{y}$, where W is orthonormal wavelet transformation matrix (Mallat 1999), which depends on the selected orthogonal wavelet basis. A large families of choices for wavelet basis functions are available for use, see, for example, Daubechies (1992). Also see Mallat (1999) for an efficient algorithm to implement DWT. In this article, the Haar transform is chosen as one way of DWT because Haar coefficients have an explicit interpretation of the changes in the profile observations. Also see Jin and Shi (2001) and Zhou et al. (2006) as examples of applying Haar transform to monitor profile samples.

For the observed p-dimension profile, $y = (y(x_1), \ldots, y(x_p))$, we consider the Haar transformation with wavelet basis functions:

$$\phi_{00}(x) = 1, x \in [0, 1] \tag{6.3}$$

$$\psi_{km}(x) = \begin{cases} 2^{\frac{k-1}{2}}, & \frac{m-1}{2^{k-1}} < x < \frac{m-1/2}{2^{k-1}} \\ -2^{\frac{k-1}{2}}, & \frac{m-1/2}{2^{k-1}} < x < \frac{m}{2^{k-1}} \\ 0, & \text{elsewhere} \end{cases} \tag{6.4}$$

where k represents the scale of Haar transform and $m = 1, 2, \ldots, 2^{k-1}$.

For simplicity, we assume $p = 2^J$ (otherwise we can add new extra zero coordinations to the original profile if needed). When Haar transform is chosen, the wavelet coefficients $\mathbf{c} = (c(1), c(2), \ldots, c(p))^T$ are often written as $(c_0^0, c_1^1, c_2^1, c_2^2, \ldots, c_J^1, \ldots, c_J^{2^{J-1}})^T$, which represent the Haar coefficients for different levels from 0 to J.

For any new observed p-dimension profile, $y = (y(x_1), \ldots, y(x_p))$, the explicit expression of these Haar coefficients is given by

$$\begin{aligned} c_0^0 &= 2^{-\frac{J}{2}} \sum_{\ell=1}^{2^J} y(x_\ell), \\ c_k^m &= 2^{\frac{J-k-1}{2}} \{s[(m-1)2^{J-k+1}+1, (m-\frac{1}{2})2^{J-k+1}] \\ &\quad -s[(m-\frac{1}{2})2^{J-k+1}+1, m2^{J-k+1}]\}, \\ &= 2^{-\frac{J-k+1}{2}} \{ \sum_{\ell=(m-1)2^{J-k+1}+1}^{(m-\frac{1}{2})2^{J-k+1}} y(x_\ell) - \sum_{\ell=(m-\frac{1}{2})2^{J-k+1}+1}^{m2^{J-k+1}} y(x_\ell)\} \end{aligned} \tag{6.5}$$

for $k = 1, \ldots, J; m = 1, 2, \ldots, 2^{k-1}$ and $s[i, j]$ is defined by $s[i, j] = \frac{1}{j-i+1} \sum_{\ell=i}^{j} y(x_\ell)$. In other words, the Haar coefficient c_0^0 is proportional to the mean of all data and the other coefficients c_k^m are proportional to the mean difference of two adjacent intervals of length 2^{J-k}.

6.3 Our Proposed Method

At the high-level, our proposed profile monitoring method is based on monitoring the mean shifts on wavelet coefficients of nonlinear profiles $y_k(x)$'s. First, we use the in-control profiles from the historical training data to estimate the pre-change distributions of the wavelet coefficients. Second, we construct local monitoring statistics for each wavelet coefficient by recursively estimating the post-change mean of the wavelet coefficients. Third, we construct global monitoring procedure based on the information of the first several largest monitoring statistics.

It is necessary to emphasize that in the literature, wavelets are usually used for dimension reduction to select significant features and filter out noise (Donoho and Johnstone 1994). Here our proposed method is constructing efficient monitoring statistic for each wavelet coefficients and then performs dimension reduction on the monitoring statistics. There are two technical challenges that need special attention. The first one is that we do not know which wavelet coefficients will be affected under the out-of-control state, and the second one is that we do not know what are the changed magnitudes or the post-change distributions for those affected wavelet coefficients. To address these two challenges, we propose a computationally efficient algorithm that can monitor a large number of wavelet coefficients simultaneously in parallel based on local recursive CUSUM procedures, and then combine these local procedures together to raise a global alarm using the order-thresholding transformation in Liu et al. (2017) to filter out those unaffected Haar coefficients. The recursive CUSUM procedure is to adaptively update the estimates of the post-change means, and it was first proposed in Lorden and Pollak (2008) for detecting a normal mean shift from 0 to some unknown, positive values. Here we extend it to the wavelet context when one wants to detect both positive and negative mean shifts of the wavelet coefficients.

For the purpose of demonstration, in the remaining of the paper, we consider Haar coefficients as an example since they can easily be calculated and interpreted. Furthermore, they can capture the local changes on the profile efficiently.

For better presentation of our proposed nonlinear profile monitoring methods, we split this section into four subsections. Section 6.3.1 focuses on estimating the in-control means of Haar coefficients, and Sect. 6.3.2 discusses how to recursively estimate possible mean shifts of Haar coefficients and constructs local monitoring statistics for each wavelet coefficient. Section 6.3.3 derives our proposed monitoring method and Sect. 6.3.4 discusses how to choose tuning parameters.

6.3.1 In-Control Estimation

In our case study and in many real-world applications, it is reasonable to assume that some in-control profiles are available for learning the process variables. Without loss of generality, assume that there are m in-control profiles before online monitoring, and denote $\mathbf{c}_\ell$ as the vector of Haar coefficients of the ℓth profile $y_\ell(x)$ under the in-control status for $\ell = -m+1, \cdots, -1, 0$. If we denote $\mathbf{c}^{(ic)}$ as the mean vector of Haar coefficients under the in-control state, then Haar coefficients under the in-control state are assumed as

$$\mathbf{c}_\ell = \mathbf{c}^{(ic)} + \mathbf{e}_\ell, \qquad \text{where} \qquad \mathbf{e}_\ell \sim N(\mathbf{0}, \Sigma_p). \tag{6.6}$$

for $\ell = -m+1, \cdots, -1, 0$. In other words, when there are no changes, the Haar coefficients $\mathbf{c}_\ell$ are i.i.d. multivariate normally distributed with in-control mean $\mathbf{c}^{(ic)}$ and diagonal covariance matrix $\Sigma_p = \text{diag}(\sigma_1^2, \ldots, \sigma_p^2)$.

It is well known that the sample mean based on the in-control Haar coefficients $\mathbf{c}_\ell$ is not always a good estimator for $\mathbf{c}^{(ic)}$ when the dimension p is large (James and Stein 1961). In the offline wavelet context, it is often assumed that the in-control p-dimensional mean vector of the Haar coefficients, $\mathbf{c}^{(ic)} = (c_1^{(ic)}, \cdots, c_p^{(ic)})$, has a sparsity structure and applying shrinkage techniques to filter out noise and obtain an accurate estimation (Donoho and Johnstone 1995, 1994). In this article, we follow the literature and apply hard shrinkage on the sample mean of in-control Haar coefficients. Specifically, let $\bar{\mathbf{c}}$ be the sample mean of m in-control Haar coefficient vectors, i.e.,

$$\bar{\mathbf{c}} = \frac{1}{m} \sum_{\ell=-m+1}^{0} \mathbf{c}_\ell.$$

Then the estimator of $\mathbf{c}^{(ic)} = (c_1^{(ic)}, \cdots, c_p^{(ic)})$ is

$$\hat{c}_i^{(ic)} = \begin{cases} \bar{c}_i^{(ic)}, & \text{if } |\bar{c}_i^{(ic)}| > \rho_1 \hat{\sigma}_i \\ 0, & \text{if } |\bar{c}_i^{(ic)}| \le \rho_1 \hat{\sigma}_i \end{cases} \tag{6.7}$$

where $\hat{\sigma}_i$ is the sample standard deviation of the i-th Haar coefficient, and ρ_1 is a crucial tuning parameter to control the sparsity of the mean vector $\mathbf{c}^{(ic)}$. The choice of ρ_1 will be discussed in detail later.

6.3.2 *Out-of-Control Estimation and Local Statistics*

In the profile monitoring context, the p-dimensional mean vector of the Haar coefficients is assumed to shift from the in-control value $\mathbf{c}^{(ic)}$ to an out-of-control value $\mathbf{c}^{(oc)} = (c_1^{(oc)}, \cdots, c_p^{(oc)})$. The difficulty is that one generally has limited knowledge about the out-of-control or fault samples in online profile monitoring, and thus one may not be able to accurately estimate the out-of-control mean $\mathbf{c}^{(oc)}$ even if we also put the sparsity constraints on $\mathbf{c}^{(oc)}$. For that reason, it makes more sense in online profile monitoring to assume that the difference vector $c^{(oc)} - c^{(ic)}$, instead of $c^{(oc)}$ itself, is sparse. To be more concrete, below we assume that only a few components of $\mathbf{c}^{(oc)} - \mathbf{c}^{(ic)}$ are non-zero, and $|c_i^{(oc)} - c_i^{(ic)}|/\sigma_i > \rho_2$ if the i-th component is affected, for some constant $\rho_2 > 0$, where σ_i is the standard deviation in (6.6).

Note that the change may affect those components with in-control value $c_i^{(ic)} = 0$, and thus one cannot simply monitor those non-zero components under the in-control state. Also, since we do not know which Haar coefficients will have mean shifts and do not know what the magnitudes of mean shift are, one intuitive idea is to adaptively and accurately estimate the post-change mean $\mathbf{c}^{(oc)}$ as we collect data for online monitoring under the sparsity assumption of $\mathbf{c}^{(oc)} - \mathbf{c}^{(ic)}$. Unfortunately, such an approach is generally computationally expensive and infeasible for online monitoring. Here we observe that the focus of profile monitoring is not necessarily on the accurate estimation of $\mathbf{c}^{(oc)}$, but on accurately raising a global alarm when there is a change. Hence, we propose a different approach that first locally monitors each component for a possible significant local mean shift, and then apply the order-thresholding technique to raise a global alarm under the sparse assumption that only a few local components are affected by the change.

When monitoring online profiles y_k's, at each time k, we first use (6.5) to derive the corresponding p-dimension Haar coefficients $\mathbf{c}_k$, and then standardize each of p components by

$$X_{i,k} = \frac{c_k(i) - \hat{c}_i^{(ic)}}{\hat{\sigma}_i}, \tag{6.8}$$

for $i = 1, \cdots, p$, where $\{\hat{c}_i^{(ic)}, \hat{\sigma}_i\}_{i=1,\cdots,p}$ are estimators of the in-control mean $\mathbf{c}^{(ic)}$ and standard deviation $\boldsymbol{\sigma}$ in (6.7) based on in-control samples.

By (6.7), rigorously speaking, the normalized coefficients $X_{i,k}$ might not be i.i.d. $N(0, 1)$ unless the tuning parameter $\rho_1 = 0$. In the context of online profile monitoring, the tuning parameter ρ_1 will often be small, and thus it is not bad to assume that the $X_{i,k}$'s satisfy the normality assumption from the practical viewpoint. Hence, the profile monitoring problem is reduced to the problem of monitoring the possible mean shifts of p-dimensional multivariate normal random vectors $\mathbf{X}_k = (X_{1,k}, \cdots, X_{p,k})$, where the means of some components may shift from 0 to some positive or negative value with magnitude of at least $\rho_2 > 0$.

If we know the exact post-change mean μ_i for the i-th component that is affected by the change, it is straightforward to develop an efficient local detection scheme, since one essentially faces the problem of testing the hypotheses in the change-point model where $X_{i,1}, \cdots, X_{i,\nu-1}$ are i.i.d. $f_0(x)$ = pdf of $N(0, 1)$ and $X_{i,\nu}, \cdots, X_{i,n}$ are i.i.d. $f_1(x)$ = pdf of $N(\mu_i, 1)$. At each time k, we repeatedly test the null hypothesis $H_0 : \nu = \infty$ (no change) against the alternative hypothesis $H_1 : \nu = 1, 2, \cdots$ (a change occurs at some finite time), see Lorden (1971). Thus the log generalized likelihood ratio statistic at time k becomes

$$W_{i,k}^* = \max_{1 \le \nu \le k} \frac{\prod_{\ell=1}^{\nu} f_0(X_{i,\ell}) \prod_{\ell=\nu+1}^{k} f_1(X_{i,\ell})}{\prod_{\ell=1}^{k} f_0(X_{i,\ell})}, \tag{6.9}$$

which can be recursively computed for normal distributions as

$$W_{i,k}^* = \max\left(W_{i,k-1}^* + \mu_i X_{i,k} - \frac{1}{2}(\mu_i)^2, 0 \right), \tag{6.10}$$

for $k = 1, \cdots$, with the initial value $W^*_{i,k=0} = 0$. In the literature, the statistic $W^*_{i,k}$ in (6.10) was first defined by Page (1954), and is called cumulative sum (CUSUM) statistics and enjoys theoretical optimality (Lorden 1971; Moustakides 1986).

In our context of profile monitoring, we do not know the value of the post-change mean μ_i except that $|\mu_i| \geq \rho_2$, thus we cannot use the CUSUM $W^*_{i,n}$ in (6.10) directly. One natural idea is to estimate μ_i from observed data, and then plug-in the estimated $\hat{\mu}_i$ into the CUSUM statistics in (6.10). For that purpose, at time k, denote by $\hat{\nu}_k$ the largest $\ell \leq k-1$ such that $W^*_{i,\ell} = 0$. Then the generalized likelihood ratio properties suggest that $\hat{\nu}_k$ is actually the maximum-likelihood estimate of the change-point ν at time k, and thus one would expect that the data between time $[\hat{\nu}_k, k]$ would likely come from the post-change distributions, which allows us to provide a reasonable estimate of the post-change mean $\hat{\mu}_i$ at time k. This idea was first rigorously investigated in Lorden and Pollak (2008) for detecting positive mean shifts of normal distributions, and here we aim to detect either positive or negative mean shifts. Specifically, at time k, for the i-th standardized Haar coefficients $X_{i,k}$'s, we define $\hat{\mu}^{(1)}_{i,k}$ and $\hat{\mu}^{(2)}_{i,k}$ as the estimates of the post-change mean of $X_{i,k}$ when restricted to the positive and negative values, respectively, under the assumption that $|\mu_i| \geq \rho_2$, with the explicit expressions as:

$$\hat{\mu}^{(1)}_{i,k} = \max\left(\rho_2, \frac{s + S^{(1)}_{i,k}}{t + T^{(1)}_{i,k}}\right) > 0, \qquad \hat{\mu}^{(2)}_{i,k} = \min\left(-\rho_2, \frac{-s + S^{(2)}_{i,k}}{t + T^{(2)}_{i,k}}\right) < 0, \quad (6.11)$$

and for $j = 1, 2$ and for any k, the sequences $(S^{(j)}_{i,k}, T^{(j)}_{i,k})$ are defined recursively

$$\begin{pmatrix} S^{(j)}_{i,k} \\ T^{(j)}_{i,k} \end{pmatrix} = \begin{cases} \begin{pmatrix} S^{(j)}_{i,k-1} + X_{i,k-1} \\ T^{(j)}_{i,k-1} + 1 \end{pmatrix} & \text{if } W^{(j)}_{i,k-1} > 0 \\ \begin{pmatrix} 0 \\ 0 \end{pmatrix} & \text{if } W^{(j)}_{i,k-1} = 0 \end{cases}. \quad (6.12)$$

Roughly speaking, for each estimate $\hat{\mu}^{(j)}_{i,k}$, if $\hat{\nu}^{(j)}_k$ is the candidate change-point, then $T^{(j)}_{i,k}$ denotes the time steps between $\hat{\nu}^{(j)}_k$ and k, whereas $S^{(j)}_{i,k}$ is the summation of all observations in the interval $[\hat{\nu}^{(j)}_k, k]$. The constants s and t in (6.11) are pre-specified, non-negative constants, and s/t can be thought of as a prior estimate of the post-change mean.

By plugging the adaptive estimations $\hat{\mu}^{(j)}_{i,k}$ of the post-change mean μ_i in the CUSUM statistics in (6.10), we can derive the local monitoring adaptive CUSUM statistics by

$$W_{i,k} = \max(W^{(1)}_{i,k}, W^{(2)}_{i,k}), \quad (6.13)$$

where $W_{i,k}^{(1)}$ and $W_{i,k}^{(2)}$ are the local detection statistics for detecting positive and negative mean shifts:

$$W_{i,k}^{(1)} = \max\left(W_{i,k-1}^{(1)} + \hat{\mu}_{i,k}^{(1)} X_{i,k} - \frac{1}{2}(\hat{\mu}_{i,k}^{(1)})^2, 0\right),$$
$$W_{i,k}^{(2)} = \max\left(W_{i,k-1}^{(2)} + \hat{\mu}_{i,k}^{(2)} X_{i,k} - \frac{1}{2}(\hat{\mu}_{i,k}^{(2)})^2, 0\right). \quad (6.14)$$

6.3.3 Global Online Monitoring Procedure

At time k, we have p local detection statistics $W_{i,k}$'s for $i = 1, \cdots, p$, one for monitoring each specific Haar coefficient locally. In general, the larger values of the $W_{i,k}$'s, the more likely the Haar coefficient is affected. Since we don't know which Haar coefficients are affected by the change, we follow Liu et al. (2017) to raise a global alarm based on the largest r values of the $W_{i,k}$'s. This allows us to filter out those non-affected Haar coefficients, and provides the list of candidate affected Haar coefficients.

Specifically, at each time k, we order p local detection statistics $W_{i,k}$'s for p Haar coefficients, say, $W_{(1),k} \geq W_{(2),k} \geq \ldots \geq W_{(p),k}$ are order statistics of $W_{i,k}$'s. Then our proposed profile monitoring scheme $N(b, r)$ is to raise an alarm at first time when the summation of the top r statistics $W_{(1),k}, \ldots, W_{(r),k}$ exceed some pre-defined threshold b, i.e.,

$$N(b, r) = \inf\{k : \sum_{i=1}^{r} W_{(i),k} \geq b\}, \quad (6.15)$$

where r is the tuning parameter that is determined by the sparsity of the post-change, b is the pre-specified constant to control false alarm.

In summary, our proposed profile monitoring scheme $N(b, r)$ in (6.15) is based on monitoring Haar coefficients. We use recursive CUSUM procedures, which can adaptively estimate unknown changes, to monitor each Haar coefficient individually, and use order-thresholding to address the sparse post-change scenario when only a few Haar coefficients are affected by the change.

It is important to emphasize that our proposed procedure $N(b, r)$ is robust in the sense that it can detect a wide range of possible changes on the profiles without requiring any knowledge on the potential failure pattern. Additionally, by the recursive formulas in (6.12) and (6.14), for a new coming profile, our proposed procedure only involves a computational complexity of order $O(p)$ to update local detection statistics for p Haar coefficients, as well as additional order of $O(p \log(p))$ to sort these p local detection statistics. Thus at each fixed time step, the overall computational complexity of our proposed methodology is of order $O(p \log(p))$. Meanwhile, for the GLR procedure in Chicken et al. (2009), the computational complexity is of order $O(t^2p^2)$ at time step t, which can be reduced to the order of $O(K^2p^2)$ if one only uses a fixed window size of K latest observations to make decisions instead of all t observations, where K often needs to be at least of order $O(\log(p))$ to be statistically efficient. Hence, as compared to the GLR procedure, our proposed procedure can be easily implemented recursively and thus is scalable when online monitoring high-dimension profile data over a long time period.

Algorithm 1 Implementation of our proposed procedure $N(b, r)$ in (6.15)

Initial parameters: ρ_1, ρ_2, s, t, and r.
In-control estimation: Using a set of m in-control p-dimensional profile samples $\mathbf{y}_1, .., \mathbf{y}_\mathbf{m}$, perform the following steps.
Step 1: get the Haar coefficients $\mathbf{c}_1, .., \mathbf{c}_\mathbf{m}$ by Eq. (6.5).
Step 2: get the estimation of standard deviation of the ith Haar coefficient $\hat{\sigma}_i$.
Step 3: get $\hat{\mathbf{c}}^{(ic)}$ by Eq. (6.7) with the threshold ρ_1.
Online monitoring:
initialize $k = 0$, and set all initial observations $X_i = 0$ and all $S_i^{(j)} = T_i^{(j)} = W_i^{(j)} = 0$, for $i = 1, \ldots, p$ and $j = 1, 2$.
While the scheme $N(b, r)$ has not raised an alarm
do 1. Update $(S_i^{(j)}, T_i^{(j)})$ via (6.12).
2. Compute the intermediate variables $\hat{\mu}_i^{(j)}$ from (6.11) which are the estimates of the post-change means.
3. Input new p-dimensional profile $\mathbf{y}$, using the estimated in-control mean $\hat{\mathbf{c}}^{(ic)}$ and standard deviation $\hat{\boldsymbol{\sigma}}$ to get the updated standardized p components $\{X_1, \ldots, X_p\}$ by (6.8).
4. For $i = 1, \ldots, p$, recompute the local monitoring statistics $W_i^{(j)}$ in (6.14) and W_i in (6.13).
5. Get the order statistics of $\{W_1, \ldots, W(p)\}$ denoted by $W_{(1)} \geq W_{(2)} \geq \ldots \geq W_{(p)}$
6. Compute the global monitoring statistics

$$G = \sum_{i=1}^{r} W_{(i)}$$

if $G \geq b$ **terminate:** Raising an alarm at time k and declaring that a change has occurred;
end the while loop

6.3.4 *Parameter Settings*

For our proposed monitoring procedure $N(b, r)$, there are two global parameters, r and b, and four local parameters, ρ_1, ρ_2, s, t. Optimal choices of these parameters will depend on the specific applications and contexts, and below we will discuss how to set the reasonable values of those parameters based on our extensive numerical experiences.

Let us first discuss the choices of two global parameters, r and b. The optimal choice of r that maximizes the detection power of the proposed procedure $N(b, r)$ is the number of truly changed Haar coefficients, which is often unknown. Based on our extensive simulations (Liu et al. 2017), when monitoring hundreds or thousands of Gaussian data streams simultaneously with an unknown number of affected local streams, the value $r \in [5, 10]$ often can reach a good balance on the detection power and the robustness to detect a wide range of possible shifts. Hence, in the case study and simulation study, we choose $r = 8$. As for the global parameter b, it controls when to stop the monitoring procedure and is often chosen to satisfy the pre-specified false alarm constraints. A standard approach in the literature is to choose b by repeatedly sampling in-control measurements either from in-control training data or from Monte Carlo in-control models, so that the monitoring procedure $N(b, r)$ will satisfy false alarm constraint.

Next, the local parameter ρ_1 in (6.7) essentially conducts a dimension reduction for in-control profiles. A good choice of the ρ_1 will depend on the characteristics of in-control profile data in specific applications, and in general the cutoff threshold ρ_1 should be chosen balance the bias-variance trade-off of estimation of the in-control mean profile. Much theoretical research has been done on how to choose ρ_1 for the single profile (Donoho and Johnstone 1994, 1998). These existing approaches focus more on the wavelet coefficient or mean profile estimation in the context of de-noising while the main objective in our context is to detect the changes of wavelet coefficients. Since we will conduct another dimension reduction at the layer of local detection statistics, it is often better to be conservative to choose a small constant $\rho_1 > 0$ value so as to keep more Haar coefficients from the in-control profiles. Also automatic or tuning-free approaches have been developed to choose the cutoff threshold such as ρ_1 adaptively in other contexts, see Zou and Qiu (2009) and Zou et al. (2015). However, such approaches are often computationally expensive, and it is unclear how to extend them to multiple profiles monitoring while keeping the proposed procedure to be scalable. In our simulation and case study, we found out that a simple choice of $\rho_1 = 0.15$ will yield significantly better results as compared with the existing methods in the literature. It remains an open problem to derive the optimal choice of ρ_1 under the general setting so that our proposed procedures are efficient in both computational and statistical viewpoints.

Finally, the local parameter ρ_2 represents the interested-smallest magnitude of mean shift of wavelet coefficients to be detected. In practice, it can be set based on the engineering domain knowledge to ensure production yield. In this paper, we set $\rho_2 = 0.25$. In addition, the local parameters, s and t in (6.11), are related to the prior distribution of the unknown post-change mean μ_i, so that the corresponding estimators of μ_i is a Bayes estimator and will be more robust than using the sample mean directly. In this paper, we follow Lorden and Pollak (2008) to choose $s = 1$ and $t = 4$.

6.4 Case Study

In this section, we apply our proposed wavelet-based methodology to a real progressive forming manufacturing process dataset in Lei et al. (2010) that includes 307 normal profiles and 5 different groups of fault profiles. Each group contains 69 samples which are collected under the faults due to missing part occurring in one of these five operations, respectively. Additionally, there are $p = 2^{11} = 2048$ measurement points in each profile.

The original research on Lei et al. (2010) focuses on the offline classification of normal and fault profile samples, while our research mainly emphasizes on the fast online detection. We will compare the performance of our proposed monitoring procedure with the other two common used procedures to illustrate the efficiency of our scheme. First one is the Hotelling's T^2 control chart based on selected wavelet coefficients (Zhou et al. 2006). The second one is based on the asymptotic maximum-likelihood test in Chicken et al. (2009). Specifically, we consider the following three procedures:

- Our proposed method $N(b, r)$ in (6.15);
- Hotelling's T^2 control chart based on the first r out of p wavelet coefficients:

$$T(b, r) = \inf\left\{j \geq 1 : w_j \geq b\right\}. \tag{6.16}$$

where

$$w_j = \sum_{i=1}^{r}(\frac{c_j(i) - \hat{c}_i^{(ic)}}{\hat{\sigma_i}^2})^2$$

- The method in Chicken et al. (2009), where the generalized likelihood ratio test was used on all p wavelet coefficients:

$$M^*(b) = \inf\left\{n \geq 1 : \max_{1 \leq i < n}\left\{[\frac{\sum_{j=i+1}^{n}\tilde{w}_j}{n-i} - \frac{\sum_{j=1}^{i}\tilde{w}_j}{i}] * \sum_{j=i+1}^{n}(\frac{w_j}{p} - 1)\right\} \geq b\right\}.$$

where

$$\tilde{w}_j = \frac{1}{\hat{\sigma}^2}\sum_{i=1}^{p}\{\max(0, |c_j(i) - \hat{c}_i^{(ic)}| - \lambda)\}^2$$

$$\lambda = \sqrt{2\frac{\log p}{p}}\hat{\sigma}.$$

Table 6.1 A comparison of the detection delays of three methods with in-control average run length equal to 200 based on 500 repetitions in Monte Carlo simulations

Method	Fault 1	Fault 2	Fault 3	Fault 4	Fault 5
N(b=73,r=8)	1(0)	1(0)	1(0)	1.51(0.03)	1.01(0.01)
T(b=23.33,r=8)	1(0)	1(0)	1(0)	17.71(0.78)	1(0)
$M^*(b = 600)$	1(0)	1(0)	1(0)	4.47(0.13)	1.22(0.02)

The standard errors of the detection delays are reported in the bracket

In order to have a fair comparison, r is chosen as 8 for our proposed method $N(b, r)$ in (6.15) and the Hotelling's T^2 control chart $T(b, r)$ in (6.16).

To evaluate the detection efficiency of those methods, we first find the appropriate values of the global threshold b such that the average run length of each scheme is 200 when the samples are collected by sampling from the 307 normal profiles with replacement. Then, using the obtained global threshold value b, we simulate the detection delay when the samples are sequentially collected by sampling from the 69 fault profiles. All Monte Carlo simulations are based on 500 repetitions. The results of detection delay and standard error are summarized in Table 6.1.

From Table 6.1, we can see all of these three methods can detect the change of Fault 1, 2, 3, and 5 very fast (on average, just need one sample to detect such change). It is necessary to emphasize that although as shown in Fig. 6.2, the difference between normal profile and the Fault 4 profile is very subtle, and our proposed method can detect the Fault 4 change much faster than the other two methods.

6.5 Simulation Study

In this section, we present the simulation study results to illustrate the efficiency of our proposed procedure. We follow the nonlinear profile monitoring literature to consider the in-control mean profile as the Mallet's piecewise smooth function in Mallat (1999) , see Fig. 6.4. This testbed curve is a complicated function with several non-differentiable points and difficult patterns, including several transient jumps, therefore cannot easily be modeled by parametric models or other nonparametric models and has been popularly used in much research to evaluate the performance of nonlinear profile monitoring procedures, see Jeong et al. (2006), Chicken et al. (2009), and Lee et al. (2012).

The out-of-control mean profile follows the same setup in the previous literature Lee et al. (2012) and assumes a local mean shift on some intervals. Specifically, the out-of-control mean profiles are designed as $f_1(x) = f_0(x)+\mu I_\delta(x)$ where the shift magnitude $\mu \in \{0.25, 0.5, 1\}$ and three different changed intervals: (1) $\delta = [0, 1]$,

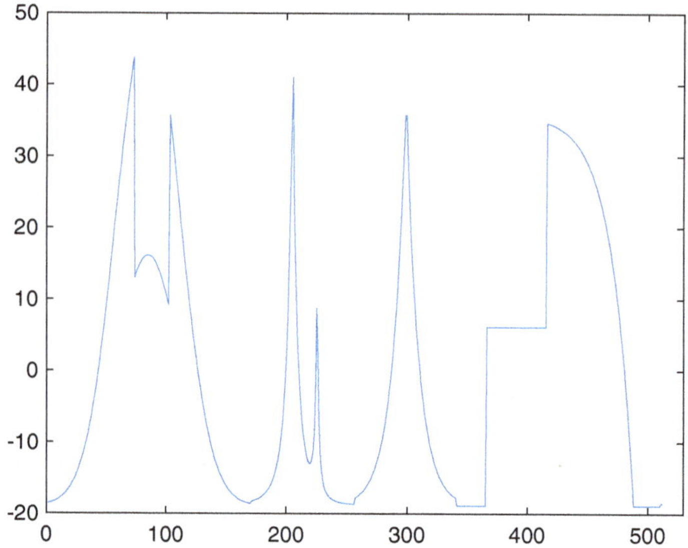

Fig. 6.4 Mallat's piecewise smooth function

which is referred as Global shift; (2) $\delta = [\frac{73}{512}, \frac{76}{512}] \cup [\frac{288}{512}, \frac{296}{512}]$, which is referred as Local shift I, and (3) $\delta = [\frac{3}{512}, \frac{15}{512}] \cup [\frac{344}{512}, \frac{347}{512}]$, which is referred as Local shift II.

Based on the mean profiles, we generate in-control and out-of-control sample profiles, which consist of a realization of $p = 512$ pairs $(x_i, y(x_i))$ with $x_1, \ldots, x_p$ equal spaced on $[0, 1]$ and $y(x_i) = f_0(x_i) + \epsilon(x_i)$ as in-control sample profile and $y(x_i) = f_1(x_i) + \epsilon(x_i)$ as out-of-control sample profile, where $\epsilon(x_i)$ is i.i.d standard normally distributed $N(0, 1)$.

We will compare the performance of our proposed method $N(b, r = 8)$ in (6.15) with the same two methods in the previous section: the method $M^*(b)$ in (6.16) and the method $T(b, r = 8)$ in (6.16). In this simulation study, we still set $\rho_1 = 0.15$, $\rho_2 = 0.25$, $s = 1$, $t = 4$ for our proposed scheme.

Specifically, based on 1000 Monte Carlo simulations, we keep the in-control average run length of those schemes as 200 and compare the detection delay under the Global shift, Local shift I, and Local shift II with different magnitudes of mean shift. The results are summarized in Table 6.2.

From Table 6.2 we can see that (1) our proposed method $N(b, r)$ yields the smallest detection delay for detecting local shifts compared with the other two methods $M^*(b)$ and $T(b, r)$; (2) a competitive results for detecting the global shifts under different magnitudes of shifts. This implies our proposed wavelet-based monitoring procedure is more robust to the unknown changes.

Table 6.2 A comparison of the detection delays of 3 methods with in-control average run length equal to 200 based on 1000 repetitions in Monte Carlo simulations

Method	μ	Global shift	Local shift I	Local shift II
	0.25	2.59(0.01)	92.38(0.52)	67.41(0.42)
$N(b = 51, r = 8)$	0.5	1(0.01)	31.63(0.18)	22.17(0.14)
	1	1(0.00)	9.46(0.05)	6.53(0.04)
T(b=21.7, r=8)	0.25	1.03(0.01)	151.82(4.68)	253.57(7.15)
	0.5	1.00(0)	144.38(4.39)	100.59(2.99)
	1	1.00(0)	79.08(2.58)	24.81(0.74)
$M^*(b = 10.1)$	0.25	8.26(0.18)	157.40(4.81)	151.55(4.73)
	0.5	1.29(0.02)	125.24(4.09)	106.31(3.58)
	1	1.00(0)	35.97(0.87)	24.55(0.55)

The standard errors of the detection delays are reported in the bracket

6.6 Conclusions

In this article, we develop a new scalable scheme for monitoring nonlinear profiles with unknown post-change distribution. This article makes three methodological contributions. First, we propose to use all wavelet coefficients to monitor the process, while the prior literature of nonlinear profile monitoring is dominated by analyzing and using just significant coefficients. Second, we propose to use two shrinkage techniques to filter out the noise introduced by using all wavelet coefficients. One is using hard shrinkage to estimate the in-control mean coefficients. The other one is to build monitoring procedure only focusing on the information of a few coefficients, which have higher likelihood to be changed. Third, we propose to utilize a recent developed adaptive-CUSUM procedure in Liu et al. (2017) to efficiently monitor the standardized wavelet coefficients without knowing the information about the post-change.

There is plenty of room for improving our proposed scheme for monitoring nonlinear profiles, calling for further research. First, this article mainly focuses on the detection of mean shift of the normal distributed profile. Although there are many applications of our proposed scheme, it is also necessary to work on the detection procedures for more generally distributed profiles. Second, this article makes an independence assumption on the noise distribution in (6.1). It will be useful to develop a more robust method that can handle different correlation structure of the profile data.

Acknowledgements This research is partially supported by NSF grant CMMI-1362876.

References

Chang, S. I., & Yadama, S. (2010). Statistical process control for monitoring non-linear profiles using wavelet filtering and b-spline approximation. *International Journal of Production Research, 48*(4), 1049–1068.

Chang, T. C., & Gan, F. F. (2006). Monitoring linearity of measurement gauges. *Journal of Statistical Computation and Simulation, 76*(10), 889–911.

Chen, S., & Nembhard, H. B. (2011). A high-dimensional control chart for profile monitoring. *Quality and Reliability Engineering International, 27*(4), 451–464.

Chicken, E., Pignatiello, J. J., Jr., & Simpson, J. R. (2009). Statistical process monitoring of nonlinear profiles using wavelets. *Journal of Quality Technology, 41*(2), 198.

Daubechies, I. (1992). Ten lectures on wavelets. Philadelphia: SIAM.

Donoho, D. L., & Johnstone, I. M. (1994). Ideal spatial adaptation by wavelet shrinkage. *Biometrika, 81*(3), 425–455.

Donoho, D. L., & Johnstone, I. M. (1995). Adapting to unknown smoothness via wavelet shrinkage. *Journal of the American Statistical Association, 90*(432), 1200–1224.

Donoho, D. L., & Johnstone, I. M. (1998). Minimax estimation via wavelet shrinkage. *The Annals of Statistics, 26*(3), 879–921.

Fan, J. (1996). Test of significance based on wavelet thresholding and Neyman's truncation. *Journal of the American Statistical Association, 91*(434), 674–688.

Fan, J., & Lin, S. K. (1998). Test of significance when data are curves. *Journal of the American Statistical Association, 93*(443), 1007–1021.

Gardner, M. M., Lu, J. C., Gyurcsik, R. S., Wortman, J. J., Hornung, B. E., Heinisch, H. H., et al. (1997). Equipment fault detection using spatial signatures. *IEEE Transactions on Components, Packaging, and Manufacturing Technology: Part C, 20*(4), 295–304.

Hall, P., Poskitt, D. S., & Presnell, B. (2001). A functional data-analytic approach to signal discrimination. *Technometrics, 43*(1), 1–9.

James, W., & Stein, C. (1961). Estimation with quadratic loss. In *Proceedings of the fourth Berkeley symposium on mathematical statistics and probability* (vol. 1, pp. 361–379).

Jeong, M. K., Lu, J. C., & Wang, N. (2006). Wavelet-based SPC procedure for complicated functional data. *International Journal of Production Research, 44*(4), 729–744.

Jin, J., & Shi, J. (1999). Feature-preserving data compression of stamping tonnage information using wavelets. *Technometrics, 41*(4), 327–339.

Jin, J., & Shi, J. (2001). Automatic feature extraction of waveform signals for in-process diagnostic performance improvement. *Journal of Intelligent Manufacturing, 12*(3), 257–268.

Kang, L., & Albin, S. L. (2000). On-line monitoring when the process yields a linear profile. *Journal of Quality Technology, 32*(4), 418.

Kazemzadeh, R. B., Noorossana, R., & Amiri, A. (2008). Phase I monitoring of polynomial profiles. *Communications in Statistics–Theory and Methods, 37*(10), 1671–1686.

Kim, M. H., & Akritas, M. G. (2010). Order thresholding. *The Annals of Statistics 38*(4), 2314–2350.

Lee, J., Hur, Y., Kim, S. H., & Wilson, J. R. (2012). Monitoring nonlinear profiles using a wavelet-based distribution-free CUSUM chart. *International Journal of Production Research, 50*(22), 6574–6594.

Lei, Y., Zhang, Z., & Jin, J. (2010). Automatic tonnage monitoring for missing part detection in multi-operation forging processes. *Journal of Manufacturing Science and Engineering, 132*(5), 051010.

Liu, K., Zhang, R., & Mei, Y. (2017). Scalable sum-shrinkage schemes for distributed monitoring large-scale data streams. *Statistica Sinica (Accepted).*

Lorden, G. (1971). Procedures for reacting to a change in distribution. *The Annals of Mathematical Statistics, 42*(6), 1897–1908.

Lorden, G., & Pollak, M. (2008). Sequential change-point detection procedures that are nearly optimal and computationally simple. *Sequential Analysis, 27*(4), 476–512.

Mallat, S. (1999). *A wavelet tour of signal processing*. London: Academic Press.

Mallat, S. G. (1989). Multifrequency channel decompositions of images and wavelet models. *IEEE Transactions on Acoustics, Speech, and Signal Processing 37*(12), 2091–2110.

Moustakides, G. V. (1986). Optimal stopping times for detecting changes in distributions. *The Annals of Statistics, 14*(4), 1379–1387.

Neyman, J. (1937). Smooth test for goodness of fit. *Scandinavian Actuarial Journal, 20*(3–4), 149–199.

Page, E. S. (1954). Continuous inspection schemes. *Biometrika, 41*(1–2), 100–115.

Paynabar, K., Zou, C., & Qiu, P. (2016). A change-point approach for phase-i analysis in multivariate profile monitoring and diagnosis. *Technometrics, 58*(2), 191–204.

Qiu, P., Zou, C., & Wang, Z. (2010). Nonparametric profile monitoring by mixed effects modeling. *Technometrics, 52(3)*, 265–277.

Zhou, C., Liu, K., Zhang, X., Zhang, W., & Shi, J. (2016). An automatic process monitoring method using recurrence plot in progressive stamping processes. *IEEE Transactions on Automation Science and Engineering, 13*(2), 1102–1111.

Zhou, S., Sun, B., & Shi, J. (2006). An SPC monitoring system for cycle-based waveform signals using Haar transform. *IEEE Transactions on Automation Science and Engineering, 3*(1), 60–72.

Zou, C., & Qiu, P. (2009). Multivariate statistical process control using lasso. *Journal of the American Statistical Association, 104*(488), 1586–1596.

Zou, C., Qiu, P., & Hawkins, D. (2009). Nonparametric control chart for monitoring profiles using change point formulation and adaptive smoothing. *Statistica Sinica, 19*(3), 1337–1357.

Zou, C., Tsung, F., & Wang, Z. (2007). Monitoring general linear profiles using multivariate exponentially weighted moving average schemes. *Technometrics, 49*(4), 395–408.

Zou, C., Wang, Z., Zi, X., & Jiang, W. (2015). An efficient online monitoring method for high-dimensional data streams. *Technometrics, 57*(3), 374–387.

Zou, C., Zhou, C., Wang, Z., & Tsung, F. (2007). A self-starting control chart for linear profiles. *Journal of Quality Technology, 39*(4), 364–375.

Chapter 7
Estimating the Confidence Interval of Evolutionary Stochastic Process Mean from Wavelet Based Bootstrapping

Aline Edlaine de Medeiros and Eniuce Menezes de Souza

7.1 Introduction

Time series data are naturally found in a range of fields such as Agriculture, Geophysics, Meteorology, Health, Economy and Social Sciences, among several others (Chatfield 2016; Wei 2006). Given a parametric space T and a probability space (Ω, A, P), a stochastic process is a family $Z = \{Z(t), t \in T\}$, such that, for each $t \in T$, $Z(t)$ is a random variable (Morettin and Toloi 2006). A time series is considered as the finite realization of a stochastic process. In other words, an observed time series is a trajectory of a stochastic process.

Indeed, $Z(t)$ is a two variable function $Z(t, w)$ wherein $t \in T$, and $w \in \Omega$. Considering $f_Z(z)$ as the probability density function of $Z(t, w)$, Fig. 7.1, adapted from Morettin and Toloi (2006), represents a stochastic process as aforementioned.

In many situations, accessing more than one observation of a phenomenon for each instant of time is impossible. In general, the function $Z(t, w)$ is assumed to follow a Gaussian distribution for each instant, and the observed time series represents the mean μ_t of the stochastic process for each $t \in T$. But, this assumption is not always true and $f_Z(z)$ can be different at each instant of time. It would be very important to estimate the uncertainty associated with μ_t, from its confidence interval.

The technique called bootstrap (Efron and Gong 1983), which is an appropriate methodology for solving a variety of inferential problems, could be a good alternative to estimate the uncertainty for μ_t. However, bootstrap is more designed

A. E. de Medeiros (✉)
Graduate Program in Biostatistics, State University of Maringa, Maringa, Brazil

E. M. de Souza
Department of Statistics, State University of Maringa, Maringa, Brazil
e-mail: emsouza@ucm.br

Y. Zhao, D.-G. Chen (eds.), *New Frontiers of Biostatistics and Bioinformatics*, ICSA Book Series in Statistics, https://doi.org/10.1007/978-3-319-99389-8_7

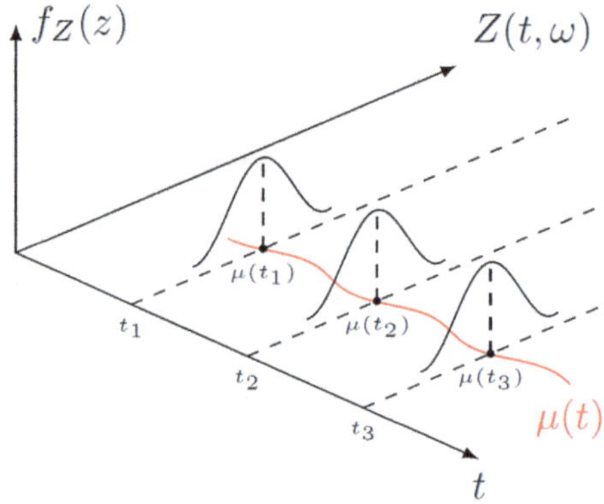

Fig. 7.1 A stochastic process represented as a family of random variables

for uncorrelated data, and not for those exhibiting short or long-range dependence as time series. Fortunately, in the last years, several methods were developed to deal with resampling time series, some of them based on wavelets, especially using DWT (Angelini et al. 2005; Golia 2002; Percival et al. 2000; Yi et al. 2007).

To estimate the confidence interval for μ_t, we aim to evaluate and implement some methods from the literature but with some modifications, and propose others involving NDWT:

M1: Naive bootstrapping based on NDWT;
M2: DWT two-step wavestrapping (TSWDWT);
M3: NDWT two-step wavestrapping (TSWNDWT);
M4: Reinflation of the bootstrap resamples of TSWDWT.
M5: Reinflation of the bootstrap resamples of TSWNDWT.

The methods M1, M2, M3, M4, and M5 are going to be applied and compared for estimation of the uncertainty for bronchiolitis hospitalization rate in Paraná State from 2000 to 2014. Bias, standard errors, and coefficients of variation can evaluate the ensembles of resampled time series.

This work is organized as follows. Section 7.2 presents a brief review of the techniques usually used to resample time series. In Sect. 7.3, we describe the methods we are using to estimate the uncertainty associated with the bronchiolitis hospitalization rate. In Sects. 7.4 and 7.5, the main results and conclusions of our study are presented, respectively.

7.2 Resampling Time Series

One of the most important characteristics of a time series is the dependence on nearby observations. Because of this correlation structure, maintaining the data order is of great importance. So, resampling time series requires appropriate techniques that consider the dependence and the order of the observations. One of the usual approaches is the Stationary Bootstrap (SB) (Politis and Romano 1994).

Considering Y_t as a strictly stationary and weakly dependent time series, the SB is a special case of blocks resampling, which consists in defining two sequences of random variables $L_1, L_2, \ldots$ and $I_1, I_2, \ldots$, both independent of each other and independent of Y_t, and such that $L_1, L_2, \ldots$ follow a geometric distribution with parameter p and $I_1, I_2, \ldots$ follow a uniform distribution on $\{1, 2, \ldots, n\}$. Then, the random blocks B_{I_i,L_i}, with random blocks length L_{i-1}, are given by

$$B_{I_i,L_i} = (Y_{I_i}, Y_{I_i+1}, \ldots, Y_{I_i+L_{i-1}}). \tag{7.1}$$

However, SB is not applicable to those time series exhibiting non-stationary and long-range dependence.

In the last years, the wavelet analysis has been standing out as a tool for resampling time series (Angelini et al. 2005; Breakspear et al. 2003; Golia 2002; Percival et al. 2000). Basically, considering we have a multiresolution analysis (MRA) (Mallat 1989), a time series $Y = (y_0, y_1, \cdots, y_{n-1})$ can be represented as a function f in terms of the scaling function ϕ and wavelet function ψ as

$$f(t) = \sum_{k=0}^{n-1} c_{J_0,k}\phi_{J_0,k}(t) + \sum_{j=J_0}^{J-1}\sum_{k=0}^{n-1} d_{j,k}\psi_{j,k} \tag{7.2}$$

where $J - 1 < \log_2 n \leq J$, $j = J_0, \cdots, J-1$ representing a multiresolution level, and $k = 0, \cdots, n-1$. The coefficients $c_{J_0,k}$ and $d_{j,k}$ are called the smooth (scaling) and detail (wavelet) coefficients, respectively (Kang and Vidakovic 2017).

When we take $\phi_{J_0,k}(t) = 2^{J_0/2}\phi(2^{J_0}t - k)$ and $\psi_{j,k}(t) = 2^{j/2}\psi(2^j t - k)$, the coefficients $c_{J_0,k}$ and $d_{j,k}$ comprise the DWT of the time series Y. On the other hand, taking $\phi_{J_0,k}(t) = 2^{J_0/2}\phi(2^{J_0}(t-k))$ and $\psi_{j,k}(t) = 2^{j/2}\psi(2^j(t-k))$, the detail and smooth coefficients represent the NDWT of the time series Y.

The DWT wavelet coefficients have less autocorrelation than the observed time series, and this allows applying bootstrap (wavestrap), even for non-stationary time series (Golia 2002; Tang et al. 2008; Yi et al. 2007). However, in conditions where translation or shift-invariance (Nason and Silverman 1995) is important, as for time series, the NDWT is a good alternative.

One difficulty in applying bootstrap on DWT is the number of wavelet coefficients which becomes smaller at each resolution level. NDWT has the same number of wavelet coefficients in each resolution level, overcoming this DWT limitation. NDWT is also more flexible with respect to the time series length, being appropriate for all those which are a multiple of two. Furthermore, NDWT has an easy implementation with more than one algorithm, including the pyramidal algorithm (Mallat 1989).

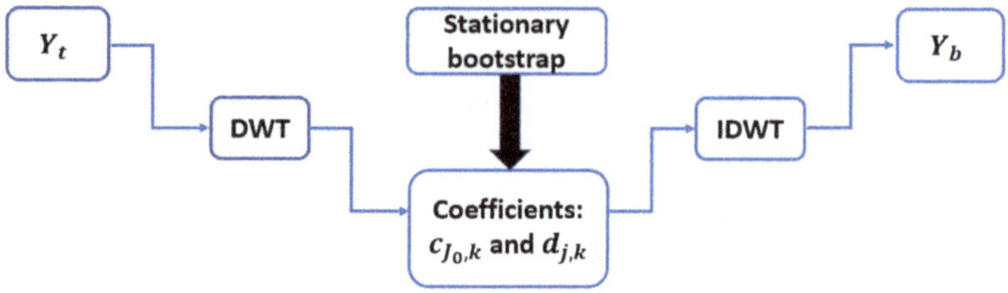

Fig. 7.2 Wavelet based stationary bootstrap

7.2.1 Bootstrap Based on Wavelets

Golia (2002) applied the stationary bootstrap to the wavelet coefficients of time series exhibiting long memory (Golia 2002). This application was possible because the wavelet coefficients are wide-sense stationary and weakly correlated in each scale (Wornell and Oppenheim 1996). In her work, she used the Daubechies wavelet with four vanishing moments and coarsest level of details equals to 4 in DWT. The results were good, however, the author comments the need of evaluating this approach for other long memory processes. The procedure of this wavelet based stationary bootstrap is described in Fig. 7.2.

Considering the SB and its use combined with wavelets, Yi et al. (2007) developed a DWT-based method called Two-Step Wavestrapping (TSW), to simulate non-stationary acceleration data in the mobile computing context (Yi et al. 2007). In this context, they intended to simulate the acceleration data collected from a group composed by one hundred twenty six undergraduate students. In this sort of data, each student provides one time series for each one of the three evaluated axes forming a group of time series. Each group of time series was divided into subgroups statistically characterized by Hurst exponents, and then TSW procedure is applied by subgroups.

Describing TSW for only one time series, the first part of the TSW consists in performing the SB in one-step DWT, which is called Stationary Parallel Bootstrapping. In other words,

1. Given a time series Y of power-of-two length, apply the DWT to generate the coarsest level of detail (J_0) and scale coefficients;
2. Resample these scaling and wavelet coefficients using the Stationary Bootstrap (Politis and Romano 1994);
3. Apply the inverse discrete wavelet transform (IDWT) to the resampled wavelet coefficients to generate a surrogate time series Y_b, wherein b indicates the performed bootstrap.

The second step consists in adjusting the trend and energy. For trend adjustment, both the time series Y and its surrogate Y_b are decomposed using the DWT. Then, the scaling coefficients c_b obtained from Y_b are surrogated by the scaling

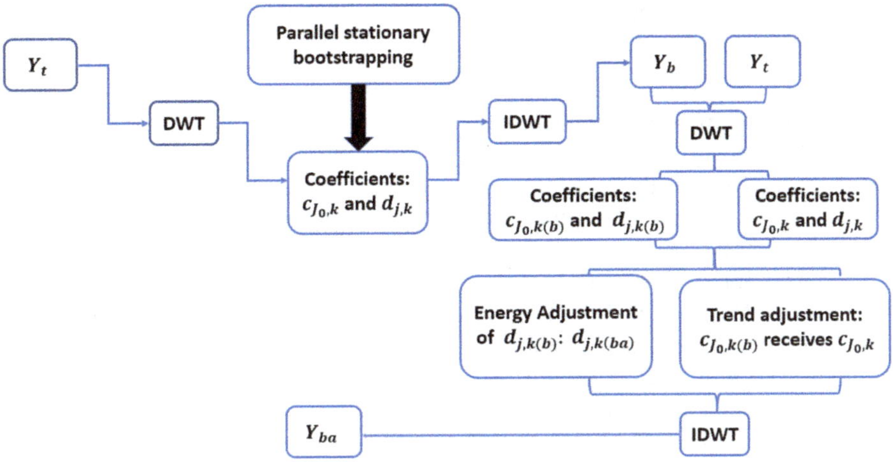

Fig. 7.3 Two step wavestrap (TSW) algorithm

coefficients c generated from Y. For the energy adjustment, the following steps can be performed:

1. Generate the average energy for each decomposition level of Y and Y_b, given by

$$\bar{e}_j = \sum_{k=0}^{2^j-1} \frac{{d_{j,k}}^2}{2^j}, \quad j = J_0, \cdots, J-1, \tag{7.3}$$

   where $d_{j,k}$ is the kth wavelet coefficient in the jth decomposition level.
2. Adjust the average energy for each decomposition level of Y_b to the average energy of the levels of Y, doing

$$d_{ba\,j,k} = d_{b\,j,k}\sqrt{\frac{\bar{e}_j}{\bar{e}_{bj}}}, \tag{7.4}$$

   where d_{ba} represents the adjustment done in each decomposition level of Y_b;

Figure 7.3 summarizes the TSW algorithm. An important contribution of this methodology is the idea of an energy adjustment in the levels to preserve the inherent variability of the original data, even after the resampling. Furthermore, each realization of this procedure provides a surrogate time series with the same feature of the original time series. Another important point is that the vertical correlation of wavelet coefficients among scale levels was taken into account, since the scaling and wavelet coefficients were resampled together.

In the next section, we present the proposed bootstrap methods, that are based on NDWT, SB, and TSW of the wavelet coefficients.

7.3 Proposed Methods

Using the statistical language R (R Core Team 2016), we implemented five methods to generate the proposed confidence interval for μ_t.

The first bootstrap technique consists in performing a decomposition of the time series using NDWT, applying the naive bootstrap to detail coefficients and then generating a surrogate time series using INDWT. Figure 7.4 presents the NDWT naive algorithm.

The next approaches were implemented following the same steps as TSW. The first one, called TSWNDWT, follows the same steps of TSW but replacing the DWT by NDWT. In the second step, we work only with NDWT coefficients that comprise the first level of details. As in TSW we also developed the trend and energy adjustment as described in Sect. 7.2.1. Figure 7.5 describes the TSWNDWT algorithm.

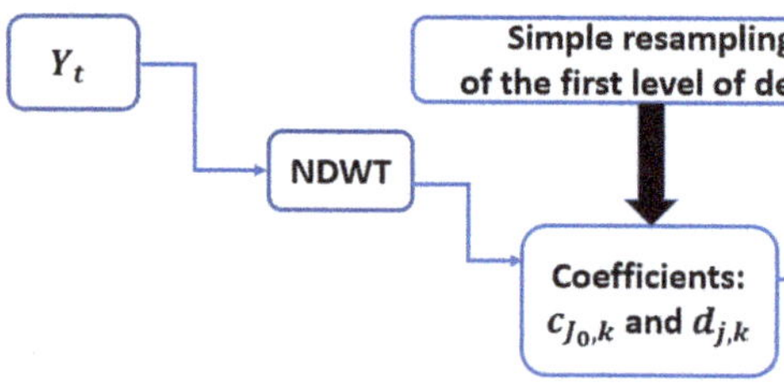

Fig. 7.4 Naive bootstrap based on NDWT

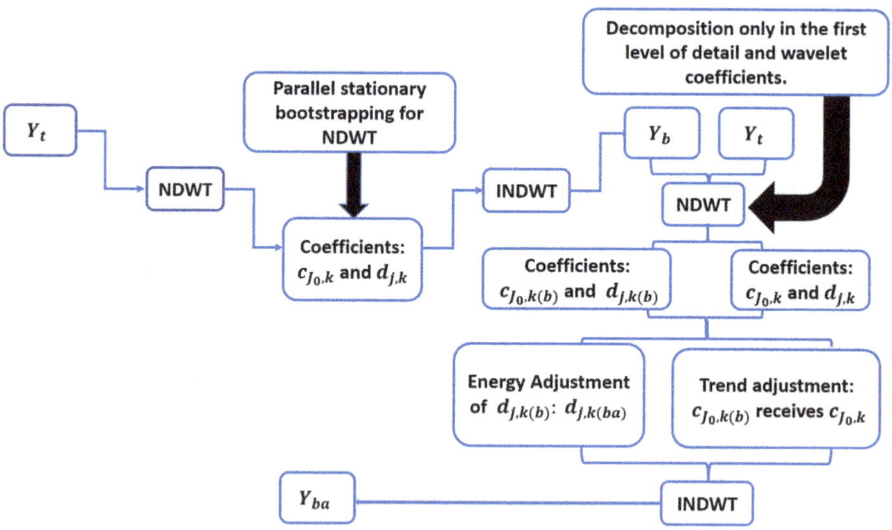

Fig. 7.5 TSWNDWT algorithm

The method called TSWDWT follows the same steps of TSWNDWT, but the decomposition and reconstruction of the time series is performed using DWT. The latest approaches consist in reinflating the surrogate time series obtained from TSWDWT and TSWNDWT. In the literature, reinflation means multiplying the correlation factor correction $\sqrt{1.1}$ to the surrogate time series (Tang et al. 2008).

To illustrate the confidence interval of the evolutionary stochastic process mean we used the month rate of bronchiolitis hospitalizations time series from the Paraná State—BR, in the period from 2000 to 2014. This time series was collected from DATASUS database and contains 180 observations.

The resampling methods based on DWT require data of power-of-two size. So, we extend the time series by reflection to 256 observations.

For each one of the proposed methods, we fixed the orthonormal Daubechies' wavelet (Daubechies 1992), with two vanishing moments ($d4$). This family of wavelets has been frequently used in similar works (Golia 2002; Tang et al. 2008). Furthermore, to obtain the mean of the stochastic process μ_t, the level mean of the time series, standard errors, and bias we resampled the time series 5000 times for each one of the proposed methods.

7.4 Results

Figure 7.6 presents the time series of the rate of monthly hospitalizations for bronchiolitis (Y), and the mean of the group of surrogate time series for Y from each presented bootstrap method. From Fig. 7.6 all the bootstrap means seems to be

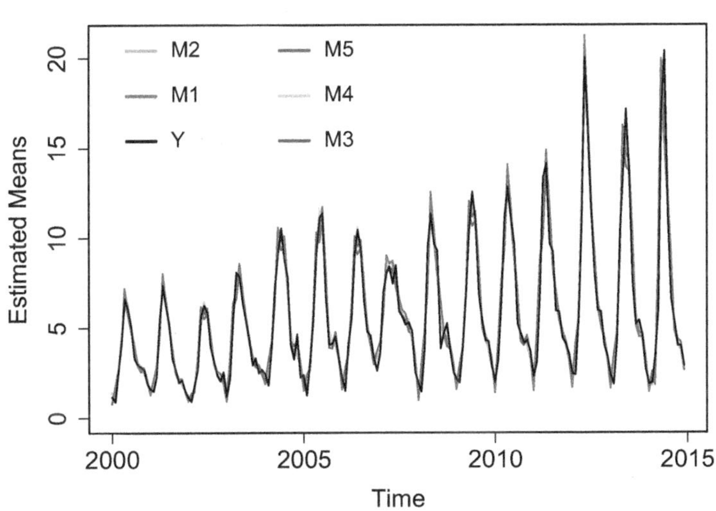

Fig. 7.6 Averages of the surrogate time series: Y—rate of bronchiolitis time series, M1—naive bootstrap based on NDWT, M2—TSWNDWT, M3—TSWDWT, M4—reinflated TSWNDWT and M5—reinflated TSWDWT

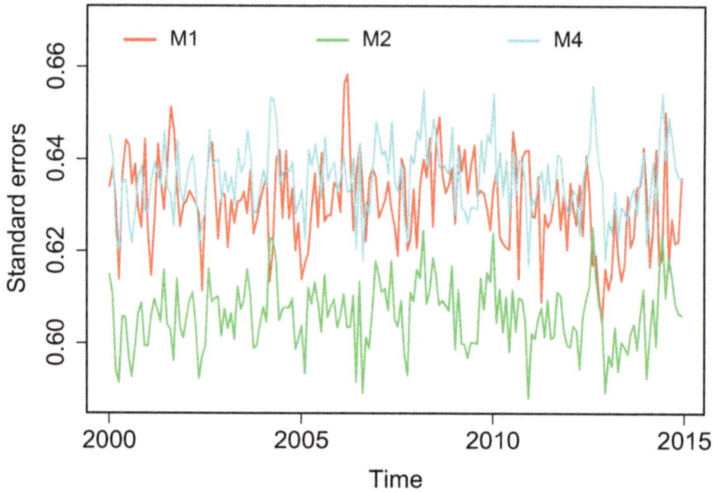

Fig. 7.7 Standard errors of surrogate time series: M1—naive bootstrap based on NDWT, M2—TSWNDWT, and M4— reinflated TSWNDWT

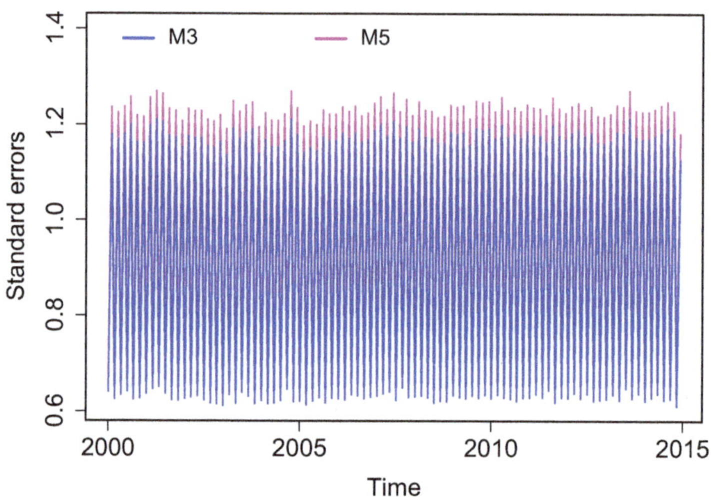

Fig. 7.8 Standard errors of surrogate time series: M3—TSWDWT, and M5—reinflated TSWDWT

similar to the observed time series. We can also generate the standard errors of the surrogate time series, as presented in Figs. 7.7 and 7.8.

We can see that the naive bootstrap based on wavelet, TSWNDWT and reinflated TSWNDWT methods presented low variability, whereas TSWDWT, and reinflated TSWDWT standard errors present a high level of oscillation. Possibly this behavior is related to the number of coefficients in each decomposition level. While the

Table 7.1 Average of standard errors (SE), coefficient of variation (CV) and bias of the surrogate time series

Classes	NDWT bootstrap	TSNDWT	TSDWT	Reinflated TSNDWT	Reinflated TSDWT
SE	0.63	0.61	0.90	0.64	0.95
CV	11.23	10.79	16.05	10.79	16.05
Bias	2.74×10^{-5}	-7.71×10^{-6}	6.15×10^{-5}	-2.74×10^{-1}	-2.74×10^{-1}

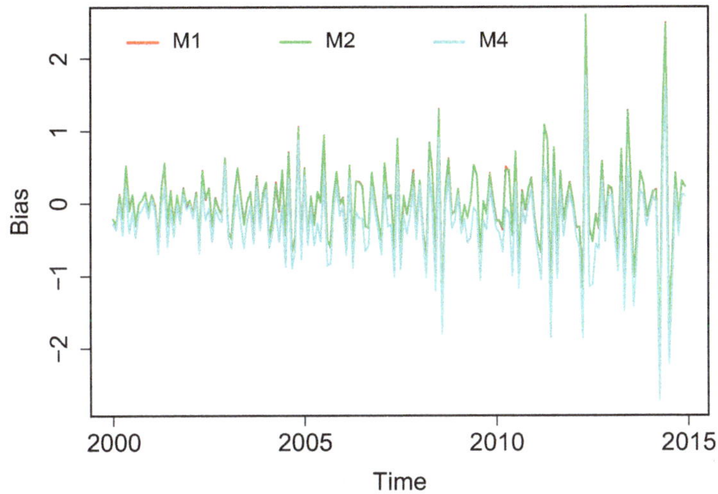

Fig. 7.9 Bias: M1—naive bootstrap based on NDWT, M2—TSWNDWT, and M4—reinflated TSWNDWT

number of coefficients in each level of NDWT remains the same as the observed time series, in DWT, the number of coefficients decreases by half in each level. In general, the TSWNDWT presented the best standard errors and coefficient of variation.

The averages of the standard errors, coefficient of variation, and bias are represented in Table 7.1. The results corroborate with the graphical analyses, pointing the TSWNDWT as the method with the smallest variability.

The results in Table 7.1 also corroborate with Fig. 7.8 indicating the largest variability for the methods that use DWT.

We also can see that the naive bootstrap based on wavelet, TSWNDWT, and reinflated TSWNDWT average bias is smaller than those for TSWDWT and reinflated TSWDWT methods. The TSWNDWT presented the best average of bias, which is about −0.000008.

Figures 7.9 and 7.10 present the bias of the mean of the surrogate time series for each one of the evaluated methods. As in the standard errors analyse, the naive bootstrap based on wavelet, TSWNDWT, and reinflated TSWNDWT methods

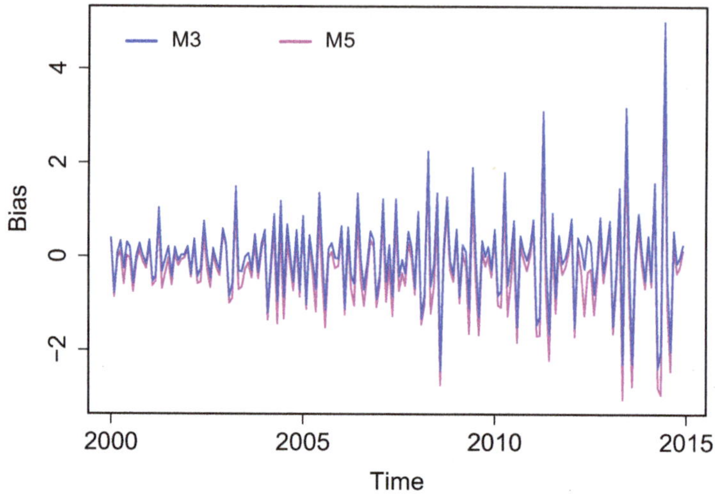

Fig. 7.10 Bias: M3—TSWDWT, and M5—reinflated TSWDWT

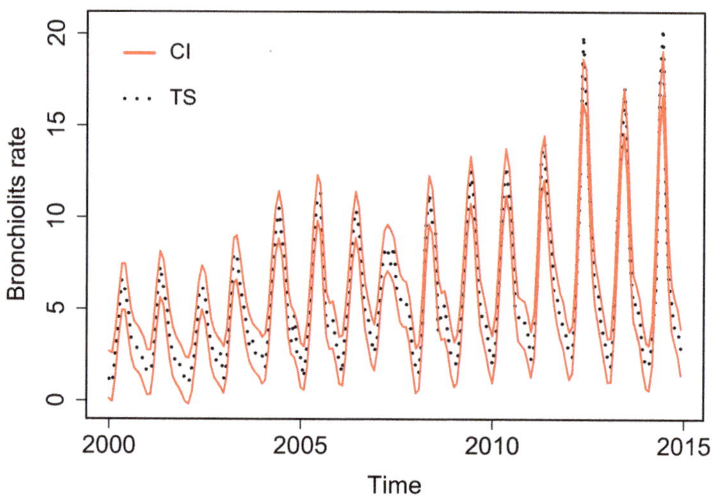

Fig. 7.11 Confidence interval (CI) obtained from naive bootstrap for the rate of bronchiolitis hospitalizations time series (TS)

presented best results. On the other hand, TSWDWT and reinflated TSWDWT bias reached largest values.

In Figs. 7.11, 7.12, 7.13, 7.14, and 7.15 the confidence interval for the rate bronchiolitis hospitalizations time series obtained from each discussed method is presented. In all graphs, the time series Y is represented as a black dotted line and the confidence interval as a red line.

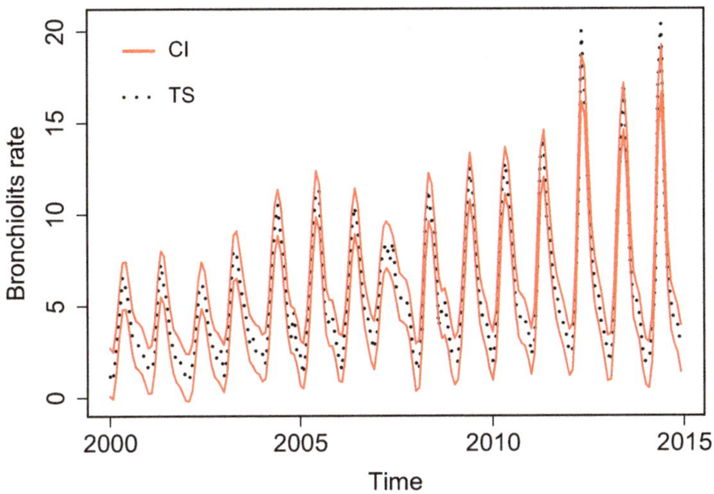

Fig. 7.12 Confidence interval obtained from TSWNDWT

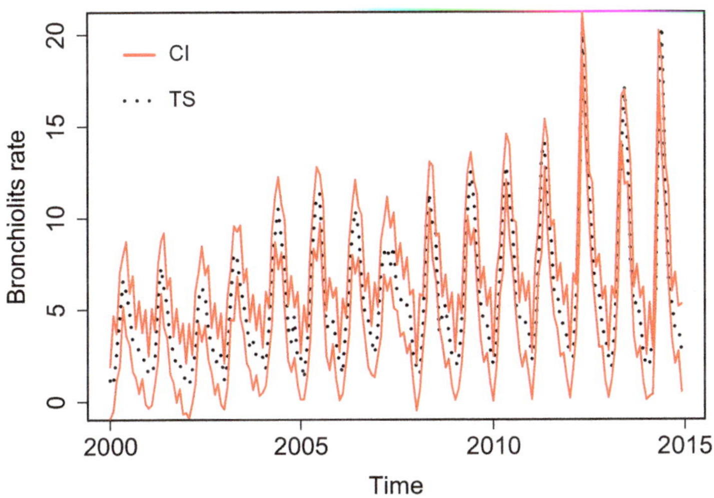

Fig. 7.13 Confidence interval obtained from TSWDWT

Figure 7.11 presents the confidence interval generated from the naive bootstrap based on NDWT. The CI constructed from this method included almost all the values of the observed time series, which represents the mean of the stochastic process.

From Fig. 7.12 one can observe the confidence interval generated from TSWNDWT. This method also includes almost all the values of that observed

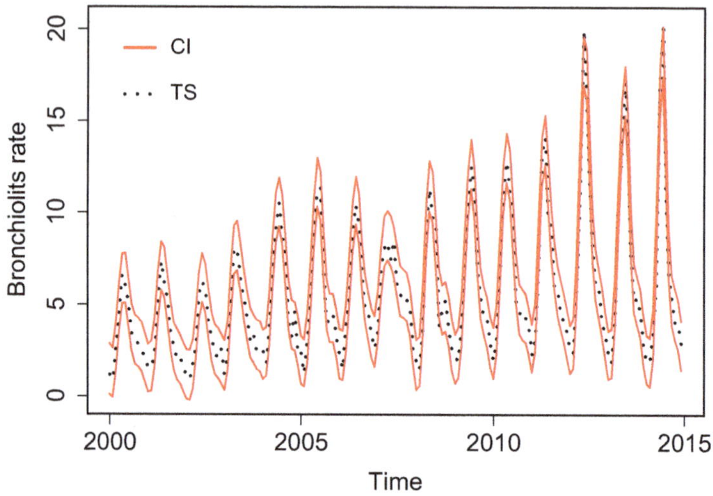

Fig. 7.14 Confidence interval obtained from reinflated TSWNDWT

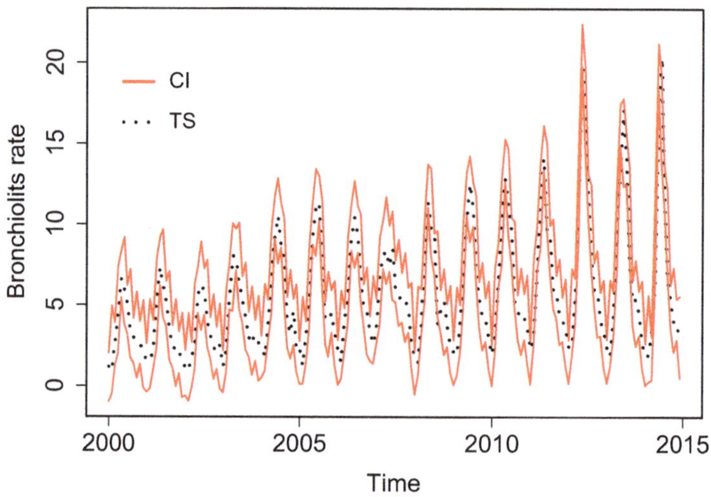

Fig. 7.15 Confidence interval obtained from reinflated TSWDWT

time series, but we can observe that this interval is little more narrower than in the confidence interval using only naive NDWT bootstrap.

Figure 7.13 presents the confidence interval generated from TSWDWT. The CI obtained from this method contains the most part of the observed values, and it seems to have less point out of the interval than the two methods already analyzed. However, each one of the time series that compose the confidence interval has more noise than those obtained from naive NDWT bootstrap and TSWNDWT.

The presence of more noise in the confidence interval generated from TSWDWT is expected since this method based on DWT seems to present more variability, bias, besides fewer wavelet coefficients in each multiresolution level to be resampled.

Figure 7.14 presents the confidence interval generated from reinflated TSWNDWT. The CI obtained from this method contains almost all the observed process values, and a few outside points. In general, we observe that the methods based on NDWT have a similar behavior.

In Fig. 7.15, the confidence interval generated from the reinflated TSWNDWT is represented. The CI obtained from this method also contains almost all the observed process values, but as in TSWDWT, the time series that compose the confidence interval is more noisy than those NDWT based methods.

In general, the built confidence intervals include almost all the time series values that represent the mean of the stochastic process. But, when the bronchiolitis hospitalizations are high producing spikes in the time series, mainly in May of 2012 and June 2014 the CI does not contain the time series values.

Although all the methods contain observed points that are not inside of the confidence intervals, those based on NDWT have less outside points. The TSWNDWT share more interesting results presenting low variability and the smaller bias.

7.5 Final Considerations

The difficulty or impossibility in accessing more than one trajectory in a stochastic process such as the monthly rate of bronchiolitis hospitalizations is well known. Providing a method to estimate the uncertainty associated with the evolutionary stochastic process mean without considering the presupposition of normality is a challenging problem. With the presented possibilities, this problem can be taken into account from wavelet-based bootstrapping.

All the evaluated methods provide a measure of the confidence interval of the mean μ_t for the monthly hospitalization rate for bronchiolitis via wavelet decomposition using the Daubechies' wavelet $d4$. At the moment, we are analyzing these methods considering different wavelet families and vanishing moments, as well as other time series with diverse behaviors and lengths. In the literature, the usual methods for resampling time series are based on DWT. In this work, we observed that NDWT provides good estimates, with the smallest standard errors and coefficient of variations.

The generation of the confidence interval for μ_t can also be used to estimate the uncertainty for wavelet regression models, since they also represent the mean of a stochastic process.

Acknowledgements The authors acknowledge and appreciate the anonymous reviewers for the valuable comments and such a positive feedback. The authors also thank the financial support of the Brazilian Federal Agency for Support and Evaluation of Graduate Education (CAPES).

References

Angelini, C., Cava, D., Katul, G., & Vidakovic, B. (2005). Resampling hierarchical processes in the wavelet domain: A case study using atmospheric turbulence. *Physica D: Nonlinear Phenomena, 207*(1), 24–40.

Breakspear, M., Brammer, M., & Robinson, P. A. (2003). Construction of multivariate surrogate sets from nonlinear data using the wavelet transform. *Physica D: Nonlinear Phenomena, 182*(1), 1–22.

Chatfield, C. (2016). *The analysis of time series: an introduction*. Boca Raton: CRC Press.

Daubechies, I. (1992). Ten lectures on wavelets || 1. The what, why, and how of wavelets. https://doi.org/10.1137/1.9781611970104.

Efron, B., & Gong, G. (1983). A leisurely look at the bootstrap, the jackknife, and cross-validation. *The American Statistician, 37*(1), 36–48.

Golia, S. (2002). Evaluating the GPH estimator via bootstrap technique. In W. Härdle, B. Rönz (Eds.), *Compstat* (pp. 343–348). Heidelberg: Physica.

Kang, M., & Vidakovic, B. (2017). MEDL and MEDLA: Methods for assessment of scaling by medians of log-squared nondecimated wavelet coefficients. arXiv preprint arXiv:1703.04180.

Mallat, S. G. (1989). A theory for multiresolution signal decomposition: The wavelet representation. *IEEE Transactions on Pattern Analysis and Machine Intelligence, 11*(7) , 674–693.

Morettin, P. A., & Toloi, C. (2006). *Análise de séries temporais*. Spindale: Blucher.

Nason, G. P., & Silverman, B. W. (1995). The stationary wavelet transform and some statistical applications. In A. Antoniadis & G. Oppenheim (Eds.), *Wavelets and statistics* (pp. 281–299). New York: Springer.

Percival, D., Sardy, S., & Davison, A. (2000). Wavestrapping time series: Adaptive wavelet-based bootstrapping. In *Nonlinear and nonstationary signal processing* (pp. 442–471). Cambridge: Cambridge University Press.

Politis, D. N., & Romano, J. P. (1994). The stationary bootstrap. *Journal of the American Statistical association, 89*(428), 1303–1313.

R Core Team (2016). *R: A language and environment for statistical computing*. Vienna: R Foundation for Statistical Computing.

Tang, L., Woodward, W. A., & Schucany, W. R. (2008). Undercoverage of wavelet-based resampling confidence intervals. *Communications in Statistics – Simulation and Computation®, 37*(7), 1307–1315.

Wei, W. S. W. (2006). *Time series analysis: univariate and multivariate methods*. Boston: Pearson Addison Wesley.

Wornell, G., & Oppenheim, A. V. (1996). *Signal processing with fractals: A wavelet-based approach*. Upper Saddle River: Prentice Hall Press.

Yi, J.-S., Jung, Y.-Y., Jacko, J., Sainfort, F., & Vidakovic, B. (2007). Parallel wavestrap: Simulating acceleration data for mobile context simulator. *Current Development in Theory and Applications of Wavelets, 1*, 251–272.

Chapter 8
A New Wavelet-Based Approach for Mass Spectrometry Data Classification

Achraf Cohen, Chaimaa Messaoudi, and Hassan Badir

8.1 Introduction

Application of new technologies of big data and statistical learning theory to mass spectrometry data classification problem can have a valuable impact on public health. This need is particularly critical in early detection and identification of cancer. Many strategies can be implemented to combat cancer such as early detection, close monitoring of the patient after initial treatment, and others (Diamandis 2004). Proteomic patterns through mass spectrometry techniques have shown a promising strategy to diagnose cancer.

Mass Spectrometry (MS) is an analytical chemistry technique that was introduced to help to identify the amount and type of chemicals present in a sample by measuring the mass-to-charge ratio and abundance of gas-phase ions. The mass spectrometers consist of three principal elements: an ion source, a mass analyzer, and an ion detection system (Aebersold and Mann 2003). The ionization is the first step in mass spectrometry analysis. The second step is the separation of the ions according to their mass to charge ratio. Finally, the compounds are detected and the relative abundance of each of the resolved ionic species is recorded. The output of the detector is a mass spectrum presented in a plot of the relative abundance or relative intensity as a function of the mass-to-charge ratio, see Fig. 8.1.

A. Cohen (✉)
Department of Mathematics and Statistics, University of West Florida, Pensacola, FL, USA
e-mail: acohen@uwf.edu

C. Messaoudi · H. Badir
National School of Applied Sciences-Tangier, ENSAT, Abdelmalek Essaadi University, Tangier, Morocco
e-mail: messaoudi@ensat.ac.ma; badir.hassan@uae.ma

Y. Zhao, D.-G. Chen (eds.), *New Frontiers of Biostatistics and Bioinformatics*,
ICSA Book Series in Statistics, https://doi.org/10.1007/978-3-319-99389-8_8

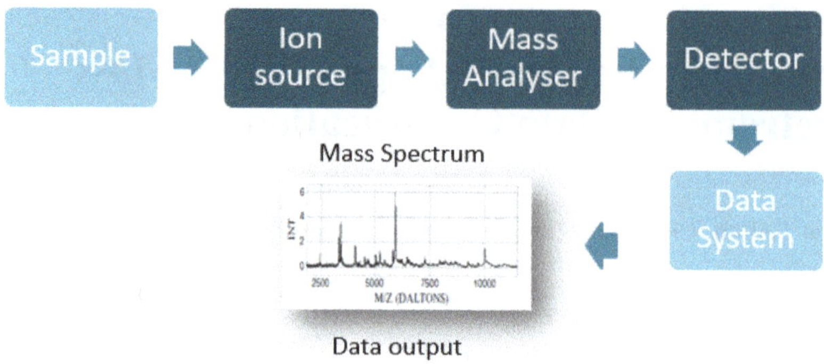

Fig. 8.1 Components of a mass spectrometer

In the last decade, MS data analysis has become an increasingly prominent field allowing the identification, quantification, and characterization of peptides and proteins in biological samples. It has been applied to discover patterns of differentially expressed protein in clinical samples such as blood serum. Especially, biomarker identification that can be used for diagnosis and monitoring of many diseases (Cravatt et al. 2007). The Matrix Assisted Laser Desorption/Ionization Time-Of-Flight (MALDI-TOF) and Surfaced-Enhanced Laser Desorption/Ionization Time-Of-Flight (SELDI-TOF) are high-throughput technologies for the acquisition of protein expression profiles from biological fluids (serum, plasma, etc.). The use of these technologies, with statistical modeling, is essential for (1) the identification of novel protein biomarkers of disease and (2) the classification of a new unseen mass spectrum.

MS data are given by the number of mass-to-charge ratios (m/z). Tens of thousands of m/z are available in the data but not necessarily all are used to MS data classification. It is reasonable to have a feature extraction procedure that is able to decrease the effects of noise, reduce dimension, and define new features to represent the data. These new features are used to develop a good classification model (Das 2001). In the last decades, MS data analysis for cancer identification has focused on two main concepts (1) selecting features from MS spectrum and (2) developing classification models for prediction. Both concepts should work together in order to provide an accurate model for classifying MS data.

The conventional method for processing an MS spectrum is to perform a number of preprocessing steps before developing any statistical models. These tools include baseline correction, normalization, and denoising (Dubitzky et al. 2007, pp. 79–102). The authors in Petricoin et al. (2002) developed a bioinformatics tool to identify ovarian cancer using self-organizing clustering analysis and genetic algorithm. In Tang et al. (2010), the authors proposed an approach for dimensionality reduction and tested it using mass spectrometry data for ovarian cancer. They used the mean, variance, skewness, and kurtosis in order to reduce the dimension. A Kernel Partial Least Squares model is then developed for ovarian cancer classification. Moreover, Li and Zeng (2016) proposed a method based on the model of uncorrelated linear discriminant analysis combined with variable selection method, applied to serum SELDI-TOF MS for ovarian cancer identification. Wu et al. (2016) proposed a classification model based on probabilistic principal component analysis and support vector machine. The model was applied to ovarian cancer. Sharma and Singh (2016) suggested the use of the neural network for diagnosis of ovarian cancer.

de Noo et al. (2006) studied colorectal cancer using the MALDI-TOF serum. In a randomized block design, pre-operative samples from 66 colorectal cancer patients and 50 controls were used, and a classification model is built using a linear discriminant analysis with double cross-validation. Another study on colorectal cancer is given in Ward et al. (2008). The authors used a logistic regression model to classify MS spectrum for 67 patients with colorectal cancer and 72 non-cancer control subjects. Lung cancer was studied in Yildiz et al. (2007), the paper investigated MS data to identify lung cancer cases from matched controls. MALDI-MS data were used with two methods of analysis: the weighted flexible compound covariate method and support vector machine.

Pancreatic cancer was the goal of the study given in Ge and Wong (2008). The authors investigated the utility of three feature selection schemas Student t-test, Wilcoxon rank sum test, and genetic algorithm. Some of the selected features were then used to classify MS Pancreatic cancer through six different decision tree classifier ensembles, such as Random forest, Adaboost, and others. Ohn et al. (2016) used 2D polyacrylamide gel electrophoresis (2D PAGE) approach to generate the 2D proteome patterns, and they then compared three classification methods: genetic algorithm combined with SVM, stepwise forward feature selection with K-NN, and random forest. These methods were applied to identify breast cancer.

Lancashire et al. (2009) presented a review of the concepts related to neural networks with their applications in mass spectrometry and focus on cancer studies. In this study (Gromski et al. 2014), the researchers compared feature selection methods with some classification approaches such as Random Forest with its variable selection techniques and SVM combined with support vector machines-recursive feature elimination, and they showed better performance is given by SVM. Awedat et al. (2016) proposed a compressive sensing sampling approach to reducing the dimension. They showed L2-algorithm with regularization terms has better performance than standalone L2-algorithm.

Wavelet analysis has been shown potential application for MS classification to (a) reducing dimension, (b) extracting features, or (c) denoising data. In Yu et al. (2005), a procedure to classify ovarian cancer based on MS data was developed. The authors combined binning, Kolmogorov-Smirnov test, wavelet analysis, and support vector machines to preprocess and develop a classification model, The authors used the *db4* wavelet. Another classification approach of proteomic MS data based on bi-orthogonal discrete wavelet transform and support vector machines was proposed in Schleif et al. (2009). The authors used *bior3.7* wavelet for denoising purposes. Du et al. (2009) proposed a workflow for MS classification based on wavelet analysis, Kolmogorov-Smirnov test with bagging predictor. The wavelet *sym8* was used to denoise the MS data. Nguyen et al. (2015) showed that combining *Haar* wavelet coefficients and genetic algorithm provides a good selection feature subset for the performance classification, but genetic algorithms require a random initialization that may lead to different results. The Wavelet-based function mixed model, which generalizes the linear mixed models to the case of functional data was used in order to analyze MS-data (Morris et al. 2006).

The goal of this chapter is to present a new approach for MS data classification. The proposed approach is original and based on a combination between principal component and wavelet analyses in addition to a new T^2 statistic. Most of the previous research using wavelet analysis did not show how they did select the wavelet family for their analysis. To this end we propose a prior study to select the best-suited wavelet for the analysis. This will help future MS research to have a subjective tool for wavelet selection. The principal component analysis is applied to six features (statistics) that are calculated on the wavelets coefficients (approximation and details). Next, we propose a new statistic $T^2 = \sqrt{T_a^2 + T_d^2}$ combining T^2 on the approximation and details coefficients, respectively. Finally, a support vector machine model is built on the new aforementioned statistic. The proposed approach shows high accuracy, specificity, and sensitivity. We provide a detailed description of each step to ensure the reproductivity of the present research work.

This chapter is organized as follows. Section 8.2 presents the proposed approach for mass spectrometry data classification. In Sect. 8.3, experiments and results are given, and Sect. 8.4 presents conclusions and some directions of research.

8.2 The Proposed Approach

We have designed and implemented a new approach to classify mass spectrometry data. The main steps of our proposed approach are illustrated in Fig. 8.2. The philosophy of the method consists of subdividing the MS sample into several windows and extracting from them some features that will help discriminate/classify the entire MS spectrum. The wavelets analysis has potential capabilities to extract features especially when noisy data is used such as the case with MS data. The principal component analysis with T^2 statistic is used in order to aggregate the features from the wavelets coefficients into one statistic.

The proposed approach can be implemented as follows:

In this approach, each MS spectrum is represented by a T^2 statistic calculated into the feature space of the principal component analysis. The latter is applied to the features (Energy, Mean, Kurtosis, Skewness, Variance, and Coefficient of Variation) of the wavelets coefficients. This approach will be applied to a real dataset in the next section.

1: Input data: MS spectrum X of length L
2: Divide X into n samples of length $N = 2^J$, and rearrange X into a data matrix $Z_{(N \times n)}$
3: Apply Discrete Wavelet Transform using *bior3.1* to each column of $Z_{(N \times n)}$; see Sect. 8.2.1
4: Compute [Energy, Mean, Kurtosis, Skewness, Variance, and Coefficient of Variation (CV)] of the approximation and details wavelets coefficients as follows:

$$Energy = \sum w_i^2; \qquad Mean = \sum w_i/m \tag{8.1}$$

$$Variance = \sum (w_i - Mean)^2/(m-1); \qquad CV = \frac{\sqrt{Variance}}{Mean} \tag{8.2}$$

$$Skewness = E\left[\left(\frac{X-\mu}{\sqrt{Variance}}\right)^3\right]; \qquad Kurtosis = E\left[\left(\frac{X-\mu}{\sqrt{Variance}}\right)^4\right] \tag{8.3}$$

The data look now as follows, for both approximation and details coefficients:

Feature =

	Energy	**Mean**	**Variance**	**Skewness**	**Kurtosis**	**CV**
$Window_1$	E_1	M_1	V_1	Sk_1	Ku_1	CV_1
$Window_2$	E_2	M_2	V_2	Sk_2	Ku_2	CV_2
$\vdots$	$\vdots$	$\vdots$	$\vdots$	$\vdots$	$\vdots$	$\vdots$
$Window_n$	E_n	M_n	V_n	Sk_n	Ku_n	CV_n

5: Apply Principal Component Analysis to $Feature_{a_j}$ and $Feature_{d_j}$ data matrices, and select a number of principal components (reduced space)
6: Compute $T_{a_j}^2$ and $T_{d_j}^2$ statistics corresponding to the approximation and details wavelets coefficients, respectively, in the reduced space, see Sect. 8.2.2
7: Develop an SVM model on the $T^2 = \sqrt{T_{a_j}^2 + T_{d_j}^2}$ statistic, see Sect. 8.2.3

where w_i are the wavelet coefficients (either approximations or details), m is the number of coefficients, L is the length of the MS, J is a natural number, and a_j and d_j are the approximations and details wavelet coefficients, respectively.

8.2.1 Wavelets Analysis

Wavelet analysis is a mathematical tool that consists of projecting data into a time-frequency representation. The theory of Multi-Resolution Analysis (MRA) has linked wavelets theory to filter analysis. It opened the door to apply wavelets to image processing and also resulted in the implementation of the Fast Wavelet Transform (FWT) algorithm (Mallat 1989; Misiti et al. 1996). Wavelets functions are grouped by families such as Haar, Daubechies, Coiflet, Symlet, and Biorthogonal (Daubechies 1992). The Continuous Wavelet Transform (CWT) is a redundant transformation since the scale and the translation parameters are changed

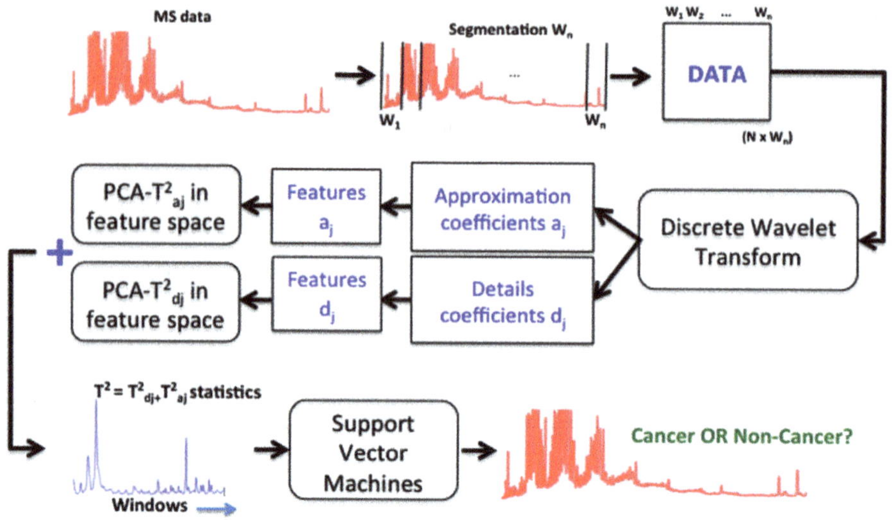

Fig. 8.2 The proposed method for MS classification

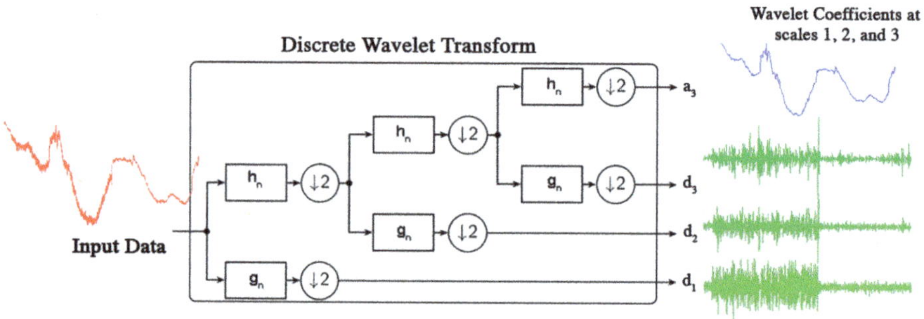

Fig. 8.3 The discrete wavelet transform through filter banks

continuously. The Discrete Wavelet Transform (DWT) is computationally efficient and can be achieved by the discretization of the scale s and translation τ parameters, as follows:

$$\psi_{j,k}(t) = 2^{j/2}\psi(2^j t - k) \tag{8.4}$$

where $s = 2^j$ and $\tau = ks$; $j, k \in \mathbf{Z}$.

These wavelet bases are orthogonal and defined in the framework of the Multi-Resolution Analysis (MRA), which provides a multiscale decomposition using orthogonal wavelets families across filter banks, see Fig. 8.3.

The wavelets coefficients of the DWT, approximations $a_j(k)$ and details $d_j(k)$, are given as follows:

$$a_j(k) = \sum_{i=0}^{l} h[i]a_{j-1}[2k - i] \tag{8.5}$$

$$d_j(k) = \sum_{i=0}^{l} g[i]a_{j-1}[2k - i] \tag{8.6}$$

where $a_0 = x$ the original signal, j represents the decomposition scale; $k \in Z$; l is the filter length; h and g are the scaling and wavelets filters, respectively.

The past research publications in bioinformatics have used the shrinkage techniques, which consist of thresholding wavelets coefficients. These techniques have shown a good performance for reducing noise in mass spectrometry data. Several thresholds have been developed, VisuShrink (Donoho and Johnstone 1994), RiskShrink, SUREShrink (Donoho and Johnstone 1995; Donoho 1995), FirmShrink (Gao et al. 1997; Gao 1998), to name a few. One of the benefits of the wavelet transform is the plenty of the wavelets functions developed over the past decades, but from such advantage arises the question of how to select a wavelet that is best suited for analyzing MS data.

There are two approaches in order to choose a wavelet for a specific signal. First, the qualitative methods such as *orthogonality*, *symmetry*, and *compact support*. Second, the quantitative measures such as *energy*, *entropy*, *mutual Information*, *conditional entropy*, and *energy-to-Shannon entropy ratio*. In this work, we used the *energy-to-Shannon entropy ratio*, which is defined as:

$$R = \frac{Energy}{Entropy} = \frac{\sum^{N} \mid wt(s,i) \mid^2}{-\sum^{N} p_i \log_2 p_i} \tag{8.7}$$

where N is the number of wavelet coefficients and wt represents the wavelets coefficients, s is the scaling parameter, and $p_i = \frac{|wt(s,i)|^2}{Energy}$.

The set of wavelets that has given a large *energy-to-Shannon entropy ratio* should be considered the candidate wavelets, one can choose the wavelets that have produced the largest *energy-to-Shannon entropy ratio*.

We conducted a preliminary study to choose which wavelet will be used to extract features. We considered 58 wavelets, and by using the Breast cancer Mass Spectrometry data presented in Sect. 8.3. The largest average *energy-to-Shannon entropy ratio* is equal to 49.88 and given by *bior3.1*, see Fig. 8.4. Therefore, we chose the biorthogonal wavelet (Cohen et al. 1992) *bior3.1* as the best-suited wavelet of the analysis.

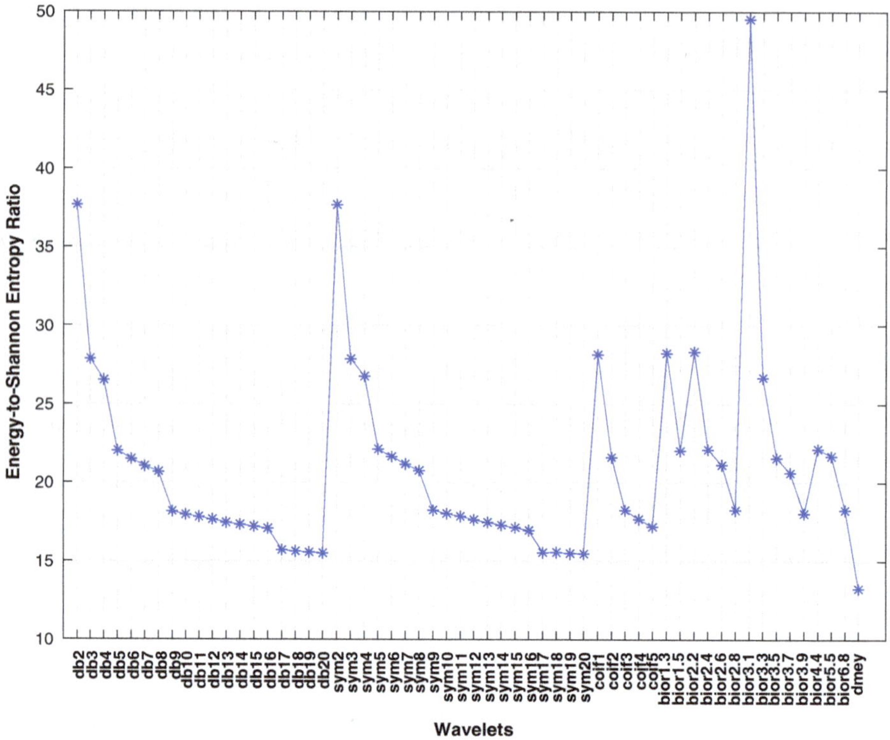

Fig. 8.4 The average energy-to-Shannon entropy ratio using 30 MS Spectrum and 58 Wavelets. dbN: Daubechies of order N; Sym: Symlet; Coif: Coiflet; bior: Biorthogonal, dmey: discrete Meyer

8.2.2 Principal Component Analysis and Hotelling T^2 Statistic

Principal component analysis (PCA) (Jolliffe 1986) is widely used for data exploration and interpretation. Principal component analysis of a data matrix provides new uncorrelated variables (principal components) whose variances are as large as possible. Consider a normalized data matrix, with p variables and N observations.

$$\mathbf{Z} = \begin{pmatrix} z_{11} & z_{12} & \cdots & z_{1p} \\ z_{21} & z_{22} & \cdots & z_{2p} \\ \vdots & \vdots & \ddots & \vdots \\ z_{N1} & z_{N2} & \cdots & z_{Np} \end{pmatrix} \tag{8.8}$$

The covariance matrix of Z can be approximated as:

$$\hat{\Sigma} = \frac{1}{N-1} Z^T Z = P \Lambda P^T \tag{8.9}$$

where $\Lambda = diag(\lambda_1, \lambda_2, \ldots, \lambda_p)$ with $\lambda_1 \geq \lambda_2 \geq \ldots, \geq \lambda_p$. λ_i are the eigenvalues and P are the eigenvectors of $\hat{\Sigma}$. According to λ_i's, P and Λ could be divided into a feature space (*feat*) and a residual space (*res*). We can then rewrite P and Λ as follows:

$$P = \begin{bmatrix} P_{feat} & P_{res} \end{bmatrix} \tag{8.10}$$

$$\Lambda = \begin{bmatrix} \Lambda_{feat} & 0 \\ 0 & \Lambda_{res} \end{bmatrix} \tag{8.11}$$

The Hotelling T^2 statistic can then be computed as follows:

$$T^2 = Z P_{feat} \Lambda_{feat}^{-1} P_{feat}^T Z^T \tag{8.12}$$

where T^2 is the Hotelling statistic calculated into the multivariate feature space of the principal component analysis, and P^T is the transpose of P. The number of components in the feature space can be determined by using techniques such as the cumulative explained variance and the scree plot. In our approach, the number of principal component in the feature space is determined by the cumulative explained variance technique.

8.2.3 *Support Vector Machines*

The Support Vector Machines (SVM) are one of the most used statistical learning methods for classification and regression. Classification using SVM can handle problems where a training set S is linearly separable or linearly non-separable. The kernel approach allows the training data to be projected into a higher dimensional feature space where the data become separable (Shawe-Taylor and Cristianini 2004; Cristianini and Shawe-Taylor 2000; Vapnik 2013). This property makes the use of SVM valuable for many applications.

Given a training set of pairs (X_i, Y_i), $i = 1, \ldots, N$ where $X_i \in R^n$ and $Y \in \{1(cancer), -1(Non-cancer)\}^N$, the support vector machines find a hyperplane that separates the two classes. The generalized optimal separating hyperplane is determined by w that minimizes the following optimization problem (Cortes and Vapnik 1995):

$$\frac{1}{2} w^T w + C \sum_{i=1}^{N} \epsilon_i \tag{8.13}$$

subject to

$$y_i(w^T \phi(x_i) + b) \geq 1 - \epsilon_i; \qquad \epsilon_i \geq 0 \tag{8.14}$$

where the training set x_i is mapped into a higher dimension space using the function ϕ. C is a positive constant (regularization parameter). It is shown that the problem presented in Eqs. (8.13)–(8.14) depends only on the inner product of x. Therefore, the inner product in the high dimensional feature space can be performed in the input space via the kernel functions. Many kernel functions exist such as polynomial and Gaussian Radial Basis Function (RBF). The Gaussian kernel has given a great attention and defined by:

$$K(x, x') = exp\Big(-\gamma \mid\mid x - x' \mid\mid^2 \Big) \tag{8.15}$$

where γ is the Kernel parameter. In order to find the best decision boundaries, the hyperparameters C and γ should be controlled. The hyperparameter optimization can be used to select C and the Kernel parameter. Methods such as grid search, random search, and Bayesian optimization are often used for such purpose. In this work, we used a grid search on $C \in \{2^{-8}, 2^{-7}, \ldots, 2^8\}$ and $\gamma \in \{2^{-10}, 2^{-9}, \ldots, 2^{10}\}$ to select the hyperparameters C and γ. We run a 50-fold cross validation on the training data set to determine the hyperparameters as follows: $C = 2$ and $\gamma = 0.0009765$

8.3 Experiments and Results

The data used are low-mass range SELDI spectra derived from patients with breast cancer and from normal controls. They can be found online at the Department of Bioinformatics and Computational Biology at the University of Texas M.D. Anderson Cancer Center (P. datasets for Breast Cancer 2004). The datasets were generated using IMAC-3 protein chip and sample application was performed using the Biomek 2000 Laboratory Automation Workstation robot (Beckman Coulter, Fullerton, CA). There are 33,885 m/z values and 156 samples where control (normal) patients contribute with 57 samples and 99 samples are cancer. The analysis was done using Matlab and R. The authors are willing to share the code used in this work if requested by e-mail. An example of a sample of cancer and a sample from normal patients is in Fig. 8.5.

In the experiment, each MS sample is subdivided into 64 windows of $2^8 = 256$ observations that is 32,768 m/z values. Since the data have 33,885 values the remaining values are not used. The data then are arranged into a matrix data of 256 rows and 64 columns. Next, the discrete wavelet transform is applied to each column (window) using the *bior3.1* wavelet as shown in Sect. 8.2.1. This results in obtaining the approximation coefficients and details coefficients for each window. Afterwards, the features presented in Sect. 8.2 are computed for each window, therefore the

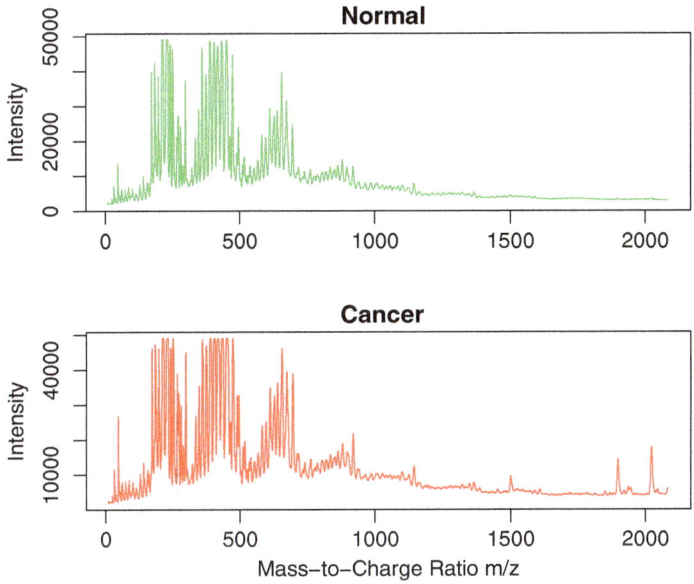

Fig. 8.5 An example of MS samples from cancer and normal patients

data are now arranged as two matrices data of 64 rows (windows) and 6 columns (features), one matrix data is based on the approximations and the other is given by the details coefficients. Each window is represented by six features of the wavelets coefficients, namely Energy, Mean, Variance, Skewness, Kurtosis, and Coefficient of Variation (CV).

Next, the principal component analysis is conducted on the two data matrices, and the number of principles components are selected based on the explained variances. We selected a number of principal components that explain at least 90% and at most 95% of the data. Then, the Hotelling $T^2_{a_j}$ and $T^2_{d_j}$ are calculated into the reduced space, see Sect. 8.2.2, for the approximation and details coefficients, respectively. Finally, $T^2 = \sqrt{T^2_{a_j} + T^2_{d_j}}$ statistic is then calculated to represent the original MS spectrum, see Fig. 8.6. Each MS sample will be given by T^2 statistic, and the classification model will be built on T^2 statistics for each patient.

8.3.1 *Results and Performance*

The classification model is built in two phases, a training phase, and a test phase. 80% of the dataset is used as the training dataset (78 Cancer, 46 Normal) and 20%

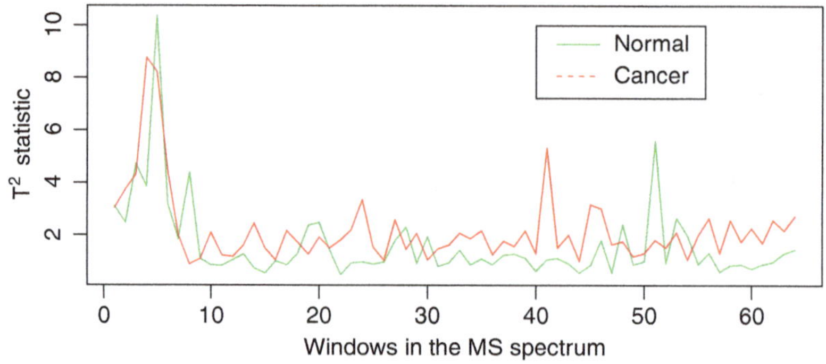

Fig. 8.6 T^2 statistic for normal and cancer MS spectrum

as the testing data set. The MS data are then classified as normal or cancer. The classification results will be given in terms of accuracy, sensitivity, and specificity, as follows:

$$Accuracy = \frac{TP + TN}{TP + TN + FP + FN}, \tag{8.16}$$

$$Sensitivity = \frac{TP}{TP + FN}, \tag{8.17}$$

$$Specificity = \frac{TN}{TN + FP}, \tag{8.18}$$

where TP is True Positive, TN True Negative, FP False Positive, and FN False Negative. The proposed framework achieves a reasonable classification performance. The combination of wavelets coefficients and higher order statistics provide useful discriminatory information about MS data.

The classification model is developed using the training set and then tested using the testing samples. The predictive procedure optimizes the model parameters to build a model that fits the training data as well as possible, which then may lead to an overfitting. Cross-validation can be used as a model validation technique for assessing how the results of a statistical analysis will generalize to an independent data set, therefore, this will help overcome the overfitting problem.

We performed a k-fold cross-validation procedure in order to avoid overfitting and evaluate the generalization capabilities of our model. The parameter k varies into {5, 10, 20, 50}. A Monte-Carlo simulation repeating the whole process including the selection of the training and testing sets is also performed. 100 simulation runs were conducted. Consequently, the results reported are the averages and standard errors of the performance measures given in Eqs. (8.16)–(8.18).

Table 8.1 shows a summary of the performance results of our proposed method. The accuracy classification is 100% on average with 0 standard error. The average sensitivity and specificity are equal to 100% for the training set and the testing set.

Table 8.1 The average (standard error) of the performance results using 100 replications of k-Fold cross validation, and the hyperparameters C = 2 and γ = 0.0009765 obtained from the grid search optimization

k ↓	Accuracy(SE) %	Sensitivity		Specificity	
		Training set	Testing set	Training set	Testing set
5	99.91 (0.12)	1 (0)	1 (0)	1 (0)	1 (0)
10	100 (0)	1 (0)	1 (0)	1 (0)	1 (0)
20	100 (0)	1 (0)	1 (0)	1 (0)	1 (0)
50	100 (0)	1 (0)	1 (0)	1 (0)	1 (0)

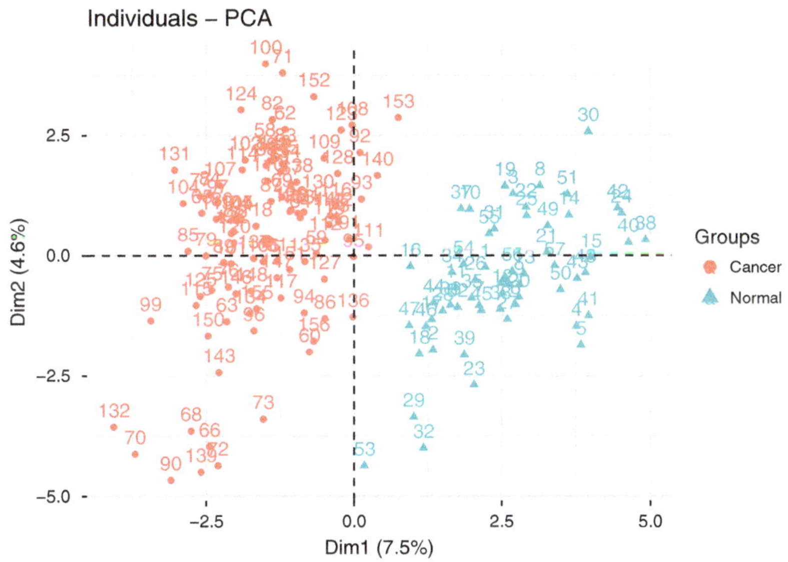

Fig. 8.7 The first two principal components on T^2 statistic for 99 cancer samples and 57 normal samples

In order to investigate the excellent performance of the proposed method, we conducted a principal component analysis (PCA) on the 156 samples of MS data (57 normal and 99 cancer). The PCA is applied to a data matrix of 156 rows (patients) and 64 columns (T^2 statistics). The results given in Fig. 8.7 show clearly that the proposed method ingeniously separates the cancer MS samples from the Normal MS samples.

This result has a valuable scientific impact on public health, especially on the early cancer detection. In fact, automatic classification of these mass spectrometry patterns will definitely help physicians in the diagnosis of diseases such as cancer. In addition, the higher classification performance we have, the more confident in the diagnosis we become.

8.4 Conclusion

The MS data are important for clinical diagnosis and health advances. Preprocessing methods and transformation such as wavelets analysis and principal component analysis can help face high dimensionality and reduce noise.The accuracy of the proposed model is 100% on average with 0 standard error. The average sensitivity and specificity are equal to 100% for the training set and the testing set. This paper contributes to the development of accurate models for MS classification. We aim at applying the proposed method to other MS data of different types of cancer.

References

Aebersold, R., & Mann, M. (2003). Mass spectrometry-based proteomics. *Nature, 422*(6928), 198–207.

Awedat, K., Abdel-Qader, I., & Springstead, J. R. (2016). Mass spectrometry sensing data for robust cancer classification. In *Electro Information Technology (EIT), 2016 IEEE International Conference on* (pp. 0258–0262). Piscataway: IEEE.

Cohen, A., Daubechies, I., & Feauveau, J.-C. (1992). Biorthogonal bases of compactly supported wavelets. *Communications on Pure and Applied Mathematics, 45*(5), 485–560.

Cortes, C., & Vapnik, V. (1995). Support vector networks. *Machine Learning, 20*(3), 273–297.

Cravatt, B. F., Simon, G. M., & Yates Iii, J. R. (2007). The biological impact of mass-spectrometry-based proteomics. *Nature, 450*(7172), 991.

Cristianini, N., & Shawe-Taylor, J. (2000). *An introduction to support vector machines and other kernel-based learning methods*. Cambridge: Cambridge University Press.

Das, S. (2001). Filters, wrappers and a boosting-based hybrid for feature selection. In *ICML* (Vol. 1, pp. 74–81).

Daubechies, I. (1992). *Ten lectures on wavelets*. Philadelphia, PA: Society for Industrial and Applied Mathematics.

de Noo, M. E., Mertens, B. J., Özalp, A., Bladergroen, M. R., van der Werff, M. P., van de Velde, C. J., et al. (2006). Detection of colorectal cancer using maldi-tof serum protein profiling. *European Journal of Cancer, 42*(8), 1068–1076.

Diamandis, E. P. (2004). Mass spectrometry as a diagnostic and a cancer biomarker discovery tool opportunities and potential limitations. *Molecular & Cellular Proteomics, 3*(4), 367–378.

Donoho, D. (1995). De-noising by soft-thresholding. *IEEE Transactions on Information Theory, 41*(3), 613–627.

Donoho, D. L., & Johnstone, J. M. (1994). Ideal spatial adaptation by wavelet shrinkage. *Biometrika, 81*(3), 425–455.

Donoho, D. L., & Johnstone, J. M. (1995). Adapting to unknown smoothness via wavelet shrinkage. *Jouranl of the American Statistical Association, 90*, 1200–1224.

Du, J., Wu, X.-M., Wang, B., Su, H.-J., Ma, K., & Zhang, H.-Q. (2009). Wavelet transform and bagging predictor approaches to cancer identification from mass spectrometry-based proteomic data. In *Bioinformatics and Biomedical Engineering, 2009. ICBBE 2009. 3rd International Conference on* (pp. 1–4). Piscataway: IEEE.

Dubitzky, W., Granzow, M., & Berrar, D. P. (2007). *Fundamentals of data mining in genomics and proteomics*. Berlin: Springer Science and Business Media.

Gao, H.-Y. (1998). Wavelet shrinkage denoising using the non-negative garrote. *Journal of Computational and Graphical Statistics, 7*(4), 469–488.

Gao, H.-Y., & Bruce, A. G. (1997). Waveshrink with firm shrinkage. *Statistica Sinica, 7*(4), 855–874.

Ge, G., & Wong, G. W. (2008). Classification of premalignant pancreatic cancer mass-spectrometry data using decision tree ensembles. *BMC Bioinformatics, 9*(1), 275.

Gromski, P. S., Xu, Y., Correa, E., Ellis, D. I., Turner, M. L., & Goodacre, R. (2014). A comparative investigation of modern feature selection and classification approaches for the analysis of mass spectrometry data. *Analytica Chimica Acta, 829*, 1–8.

Jolliffe, I. T. (1986). Principal component analysis and factor analysis. In *Principal component analysis* (pp. 115–128). Berlin: Springer.

Lancashire, L. J., Lemetre, C., & Ball, G. R. (2009). An introduction to artificial neural networks in bioinformatics—application to complex microarray and mass spectrometry datasets in cancer studies. *Briefings in Bioinformatics, 10*, 315–329. https://doi.org/10.1093/bib/bbp012.

Li, Y., & Zeng, X. (2016). Serum seldi-tof ms analysis model applied to benign and malignant ovarian tumor identification. *Analytical Methods, 8*(1), 183–188.

Mallat, S. G. (1989). A theory for multiresolution signal decomposition: The wavelet representation. *IEEE Transactions on Pattern Analysis and Machine Intelligence, 11*(7), 674–693.

Misiti, M., Misiti, Y., Oppenheim, G., & Poggi, J. (1996). *Wavelet toolbox*. Natick, MA: The MathWorks Inc.

Morris, J. S., Brown, P. J., Baggerly, K. A., & Coombes, K. R. (2006). Analysis of mass spectrometry data using bayesian wavelet-based functional mixed models. In *Bayesian inference for gene expression and proteomics* (pp. 269–288). Cambridge: Cambridge University Press.

Nguyen, T., Nahavandi, S., Creighton, D., & Khosravi, A. (2015). Mass spectrometry cancer data classification using wavelets and genetic algorithm. *FEBS Letters, 589*(24), 3879–3886.

Ohn, S.-Y., Chi, S.-D., & Heo, C. (2016). Identification of breast cancer by classification of proteome patterns. *International Journal of Modeling, Simulation, and Scientific Computing, 7*(04), 1643004.

P. Datasets for Breast Cancer (2004). http://bioinformatics.mdanderson.org/pubdata.html.

Petricoin, E. F., Ardekani, A. M., Hitt, B. A., Levine, P. J., Fusaro, V. A., Steinberg, S. M., et al. (2002). Use of proteomic patterns in serum to identify ovarian cancer. *The Lancet, 359*(9306), 572–577.

Schleif, F.-M., Lindemann, M., Diaz, M., Maaß, P., Decker, J., Elssner, T., et al. (2009). Support vector classification of proteomic profile spectra based on feature extraction with the bi-orthogonal discrete wavelet transform. *Computing and Visualization in Science, 12*(4), 189–199.

Sharma, A., & Singh, S. (2016). Neural network for diagnosis of ovarian cancer based on proteomic patterns in serum. *Journal of Scientific and Technical Advancements, 2*(2), 25–27.

Shawe-Taylor, J. & Cristianini, N. (2004). *Kernel methods for pattern analysis*. Cambridge: Cambridge University Press.

Tang, K.-L., Li, T.-H., Xiong, W.-W., & Chen, K. (2010). Ovarian cancer classification based on dimensionality reduction for seldi-tof data. *BMC Bioinformatics, 11*(1), 109.

Vapnik, V. (2013). *The nature of statistical learning theory*. Berlin: Springer Science and Business Media.

Ward, D. G., Nyangoma, S., Joy, H., Hamilton, E., Wei, W., Tselepis, C., et al. (2008). Proteomic profiling of urine for the detection of colon cancer. *Proteome Science, 6*(1), 19.

Wu, J., Ji, Y., Zhao, L., Ji, M., Ye, Z., & Li, S. (2016). A mass spectrometric analysis method based on ppca and svm for early detection of ovarian cancer. *Computational and Mathematical Methods in Medicine, 2016*, 6169249.

Yildiz, P. B., Shyr, Y., Rahman, J. S., Wardwell, N. R., Zimmerman,L. J., Shakhtour, B., et al. (2007). Diagnostic accuracy of maldi mass spectrometric analysis of unfractionated serum in lung cancer. *Journal of Thoracic Oncology, 2*(10), 893–901.

Yu, J., Ongarello, S., Fiedler, R., Chen, X., Toffolo, G., Cobelli, C., et al. (2005). Ovarian cancer identification based on dimensionality reduction for high-throughput mass spectrometry data. *Bioinformatics, 21*(10), 2200–2209.

Part III
Clinical Trials and Statistical Modeling

Chapter 9
Statistical Power and Bayesian Assurance in Clinical Trial Design

Ding-Geng Chen and Jenny K. Chen

9.1 Introduction

A well-known practice in clinical trial design is to determine the appropriate number of patients (i.e., sample size) needed for adequate statistical power and for addressing clinical objectives. Based on this common knowledge, most clinical trials are "powered" at 80% to ensure the "success" of the planned clinical trials. However, it is not uncommon that a failed trial could be found even when all the protocols are followed. What could be wrong then? We often hear the statement, "my clinical trial is powered at 80%, so I have an 80% chance of being successful." Intuitively this seems correct, but in reality this is unlikely, and the actual chance for a successful trial could be much lower. In fact, statistical power is often not the same as the probability of success, even if we purely define "success" as seeing a statistically significant treatment effect.

Let's consider a typical superiority trial, which provides evidence that a new drug treatment is better than a current standard placebo treatment. In designing this trial, we usually aim for a statistical power of 80%. Does this mean that the new trial has an 80% chance of showing the better efficacy of the new drug treatment? Unfortunately no, as this is not true for many reasons. For example, one of the many reasons is how we perform power calculations. In order to calculate statistical

D.-G. Chen (✉)
Department of Biostatistics, Gillings School of Global Public Health, University of North Carolina at Chapel Hill, Chapel Hill, NC, USA

Department of Statistics, University of Pretoria, Pretoria, South Africa
e-mail: dinchen@email.unc.edu

J. K. Chen
Department of Biostatistics, Gillings School of Global Public Health, University of North Carolina at Chapel Hill, Chapel Hill, NC, USA

Y. Zhao, D.-G. Chen (eds.), *New Frontiers of Biostatistics and Bioinformatics*, ICSA Book Series in Statistics, https://doi.org/10.1007/978-3-319-99389-8_9

power, we have to have prior knowledge of the expected treatment difference, the underlying inter-patient variability, the accepted type-I error rate (alpha), and the sample size. However, we don't know the true value of the expected treatment effect, so we resort to *estimates* instead, estimates that are mostly based on previous experiences. These estimates, however, can have different levels of uncertainty. The true treatment effect may be smaller than the estimate, which would then give a smaller statistical power. As a result, the statistically powered trial may not provide a good assurance of success for the trial being planned.

Conceptually, the statistical power is defined as the probability of rejecting the null (such as when the new drug has similar efficacy to the placebo) given that the true clinical trial treatment effect equals a predetermined value. Therefore, the statistical power is a conditional probability to this unknown predetermined value, which could be very different from the observed treatment effect estimate. Whenever this prior estimate is far away from the truth, the calculated statistical power could be very misleading. This could lead to a scenario where a well-designed clinical trial is ineffective and a failure.

By incorporating the uncertainties of prior estimates on treatment effects and other clinical parameters, a Bayesian assurance has been developed as an alternative to statistical power, which is a paradigm change in clinical trial design. This is seen in O'Hagan and Stevens (2001), and O'Hagan et al. (2005). Bayesian assurance calculates the unconditional probability that a trial will lead to the desired outcome, which is different from statistical power, the conditional probability based on the assumed unknown treatment effect. More development are followed by Chuang-Stein (2006), and Chen and Ho (2017). Even though this new concept is intuitively important, it still remains new to most clinical trialists. As a result, further promotion of this concept is needed, and thus leads to this chapter. In this chapter, we outline the concept of assurance and discuss the computations of assurance using a Monte-Carlo simulation-based approach.

In Sect. 9.2, we outline the concept of paradigm change from conventional statistical power to Bayesian assurance. In Sect. 9.3, we show the Monte-Carlo simulation-based approach to calculate statistical power and assurance. Finally, in Sect. 9.4, a discussion is provided.

9.2 A Paradigm Change from Statistical Power to Bayesian Assurance

9.2.1 Conventional Statistical Power and Its Limitations

Every clinical trial compares whether a new drug is better than the placebo. In order to demonstrate the efficacy of a new drug, a large enough sample of patients is needed so that statistical tests (such as t-test and χ^2-test) can be used to statistically test whether the new drug is more effective than the placebo.

Expressing this in statistical terms, we define the null hypothesis, H_0, as the new drug and the placebo having no difference in efficacy. The alternative hypothesis, H_a, is that the new drug is more effective than the placebo. The hypothesis test will test whether or not there is a statistically significant drug effect. Associated concepts include type-I error and type-II error, where type-I error (α, typically controlled at 5%) is defined as the probability of rejecting the null hypothesis when it is true, and type-II error (β) as the probability of failing to reject the null hypothesis when it is false. Statistical power (π) is then defined as the probability of rejecting the null hypothesis when it is false (i.e., $\pi = 1 - \beta$), which is typically set between 0.8 and 0.9 in most clinical trials. The associated sample size is then determined based on this power and the type-I error rate.

Following the notations from O'Hagan et al. (2005), we denote R as rejecting the null hypothesis. The conventional definition of statistical power can then be expressed as

$$\pi(\theta) = P(R|\theta) \tag{9.1}$$

where $\pi(.)$ is the power function and θ is a vector of the assumed parameters, such as the treatment effect, sampling variance, and possible others. It can be seen that the statistical power equation (9.1) is a *conditional* probability of R conditioned on the unknown parameter vector θ. The value of this power and the associated sample size calculation is then dependent on the unknown parameter vector θ.

However, θ cannot be obtained precisely in real clinical trials as pointed out in O'Hagan et al. (2005) and others. It is rare that the estimate of θ from previous clinical trials will be close to the predetermined parameter value, which then leads to overpowered or underpowered clinical trials.

To illustrate the limitations of conventional statistical power in designing a confirmatory trial on a promising new drug, let's say that we had a small trial with $n_0 = 20$ (for each treatment) patients against a matching placebo which yielded an estimated treatment effect of $d = 2.5$ units and standard deviation of $\sigma = 8$ units. The effect size is calculated to be 0.3125 (=2.5/8) with a t-statistic of 0.988 and a one-sided p-value of 0.165. This p-value is larger than the one-sided significance level of 0.025, so the trial is not statistically significant. However, it is promising for further study due to the moderate effect size of 0.3125.

Based on this finding, a new confirmatory trial is planned with a one-sided test at $\alpha = 2.5\%$, power = 0.80, and a sample size of 162 per group.

However, the estimated treatment effect of $d = 2.5$ and standard deviation of $\sigma = 8$ are point estimates and are subject to errors. It is then more practical to say that both the treatment effect and the population standard deviation are in a range where the treatment effect is mostly in the range of (1.5, 3.5) and the standard deviation is in the range of (6, 10). With these guesstimates, we can further assume that the treatment effect is distributed normally with a mean of the observed 2.5 and a standard deviation of 0.5, i.e., $d \sim N(2.5, 0.5)$. The standard deviation would be distributed as an inverse-gamma distribution where the two parameters can be determined from the observed $\sigma = 8$ and 1 standard deviation away from the true σ. With these data from the small trial, if $d = 2$, the power would be 0.61 (i.e., underpowered confirmative trial) with a sample size of 162. If $d = 3$, the power would be 0.92 (i.e., overpowered confirmative trial). Similarly, instead of $\sigma = 8$ estimated from the small trial, if $\sigma = 7$, the power would be 0.89 (i.e., overpowered confirmative trial) and if $\sigma = 9$, the power would be 0.70 (i.e., underpowered confirmative trial).

In practice, we rarely know the truth of d and σ as well as other design parameters. Therefore, the resulting power calculation and sample size are subject to all these uncertainties and seriously limit the success of designed clinical trials.

9.2.2 Bayesian Assurance in Clinical Trials

To eliminate these limitations of conventional statistical power, O'Hagan and Stevens (2001) advocated the use of "assurance" (denoted by γ), which is defined as an *unconditional* probability to reject the null hypothesis, i.e., $\gamma = P(R)$, where R is rejection of the null hypothesis. Using the Bayesian framework, the assurance can be expressed as the expected power to the parameter vector space of θ. It can be seen that

$$\gamma = P(R) = \int P(R, \theta)\, d\theta = \int P(R|\theta)\, P(\theta)\, d\theta = E_{\theta}\left(\mathrm{P}\left(\mathrm{R}|\theta\right)\right) \qquad (9.2)$$

where the expectation is to the (prior) probability distribution of parameter vector space of θ.

With this definition, the "Bayesian assurance" provides a bridge between the frequentists' approach with statistical power and the Bayesian paradigm of averaging or integrating out the conditional statistical power with all possible (prior) values of the parameter vector space of θ. Bayesian assurance can then provide an unconditional probability or evidence to assess the success of a clinical trial and is therefore more realistic and robust than that of conventional statistical power.

To our experience and knowledge in clinical trials, it is very reasonable to use this hybrid frequentist-Bayesian approach in study design since prior information has always been used to calculate sample size. Whenever this prior information for the unknown parameter θ, (such as treatment effect) is sufficiently strong, the prior variance would approach zero, and the assurance equation (9.2) would approach the conventional statistical power defined in Eq. (9.1). On the other hand, if the prior information is weak, the prior variance would be large and the assurance defined in Eq. (9.1), which averages all the potential values of this vague prior distribution, would be more appropriate than the conventional statistical power to assess the probability of a successful trial.

9.3 Computational Implementation on Bayesian Assurance

With the conceptual paradigm change from the conventional statistical power defined in Eq. (9.1) to the Bayesian assurance defined in Eq. (9.2), the next logical step to actually show how to calculate the Bayesian assurance. It can be seen that the Bayesian assurance defined in Eq. (9.2) is the expected power of the parameter vector space of θ. θ can be high dimensional, such as the treatment effect or the associated variance from normally distributed data. In this situation, the computation in Eq. (9.2) would involve high-dimensional integration which would be impractical to implement in statistical software. This, however, can be resolved with a Monte-Carlo (MC) simulation-based statistical computing algorithm.

As proposed in O'Hagan et al. (2005) for assurance calculation, the general principle is to incorporate sampling from the prior distribution of θ before sampling from the data. Specifically, the general algorithm to compute the Bayesian assurance with MC of outcomes $A_1, A_2, \ldots, A_k$ is as follows:

1. Define counters I for iteration and $T_1, T_2, \ldots, T_k$ for the assurances, and set all counters to 0. Set the required number, N. Set I = 0 and start looping.
2. Sample θ from the prior distribution.
3. Sample the data and calculate the sufficient statistics using the model and the sampled value of θ from step 2.
4. For $j = 1, 2, \ldots,$ k, increment T_j by 1 if the outcome A_j has occurred.
5. Increment I: If I < N; go to step 2.
6. For j = 1, 2, …, k, estimate assurance $\gamma_j = P(A_j)$ by T_j/N.

Chen and Ho (2017) detailed the MC for several types of data and provided the R program for implementation. In this chapter, we illustrate this MC for normally distributed data due to its extensive use in clinical trials.

For normally distributed data, we typically consider two situations depending on whether the variance parameter is known or not. In the situation where the variance is assumed to be known (although this is rare in reality), the computational implementation of Bayesian assurance would be very straightforward since an analytical formula can be derived from Eq. (9.2) as seen in Chen and Ho (2017). A more realistic situation is that the variances are unknown. In this situation, the commonly used test statistic is the Student t with the test statistic formulated as $t = \frac{\overline{x}_2 - \overline{x}_1}{\widehat{\sigma}\sqrt{\frac{1}{n_1} + \frac{1}{n_2}}}$ which follows the Student t-distribution with degrees of freedom, $df = n_1 + n_2 - 2$, where $\widehat{\sigma}$ is the estimated pooled standard deviation (see Chen et al. 2017, for details).

The standard test for superiority of a new drug has a null hypothesis of no treatment difference H_0: $\delta = 0$, against the one-sided alternative H_a: $\delta > 0$, which is to reject the null hypothesis if $t > t_{\alpha, df}$ with $\alpha = 0.025$. The statistical power can then be calculated based on this t-distribution.

Based on this distribution, the calculation of Bayesian assurance can also be done by using two-dimensional numerical integration over the parameter space of δ and

σ^2 with the non-central t-distribution. However, the Monte-Carlo simulation-based approach can be easily implemented in the following steps:

1. Set counter I = 0 and the number of simulations, N (say, N = 1,000,000).
2. Sample δ and σ^2 from their prior distributions.
3. Sample $\overline{x}_2 - \overline{x}_1 \sim N\left(\delta, \left(n_1^{-1} + n_2^{-1}\right)\sigma^2\right)$ and $(n_1 + n_2 - 2)\widehat{\sigma}^2 / \sigma^2 \sim \chi^2_{df}$, calculate the t-test statistic and statistical power.
4. Estimate the assurance with the average of the resulted sample of N statistical powers.

This MC approach can be easily implemented in R as follows:

```
# Function for Bayesian Assurance for Normal Data with unknown sigma (ANDus)
    where:
# nsimu = the number of MC simulations, recommended to be >1,000,000
# prior.mean = mean value for prior distribution
# prior.sd = standard deviation for prior distribution
# prior.size = sample size from the prior clinical trial
# post.size = sample size from planned new trials so Bayesian assurance will be
    calculated
ANDus = function(nsimu,prior.mean,prior.sd,prior.size,post.size){
sim.pow = rep(0, nsimu) # temp-holder for the statistical power
# for-loop to calculate the statistical power
for(i in 1:nsimu){

        # sample chisq for sigma since (n-1)*s^2/sigma^2 ~chisq(n-1)
        sd = sqrt((prior.size-1)*prior.sd^2/rchisq(1,df=prior.size-1))
        # calculate the standard deviation for the mean
        prior.sdm = sqrt(2/prior.size)*sd # prior sd for the mean
        post.sdm = sqrt(2/post.size)*sd # posterior sd for the mean
        # sample the prior
        Delta = rnorm(1, prior.mean,prior.sdm)
        # with the sampled prior, calculate the power
        sim.pow[i] = pnorm(1.96*post.sdm,mean=Delta,
                         sd=post.sdm,lower.tail=FALSE,log.p=FALSE)
          } # end of i-loop

# Bayesian assurance is the average of simulated power
mean(sim.pow)
} # end of "ANDus" function
```

We can then call this function to calculate the Bayesian assurance using

```
> ANDus(1000000,2.5,8,20,162)
```

which produces the Bayesian assurance of 0.605. This is smaller than the conventional statistical power of 0.8.

Table 9.1 The conventional statistical power and Bayesian assurance for different sample size per treatment with the setup of mean difference of 2.5, standard deviation of 8, and sample size of 20 from prior clinical trial

Sample size	Statistical power	Bayesian assurance
10	0.097	0.151
40	0.281	0.370
70	0.451	0.475
100	0.594	0.535
130	0.709	0.576
160	0.796	0.604
200	0.876	0.631
250	0.937	0.656
300	0.969	0.673

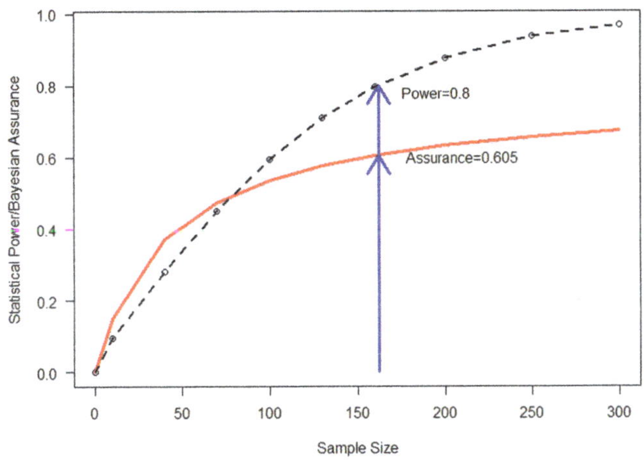

Fig. 9.1 Statistical power and Bayesian assurance for a range of sample sizes

For further illustration, we calculated the conventional statistical power and Bayesian assurance for a series of different sample sizes in Table 9.1.

It can be seen from Table 9.1 that the statistical power is generally larger than the Bayesian assurance for larger sample sizes (>80 in this simulation) since Bayesian assurance is an average of the 1,000,000 MC simulated statistical powers. However, for small sample sizes (<80 in this simulation), the Bayesian assurance is larger than the conventional statistical power due to the incorporation of prior information. It is observed that the statistical power is equal to 0.8 for sample size 162 per group due to the design. These properties can be graphically seen in Fig. 9.1, where the dashed line indicates the conventional statistical power and the solid line indicates Bayesian assurance. The vertical arrows indicate the associated statistical power at 0.8 and the Bayesian assurance of 0.605 for the designed sample size of 162.

9.4 Discussion

In this chapter, we discussed the transition from conventional statistical power to Bayesian assurance proposed in O'Hagan and Stevens (2001) in designing clinical trials. The conventional statistical power is the probability of rejecting the null hypothesis conditional on the specified treatment effect, whereas the Bayesian assurance is the unconditional probability of a successful clinical trial averaged over the parameter space of this pre-specified treatment effect. The calculation of assurance involves a high-dimensional integration that would have to resort to numerical integration. We promote the Monte-Carlo simulation-based approach and illustrated its implementation in R for clinical trials with normally distributed data.

It is common knowledge that a traditionally powered clinical trial at 80% does not guarantee 80% probability of success since the power calculation is based on a pre-specified, fixed treatment effect that will most likely be different from the true treatment effect. Typically, the Bayesian assurance is lower than the statistical power for a sufficient sample size, even though we observe that the Bayesian assurance could be higher than the statistical power for underpowered clinical trials as seen in Table 9.1 and Fig. 9.1. When compared to traditional power, we believe that Bayesian assurance can provide a more realistic and robust measure of probability of success. Therefore, we recommend the application and implementation of Bayesian assurance in clinical trial design.

References

Chen, D. G., & Ho, S. (2017). From statistical power to statistical assurance: It's time for a paradigm change in clinical trial design. *Communications in Statistics - Simulation and Computation., 46*(10), 7957–7971.

Chen, D. G., Peace, K. E., & Zhang, P. (2017). *Clinical trial data analysis using R and SAS*. Boca Raton, FL: Chapman & Hall/CRC Biostatistics Series.

Chuang-Stein, C. (2006). Sample size and the probability of a successful trial. *Pharmaceutical Statistics, 5*, 305–309.

O'Hagan, A., & Stevens, J. W. (2001). Bayesian assessment of sample size for clinical trials of cost-effectiveness. *Medical Decision Making, 21*, 219–230.

O'Hagan, A., Stevens, J. W., & Campbell, M. (2005). Assurance in clinical trial design. *Pharmaceutical Statistics, 4*, 187–201.

Chapter 10
Equivalence Tests in Subgroup Analyses

A. Ring, M. Scharpenberg, S. Grill, R. Schall, and W. Brannath

10.1 Introduction

10.1.1 Why Are Subgroup Analyses Important Within the Framework of Evidence-Based Medicine?

When aiming to receive marketing authorisation for a new drug product, the critical step in drug development is the performance of the drug in confirmatory Phase 3 trials, in the general patient population of the relevant therapeutic area. Within the clinical development programme, these confirmatory trials are typically larger than earlier, exploratory Phase 1/2 trials, and include a wider patient population (Friedman et al. 2010; Machin and Campbell 2005).

A. Ring (✉)
Department of Mathematical Statistics and Actuarial Science, University of the Free State, Bloemfontein, South Africa

medac GmbH, Wedel, Germany
e-mail: ringa@ufs.ac.za

M. Scharpenberg · W. Brannath
Faculty of Mathematics/Computer Sciences, Competence Center for Clinical Trials Bremen, University of Bremen, Bremen, Germany

S. Grill
Faculty of Mathematics/Computer Sciences, Competence Center for Clinical Trials Bremen, University of Bremen, Bremen, Germany

Leibniz Institute for Prevention Research and Epidemiology - BIPS, Bremen, Germany

R. Schall
Department of Mathematical Statistics and Actuarial Science, University of the Free State, Bloemfontein, South Africa

IQVIA Biostatistics, Bloemfontein, South Africa

Y. Zhao, D.-G. Chen (eds.), *New Frontiers of Biostatistics and Bioinformatics*, ICSA Book Series in Statistics, https://doi.org/10.1007/978-3-319-99389-8_10

The primary objective of confirmatory Phase 3 trials is to demonstrate superior (or sometimes non-inferior) efficacy of the new treatment over the control. Therefore, these trials are randomised controlled trials which are powered adequately. It is generally expected that the patient characteristics are balanced among the trial treatments due to the randomisation; however, it is possible that the treatment effect may vary within subgroups.

The statistical analysis is first performed for the primary endpoint, a pre-specified outcome variable, for which average effects in each treatment group are compared. However, both random variability and the impact of non-random variables could have an impact on the outcome.

The magnitude of random variability (e.g. measurement errors) will affect the precision of the estimates of treatment effect, and the sample size is selected at the design stage to accommodate this variability. If non-random variables—patient characteristics such as gender, age, ethnicity or metabolic genetic variations, or comorbidities, such as previous experience of stroke, or exacerbations—have an impact on the outcome, the average treatment effect in the total study population might not be representative of the treatment effect in subgroups of the patient population. Ultimately, it is important both from the public health perspective and in the treatment of individuals to identify treatment effect modifiers in order to select the best treatment for each condition (Varadhan and Seeger 2013). Therefore, assessment of consistency of the treatment effect across subgroups is an important part of Phase 3 clinical trials, which is also requested by regulatory guidelines (EMA 2013).

Confirmatory Phase 3 trials collect and report a large number of patient characteristics. It is not uncommon that published trial reports present up to 20 or 30 analyses aimed to statistically detect signals of heterogeneity for the overall interpretation of the benefit-risk profile—including efficacy and safety—of the new drug (examples are the empagliflozin outcome trial (Zinman et al. 2015) or the dabigatran RE-LY trial (Dans et al. 2013)). As these subgroup analyses are typically uncontrolled for multiplicity, signals may be generated due to the sheer volume of data and the applied analyses that are generated in such trials, and due to the multiple medical and demographic conditions that can be investigated for their impact; multiple testing invariably inflates the type-I error for generating such findings.

A COCHRANE evaluation of gender-by-treatment interaction in meta-analyses of randomised clinical trials found that "statistically significant ... interactions are only slightly more frequent than what would be expected by chance and there is little evidence of subsequent corroboration or clinical relevance of sex-treatment interactions" (Wallach et al. 2016). Thus almost all statistically significant gender-by-treatment interaction effects represent type-I errors. However, there might be other subgroups which could be more prone to interactions (e.g. comorbidities). This problem leads to the need to separate the real subgroup differences from the apparent subgroup differences (significant differences due to type-I error), and more importantly, relevant from non-relevant heterogeneities (that is, heterogeneities that are not only statistically significant but also clinically important). This second question—namely to identify heterogeneity of drug efficacy of a relevant magnitude within the therapeutic area—has often been neglected.

10.1.2 Approaches to Perform and Interpret Subgroup Analyses

Many strategies for subgroup analyses have been developed and discussed, summarised in the reviews by Varadhan et al. (2013), Hemmings (2014) and Dmitrienko et al. (2016).

- ***Descriptive subgroup analyses*** which only provide the estimate of the treatment effect in each subgroup in addition to the average overall treatment effect, without statistical testing.
- ***Exploratory subgroup analyses*** which aim to detect heterogeneity by performing statistical interaction tests for a large number of subgroups. These types of analyses should be interpreted in a non-confirmatory way but could be employed to generate signals for further confirmatory investigation.
- ***Confirmatory subgroup analyses*** which test and confirm hypotheses that affect subgroup effects. Few pre-specified covariates (e.g. identified in previous trials or by pre-clinical or epidemiological plausibility) are examined regarding their (statistically significant) impact on the endpoint.

Sometimes, challenges may arise because the trial data is reviewed from a different perspective than originally planned during the design of the trial. Signals of subgroup heterogeneity often only arise when large patient populations are studied, which is during Phase 3. In the absence of information from pre-clinical studies or medical knowledge about typical subgroup heterogeneity in a particular therapeutic area, the sponsor designs Phase 3 trials for meeting the primary objective, and hence plans subgroup analyses with the purpose to describe the data and to perform exploratory subgroup analyses. Any finding of potential subgroup heterogeneity would often not be resolved before applying for marketing authorisation unless these signals are strong. When the data are later reviewed, e.g. by health authorities searching for generalisability of the results and hence epidemiological evidence, those authorities might interpret findings of subgroup heterogeneity in a confirmatory way when they would need to decide on the implementation and reimbursement of new health technologies (Kent et al. 2010).

One of the fundamental issues is that the subgroup-by-treatment interaction (see Sect. 10.2) as a covariate of the analysis is tested for statistical significance—that is, one determines whether the hypothesis of subgroup homogeneity can be rejected.

Here we present an alternative, rather contrary approach: We propose to use tests that allow one to confirm the homogeneity of treatment effects across subgroups by rejecting heterogeneity. Thus we answer the question about the relevance of apparent heterogeneity of treatment effects across subgroups by performing equivalence tests (which in the present context we call *consistency tests*). Using an appropriate characteristic, consistency of treatment effects across subgroups is shown when the confidence interval for the characteristic is fully included within pre-defined consistency margins. This approach allows one to demonstrate the absence of relevant differences in treatment effects between the subgroups. Conventional tests of subgroup-by-treatment interaction cannot do that: the conventional test can neither confirm homogeneity of treatment effects (absence of differences in treatment effects between the subgroups), nor does it address the clinical relevance of such differences (Ting 2017).

Importantly, this investigation is based on the scenario when the primary objective, e.g. the demonstration of the superiority of one treatment over the other, has already been accomplished. There are other methods which aim to estimate treatment effects in promising subgroups when the trial failed to demonstrate the primary objective (Tanniou et al. 2017), as well as adaptive designs which seek to identify promising subgroups during the conduct of the trial and recruit more patients in the subgroups in question (enrichment designs (Wassmer and Dragalin 2015)). These methods are not considered here because their aim is to utilise heterogeneity between subgroups when it exists, while we aim to demonstrate the consistency of subgroup effects.

10.1.3 Objectives and Organisation of This Chapter

This investigation has two aims:

1. To present a framework of using equivalence tests for judging the consistency of treatment effects across subgroups; and to apply this framework to both normally distributed and binary endpoints.
2. To develop considerations on the selection of appropriate equivalence margins for such tests.

In Sect. 10.2, the general concept of those tests is developed, and the generalised linear model as the basis for subgroup investigations is presented. In addition, the setup of the simulations is described.

In Sect. 10.3, the tests are applied to models with quantitative (normally distributed) endpoints, and results on the performance of the tests in relation to the chosen margins are shown. Similarly Sect. 10.4 is devoted to binary endpoints. The final Sect. 10.5 discusses similarities and differences of subgroup investigations for the two types of endpoint, and summarises the recommendations for the selection of the equivalence margins.

Throughout we assume that one randomised clinical trial with two treatments has been performed, which aims to demonstrate the superiority of one treatment over the other. For simplicity we assume that both treatments and the two subgroups are balanced independently so that each combination of treatment and subgroup comprises 25% of the trial subjects.

10.2 The Concept of Testing Equivalence of Subgroup Outcomes

10.2.1 Generalised Linear Model

The generalised linear model underlying the developments in this chapter is as follows:

$$h\left(E\left(Y_i|X_{iT}, X_{iS}\right)\right) = \beta_0 + \beta_T X_{iT} + \beta_S X_{iS} + \beta_{TS} X_{iTS}, \tag{10.1}$$

where $E(Y_i| X_{iT}, X_{iS})$ is the expected value of the response variable for the i-th individual, h is a measurable function (link function),

$$\begin{aligned} X_{iT} &= \begin{cases} 1 \text{ if subject } i \text{ is in treatment group} \\ 0 \text{ if subject } i \text{ is in control group} \end{cases} \\ X_{iS} &= \begin{cases} -0.5 \text{ if subject } i \text{ belongs to subgroup 1} \\ 0.5 \text{ if subject } i \text{ belongs to subgroup 2} \end{cases} \end{aligned} \tag{10.2}$$

and $X_{iTS} = X_{iT}X_{iS}$. That is, β_T represents the treatment effect on $h(E(Y))$, β_S is the subgroup effect, and β_{TS} is the subgroup-by-treatment interaction. The above parameterisation of X_{iS} ensures that $E(X_{iS}) = 0$ for the case of equally sized subgroups considered here. Model (10.1) will be applied to the case of normally distributed responses in Sect. 10.3, and to the case of binary responses in Sect. 10.4.

In further discussions, the subgroup specific treatment effects will be of interest. They are defined as

$$\begin{aligned} \delta_1 =& h\left(E\left(Y_i|X_{iT}=1, X_{iS}=-0.5\right)\right) - h\left(E\left(Y_i|X_{iT}=0, X_{iS}=-0.5\right)\right) \\ =& \beta_T - 0.5\beta_{TS} \end{aligned} \tag{10.3}$$

and

$$\begin{aligned} \delta_2 =& h\left(E\left(Y_i|X_{iT}=1, X_{iS}=0.5\right)\right) - h\left(E\left(Y_i|X_{iT}=0, X_{iS}=0.5\right)\right) \\ =& \beta_T + 0.5\beta_{TS}. \end{aligned} \tag{10.4}$$

From (10.3) and (10.4) it is easy to see that the difference in subgroup specific treatment effects, $\delta_2 - \delta_1$, equals the interaction parameter β_{TS}.

Furthermore, we define the overall treatment effect as

$$\Delta = E\left(h\left(E\left(Y_i|X_{iT}=1\right)\right) - h\left(E\left(Y_i|X_{iT}=0\right)\right)\right) = \beta_T. \tag{10.5}$$

The parameter Δ thus quantifies the average effect of treatment on the expected outcome on the linear predictor scale.

In order to characterise the magnitude of the interaction, the following parameter is defined:

$$\phi = 1 - \frac{\min(\delta_1, \delta_2)}{\max(\delta_1, \delta_2)}. \tag{10.6}$$

The parameter ϕ provides a simple characteristic of treatment effect heterogeneity in the subgroups. It tells us by which percentage the smaller subgroup specific treatment effect is below the larger one. The parameter equals zero if the treatment effect is homogenous across subgroups, and is equal to 1 if there is no treatment effect in one of the subgroups. A value of $\phi = 0.5$ indicates that the effect in one of the subgroups is half of that in the other subgroup, which is a convenient interpretation. The parameter ϕ is bounded by 2 when the treatment effect is non-negative. Values of $\phi > 1$ indicate a qualitative interaction (the treatment effects in the two subgroups have different signs). Since qualitative interactions require typically a rather complex biologic action (a reverse effect in one of the subgroups, e.g. Mok et al. 2009) and values of $\phi \leq 1$ are sufficient to investigate the statistical properties of the consistency test, only quantitative interactions are considered here.

Brookes et al. (2001, 2004) defined another parameter for the quantification of the interaction in this context. This is done by regarding the subgroup-by-treatment interaction $(\delta_2 - \delta_1)$ relative to the overall treatment effect Δ, which leads to the following ratio denoted by ψ:

$$\psi = \frac{\delta_2 - \delta_1}{\Delta} = \frac{\beta_{TS}}{\beta_T}. \tag{10.7}$$

While the formula for ψ is simpler than that of ϕ, its values are less convenient to interpret. Again, $\psi = 0$ corresponds to no difference between the subgroups (because $\beta_{TS} = 0$), while given the overall treatment effect is positive ($\beta_T > 0$), the case of no effect in subgroup 1 leads to $\psi = 2$ (because the subgroups are balanced), the case of subgroup 1 having half the effect of the other leads to $\psi = 2/3$. If the treatment effect in subgroup 1 exceeds that in subgroup 2, we obtain negative values of the same magnitude for ψ.

For further considerations in this section we assume that $\delta_2 \geq \delta_1$, i.e. the treatment effect in subgroup 1 does not exceed that in subgroup 2. This can always be achieved by reordering the subgroups considered and can be formalised by $\beta_{TS} \geq 0$. For $\delta_2 < \delta_1$ the sign of ψ respectively β_{TS} in the formulas (10.8) and (10.9) changes.

Both parameters can be transformed into the other. The following equations hold:

$$\phi = \frac{2\psi}{2 + \psi} \quad \text{and} \quad \psi = \frac{2\phi}{2 - \phi}. \tag{10.8}$$

Figure 10.1 shows the relationship between ϕ and ψ. Since the parameter ϕ has the above-mentioned interpretation on the interval (0, 1) we will use ϕ in the rest of

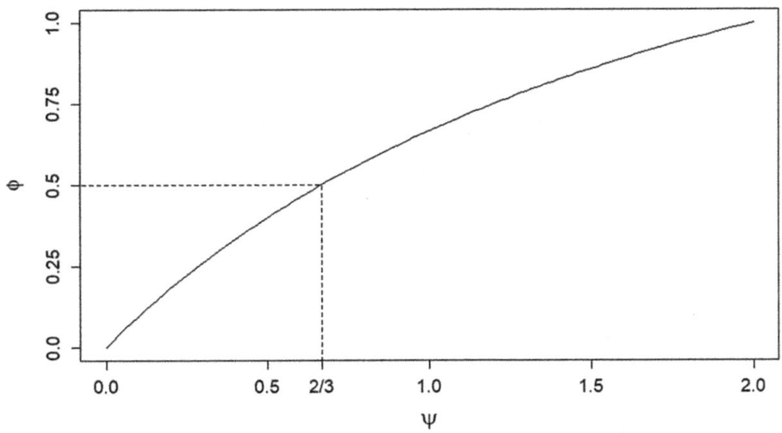

Fig. 10.1 Relation between the two alternative parameters ψ and ϕ that quantify the magnitude of the treatment-subgroup interaction

this chapter. An alternative formula which can be obtained from the transformations above is given by:

$$\phi = \frac{2\,\beta_{TS}}{2\beta_T + \beta_{TS}}. \tag{10.9}$$

This shows, how ϕ depends on the overall treatment effect and the interaction term, which as stated above is the difference in subgroup specific treatment effects. Solving for β_{TS} in (10.9) shows how the interaction parameter depends on the magnitude of interaction measured by ϕ:

$$\beta_{TS} = \beta_T \frac{2\phi}{2-\phi}. \tag{10.10}$$

Finally, we want to show how the magnitude of interaction affects the subgroup specific treatment effects δ_1 and δ_2:

$$\delta_1 = \beta_T\left(1 - \frac{\phi}{2-\phi}\right) \quad \text{and} \quad \delta_2 = \beta_T\left(1 + \frac{\phi}{2-\phi}\right). \tag{10.11}$$

10.2.2 Equivalence Tests

Traditionally, interaction effects are tested using a null hypothesis of the type H_0: $\beta_{TS} = 0$. The objective of such a test is the rejection of the null hypothesis with a pre-specified type-I error, in order to claim that there is a "significant difference". When the null hypothesis can be rejected, the difference is demonstrated (by the level of significance). When the null hypothesis cannot be rejected, no sufficient evidence against the null hypothesis has been seen, but this does indeed not imply that the null hypothesis is true.

For the evaluation of subgroup homogeneity such test of H_0: $\beta_{TS} = 0$ is of little relevance. When the overall trial result was positive, a minor difference in the treatment effect between two subgroups might be clinically acceptable and hence unimportant for the interpretation of the benefits or risks of the new therapy. Furthermore, the test outcome might just depend on the sample size of the trial: A larger sample size might have high power to confirm a particular difference, although such a difference might not be clinically relevant; in contrast, a small sample size will be associated with low power to reject the null hypothesis so that a potentially clinically relevant difference between the subgroups might not be detected. This setup is neither in the interest of the trial sponsor nor of the patients, since differences detected as statistically significant might not be clinically important, and vice versa, statistically non-significant differences might be clinically important.

Equivalence tests address the above-mentioned problems directly: By defining a medically relevant margin θ_c—a value which should not be exceeded as the "heterogeneity" between the two subgroups—an appropriately specified null hypothesis and test can potentially demonstrate the similarity of the effects between the two subgroups (Fig. 10.2). There are two null hypothesis, which claim that there is a relevant difference to either direction:

$$H_{0,1} : -\theta_c \geq x \quad \text{and} \quad H_{0,2} : x \geq \theta_c \tag{10.12}$$

where x is the parameter of interest. Rejection of both one-sided hypotheses at level α would allow one to conclude

$$-\theta_c < x < \theta_c \tag{10.13}$$

and thus the consistency of the treatment effect across the subgroups. This way of specifying the null and alternative hypotheses is called the Two One-Sided Test procedure (TOST—Schuirmann (1987)) or interval inclusion principle (Ocaña et al. 2008). The test decision is illustrated in Fig. 10.2. Figure 10.3 indicates the general principle, showing possible scenarios for treatment contrasts within each subgroup, which should ultimately lead to decisions similar to those depicted in Fig. 10.2.

The key issue for the equivalence test is the selection of an appropriate margin θ_c. While there will always be some statistical considerations to determine the margin, the main input would have to be provided by medical experts, to address the question which difference would just be large enough to be considered relevant in the therapeutic area of interest. The relevance would be determined based on the current experience with regard to therapy benefits in this area, and the efficacy that is expected on average in the whole population. The ICH E10 2000 guideline describes general considerations when selecting equivalence margins, which would need to be adapted for the case of subgroup heterogeneity.

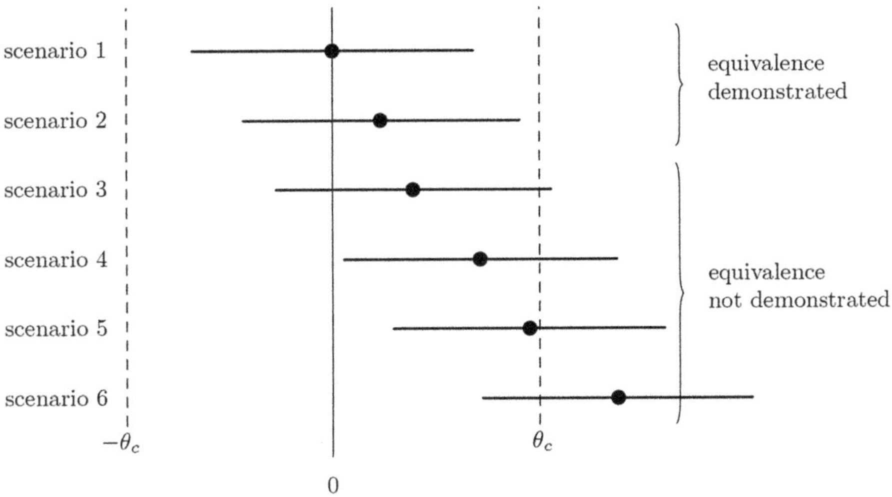

Fig. 10.2 Concept of equivalence tests illustrated for hypothetical scenarios with different magnitudes of the estimated characteristic and confidence intervals with equivalence margins $-\theta_c$ and θ_c

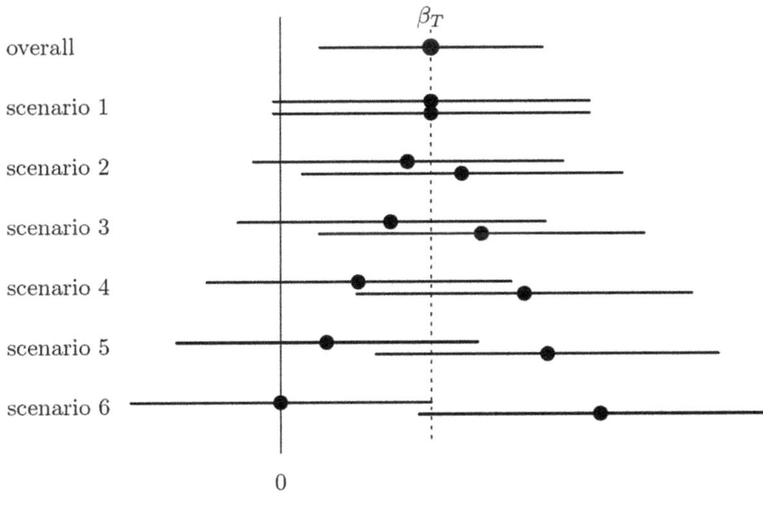

ϕ	ψ
0	0
0.29	$\frac{1}{3}$
$\frac{1}{2}$	$\frac{2}{3}$
$\frac{2}{3}$	1
$\frac{6}{7}$	$\frac{3}{2}$
1	2

Fig. 10.3 Schematic representation of subgroup heterogeneity when both subgroups are of equal size. The value of ϕ describes the magnitude of the heterogeneity. For each subgroup, the mean estimate for the treatment contrast and its confidence interval are given. As the sample size in each subgroup is half the total sample size, the confidence intervals are wider than that of the overall average treatment effect

In the rest of the chapter, we will describe statistical properties of the developed equivalence tests for subgroup heterogeneity for both normally distributed and binary endpoints. While the technical details for the two types of endpoint are different (which will be developed separately in Sects. 10.3 and 10.4), the key ideas are similar: To develop an appropriate statistic for each type of endpoint which allows one to test heterogeneity using a TOST, and to assess different equivalence margins with respect to reject the null hypotheses of inequivalence.

10.2.3 Outline of the Simulations

The key objective of Sects. 10.3 and 10.4 is to apply the equivalence tests for quantitative and binary endpoints and evaluate their performance in relation to the selected equivalence margin.

All simulations were set-up such that the treatments (active and placebo) are balanced, and the two subgroups are balanced independently. Hence the total number of subjects in each simulated trial was divisible by 4.

The Monte-Carlo simulation was performed by individually simulating and analysing clinical trials. For a given set of simulation parameters (e.g. model parameters of (10.1), power for the test of the overall treatment effect), the following steps were performed:

1. Determine the number of subjects to achieve the specified power for demonstrating overall superiority using a two-sided significance level of $\alpha = 0.05$ (based on an average effect size).
2. Derive the effect size in each subgroup from the selected value of the subgroup difference ϕ.
3. Generate data according to the underlying model derived from (10.1) (for more details, see Sects. 10.3 and 10.4).
4. Analyse the data using the statistical model that accounts for treatment, subgroup and their interaction and capture the selected test statistic to test the subgroup heterogeneity for each type of endpoint and its 90% confidence interval (CI).
5. If the CI is fully included within the given margins, the TOST hypotheses are rejected and the consistency test is declared positive.

The above procedures were repeated 10,000 times for each case, which yielded sufficient precision of the estimated power for the consistency test, as the percentage of trials which declare similarity would be determined with a standard error of less than 0.5% points (Koehler et al. 2009).

Note that for large values of ϕ rejection of the hypotheses in (10.12), i.e. concluding treatment effect homogeneity across subgroups, is not favourable since it constitutes a type-I error for the question at hand. However, we will nevertheless speak of "power" in these cases.

The analyses were performed in R (R Development Core Team 2008), version 3.4.3, with additional packages `lsmeans` (Russell 2015), `snow` and `snowfall`. For the Monte-Carlo simulations and analyses of binary data, the packages `snow and snowfall` were used for parallel computing.

10.3 An Equivalence Test for Consistency of Subgroup Effects with a Quantitative Endpoint

10.3.1 Definition of the Test

In the case of a quantitative (normally distributed) endpoint Y, the general linear model in Eq. (10.1) with identity link function becomes

$$E\left(Y_i|X_{iT}, X_{iS}\right) = \beta_0 + \beta_T X_{iT} + \beta_S X_{iS} + \beta_{TS} X_{iTS}. \tag{10.14}$$

Therefore, we can write

$$Y_i = \beta_0 + \beta_T X_{iT} + \beta_S X_{iS} + \beta_{TS} X_{iTS} + \epsilon_i \tag{10.15}$$

where β_T is the effect of treatment vs. control, β_S is the difference between the subgroups, β_{TS} is the interaction between treatment and subgroup, and ϵ_i is the common error term for subject i, normally distributed with mean 0 and common standard deviation σ. The effect size is denoted as $\Delta/\sigma = \beta_T/\sigma$.

As noted in Sect. 10.2 in Eqs. (10.3) and (10.4), the treatment contrasts in the two subgroups are

$$\delta_1 = \beta_T - 0.5\beta_{TS} \quad \text{and} \quad \delta_2 = \beta_T + 0.5\beta_{TS}. \tag{10.16}$$

The difference in subgroup specific treatment contrasts is $\delta_2 - \delta_1 = \beta_{TS}$, which corresponds to the interaction coefficient of the linear model in Eq. (10.14).

The key idea for the equivalence test with normally distributed data is to relate the interaction term to the residual: A small residual variability implies that the interindividual variance is small; therefore the therapy effect within both subgroups should not be allowed to deviate too much from each other. However, when the overall variability is large, a larger difference in subgroup effects would be acceptable, as is the (random) difference between two arbitrary subjects in the overall population. Hence, the aim of the consistency test is to judge the relevance of the subgroup differences relative to the underlying variability of the quantitative endpoint. This leads to the construction of a characteristic through scaling the subgroup difference by standard deviation σ. This ratio—the contrast of the treatment effects in the two subgroups, divided by the overall residual variability—is called the "consistency ratio":

$$CR = \frac{\beta_{TS}}{\sigma}. \tag{10.17}$$

It has been shown (Ring et al. 2018, using Schall 1995), that a (1-α)*100% confidence interval for this consistency ratio can be derived based on the non-central t-distribution as follows (LCL and UCL are the lower and upper confidence limit):

$$[LCL, UCL] = \left[m \cdot Q_\nu\left(1 - \alpha/2; \frac{\widehat{\beta_{TS}}}{m\widehat{\sigma_r}}\right), m \cdot Q_\nu\left(\alpha/2; \frac{\widehat{\beta_{TS}}}{m\widehat{\sigma_r}}\right)\right]. \tag{10.18}$$

Here $Q_\nu(t;x)$ is the inverse of F_ν when $F_\nu(x;\lambda)$ is viewed as a function of λ for fixed x, and $F_\nu(x;\lambda)$ denotes the value of the cumulative distribution function of the t-distribution with $\nu = n - 4$ degrees of freedom (n is the total sample size) and non-centrality parameter λ, evaluated at x.

Furthermore,

$$m = \sqrt{\frac{1}{n_{A1}} + \frac{1}{n_{A2}} + \frac{1}{n_{B1}} + \frac{1}{n_{B2}}} \tag{10.19}$$

with n_{Xj} denoting the sample size of treatment j in subgroup X (and $n = n_{A1} + n_{A2} + n_{B1} + n_{B2}$). As we assume that the subgroups and treatments are completely balanced, all n_{Xi} fulfil $n_{Xi} = n/4$, so that $m = 4/\sqrt{n}$.

This leads to an equivalence test, which compares the confidence interval of the consistency ratio against the pre-defined equivalence margin as

$$-\theta_c < \frac{\delta_1 - \delta_2}{\sigma_r} < \theta_c. \tag{10.20}$$

Hence the two one-sided null hypotheses to be tested are $H_{0,1}: -\theta_c \geq \frac{\delta_1-\delta_2}{\sigma_r}$ and $H_{0,2}: \frac{\delta_1-\delta_2}{\sigma_r} \geq \theta_c$.

In order to investigate the impact of particular equivalence margins on the performance of the test, we apply formulas (10.6) and (10.9) to the quantitative case. The introduction of the residual variability into (10.15) does not alter these formulas, so that we have

$$\phi = 1 - \frac{\min(\delta_1, \delta_2)}{\max(\delta_1, \delta_2)} = \frac{2\beta_{TS}}{2\beta_T + \beta_{TS}}.$$

As before, ϕ is varied between 0 (no differential subgroup effect) and 1 (no treatment effect in the second subgroup).

10.3.2 Performance of the Consistency Test with Respect to the Equivalence Margins

The Monte-Carlo simulations have been performed as outlined in Sect. 10.2.3, aiming to estimate the power of the equivalence test in relation to the subgroup heterogeneity ϕ for various values of the effect size (Δ/σ) and selected values of the equivalence margins θ_c. The true overall drug effect was selected to be $\Delta = 0.4$. As the overall variability is a scaling factor, one of the key objectives was to investigate the impact of smaller vs. larger effect sizes (0.3–0.6) by choosing different residual variabilities (σ). These effect sizes cover a typical range as seen in clinical trials (for examples, see Sect. 10.3.3).

Three investigations were performed. The first two cases examine the relationship between subgroup heterogeneity ϕ and the power of the consistency test. In the first case, the sample size was fixed across all effect sizes, in the second case the power to demonstrate superiority was fixed (leading to higher sample sizes when the effect size was smaller). The third investigation analysed the relationship between the pre-selected margins and the power of the consistency test.

When the sample size was fixed (Fig. 10.4), then the power curves of the equivalence test are quite similar for a given margin. For example with a margin of $\theta_c = 0.5$, the power for the equivalence test with a sample size of 400 (blue curve) is about 82% when there is no subgroup divergence ($\phi = 0$) across all effect sizes.

As expected, the equivalence test is more powerful when ϕ is small. Up to a value of about $\phi = 0.3$, the power of the test is only reduced slightly; this appears to be consistent for many combinations of the other parameters (effect sizes and margins). On the other end of the scale, when ϕ approaches 1 (where there is no treatment effect in one of the subgroups), the probability of successfully demonstrating equivalence is around 5%. The steepest decrease in power (inflection point) in the cases that have been investigated is reached with ϕ of 0.7–0.8. (As indicated, the term "power" might not be the best for large values of ϕ, as they specify a clear heterogeneity.)

Despite these similarities, there are remarkable quantitative differences for intermediate subgroup differences of around $\phi = 0.5$. For a larger effect size of 0.6 and a margin of 0.5, the power is about 40% for the whole range of sample sizes. For a smaller effect size such as 0.3, the power of the equivalence test is between 55 and 75%, depending on the sample size. When the effect size increases (while the same margin is selected), then the power for demonstrating equivalence decreases, because a reduction of the variability σ leads to a larger consistency ratio. This finding implies that the S-shaped power curve is flatter for smaller values of ϕ and steeper for larger values when the residual variability is larger (and hence the effect size is smaller).

This result is in line with the aims for constructing the test (Sect. 10.3.1): In the consistency ratio *CR*, the residual variance σ is a scaling factor for the observed subgroup differences. A larger residual variability means that the differences between individuals within each subgroup are larger, and hence it might be more acceptable when the efficacy between subgroups differs more.

The impact of the chosen margin among those margins that have been investigated is merely proportional—a larger margin leads to more power to demonstrate equivalences, but the shape of the power curve is quite similar. Hence a particular margin should be selected with the whole potential spectrum of subgroup differences (ϕ) in mind, and which medical relevance such a difference would have. The benefits and risks of such differences with respect to the given drug, its therapeutic area and the particular endpoint studied need to be balanced when determining the equivalence margin for a clinical trial.

The case of fixed sample sizes for various effect sizes might, however, not be fully realistic, because clinical trials will usually not primarily be powered to

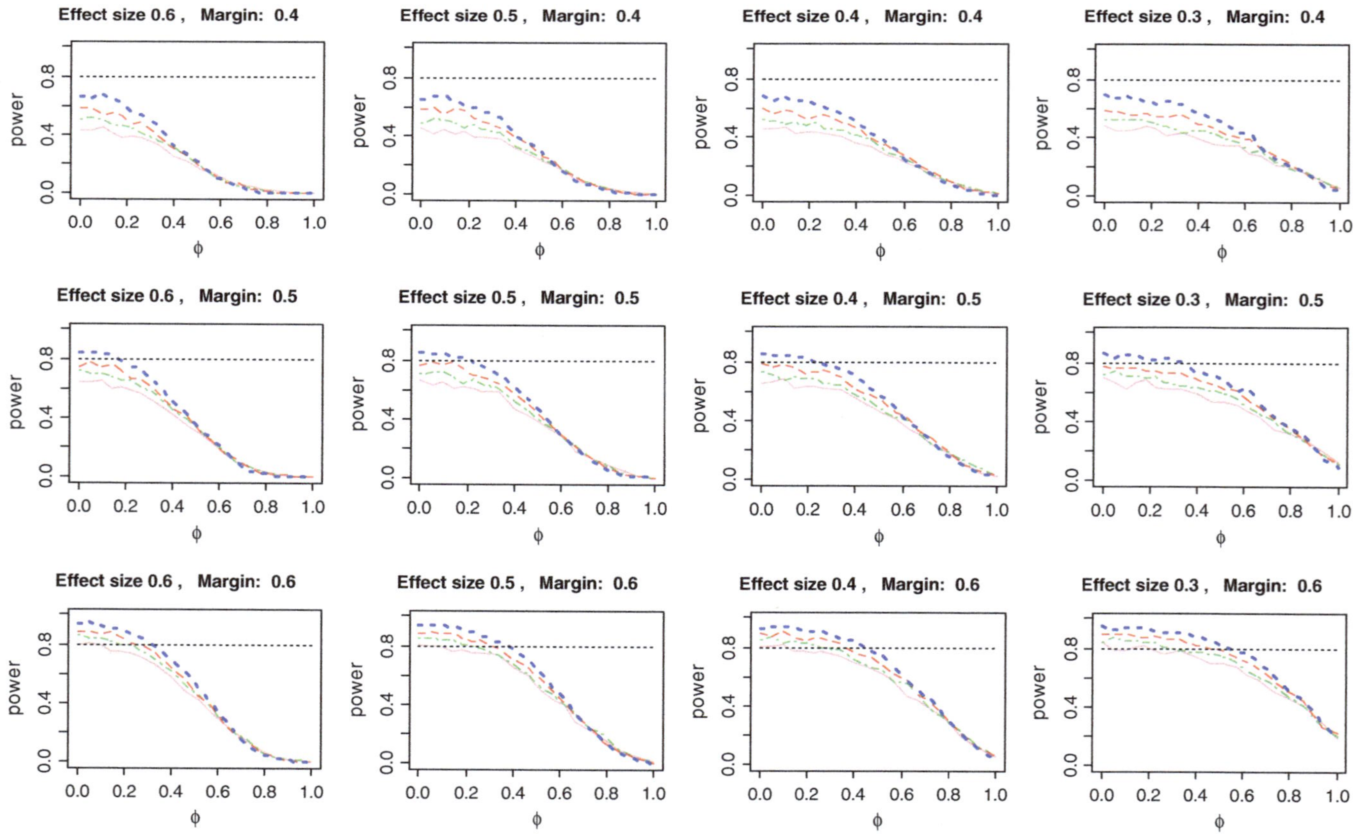

Fig. 10.4 Power of the consistency test in relation to the subgroup differences ϕ, using various cases for the effect size (0.3–0.6) and different equivalence margins θ_c (0.4–0.6) **for fixed sample sizes**. The colours of the lines correspond to the total sample size of: **pink 100, green 140, red 200 and blue 400 patients**

demonstrate equivalence between subgroups. Instead, clinical trials are typically designed with regard to a primary objective like demonstrating superiority of one treatment over the other, and the trial would be powered for this objective. Showing equivalence between subgroups would rather be a secondary objective. In the cases above, the power to demonstrate superiority was above 95% for all sample sizes when the effect size was 0.6, while it was below 80% when the effect size was only 0.3.

In the second case (Fig. 10.5) the power for the test of superiority was fixed at 80%, 85%, 90% and 95% and the sample size was calculated accordingly. Otherwise the analysis was similar. (Of note, the sample size in Fig. 10.4 was the same for the effect size of 0.5, so that these results would overlap between both figures.)

The general S-shape of the curves is similar for Figs. 10.4 and 10.5 with parts that are flat for rather small ($\phi \leq 0.3$) and steeper for rather large ($\phi \geq 0.8$) subgroup differences.

However, it can also be seen that the power of the consistency test strongly depends on the effect size for a given power for demonstrating superiority. When the effect size is rather large (0.6), the sample size to demonstrate superiority is moderate (44–74 to obtain a power of 80–95% for the superiority test). In this case, the consistency test would only have a power of less than 50%, when the margin was selected with a value of 0.5. In order to reach acceptable power for the consistency test, a much larger sample size would be required for the selected margin.

For smaller effect sizes in the range of up to 0.3, the sample size for demonstrating superiority must be about four times as large as for an effect size of 0.6, which leads to more power for the consistency test at any given margin. With a margin of 0.5, the power for the consistency test would be about 80% or more for values of the subgroup divergence ϕ of up to 0.5, when the trial is powered with at least 80%. For ($\phi \leq 0.3$), the power for the consistency test is more than 90%.

Hence, the effect size (or the residual variability for a given treatment effect) has a "double effect" on the power to demonstrate equivalence between subgroups: on one hand, a lower variability increases the consistency ratio and hence reduces the power of the consistency test; on the other hand, a lower variability requires less sample size to demonstrate superiority, but the lesser sample size also reduces the power for the consistency test.

The expected effect, that larger residual variability leads to better chance of demonstrating equivalence, could be demonstrated. But when equivalence hypotheses are tested subsequently to superiority hypotheses only based on their sample sizes, the power of the equivalence test might largely deviate from the range of 80–95%, which is typically selected for the power.

As before in Fig. 10.4, the choice of the equivalence margin has merely a proportional effect on the power. However, it seems that the selection of a margin could compensate for small or large sample sizes. When the sample size is already fixed by the trial, the margin could be selected to get the statistical properties of the equivalence test aligned with the medical relevance of actual subgroup differences. While this procedure sounds meaningful, such an approach would in other circumstances generally be discouraged: Selecting an equivalence (or non-inferiority) margin or other design element (such as the power) of a clinical trial to account for a given sample size would be a reversion of cause and outcome.

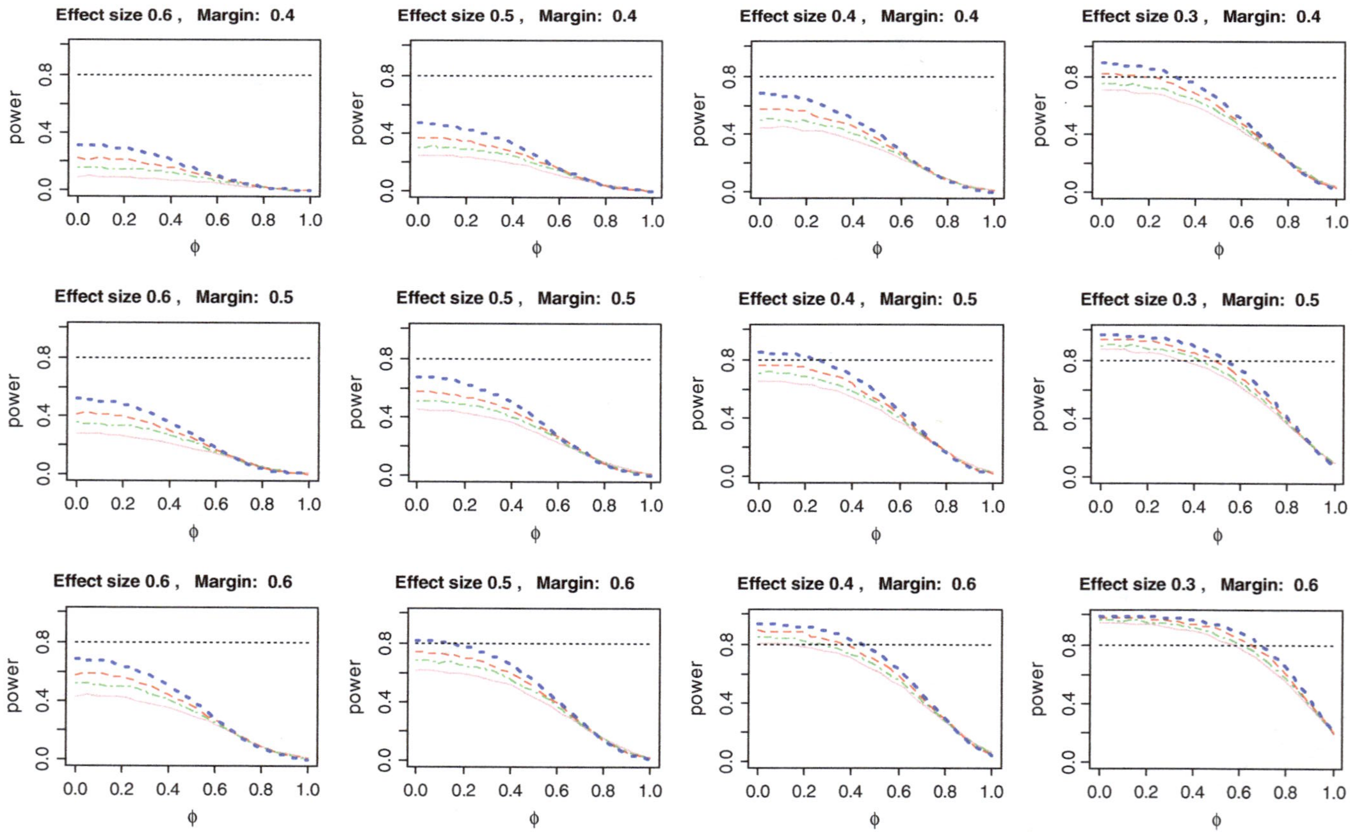

Fig. 10.5 Power of the consistency test in relation to the subgroup differences ϕ **for fixed power** of the superiority test, using various cases for the effect size (0.3–0.6) and with different equivalence margins θ_c (0.4–0.6). The colours of the lines correspond to the overall power of the trial to demonstrate superiority: **pink 80%, green 85%, red 90% and blue 95%**

Whether such a procedure would be acceptable for a (secondary) test of subgroup equivalence will need to be further discussed. This discussion should explicitly include the option of "not testing" for subgroup equivalence, because the statistical properties of such a test might not be acceptable.

The third investigation was performed similarly, but the independent axis (Fig. 10.6) is now the equivalence margin θ_C (between 0.1 and 1.0), and 4 different values for the subgroup divergence ϕ (0.0, 0.33, 0.66 and 1.0) were used. As before, the sample size was adapted to the power of the superiority test of 80 or 90%.

Again, the power for demonstrating equivalence is very similar for both smaller values of ϕ, as the red and the green line almost overlap for any value of the equivalence margins. With larger divergence, the power is largely reduced. However, when the effect size is small (and hence the sample size was larger), the power of the equivalence test becomes quite large, up to 80% and more, even when ϕ approaches 1.

10.3.3 Implications of the Variance Scaling

The main concept for the development of the consistency test for normally distributed endpoints was the scaling of the treatment contrast by the residual variability. The idea is to judge the relevance of the between-group variation relative to the between-subject variation: The larger the variability between subjects (after accounting for the subgroup covariate), the higher the variability between the subgroups that might be acceptable.

The simulations have shown that there might not be a universally acceptable equivalence margin θ_C for the proposed consistency test: If the margin is small, then the power for demonstrating equivalence is below 50% in some cases, even when there is no difference between the subgroups. If the margin is large, then large heterogeneity up to the point of no effect in one of the subgroups is accepted by the test in many cases, if only the residual variability is large as well.

The latter outcome is indeed implied by the construction of the test: as the value of σ is not limited in the consistency ratio, its magnitude could be large, hence reducing the value of the consistency ratio (as it is the denominator). Furthermore, as the sample size usually increases with increasing variability, the confidence intervals around the estimated consistency ratio decrease and hence the consistency tests get again more powerful.

It is not uncommon to observe variabilities which exceed the magnitude of the clinically relevant effects of the new medication (having effect sizes of less than 1). For example, drugs for Type-II diabetes should demonstrate a medically relevant decrease of HbA1c in the range between 0.5% and 0.7% compared to placebo, with a standard deviation of about 1.0% (Forst et al. 2010). In the area of hypertension, a placebo adjusted reduction in the range of 6–8 mmHg of systolic blood pressure (SBP) and 3–4 mmHg of diastolic blood pressure (DBP) might be considered clinically relevant, while these endpoints have a σ of 17 mmHg (SBP)

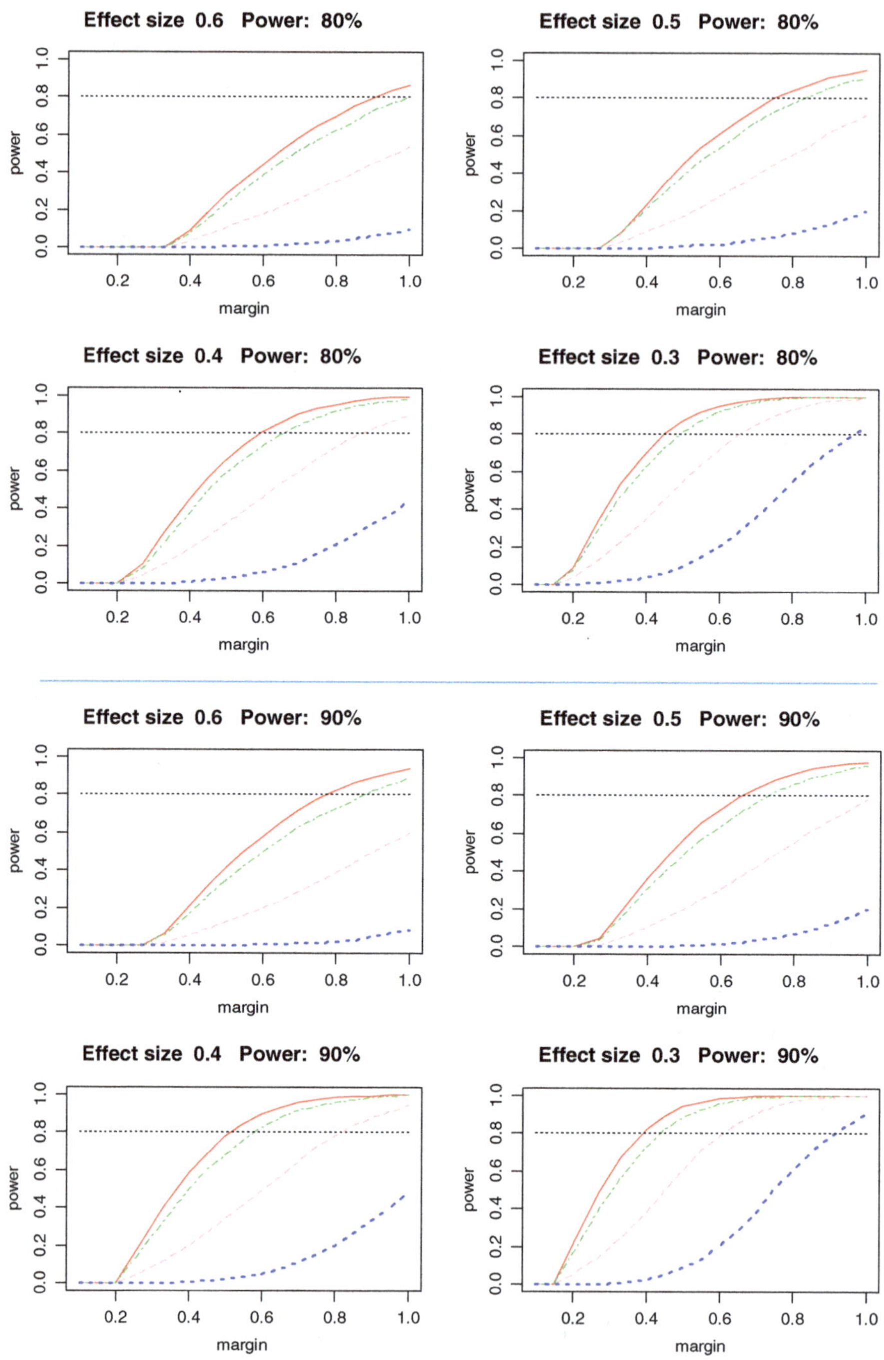

Fig. 10.6 The power of the consistency test related to the selected margin θ_c for various effect sizes and a power for the overall superiority test of 80 or 90%. The line colours show the subgroup difference ϕ with values of **0 (red), 0.33 (green), 0.66 (pink) and blue (1.0)**

or 11 mmHg (DBP), respectively (Plavnik and Ribeiro 2002; Bath et al. 2009). Finally, the minimal clinically important differences of the key endpoint for chronic obstructive pulmonary disease (COPD), the forced expiratory volume in 1 s (FEV1), should be increased by 60–100 mL in comparison with placebo. This endpoint typically has a standard deviation in the range between 150 and 190 mL (Beeh et al. 2015; Donohue 2005). These magnitudes of the effect sizes have been reflected in the simulations above.

Another aspect of variance scaling in the consistency ratio is that the observed variance in a trial combines different sources of variation. Some of these sources are "natural variability" of the endpoint in question in the patient population which is independent of the trial or study, and includes between-subject and within-subject variability. These variance components should preferably be involved in the consistency ratio.

However, some additional variability of the endpoint might be introduced by the design of the trial, how the measurements are taken, or how the data quality is monitored. This is a limitation of the method, as study specific parts of the variability can most often not be quantified. In general, it is in the interest of the sponsor and the principal investigator to minimise these types of variability, as variability reduces the overall power for the primary objective of the trial (e.g. to demonstrate a particular treatment difference). In the light of the performance of "pragmatic trials," which aim to demonstrate the efficacy and effectiveness of drugs in the "real world" situation, the variability could be inflated compared to standard trials with more controlled processes and measurements. This inflation is typically compensated by an increase of sample size, but for the consistency test, this would actually be counterproductive, as shown by the above simulations. In other words, the consistency ratio is for the parameter β_{TS} what the effect size is for β_T, as both have the variance as denominator. But when the treatment effect is evaluated for superiority, while the interaction effect is evaluated for equivalence, then the impact of an inflation of the variance is opposed for both tests.

Finally, the new drug itself could introduce additional variability of the endpoint relative to existing therapies. This element could be controlled, e.g. in repeated-measurement trials, by using the within-subject variance within the control group as the denominator of the consistency ratio, which is a solution that is used in the field of scaled average bioequivalence (Haidar et al. 2008). We will continue the discussion on modifications of or alternatives to the consistency test in Sect. 10.5.

10.3.4 Example for Planning the Study Design

An example shall illustrate how clinical and statistical team members could discuss and select appropriate equivalence margins:

Suppose a Phase 3 trial is to be designed with the primary objective to demonstrate the superiority of treatment A over control C using a normally distributed endpoint. The trial is planned for an expected treatment effect of 0.8, with a standard deviation σ of 1.6, leading to an effect size of 0.5. In order to demonstrate the primary hypothesis with a two-sided significance level $\alpha = 0.05$ with a power of 90%, a sample size of 86 subjects per group is necessary. While *a-priori* there might be no indication of effect heterogeneity between the two genders ($\phi = 0$), the team would like to perform an appropriate consistency test as a secondary hypothesis, aiming to reject the null hypothesis of non-consistency using $\alpha = 0.05$ with a power of 75%. To be conservative, they would power this test using a value of ϕ of at most 0.2 (to allow for contingency).

When reviewing Fig. 10.5 for the effect size of 0.5, consistency margin $\theta_c = 0.6$ leads to the desired power for the consistency test (the red graph corresponds to a trial that has a sample size for the rejection of the primary hypothesis of 90%). The same graph also indicates that such a test would demonstrate consistency with a power of 50% when $\phi = 0.5$, and still with a probability of about 30% when $\phi = 0.7$. Similar conclusions can be drawn from Fig. 10.6, the panel with effect size of 0.5 and overall power of 90%.

A medical consideration in this trial could be that differential subgroup effects which are twice as large in one subgroup compared to the other would be acceptable ($\phi = 0.5$). With an overall treatment effect of 0.8, the subgroup effects would be about 0.54 and 1.08.

If the team would feel that this margin (0.6) would be too liberal for larger values of heterogeneity ϕ, they could discuss using a lower value, such as 0.5. However, as can be seen from the corresponding panel in Fig. 10.5, the consistency test would not get the desired power, given the sample size for the primary hypothesis.

In this case, if the test for consistent effects for both genders would be important, a solution would be to increase the overall sample size and to select a lower equivalence margin. For example, a sample size of 104 subjects per group would lead to 95% overall power, and a consistency margin of about 0.55 would lead to 75% power for $\phi = 0.2$, about 55% power for $\phi = 0.5$, and about 25% power for $\phi = 0.7$. Of course, whether the additional costs could be justified against the gain of information would need to be discussed. Further sensitivity analyses should be performed, in particular for the case when σ is slightly smaller than expected, as this would reduce the power of the consistency test for larger values of ϕ.

To understand the analysis for the subgroup equivalence, let's assume that the trial has been performed with a pre-specified margin $\theta_c = 0.6$. The subgroups of interest are defined by the genders. The randomisation might have been stratified by gender, so that all groups are fully balanced. The simulated results are shown in Table 10.1.

The parameters of interest are estimated as $\widehat{\beta}_{TS} = \widehat{\delta}_1 - \widehat{\delta}_2 = -0.275$ and $\sigma = 1.092$. As all subgroups were fully balanced, formula (10.19) leads to $m = \frac{4}{\sqrt{176}}$. Hence applying formulas (10.17) and (10.18) using $\alpha = 0.05$ leads to $\widehat{CR} = \frac{\widehat{\beta}_{TS}}{\widehat{\sigma}} = 0.252$ and its confidence limits $LCL = 0.002$ and $UCL = 0.499$. Notably, the p-value for the interaction effect β_{TS} was 0.009, hence well below the α-level.

The conventional interpretation for such a trial could have been that an overall significant treatment effect has clearly been demonstrated. However, the p-value for

Table 10.1 Fictional outcome of a randomised controlled trial in 176 patients, with balanced treatment groups and subgroups

	Comparison A–C
Total	
Number of subjects	176
Adj. mean (SE)	0.581 (0.165)
95% CI	0.256, 0.906
p-value	<0.001
Men	
Number of subjects	88
Adj. mean (SE)	0.443 (0.233)
95% CI	−0.016, 0.903
p-value	0.058
Women	
Number of subjects	88
Adj. mean (SE)	0.718 (0.233)
95% CI	0.258, 1.177
p-value	0.002

the gender interaction effect was quite small, raising concerns that the effect might be heterogeneous across both genders.

An analysis within the subgroups would lead to the result that the superiority of the treatment could not have been demonstrated in men. The pre-specification of the equivalence test with a margin of 0.6 for the consistency ratio allows rejecting the heterogeneity of the treatment effect between genders, so that the homogeneity of the subgroup effects was demonstrated.

10.4 An Equivalence Test for the Consistency of Subgroup Effects with a Binary Endpoint

10.4.1 Definition of the Test

We now apply the ideas of Sect. 10.2 to binary data. When considering binary data, the expectation of the target variable is given by the event probability p. A popular model for binary data is the logistic regression model where the link function h of Eq. (10.1) is given by

$$h\colon (0,1) \to \mathbb{R}, \quad p \mapsto \operatorname{logit}(p). \tag{10.21}$$

The model formulation in (10.1) then yields

$$\operatorname{logit}(P(Y_i = 1 | X_{iS}, X_{iT})) = \beta_0 + \beta_T X_{iT} + \beta_S X_{iS} + \beta_{TS} X_{iTS}. \tag{10.22}$$

Again, the subgroup specific treatment effects are the basis of the assessment of treatment effect homogeneity between subgroups

$$\delta_1 = \beta_T - 0.5\beta_{TS} \quad \text{and} \quad \delta_2 = \beta_T + 0.5\beta_{TS}. \tag{10.23}$$

In the logistic regression model (10.22) considered here, the subgroup specific treatment effects coincide with the subgroup-wise log odds ratios, i.e. the log of the odds ratio of having the event under consideration of treatment versus control group, under the condition that the subject is in the respective subgroup:

$$\begin{aligned}
\delta_1 &= \log\left(\tfrac{P(Y=1|X_{iS}=-0.5,X_{iT}=1)}{P(Y=0|X_{iS}=-0.5,X_{iT}=1)} / \tfrac{P(Y=1|X_{iS}=-0.5,X_{iT}=0)}{P(Y=0|X_{iS}=-0.5,X_{iT}=0)}\right) = \log\left(OR_1\right), \\
\delta_2 &= \log\left(\tfrac{P(Y=1|X_{iS}=0.5,X_{iT}=1)}{P(Y=0|X_{iS}=0.5,X_{iT}=1)} / \tfrac{P(Y=1|X_{iS}=0.5,X_{iT}=0)}{P(Y=0|X_{iS}=0.5,X_{iT}=0)}\right) = \log\left(OR_2\right).
\end{aligned} \tag{10.24}$$

Therefore δ_k ($k = 1, 2$) is the usual treatment effect of the logistic regression model, when only subjects from subgroup k are considered.

In the case of normally distributed endpoints of Sect. 10.3, the consistency test was based on the idea to scale the difference between the subgroup specific treatment effects by the residual variance. The variance scaling was necessary for normally distributed endpoints to obtain a scale free effect size measurement. In the case of binary endpoints, we already have a scale free effect size measure at hand—the odds ratio. Therefore, the equivalence test for binary endpoints is based on a comparison of the odds ratios across subgroups. The odds ratio of treatment is homogeneous across subgroups, if and only if $OR_2/OR_1 = 1$. This ratio of odds ratios can be expressed as the exponential of the difference of the subgroup specific treatment effects $e^{\delta_2-\delta_1} = e^{\beta_{TS}}$. Therefore, the equivalence test can be based on the interaction parameter coefficient β_{TS}, which equals zero if and only if the odds ratio of treatment is homogeneous across treatment groups. To assess homogeneity across subgroups we need to reject the null hypotheses

$$H_{0,1}\colon -\theta_c \geq \beta_{TS} \quad \text{and} \quad H_{0,2}\colon \beta_{TS} \geq \theta_c \tag{10.25}$$

for the pre-specified consistency margin θ_c. As before, we want to test both hypotheses simultaneously by computing a confidence interval for β_{TS} and rejecting the hypotheses of a heterogeneous treatment effect across subgroups, when this confidence interval is included within the consistency margins.

An approximate confidence interval for the interaction coefficient β_{TS} can be based on the maximum likelihood estimator of the log odds ratio, and is given by

$$CI = \left[\widehat{\beta}_{TS} \mp z_{1-\alpha}\widehat{\sigma}_{\widehat{\beta}_{TS}}\right] \tag{10.26}$$

where $z_{1-\alpha}$ is the upper α quantile of the standard normal distribution and $\widehat{\sigma}_{\widehat{\beta}_{TS}}$ is an estimate of the standard deviation of the maximum likelihood estimator $\widehat{\beta}_{TS}$, which can be deduced from the inverse fisher matrix $F^{-1}\left(\widehat{\beta}_{TS}\right)$ of the logistic regression model (Hosmer Jr. et al. 2013). Of course any other confidence interval for β_{TS}, like the profile likelihood interval (Venzon and Moolgavkar 1988), could

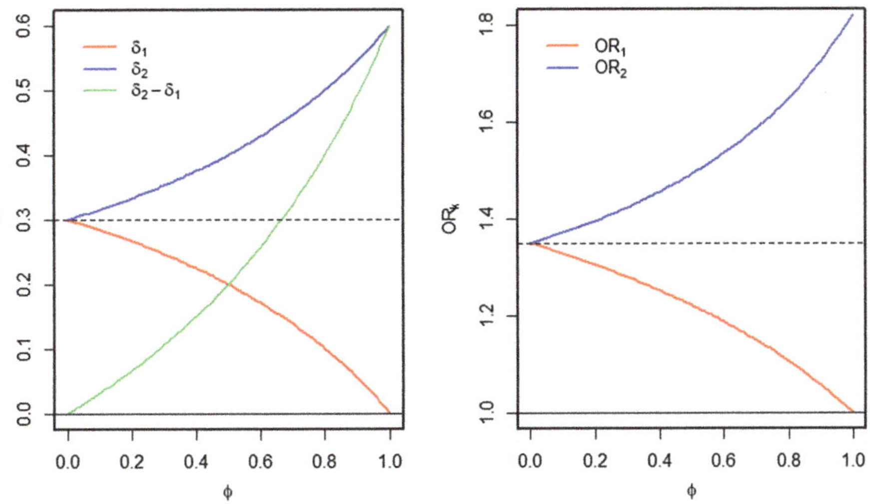

Fig. 10.7 Subgroup specific treatment effects and odds ratios for given ϕ. The overall treatment effect β_T was chosen as 0.3 (corresponding to the horizontal dashed line)

be used as basis for the test decision. However, in the simulations performed here we used the interval in (10.26).

10.4.2 Quantification of the Interaction

We again consider the parameter ϕ, which describes the magnitude of the interaction. In the setup of the logistic regression model considered here, combining (10.6) and (10.24), ϕ results in

$$\phi = 1 - \frac{\min\left(\log\left(OR_1\right), \log\left(OR_2\right)\right)}{\max\left(\log\left(OR_1\right), \log\left(OR_2\right)\right)} \tag{10.27}$$

As before ϕ will be varied between 0 (no interaction) and 1 (no treatment effect in the second subgroup). As a first step the influence of ϕ on the subgroup specific treatment effects δ_1,δ_2 and odds ratios is illustrated (Fig. 10.7). From Sect. 10.2 we know that for positive interaction terms β_{TS}

$$\delta_1 = \beta_T\left(1 - \frac{\phi}{2-\phi}\right), \quad \delta_2 = \beta_T\left(1 + \frac{\phi}{2-\phi}\right)$$

holds. Furthermore, from Eq. (10.24) we know that $OR_k = e^{\delta_k}$ $(k = 1, 2)$.

10.4.3 Simulation Setup

In the following simulation study, the influence of the consistency margin θ_c on the power of the homogeneity test derived in Sect. 10.4.1. is investigated. The model underlying the simulation uses the same event probability, denoted by p_C, in both subgroups for the control treatment. This can be formalised by setting $\beta_S = 0$ in (10.22). From (10.22), together with the constraint $\beta_S = 0$, it follows that the event probability in the control group is

$$p_C = P(Y_i = 1|X_{it} = 0) = \frac{e^{\beta_0}}{1 + e^{\beta_0}}. \tag{10.28}$$

Hence, different values of the event probability in the control group can be achieved by varying the intercept β_0 of the regression model. In the simulations β_0 is chosen such that p_C takes values 0.12, 0.27 and 0.5.

For the further specification of the model parameters, the overall (average) event probability in the treatment group which according to the parameterisation chosen (we have $E(X_{iTS}) = E(X_{iS}) = 0$ in (10.22)) can be calculated as

$$p_T = logit^{-1}(E(P(Y_i = 1|X_{it} = 1))) = e^{\beta_0+\beta_T} / \left(1 + e^{\beta_0+\beta_T}\right) \tag{10.29}$$

is considered.

For each β_0 the corresponding treatment effect coefficient β_T is determined such that the χ^2 test comparing the average event probabilities of the treatment groups has a power of 80% with a sample size of N = 100. Furthermore, for those parameters β_0 and β_T, the sample size is adjusted to achieve a power of 80–95%. The parameter ϕ is varied over a grid from 0 to 1, and the corresponding β_{TS} is calculated using (10.10). For each combination of parameters, 10,000 simulation runs were performed to determine the power of the homogeneity test depending on different values of the consistency margin θ_c.

The parameters chosen in the simulations reflect treatment effects and associated odds ratios that are observed in a variety of oncological indications. A common endpoint in cancer trials is the objective response rate (ORR), which is "the proportion of patients with tumour size reduction of a pre-defined amount and for a minimal time period" (CDER 2007).

In different types of non-small cell lung cancer (NSCLC), ORRs ranging from 12 to 65% are observed for various treatments in randomised controlled Phase 3 trials (Natale et al. 2011; Khozin et al. 2014). For metastatic renal-cell carcinoma, the observed ORRs were 31% and 6% for different treatment groups by Motzer et al. (2007) leading to an odds ratio of 7.04. Odds ratios observed in other Phase 3 trials ranged from 1.4 for 1 year survival in patients with ALK-positive lung cancer to 9.75 in a special type of NSCLC (Khozin et al. 2014). These ranges of event probabilities and odds ratios were covered by the simulations.

10.4.4 Power of the Consistency Test for Binary Endpoints

As for normally distributed endpoints of Chap. 3, two cases have been examined: First, the sample size was fixed across all values of the event probability in the control group. Second, the power to demonstrate superiority was fixed (leading to higher sample sizes when the relative treatment effect was smaller).

When the sample size was fixed (Fig. 10.8), then, for a given consistency margin, the power curves of the equivalence test are quite similar for event probabilities of 0.5 and 0.27 in the control group. However, when the event probability in the control group is smaller (0.12) the power of the equivalence test drops noticeably for small margins.

As expected, the equivalence test is more powerful when ϕ is small. Up to a value of about $\phi = 0.3$, the power of the consistency test is only reduced slightly. This appears to be consistent for many combinations of the other parameters (treatment effects and margins). For larger margins, the power of the test remains stable up to $\phi = 0.4$ and even up to $\phi = 0.6$. On the other end of the scale, when ϕ approaches 1 (meaning one of the subgroups has not any effect), the probability of declaring equivalence, which would correspond to a type-I error, is around 5% for the smaller margins and up to 30% for a margin of 3. The steepest decrease in power (inflection point) in the cases that have been investigated is reached when ϕ is between 0.6 and 0.8.

The dependence of the power of the consistency test on the chosen sample size is more pronounced for smaller consistency margins. For example, for an event probability of 0.5 and a margin of 3 the power of the consistency test varies between 0.85 and 0.99 for $\phi = 0$. For the same event probability and value for ϕ, but a margin of 1.5, the power of the consistency test varies between approximately 0.05 and 0.95.

Furthermore, the dependence of the power of the consistency test on the sample size depends non-monotonically on the event probability in the control group in the scenarios considered here. For example, with a margin of 2 and $\phi = 0$, the power varies between 0.4 and 0.99 for an event probability of 0.5, between 0.45 and 0.99 for an event probability of 0.27 and between 0.15 and 0.95 for an event probability of 0.12.

In the second case (Fig. 10.9) the power for the test of superiority was fixed and the sample size was adapted to simulate trials with 80%, 85%, 90% and 95% power to demonstrate superiority. Otherwise, the analysis was similar. Note that the red curves are the same in Figs. 10.8 and 10.9 which makes the results more comparable.

The shape of the S-curves is very similar between Figs. 10.8 and 10.9. They are flat for rather small ($\phi \leq 0.3$) and steeper for rather large ($\phi \geq 0.8$) subgroup differences.

The required sample size to reach adequate power for the consistency test depends strongly on the event probability in the control group. Again this behaviour is non-monotonic for fixed ϕ and fixed margin. For example, with a margin of 2.5 the sample size to achieve a power of 80% to show superiority in the trial (red

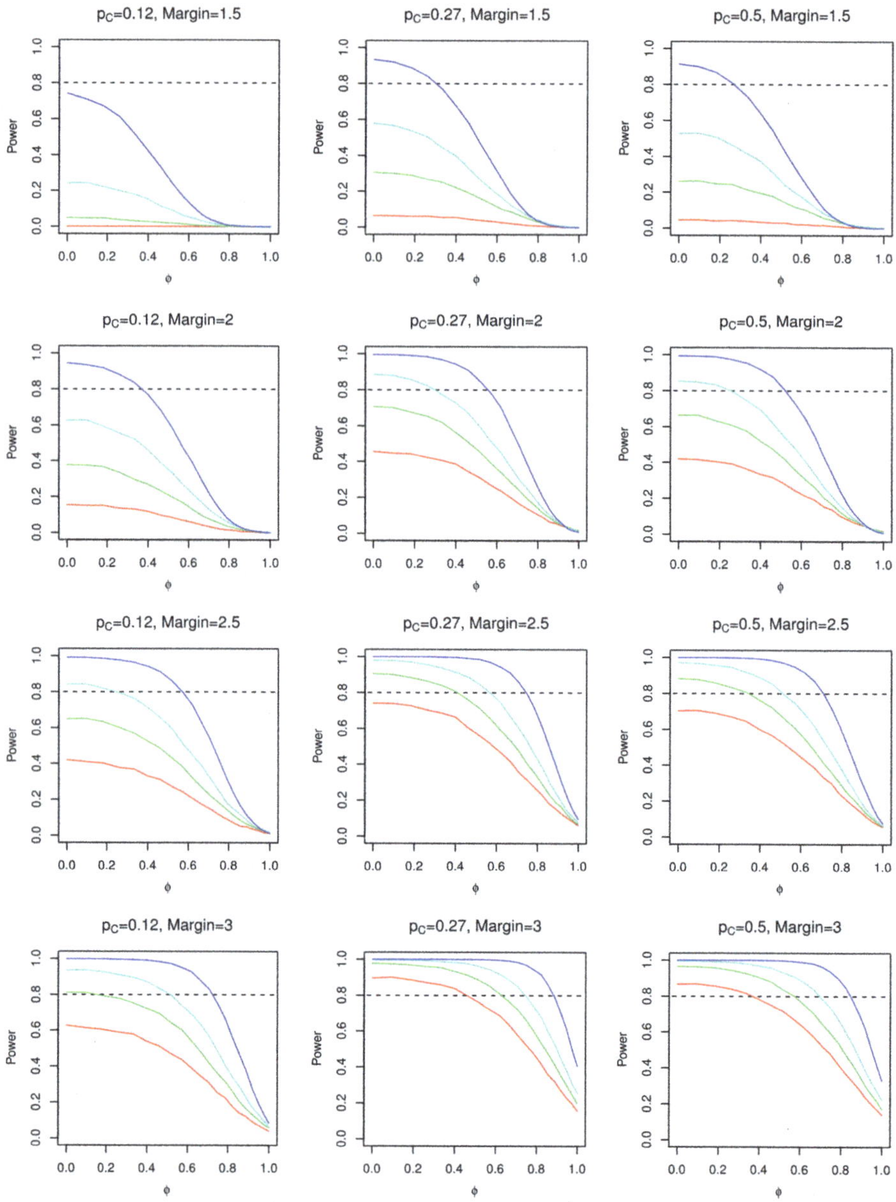

Fig. 10.8 Power of the consistency test as a function of the subgroup differences ϕ, using various event probabilities in the control group (0.12–0.5) and consistency margins (1.5–3.0) **for fixed sample sizes**. The colours of the lines depict the total sample size of: **red 100, green 140, cyan 200 and blue 400 patients**. The event probability in the treatment group is chosen such that with a sample size of $N = 100$ the test for the average event probability between treatment groups has the specified power of 80%

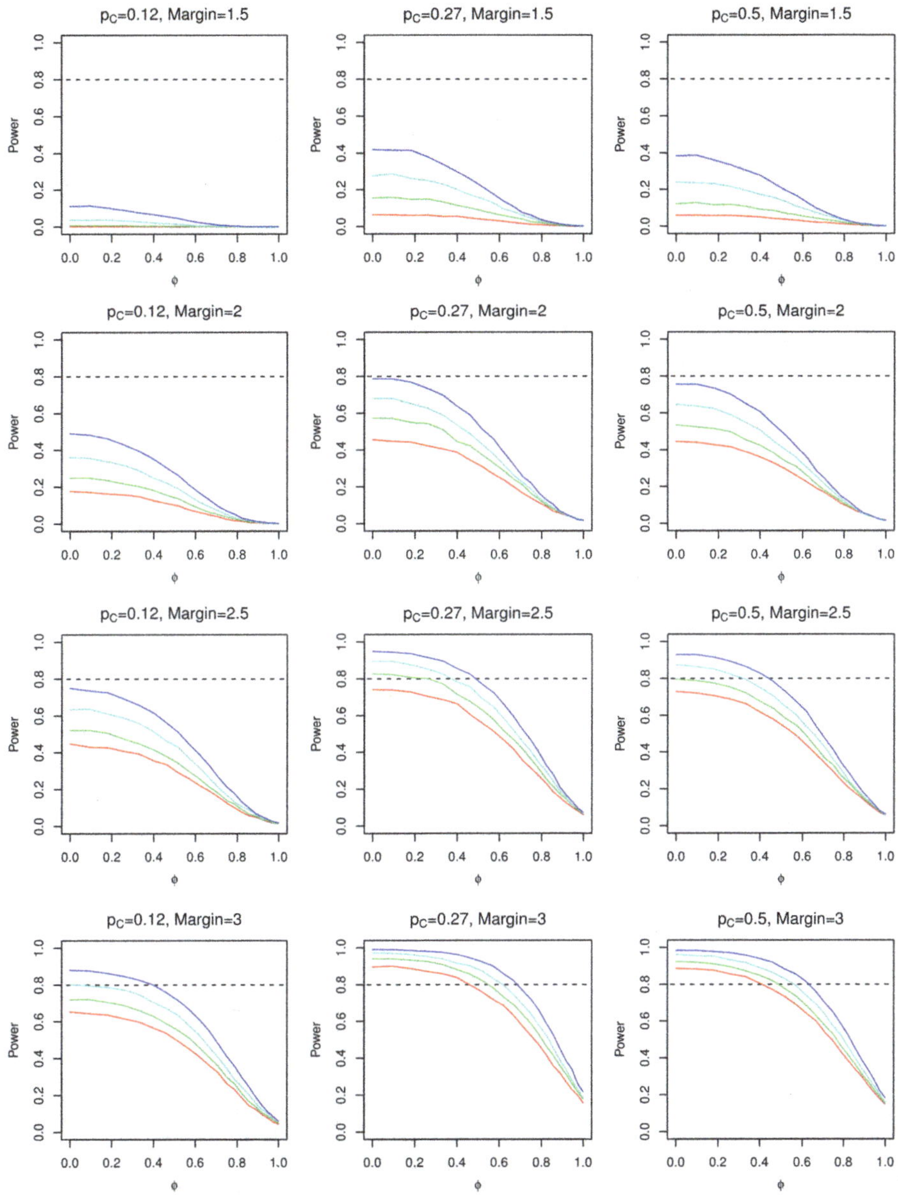

Fig. 10.9 Power of the consistency test as a function of the subgroup differences ϕ, using various event probabilities in the control group (0.12–0.5) and consistency margins (1.5–3.0). The colours of the lines depict the overall power of the trial to demonstrate superiority: **red 80%, green 85%, cyan 90% and blue 95%**

Table 10.2 Sample size used for the simulations of Fig. 10.10

			Power		
p_T	OR	β_T	0.8	0.85	0.9
0.3	1.3	0.25	1504	2864	3348
0.4	2.0	0.69	304	348	408
0.5	3.0	1.10	116	132	156
0.6	4.5	1.50	64	72	80
0.7	7.0	1.95	40	44	48

curves) leads to a power of approximately 70% for the consistency test (for $\phi = 0$) when the event probability is 0.5 in the control group. With the same margin and an event probability of 0.12, the sample size to reach 80% power in the overall test to show superiority leads to a power of approximately 40% for the consistency test (with $\phi = 0$).

Similarly, a third investigation was conducted where the event probability in the control group was fixed at 0.25, and different values for the event probability in the treatment group were examined. The independent axis (Fig. 10.10) was the consistency margin (between 1.0 and 3.0), and five different values for the subgroup divergence ϕ (0.0, 0.25, 0.50, 0.75 and 1.0) were considered. As before, the sample size was determined such that the power of the superiority test was 80–90%. In Tables 10.2 and 10.3 the parameters underlying the simulations, as well as resulting odds ratios, are given. Since for all simulations $p_C = 0.25$ was fixed, the intercept of the regression model (10.22) was $\beta_0 = -1.099$ in all cases.

The power for demonstrating equivalence is very similar for the two smallest values of ϕ, since the red and the yellow curves nearly overlap over the whole range of consistency margins for all parameter combinations. With increasing divergence, the power is reduced as expected. When the event probability in the treatment group is close to that of the control group (and hence the sample size is large), the power for demonstrating equivalence is close to one even for ϕ close to 1.

10.4.5 Discussion

The results of the simulation studies presented in Sect. 10.4.4 show that, as expected, the power of the consistency test depends on the magnitude of the subgroup-by-treatment interaction (as characterised by ϕ). For small values of ϕ the power is highest, and remains relatively stable for values smaller than 0.3–0.4, depending on the chosen margin. For higher values of ϕ the power drops considerably and flattens out as ϕ approaches 1, resulting in an S-shaped curve.

For small consistency margins a stronger dependence of the power of the consistency test on the chosen sample size was observed than for larger margins. Furthermore, the power of the consistency test depends non-monotonically on the event probabilities in the treatment groups, also when fixing the magnitude of interaction ϕ. This is partly because β_{TS} depends non-monotonically on the event

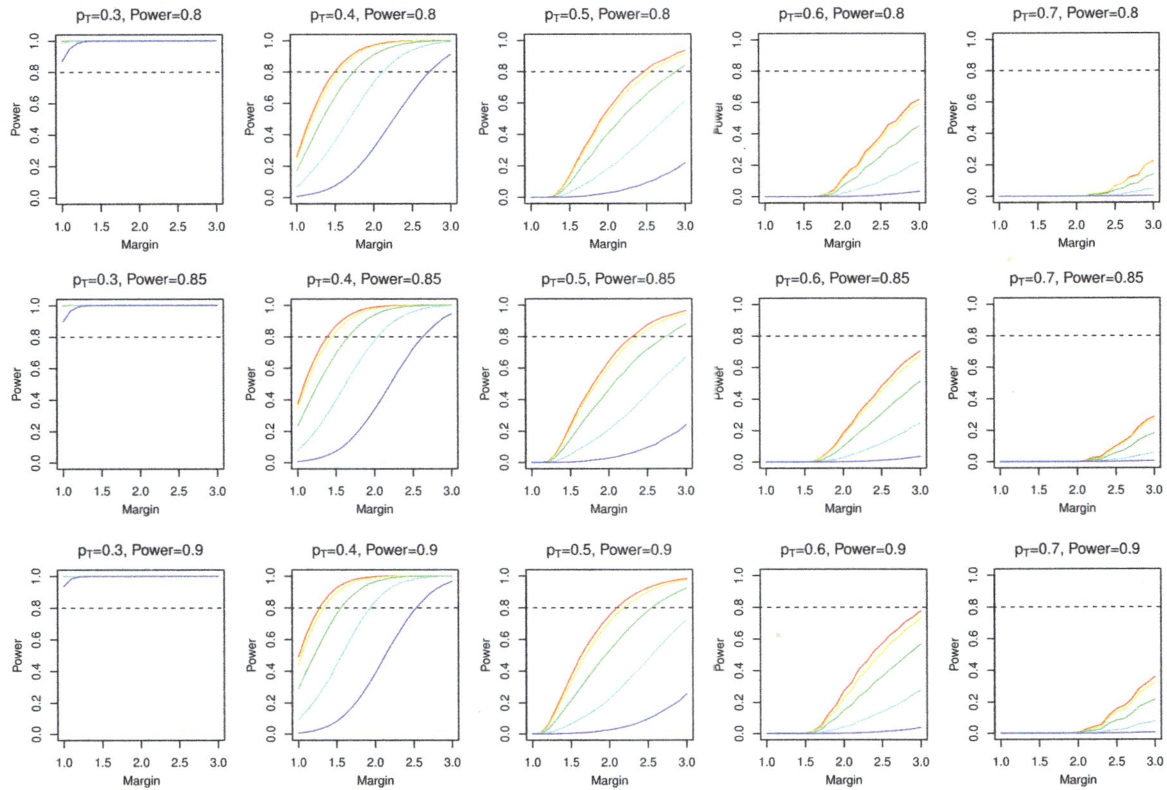

Fig. 10.10 Power of the consistency test as a function of the consistency margin for an event probability in the control group of $p_C = 0.25$ for various event probabilities in the treatment group and a power for the overall superiority test of 80–90%. The line colours depict the subgroup difference ϕ with values of **0 (red), 0.25 (yellow), 0.5 (green), 0.75 (cyan) and 1.0 (blue)**

Table 10.3 Model parameters, resulting event probabilities and odds ratios for the simulations of Fig. 10.10

p_T	OR	β_T	ϕ	β_{TS}	p_{T1}	p_{T2}	OR_1	OR_2
0.3	1.3	0.25	0	0	0.30	0.30	1.3	1.3
			0.25	0.07	0.29	0.31	1.2	1.3
			0.5	0.17	0.28	0.32	1.2	1.4
			0.75	0.30	0.27	0.33	1.1	1.5
			1	0.50	0.25	0.36	1.0	1.7
0.4	2.0	0.69	0	0	0.40	0.40	2.0	2.0
			0.25	0.20	0.38	0.42	1.8	2.2
			0.5	0.46	0.35	0.46	1.6	2.5
			0.75	0.83	0.31	0.50	1.3	3.0
			1	1.39	0.25	0.57	1.0	4.0
0.5	3.0	1.10	0	0	0.50	0.50	3.0	3.0
			0.25	0.31	0.46	0.54	2.6	3.5
			0.5	0.73	0.41	0.59	2.1	4.3
			0.75	1.32	0.34	0.66	1.6	5.8
			1	2.20	0.25	0.75	1.0	9.0
0.6	4.5	1.50	0	0	0.60	0.60	4.5	4.5
			0.25	0.43	0.55	0.65	3.6	5.6
			0.5	1.00	0.48	0.71	2.7	7.4
			0.75	1.80	0.38	0.79	1.8	11.1
			1	3.01	0.25	0.87	1.0	20.3
0.7	7.0	1.95	0	0	0.70	0.70	7.0	7.0
			0.25	0.56	0.64	0.75	5.3	9.2
			0.5	1.30	0.55	0.82	3.7	13.4
			0.75	2.34	0.42	0.88	2.2	22.5
			1	3.89	0.25	0.94	1.0	49.0

The fixed $p_C = 0.25$ leads to $\beta_0 = -1.099$ in all cases

probability in the control group, when the power of the test of overall treatment effect (or the sample size) is fixed.

Considering the relationship between the interaction term β_{TS} of the logistic regression model and the parameter ϕ, for a fixed event probability in the control group as done in Table 10.3, it becomes evident that for fixed values of ϕ the value of β_{TS} varies with the event probability in the treatment group. For example, when the event probability in the control group p_C is fixed at 0.25, a value of $\beta_{TS} = 0.3$ indicates a strong interaction ($\phi = 0.75$) when the event probability in the treatment group is $p_T = 0.3$. For a higher event probability in the treatment group ($p_T = 0.5$), a β_{TS} of about 0.3 corresponds to a moderate interaction ($\phi = 0.25$).

Since the consistency test for binary data presented here makes statements about β_{TS}, its results cannot be translated directly to the ϕ-scale without knowledge of the event probabilities observed in the study. Hence, the choice of consistency margin should be based on the specific event probabilities expected for each trial, to ensure consistent results and interpretations of the consistency test across different trials.

10.4.6 Example for Planning the Study Design

As for the case of normally distributed endpoints, an example shall illustrate how clinical and statistical team members could discuss and select appropriate equivalence margins:

Suppose a Phase 3 trial is to be designed with the primary objective to demonstrate superiority of treatment A over control C using a binary endpoint. The trial is planned for an expected treatment effect (overall OR) of 3.95, with an event probability of 0.12 in the control group, leading to an event probability of 0.35 in the treatment group. Aiming to demonstrate the primary hypothesis with a two-sided $\alpha = 0.05$ with a power of 90% leads to a sample size of 68 subjects per group. While *a-priori* there might be no indication of a gender-dependent subgroup heterogeneity ($\phi = 0$), the team would like to perform an appropriate consistency test as a secondary hypothesis, aiming to reject the null hypothesis of non-consistency using $\alpha = 0.05$ with a power of 80%. To be conservative, they would power this test using a value of ϕ of at most 0.2 (to allow for contingency).

When reviewing Fig. 10.9 for event probability of 0.12 in the control group, a consistency margin of 3 leads to the desired power for the interaction test (the cyan curve shows the graph for a trial that has a sample size for the rejection of the primary hypothesis of 90%). The same graph also indicates that such a test would demonstrate consistency with a power of 65% when $\phi = 0.5$, and still with a power of about 40% when $\phi = 0.7$.

A medical consideration in this trial could be that differential subgroup effects which are twice as large (on the log-scale) in one subgroup compared to the other would be acceptable. With an overall odds ratio of 3.95, this leads to $\phi = 0.5$, and hence subgroup specific odds ratios of about 2.53 and 6.49.

If the team would feel that such a margin would be too liberal for larger values of heterogeneity ϕ, they could discuss using a lower consistency margin, such as 2.5. However, as can be seen from this panel in Fig. 10.9, in this case the consistency test would not get the desired power, given the sample size for the primary hypothesis.

In this case, if the test for consistent effects for both genders would be important, a solution would be to increase the overall sample size and to select a lower equivalence margin. Of course, whether the additional costs could be justified against the gain of information would need to be discussed.

10.5 Discussion

10.5.1 Subgroup-by-Treatment Interaction in the General Linear Model

The objective of the investigations in this chapter was to find an alternative to the interaction test of the subgroup-by-treatment interaction for various types of endpoints. A significant interaction term is generally understood to make the overall trial result less interpretable so that the outcome of each individual subgroup should be interpreted separately. However, when numerous cofactors are tested, apparent interactions might arise, and they would need to be investigated further. The best way of doing so would be to repeat the trial in order to confirm or reject the finding, but the costs would generally not outweigh the gain in information.

The main drawback of the interaction test is that it is only powerful when the heterogeneity is very large. In fact, the power of the interaction test only reaches the power of the overall test of treatment superiority when $\phi = 1$. With intermediate heterogeneity of up to $\phi = 0.7$, its power is much smaller (Ring et al. 2018).

The null hypothesis of the interaction test is homogeneity. When this hypothesis is rejected, then interaction is claimed, without considering the magnitude of the interaction. When the null hypothesis cannot be rejected, no claim at all can be made. In addition, the test of interaction does not address the question whether the difference of the effects between subgroups would be relevant.

The equivalence test could be a potential solution, because it is based on a pre-defined (medically and statistically relevant) margin to judge the estimated differences. Within the limits of these margins, differences of the subgroups could occur, but would not alter the overall interpretation of the trial, because the differences are considered small enough when the new treatment option is to be implemented.

We have investigated two different types of endpoints, binary and normally distributed, which lead to slightly different equivalence hypotheses. The starting point was the generalised linear model and its interaction term β_{TS}. For the normally distributed endpoints, the consistency ratio—the difference between the subgroup specific treatment effects divided by the residual variability—was defined, aiming to relate the magnitude of the explained subgroup heterogeneity to the unexplained variability. For the investigation of binary endpoints the interaction term was used directly for the test, as the model does not contain a term for residual variability.

10.5.2 Selection of the Equivalence Margin

The selection of appropriate equivalence margins is a critical step during the design of a clinical trial. The margin must be stated in the clinical trial protocol with sound scientific reasoning. ICH E10 requests: "The determination of the margin in a non-inferiority trial is based on both statistical reasoning and clinical judgment, should reflect uncertainties in the evidence on which the choice is based, and should be suitably conservative."

A challenge for the selection of the margin is that the consistency test will generally be a secondary hypothesis, which will be evaluated after successful demonstration of the primary hypothesis. Indeed, the test can be imbedded into a multiple-testing strategy (Bretz et al. 2009), and the sample size could be determined accordingly. Based on the results of Sects. 10.3 and 10.4, the sample sizes could become much larger than those required for the primary hypothesis. The implementation of such statistical considerations into the selection of the margin could be a potential solution, as indicated in the examples in Sects. 10.3.4 and 10.4.6.

The simulations for the power of the consistency tests covered a broad range of parameters (effect sizes or event probabilities and the heterogeneity parameter ϕ), and one of the main results may be that a universally acceptable margin cannot be recommended across all parameter ranges. Based on the expected outcomes of the trial and the medically acceptable deviations from homogeneity in the therapeutic area, simulations may be required to provide sensible margins for the consistency test.

There were a number of common elements in the simulations of both types of endpoints. As expected, the probability to reject the hypothesis of non-consistency decreases with increasing values of the heterogeneity parameter ϕ. This decrease was small within the interval between 0 and 0.3, which is the range in which there is rather little effect discrepancy between both subgroups. However, when the margin was rather small, the power of the consistency test was less than 50% for sample sizes which provide 90% power for the superiority hypothesis. If a larger margin was selected, then the power was appropriate for small ϕ; however when ϕ was large, the probability for rejecting the heterogeneity was quite large for some cases, which is not desirable since it constitutes a type-I error.

A limitation of these investigations was that the treatments and the subgroups were fully balanced so that each combination of treatments and subgroups was present in a quarter of the total number of subjects. This may be true when the randomisations were stratified by the subgroup, but in general the subjects are not balanced in subgroups. For example, gender subgroups rarely have the exactly balanced distributions, even if the treated disease was balanced (e.g. 60% male subjects in clinical trials compared to 50% in registries for atrial fibrillation (Tanislav et al. 2015)). For other types of subgroups—such as age groups, presence or non-presence of comorbidities or subjects with disease specific criteria, the imbalance might be even stronger.

The proposed consistency tests will generally have less power with unequal distribution (for example in the normal case, the confidence interval for the consistency ratio will be larger due to formula (10.19)). Alternative tests which have been developed (Grill 2017) could not be discussed here, but generally show similar properties to the presented tests.

10.5.3 Considerations for Improvement of the Consistency Test

While we feel that an equivalence test would be a good method to assess the relevance of subgroup heterogeneity, the currently examined characteristics—the consistency ratio for normally distributed values and the β_{TS} for binary endpoints—do not completely fulfil our expectations. The consistency test has the aim of judging the relevance of heterogeneity in contrast to the pure significance of the interaction test by reversing the null hypothesis towards a TOST procedure. However, the statistical properties of both tests are somewhat similar in that they require rather large sample sizes to reject either homogeneity or heterogeneity, respectively.

When the consistency test for the normally distributed endpoints was developed, the following objectives were considered:

- There should be adequate power (e.g. > 75%) to demonstrate equivalence when the subgroup heterogeneity ($\phi \leq 0.3$) is small.
- The test should lead to rather little probability (< 25%) to demonstrate equivalence when subgroup heterogeneity is large ($\phi \geq 0.7$).
- The power curve with respect to ϕ should be S-shaped (with decreasing probability for increasing ϕ), with an inflection point that depends on the effect size, but which would be have a value within the interval of 0.3 and 0.7.

The third item shall particularly address the question of the relevance of subgroup heterogeneity, as a smaller effect size means a relatively large variability, and in this case the effect difference between subgroups may be somewhat larger as well.

While some curves in the simulations in Sects. 10.3 and 10.4 fulfil these conditions, there was no universal equivalence margin that would fulfil all three conditions. To improve the test, a number of considerations could be raised.

The variance scaling for the normally distributed consistency test could be improved by limiting the value of the common standard deviation σ with a maximum value σ_0. This would imply that the consistency ratio would not be arbitrarily reduced, in particular due to quality issues that would be in control of the study sponsor. Such a limitation is also suggested in the field of bioequivalence: While the European bioequivalence guideline allows for scaled bioequivalence methods, it limits the method to variabilities which are smaller than the geometric coefficient of variation of 50%. Such a modification needs sophisticated methods in order to reduce the risk for bias, as the estimated variability might be different from the actual, but these methods have already been developed (Tothfalusi 2017).

Another option for the normally distributed endpoints could be to evaluate the unscaled interaction term β_{TS}, as this would remove the dependence from the variability. However this would fail to address the third item so that the relevance of the heterogeneity would be judged less. A compromise might be to involve a suitable function $g(\sigma)$, which would reduce, but not completely ignore the impact of the variability.

For both the binary and the normally distributed endpoints it could be considered to develop a test involving the heterogeneity parameter ϕ instead of β_{TS}. This way the dependence of the equivalence test on the overall treatment effect might be reduced. Furthermore, the test results might be better comparable across different trials.

Finally, it could also be discussed whether the use of the same type-I error α for the primary (e.g. superiority) test and the consistency test would always be a requirement. As mentioned in Sect. 10.1, the aims of subgroup evaluations vary between descriptive, exploratory and confirmatory investigations. The investigation of Ocaña et al. (2015) showed that the relevance of carry-over effects in crossover trials can be evaluated using equivalence tests, and these tests could define their magnitude of type-I error independently from other study hypotheses. This idea could be extended towards balancing the power and type-I error for the sample size that has been given for the primary objective, similar to the proposal by Ioannidis et al. (2013).

10.5.4 Future Developments

Evaluations of subgroup effects are not restricted to the settings that have been described in this chapter, and extensions to those areas can be imagined.

The application of the generalised linear model to normally distributed and binary endpoints could be expanded to survival endpoints, and similar results as above could be obtained.

The analysis has been restricted to a single subgroup with two categories. Evaluations of more than two categories would be particularly important when analysing comorbidities. For example, a trial in subjects with cardiovascular diseases could recruit patients with various conditions, such as stroke, myocardial infarction or heart failure, and the clinical outcomes of a new treatment could potentially be different. One option could be to merge some categories together so that only two categories are analysed using the consistency test. However it might be difficult to judge clinically which categories should be merged together so that alternative methods to assess consistency more directly are warranted.

Another challenge is that there would often be more than one subgroup for which evaluations of effect consistency shall be made. The consistency test could be imbedded into a multiple-testing procedure and hence several subgroups could be tested with appropriate adjustment for multiplicity. This would still be limited to a few subgroups, and extensions towards more exploratory procedures might be a better option to address multiple evaluations simultaneously.

A promising approach for the evaluation of subgroups is the Bayesian shrinkage method (Henderson et al. 2016). This method estimates within-subgroup effects based on the overall global treatment effect. The underlying idea is that (smaller) subgroups might be more variable and hence less reliable than the whole population so that the subgroup estimates are moved towards the global mean. Although this method implicitly relies on the assumption of full exchangeability of the study subjects, it may serve a starting point for further analyses.

Bayesian analyses might be particularly valuable when prior data have been obtained before the confirmatory Phase 3 trial is performed, as it is most often the case. Pre-clinical and early clinical data could be summarised and evaluated. Even if these trials had been performed in much smaller populations, they could provide signals for subgroup heterogeneity that are worth exploring further.

Finally, the performance of well-designed meta analyses could provide more insight into subgroup effects. In general, at least two Phase 3 trials are required for regulatory approval of new health technologies, and a combined analysis of all confirmatory trials of a Phase 3 programme —while accounting for their inherent design heterogeneity—would often directly rule out or strengthen signals of non-consistent subgroup effects, because of the larger amount of individual data. This analysis would need additional assumptions, for example that there is no study-by-subgroup interaction, which however is generally plausible.

10.5.5 Conclusion

Evidence-based medicine requires detailed evaluations of risks and benefits of new treatment options, not only for the whole population, but also for individual subjects with their demographic factors and medical history. The evaluation of the consistency of subgroup effects is an important part of this evaluation, and the identification of real vs. apparent heterogeneity is often difficult. The application of equivalence tests to assess subgroup consistency might offer a new solution. The equivalence margins for such tests need to be evaluated carefully with medical and statistical evaluations for each individual trial, and simulation studies provide insight into the statistical properties of these tests.

References

Bath, P. M., Martin, R. H., Palesch, Y., Cotton, D., Yusuf, S., Sacco, R., Diener, H. C., Toni, D., Estol, C., & Roberts, R. (2009). Effect of telmisartan on functional outcome, recurrence, and blood pressure in patients with acute mild ischemic stroke: A PRoFESS subgroup analysis. *Stroke., 40*(11), 3541–3546. https://doi.org/10.1161/STROKEAHA.109.555623.

Beeh, K. M., Westerman, J., Kirsten, A. M., Hébert, J., Grönke, L., Hamilton, A., Tetzlaff, K., & Derom, E. (2015). The 24-h lung-function profile of once-daily tiotropium and olodaterol fixed-dose combination in chronic obstructive pulmonary disease. *Pulm Pharmacol Ther., 32*, 53–59. https://doi.org/10.1016/j.pupt.2015.04.002.

Bretz, F., Maurer, W., Brannath, W., & Posch, M. (2009). A graphical approach to sequentially rejective multiple test procedures. *Stat Med., 28*(4), 586–604. https://doi.org/10.1002/sim.3495.

Brookes, S. T., Whitely, E., Egger, M., Smith, G. D., Mulheran, P. A., & Peters, T. J. (2001). Subgroup analyses in randomised controlled trials: Quantifying the risks of false-positives and false-negatives. *Health Technol Assess, 5*(33), 1–56.

Brookes, S. T., Whitely, E., Egger, M., et al. (2004). Subgroup analyses in randomized trials: Risks of subgroup-specific analyses; power and sample size for the interaction test. *J Clin Epidemiol, 57*(3), 229–236.

CDER. Center for Drug Evaluation and Research (CDER), Center for Biologics Evaluation and Research (CBER). (2007, May). *Guidance for industry: Clinical trial endpoints for the approval of cancer drugs and biologics*. Retrieved January 24, 2018, from https://www.fda.gov/downloads/Drugs/Guidances/ucm071590.pdf.

Dans, A. L., Connolly, S. J., Wallentin, L., Yang, S., Nakamya, J., Brueckmann, M., Ezekowitz, M., Oldgren, J., Eikelboom, J. W., Reilly, P. A., & Yusuf, S. (2013). Concomitant use of antiplatelet therapy with dabigatran or warfarin in the Randomized Evaluation of Long-Term Anticoagulation Therapy (RE-LY) trial. *Circulation., 127*(5), 634–640. https://doi.org/10.1161/CIRCULATIONAHA.112.115386.

Dmitrienko, A., Muysers, C., Fritsch, A., & Lipkovich, I. (2016). General guidance on exploratory and confirmatory subgroup analysis in late-stage clinical trials. *J Biopharm Stat., 26*(1), 71–98. https://doi.org/10.1080/10543406.2015.1092033.

Donohue, J. F. (2005). Minimal clinically important differences in COPD lung function. *COPD., 2*(1), 111–124.

EMA. (2013). *Draft guideline on the investigation of subgroups in confirmatory clinical trials, EMA/CHMP/539146/2013*. Draft for consultation.

Forst, T., Uhlig-Laske, B., Ring, A., Graefe-Mody, U., Friedrich, C., Herbach, K., Woerle, H. J., & Dugi, K. A. (2010). Linagliptin (BI 1356), a potent and selective DPP-4 inhibitor, is safe and efficacious in combination with metformin in patients with inadequately controlled Type 2 diabetes. *Diabet Med., 27*(12), 1409–1419. https://doi.org/10.1111/j.1464-5491.2010.03131.x.

Friedman, L. M., Furberg, C. D., & DeMets, D. (2010). *Fundamentals of clinical trials*. Springer.

Grill, S. (2017). *Assessing consistency of subgroup specific treatment effects in clinical trials with binary endpoints*. MSc thesis, University of Bremen.

Haidar, S. H., Davit, B., Chen, M. L., Conner, D., Lee, L., Li, Q. H., Lionberger, R., Makhlouf, F., Patel, D., Schuirmann, D. J., & Yu, L. X. (2008). Bioequivalence approaches for highly variable drugs and drug products. *Pharm Res., 25*(1), 237–241.

Hemmings, R. (2014). An overview of statistical and regulatory issues in the planning, analysis, and interpretation of subgroup analyses in confirmatory clinical trials. *J Biopharm Stat., 24*(1), 4–18. https://doi.org/10.1080/10543406.2013.856747.

Henderson, N. C., Louis, T. A., Wang, C., & Varadhan, R. (2016). Bayesian analysis of heterogeneous treatment effects for patient-centered outcomes research. *Health Serv Outcomes Res Method, 16*, 213–233. https://doi.org/10.1007/s10742-016-0159-3.

Hosmer Jr., D. W., Lemeshow, S., & Sturdivant, R. X. (2013). *Applied logistic regression* (3rd ed.). Hoboken, NJ: Wiley.

ICH E10. (2000). *Choice of Control Group and Related Issues in Clinical Trials*.

Ioannidis, J. P., Hozo, I., & Djulbegovic, B. (2013). Optimal type I and type II error pairs when the available sample size is fixed. *J Clin Epidemiol., 66*(8), 903–910.e2. https://doi.org/10.1016/j.jclinepi.2013.03.002.

Kent, D. M., Rothwell, P. M., Ioannidis, J. P., Altman, D. G., & Hayward, R. A. (2010). Assessing and reporting heterogeneity in treatment effects in clinical trials: a proposal. *Trials., 11*, 85. https://doi.org/10.1186/1745-6215-11-85.

Khozin, S., Blumenthal, G. B., Jiang, X., et al. (2014). U.S. Food and Drug Administration approval summary: Erlotinib for the first-line treatment of metastatic non-small cell lung cancer with epidermal growth factor receptor exon 19 deletions or exon 21 (L858R) substitution mutations. *The Oncologist, 19*, 774–779.

Koehler, E., Brown, E., & Haneuse, S. J. P. A. (2009). On the assessment of Monte Carlo error in simulation-based statistical analyses. *Am Stat., 63*(2), 155–162. https://doi.org/10.1198/tast.2009.0030.

Machin, D., & Campbell, M. J. (2005). *Design of studies for medical research*. Chichester: Wiley.

Mok, T. S., Wu, Y. L., Thongprasert, S., Yang, C. H., Chu, D. T., Saijo, N., Sunpaweravong, P., Han, B., Margono, B., Ichinose, Y., Nishiwaki, Y., Ohe, Y., Yang, J. J., Chewaskulyong, B., Jiang, H., Duffield, E. L., Watkins, C. L., Armour, A. A., & Fukuoka, M. (2009). Gefitinib or carboplatin-paclitaxel in pulmonary adenocarcinoma. *N Engl J Med., 361*(10), 947–957. https://doi.org/10.1056/NEJMoa0810699.

Motzer, R. J., Hutson, T. E., Tomczak, P., et al. (2007). Sunitinib versus interferon alfa in metastatic renal-cell carcinoma. *N Engl J Med, 356*, 115–124.

Natale, R. B., Thongprasert, S., Greco, A., et al. (2011). Phase III trial of Vandetenib compared with Erlotinib in patients with previously treated advanced non-small-cell lung cancer. *Journal of Clinical Oncology, 29*(8), 1059–1066.

Ocaña, J., Sánchez, M. P., Sánchez, A., & Carrasco, J. L. (2008). On equivalence and bioequivalence testing. *Statistics & Operations Research Transactions, 32*(2), 151–176. Retrieved from www.idescat.net/sort.

Ocaña, J., Sanchez, M. P., & Carrasco, J. L. (2015). Carryover negligibility and relevance in bioequivalence studies. *Pharm Stat., 14*(5), 400–408. https://doi.org/10.1002/pst.1699.

Plavnik, F. L., & Ribeiro, A. B. (2002). A multicenter, open-label study of the efficacy and safety of telmisartan in mild to moderate hypertensive patients. *Arq Bras Cardiol., 79*(4), 339–350.

R Development Core Team. (2008). *R: A language and environment for statistical computing*. R Foundation for Statistical Computing. Retrieved from https://www.R-project.org/.

Ring, A., Day, S., & Schall, R. (2018). Assessment of consistency of treatment effects in subgroup analyses. Submitted.

Russell, L. (2015). *Lsmeans: Least-Squares Means*. R package version 2.20-2. Retrieved from http://CRAN.R-project.org/package=lsmeans.

Schall, R. (1995). Assessment of individual and population bioequivalence using the probability that bioavailabilities are similar. *Biometrics, 51*(2), 615–626.

Schuirmann, D. J. (1987). A comparison of the two one-sided test procedure and the power approach for assessing the equivalence of average bioavailability. *J Pharmacokin. Biopharm., 15*, 657–680.

Tanislav, C., Milde, S., Schwartzkopff, S., Misselwitz, B., Sieweke, N., & Kaps, M. (2015). Baseline characteristics in stroke patients with atrial fibrillation: Clinical trials versus clinical practice. *BMC Res Notes., 8*, 262. https://doi.org/10.1186/s13104-015-1237-2.

Tanniou, J., van der Tweel, I., Teerenstra, S., & Roes, K. C. B. (2017). Estimates of subgroup treatment effects in overall nonsignificant trials: To what extent should we believe in them? *Pharm Stat., 16*(4), 280–295. https://doi.org/10.1002/pst.1810.

Ting, N. (2017). Statistical interactions in a clinical trial. *Ther Innov Regulat Sci, 52*(1), 14–21.

Varadhan, R., & Seeger, J. D. (2013, January). Estimation and reporting of heterogeneity of treatment effects. In P. Velentgas, N. A. Dreyer, P. Nourjah, S. R. Smith, & M. M. Torchia (Eds.), *Developing a protocol for observational comparative effectiveness research: A user's guide*. AHRQ Publication No. 12(13)-EHC099. Agency for Healthcare Research and Quality.

Varadhan, R., Segala, J. B., Boyda, C. M., Wua, A. W., & Weiss, C. O. (2013). A framework for the analysis of heterogeneity of treatment effect in patient-centered outcomes research. *J Clin Epidemiol., 66*(8), 818–825. https://doi.org/10.1016/j.jclinepi.2013.02.009.

Venzon, D. J., & Moolgavkar, S. H. (1988). A method for computing profile-likelihood based confidence intervals. *Applied Statistics, 37*, 87–94.

Wallach, J. D., Sullivan, P. G., Trepanowski, J. F., Steyerberg, E. W., & Ioannidis, J. P. (2016). Sex based subgroup differences in randomized controlled trials: Empirical evidence from Cochrane meta-analyses. *BMJ., 24*(355), i5826. https://doi.org/10.1136/bmj.i5826.

Wassmer, G., & Dragalin, V. (2015). Designing issues in confirmatory adaptive population enrichment trials. *J Biopharm Stat., 25*(4), 651–669. https://doi.org/10.1080/10543406.2014.920869.

Zinman, B., Wanner, C., Lachin, J. M., Fitchett, D., Bluhmki, E., Hantel, S., Mattheus, M., Devins, T., Johansen, O. E., Woerle, H. J., Broedl, U. C., Inzucchi, S. E., & EMPA-REG OUTCOME Investigators. (2015). Empagliflozin, cardiovascular outcomes, and mortality in type 2 diabetes. *N Engl J Med., 373*(22), 2117–2128. https://doi.org/10.1056/NEJMoa1504720.

Chapter 11
Predicting Confidence Interval for the Proportion at the Time of Study Planning in Small Clinical Trials

Jihnhee Yu and Albert Vexler

11.1 Introduction

Frequently, the goal of a limited accrual clinical trial is a confidence interval estimate of a success rate (e.g., efficacy or safety rate of a drug) rather than conducting a hypothesis test. Such trials are carried out with very small sample sizes (e.g., 25 subjects for Bougnoux et al. (2009), 20 subjects for Martin-Schild et al. (2009), 10 subjects for Schiffer et al. (2009), and 19 subjects for Rino et al. (2010). Small sample sizes may be a result of a priori limitations such as the budget, study duration, institution size, and difficulty of accruing subjects due to the rareness of disease and eligibility criteria. Nevertheless, these small clinical trials contribute highly to investigations of novel treatments of diseases and play an important part in the medical literature.

In the study protocols for such trials, a statement of the accuracy of the estimation is more relevant than the study power. In such cases, the confidence interval width is used as the precision of the parameter estimate that will be computed at the end of the study.

Suppose that $Y_1, \ldots, Y_n$ is a random sample of Bernoulli random variables. That is, the Y_is are independent random variables each assuming the value 0 or 1, where $P(Y_i = 1) = p$, $0 < p < 1$, and $i = 1, 2, \ldots, n$. Notice that $X = \sum_{i=1}^{n} Y_i$ is a binomial random variable based on a sample of size n with success probability p. The realization of the random variable X is denoted as x.

J. Yu (✉) · A. Vexler
Department of Biostatistics, University at Buffalo, State University of New York, Buffalo, NY, USA
e-mail: jinheeyu@buffalo.edu; avexler@buffalo.edu

Y. Zhao, D.-G. Chen (eds.), *New Frontiers of Biostatistics and Bioinformatics*, ICSA Book Series in Statistics, https://doi.org/10.1007/978-3-319-99389-8_11

At a planning stage of a study, the lower and upper bounds, respectively, of a confidence interval may be defined as $l(Y_1, \ldots, Y_n)$ and $u(Y_1, \ldots, Y_n)$, functions of the "future" sample of Bernoulli random variables. Throughout this chapter, we discuss predicting values of $l(Y_1, \ldots, Y_n)$ and $u(Y_1, \ldots, Y_n)$, i.e., prediction of the "future confidence interval" and relevant sample size calculation. We use the term "future" emphasizing that the sample is not obtained yet. In this chapter, we discuss probabilistic approaches for the prediction of the future confidence interval and compare approaches to predicting the width of a future confidence interval and the corresponding sample size calculation for limited accrual clinical trials.

There are many methods available to obtain a confidence interval for a binomial proportion, p (see Newcombe and Vollset 1994; Brown et al. 2001; Pires and Amado 2008; Vollset 1993). When the sample size is small and/or p is an extreme value (e.g., less than 0.1 or greater than 0.9), the performances of many of these confidence interval estimates are not satisfactory because of actual coverage probabilities below the confidence level (Brown et al. 2001; Pires and Amado 2008). When the sample size is small and/or p is an extreme value, confidence intervals that assure the user-specified confidence level need to be used such as the exact confidence interval (e.g., Blaker 2000; Blyth and Still 1983; Wang 2014).

This chapter explains the prediction of the future interval and relevant sample size calculations for small sample size and extreme p based on the Clopper-Pearson exact method (Clopper and Pearson 1934). We propose three approaches to assess the future exact confidence interval, namely Simple plugging-in approach, Hypothesis testing approach, and Expected confidence interval approach. We note that the proposed approaches can be applied to other exact confidence interval strategies, e.g., Blaker (2000), Blyth and Still (1983) and Wang (2014), or other approximated methods.

In particular, we will also discuss the prediction of the confidence interval strategies based on the angular transformation $Z_i = \arcsin\left(\sqrt{X/n}\right)$ (henceforth, referred to as arcsine intervals), which has the coverage probability above the confidence level with small sample sizes (Pires and Amado 2008). This statistic has the favorable property that its variance, $1/(4n)$, does not depend on the success probability.

This chapter has the following structure. In Sect. 11.2, we discuss ways to predict future confidence intervals based on the Clopper-Pearson exact method (henceforth, referred to as the exact method). In Sect. 11.3, comparisons among these approaches are given. In Sect. 11.4, we discuss the prediction of the future confidence interval based on the arcsine transformation. In Sect. 11.5, we conclude with some remarks regarding the proposed approaches, provide some examples, and briefly discuss a Bayesian strategy. Computer codes in R (http://www.r-project.org) to calculate predicted widths are provided in the Appendix.

11.2 Predictions of Future Exact Confidence Intervals

In the consideration of the small sample size, we first discuss the prediction of the exact confidence interval (Clopper and Pearson 1934). This interval is called exact because its general form is based on the exact binomial probability distribution. The exact confidence interval is conservative, having a coverage probability that is greater than the user-specified confidence level, even in cases with small sample sizes and/or extreme values of p (Pires and Amado 2008). Although the exact method has been criticized as being too conservative, its guaranteed coverage probability is very appealing, particularly for small sample sizes and/or extreme p. Let us define the width of a confidence interval w as its upper bound minus its lower bound. At the planning stage, p is often derived from existing information based on accumulated data (i.e., previous trials with a similar treatment based on different diseases or available literature). Then, one may calculate the sample size based on the predicted width of the confidence interval by using a confidence interval estimation strategy given p. Three probabilistic approaches, i.e., simple plugging in, hypothesis testing approach, and the expected confidence interval are considered as follows.

11.2.1 Approach 1: Simple Plugging In

We can simply plug the value p into sample estimation of the proportion (say, $\widehat{p}$) in a confidence interval formula. This approach is more in the context of a "common" practice of the sample size calculation. For instance, suppose that an investigator assumes that the safety rate of a study drug is $p = .775$. Based on the Wald interval formula with a 95% confidence level (i.e., $\widehat{p} \pm 1.96\sqrt{\widehat{p}\,(1-\widehat{p})\,/n}$), the future confidence interval is predicted as (0.659, 0.891) with $n = 50$ by simple plugging in, producing $w = 0.232$. The sample size for the desired width w^* is obtained by solving $w^* = 2(1.96)\sqrt{p\,(1-p)\,/n}$. When the confidence interval formula includes the observed value x instead of $\widehat{p}$ (e.g., Score interval in Pires and Amado 2008), one may use $E(X|\,p) = np$ in place of x.

Once the observation x is made, the lower and upper bounds of the exact confidence interval may be expressed as

$$\pi_L = \left[1 + \frac{n-x+1}{x F_{2x,2(n-x+1),1-\alpha/2}}\right]^{-1}, \text{ and } \pi_U = \left[1 + \frac{n-x}{(x+1)\ F_{2(x+1),2(n-x),\alpha/2}}\right]^{-1}, \tag{11.1}$$

for $x = 1, \ldots, n-1$ (Agresti and Coull 1998), where $F_{a,\,b,\,c}$ denotes the $1-c$ quantile of an F-distribution with degrees of freedom a and b. If $x = 0$, $\pi_L = 0$, and if $x = n$, $\pi_U = 1$. Let L and U denote the prediction of the lower and upper bounds of the future confidence interval. Using simple plugging in, one may use the expected value of X in the form of

$$L = \left[1 + \frac{n - E(X|p) + 1}{E(X|p)\, F_{2E(X|p),2(n-E(X|p)+1),1-\alpha/2}}\right]^{-1} \text{ for } p \in (0,1]\,, \text{ and}$$

$$U = \left[1 + \frac{n - E(X|p)}{(E(X|p) + 1)\;\, F_{2(E(X|p)+1),2(n-E(X|p)),\alpha/2}}\right]^{-1} \text{ for } p \in [0,1)\,,$$

where $E(X|\,p) = np$. Also, $L = 0$ if $p = 0$ and $U = 1$ if $p = 1$.

11.2.2 Approach 2: Hypothesis Testing Approach

The direct use of hypothesis testing may provide a valid approach for future interval prediction. Consider testing the null and alternative hypotheses $H_0 : p = \pi$ vs. $H_1 : p \neq \pi$ at the significance level α. We define the acceptance region, $A(\pi)$, and the collection of the parameters corresponding to the acceptance region, $C(X) = \{\pi\,|\,X \in A(\pi)\}$. Then, similar to the statement by Chang and O'Brien (1986),

$$\pi \in C(X) \qquad \Longleftrightarrow \qquad X \in A(\pi)\,. \tag{11.2}$$

Note that Chang and O'Brien (1986) stated Eq. (11.2) in the context of group sequential designs. In a typical confidence interval estimation, the focus is finding the boundaries of π corresponding to the confidence level based on the observed value of X. Here, based on the equivalent expressions in Eq. (11.2), we obtain the boundaries of X based on the parameter π. For example, suppose that the test statistic $t(X|\,p)$ has the standard normal distribution under the assumed proportion p. The typical confidence interval is obtained by solving

$$t(X|p) = -z_{\alpha/2}, \quad \text{and } t(X|p) = z_{\alpha/2}, \tag{11.3}$$

with respect to p where z_c is $1 - c$ quantile of the standard normal distribution. For the prediction of the future confidence interval, we solve for X instead of p and the sample size calculation can be carried out based on the predicted width. Note that X values which satisfy Eq. (11.3) are the boundaries of $A(p)$. With the Wald confidence interval, solving for X (or equivalently $\widehat{p}$) based on Eq. (11.3) provides the confidence interval prediction of $(p - z_{\alpha/2}\sqrt{p(1-p)/n}$, $p+z_{\alpha/2}\sqrt{p(1-p)/n})$, which is, in this case, the same as the plugging in approach.

Approach 2 gives rise to two additional approaches using the exact confidence interval, namely a discrete version (Approach 2-1) and a continuous version (Approach 2-2). The discrete version is constructed in the following manner.

11.2.2.1 Approach 2-1: Discrete Hypothesis Testing Approach

We will reject H_0 if the outcome x results in either $P\{X \leq x|\,\pi\}$ or $P\{X \geq x|\,\pi\}$ being less than or equal to $\alpha/2$. Let x_L and x_U denote the predicted lower and upper bounds of X for the future confidence interval, respectively. Being consistent with exact hypothesis testing, x_L and x_U that guarantee the user-specified significance level should satisfy

$$P\left\{X \leq x_L|\pi\right\} = \sum\nolimits_{k=0}^{x_L} \binom{n}{k} \pi^k (1-\pi)^{n-k} \leq \alpha/2, \quad x_L \in [0, n)$$

and

$$P\left\{X \geq x_U|\pi\right\} = \sum\nolimits_{k=x_U}^{n} \binom{n}{k} \pi^k (1-\pi)^{n-k} \leq \alpha/2, \quad x_U \in (0, n]. \tag{11.4}$$

According to Eq. (11.4) and incorporating a continuity adjustment, we can define L and U specifically as

$$\begin{aligned} nL &= \max\left\{x|P\left\{X \leq x|p\right\} \leq \alpha/2\right\} + 0.5, \\ nU &= \min\left\{x|P\left\{X \geq x|p\right\} \leq \alpha/2\right\} - 0.5, \end{aligned} \tag{11.5}$$

and $nL = 0$ if the maximum in Eq. (11.5) does not exist, and $nU = n$, if the minimum in Eq. (11.5) does not exist.

11.2.2.2 Approach 2-2: Continuous Hypothesis Testing Approach

For a continuous version of Approach 2, Formula (11.4) is expressed using the incomplete beta function

$$I_y\left(a, b\right) = \frac{\Gamma\left(a+b\right)}{\Gamma(a)\Gamma(b)} \int_0^y t^{a-1} (1-t)^{b-1} dt.$$

Then, we obtain solutions of x_L and x_U satisfying

$$I_p\left(x_L + 1, n - x_L\right) = 1 - \alpha/2 \text{ for } p \in (0, 1], \text{ and}$$

$$I_p\left(x_U, n - x_U + 1\right) = \alpha/2 \text{ for } p \in [0, 1), \tag{11.6}$$

where x_L and x_U are real number solutions in $[0, n]$, and solution x_L is replaced by 0 if $x_L < 0$, and solution x_U is replaced by n if $x_U > n$. Subsequently, the upper and lower bounds of the confidence interval are $L = x_L/n$ and $U = x_U/n$.

11.2.3 Approach 3: Expected Confidence Interval

Considering that X is a random variable, we can use the expectation of the lower and upper bounds of the confidence interval. To see how this strategy affects the interval width prediction, again, assume $n = 50$ and $p = .775$ as the example in Approach 1. Applying Approach 3 to the Wald confidence interval, we obtain (0.661, 0.889), which is slightly different than the future confidence interval prediction that we obtained based on simple plug in approach (Approach 1).

Based on the exact confidence interval Eq. (11.1), the lower and upper bounds of the future interval can be predicted as

$$L = E\left(\pi_L(X)|p\right), \quad \text{and } U = E\left(\pi_U(X)|p\right), \tag{11.7}$$

where $\pi_L(X) = \left[1 + \frac{n-X+1}{XF_{2X,2(n-X+1),1-\alpha/2}}\right]^{-1}$ for $X \in \{1, 2, \ldots, n\}$, and $\pi_U(X) = \left[1 + \frac{n-X}{(X+1)\,F_{2(X+1),2(n-X),\alpha/2}}\right]^{-1}$ for $X \in \{0, 1, \ldots, n - 1\}$. Also, $\pi_L(0) = 0$ and $\pi_U(1) = 1$.

The values of L and U are found by summing the quantities inside parenthesis of Eq. (11.7) over $X = 0, 1, \ldots, n$ with corresponding probabilities. This approach is easily understandable, and it predicts the future interval without approximation or solving certain equations.

Using Taylor's expansion, it can be shown that Approach 1 approximates Approach 3 with an order of $O(n^{-1})$. Since $F_{2X,\,2(n-X+1),\,1-\alpha/2}$ in $\pi_L(X)$ and $F_{2(X+1),\,2(n-X),\,\alpha/2}$ in $\pi_U(X)$ are concave and convex functions in X, respectively, around the reasonable values of α (e.g., 0.01, 0.05, or 0.10), Approach 1 produces a smaller lower bound and a greater upper bound than Approach 3 using Jensen's inequality (Breiman 1992), implying that Approach 1, in general, provides a wider confidence width than Approach 3 (Table 11.1). However, when the sample size increases, the differences between these methods decrease.

11.3 Sample Size Calculation Based on the Future Exact Confidence Interval Prediction

Examples of confidence interval predictions based on the different approaches are shown for the various sample sizes and p in Table 11.1. Results in Table 11.1 demonstrate that Approach 1 indeed provides wider confidence intervals than Approach 3, but they can be very close when the sample sizes increase. While Approaches 2-1 and 2-2 both use the concept of the exact hypothesis test, because of the discreteness of the binomial distribution, there are sizable differences in interval assessments between the two approaches. It is shown that the confidence interval prediction based on Approach 2 can be very different than that of Approach 1 or

Table 11.1 Comparisons of the prediction of the future 95% exact confidence intervals by Approaches (App.) 1 to 3 for various proportions (p) and sample sizes (n). For each cell of the table, the values of L and U are presented as $L-U$

		p			
n	App.	0.05	0.1	0.3	0.5
10	1	0–0.3813	0.0025–0.445	0.0667–0.6525	0.1871–0.8129
	2-1	0–0.2500	0–0.3500	0–0.6500	0.1500–0.8500
	2-2	0–0.2813	0–0.3631	0–0.6457	0.1470–0.8521
	3	0.0035–0.3740	0.0114–0.4347	0.0811–0.6400	0.2001–0.7999
20	1	0.0013–0.2487	0.0123–0.317	0.1189–0.5428	0.2720–0.7280
	2-1	0–0.1750	0–0.2750	0.0750–0.5250	0.2750–0.7250
	2-2	0–0.1864	0–0.2713	0.0870–0.5326	0.2585–0.7415
	3	0.0057–0.2437	0.0193–0.3112	0.1247–0.5377	0.2770–0.7230
30	1	0.0036–0.1971	0.0211–0.2653	0.1473–0.4940	0.3130–0.6870
	2-1	0–0.1500	0–0.2500	0.1167–0.4833	0.3167–0.6833
	2-2	0–0.1557	0–0.2343	0.1275–0.4855	0.3058–0.6942
	3	0.0077–0.1934	0.0259–0.2616	0.1505–0.4910	0.3158–0.6842
100	1	0.0164–0.1128	0.0490–0.1762	0.2124–0.3998	0.3983–0.6017
	2-1	0.0500–0.1050	0.0450–0.1650	0.2050–0.3950	0.3950–0.6050
	2-2	0.0074–0.1014	0.0410–0.1672	0.2074–0.3965	0.3972–0.6078
	3	0.0176–0.1119	0.0499–0.1755	0.2129–0.3993	0.3988–0.6012

Approach 3 in cases with small sample sizes and extreme values of p. When the sample size increases and p is closer to 0.5, the differences in the confidence interval prediction between different approaches decrease.

To find the sample size, first we set the desired confidence interval width, $w*$. Let $w(n)$ denote the future confidence interval width prediction as a function of n. Given n, the interval width is easily obtained for each approach. For Approaches 1, 2-2, and 3, the sample size is obtained as $\min\{n : w(n) \leq w*\}$. For Approach 2-1, the sample size is $\min\{n : w(n) \leq w*$, and $w(n + i) \leq w*$ for $i = 1, 2, \ldots\}$ where the second condition is required since the width is not a monotone decreasing function of the sample size due to the discreteness of the binomial distribution.

Table 11.2 shows that the sample size calculations to achieve the desirable confidence interval width (upper bound minus lower bound) based on each of the four approaches. For a full investigation, various desired widths and the values of p are used although some sample size calculations are more relevant to a large clinical trial. Table 11.2 also presents the simulated exact confidence interval width using a Monte Carlo study (10,000 simulations) based on each calculated sample size.

When p is an extreme value (e.g., 0.05), the sample size calculations by Approaches 2-1 and 2-2 are too small, giving simulated widths that are far greater than the desired width. This strongly suggests that the hypothesis testing approach (Approach 2) may not be appropriate to calculate the sample size when p is an extreme value. The differences between the sample size calculations decrease when

Table 11.2 The sample sizes (n) to achieve the desired confidence interval width ($w*$) and the Monte-Carlo simulated confidence interval width based on the 95% exact confidence interval and each calculated sample size (n) for each of the four approaches (App.)

		p									
		0.05		0.1		0.15		0.3		0.5	
$w*$	App.	n	width	n	width	n	width	n	width	n	width
0.2	1	29	0.1891	44	0.1923	58	0.1951	89	0.1977	104	0.1984
	2-1	18	0.2541	45	0.1910	56	0.1989	86	0.2012	101	0.2014
	2-2	18	0.2537	43	0.1951	58	0.1954	90	0.1967	106	0.1965
	3	27	0.1975	42	0.1981	56	0.1989	87	0.2001	103	0.1994
0.25	1	20	0.2378	29	0.2397	38	0.2429	58	0.2458	67	0.2478
	2-1	10	0.3706	26	0.2535	36	0.2493	57	0.2477	65	0.2515
	2-2	12	0.3298	25	0.2596	39	0.2394	59	0.2435	69	0.2441
	3	19	0.2462	27	0.2493	36	0.2489	56	0.2498	66	0.2497
0.3	1	15	0.2853	21	0.2849	27	0.2881	40	0.2957	47	0.2957
	2-1	9	0.3962	15	0.3398	24	0.3057	40	0.2955	47	0.2957
	2-2	10	0.3700	16	0.3297	28	0.2829	42	0.2886	49	0.2897
	3	14	0.2993	19	0.2991	25	0.2984	39	0.2995	46	0.2989
0.35	1	12	0.3293	16	0.3289	20	0.3350	30	0.3407	35	0.3419
	2-1	8	0.4269	10	0.4232	16	0.3730	35	0.3159	35	0.3419
	2-2	8	0.4284	11	0.4030	20	0.3340	32	0.3300	37	0.3327
	3	11	0.3494	15	0.3399	19	0.3431	29	0.3462	34	0.3468

p becomes closer to 0.5 or smaller widths are desirable (so that a large sample is required).

Approaches 1 and 3 provide a simulated confidence interval width smaller than the target confidence interval width, a characteristic that we may consider desirable; however, Approach 3 provides consistently smaller sample sizes while keeping the simulated width close to the target confidence interval width. When a small study is planned, this reduction of the sample size can be meaningful. Thus, in the consideration of the sample size and the interval width prediction, we recommend Approach 3 to be the best way to calculate the sample size for the desired interval width without wasting study resources.

The results in Table 11.2 also demonstrate that the required sample size can vary dramatically as a function of p. For example, to achieve the same confidence interval width of 0.25, the sample size requirements based on Approach 3 are 19 for $p = 0.05$, and 66 for $p = 0.5$. This shows that, for the sample size calculation, using $p = 0.5$ a most conservative approach may result in an unnecessarily large sample size if the true p is, in fact, a much smaller value than 0.5 (e.g., 0.1). This emphasizes an importance of a proactive use of existing information for the sample size calculation.

11.4 Prediction of the Arcsine Confidence Interval

In this section, we discuss confidence intervals based on normal approximation focusing primarily on arcsine intervals. Even if arcsine intervals are based on asymptotic results, Pires and Amado (2008) argue that these intervals generally keep the coverage probability above the confidence level even with samples sizes as small as 10. We note that our approach can be easily applied to other normal approximation confidence interval strategies. If strict conservativeness matters, Pires and Amado (2008) recommend an arcsine interval with Anscombe's correction as well as the exact confidence interval.

The basic idea of the arcsine interval is that, through the arcsine transformation ($f(p) = \arcsin\sqrt{p}$), the tails of the distribution of the proportion are expanded while the middle part of the distribution compresses (Sokalr and Rohlf 1981). Based on the delta method (Rao 1973), it can be shown that the asymptotic variance of the confidence interval is independent of p. To improve the performance of the approximation of the arcsine transformation at the extreme values, Anscombe (1948) and Freeman and Tukey (1950) propose the arcsine transformation with some corrections (Sahai and Ageel 2000). In this chapter, our method is based on Anscombe's correction:

$$\pi_L = \sin^2\left(\arcsin\sqrt{\frac{x+3/8}{n+3/4}} - \frac{Z_{\alpha/2}}{2\sqrt{n}}\right), \text{ and}$$

$$\pi_U = \sin^2\left(\arcsin\sqrt{\frac{x+3/8}{n+3/4}} + \frac{Z_{\alpha/2}}{2\sqrt{n}}\right) \tag{11.8}$$

Let $\pi_L(X)$ and $\pi_U(X)$ denote π_L and π_U in (11.8), respectively, where x is replaced by the random variable X. For Approach 1, the prediction of the future confidence interval can be obtained by using $E(X|\,p)$ in place of X.

For Approach 2, the distribution of X (or $\widehat{p}$) is approximated by a normal distribution so that the discreteness of the random variable is no longer a problem. Applying Approach 2, the prediction of the arcsine interval based on Anscombe's correction is

$$L = \max\left\{\sin^2\left(\arcsin\sqrt{p} - \frac{Z_{\alpha/2}}{2\sqrt{n}}\right)\left(1+\frac{3}{4n}\right) - \frac{3}{8n},\quad 0\right\}, \text{ and}$$

$$U = \min\left\{\sin^2\left(\arcsin\sqrt{p} + \frac{Z_{\alpha/2}}{2\sqrt{n}}\right)\left(1+\frac{3}{4n}\right) - \frac{3}{8n},\quad 1\right\}. \tag{11.9}$$

Note that the minimum and maximum in Eq. (11.9) are used to restrict L and U to be from 0 to 1, inclusive.

Table 11.3 The sample sizes (n) to achieve the desired 95% confidence interval width (w^*) and the Monte-Carlo simulated confidence interval width based on the arcsine interval using Approaches (App.) 1 to 3

		p									
		0.05		0.1		0.15		0.3		0.5	
w^*	App.	n	width	n	width	n	width	n	width	n	width
0.2	1	22	0.1849	36	0.1926	50	0.1947	80	0.1982	95	0.1987
	2	15	0.2291	35	0.1951	50	0.1945	81	0.1968	97	0.1967
	3	20	0.1950	34	0.1982	48	0.1981	79	0.1994	94	0.1997
0.25	1	15	0.2299	24	0.2338	32	0.2404	51	0.2463	61	0.2463
	2	10	0.2908	22	0.2435	32	0.2403	52	0.2441	62	0.2444
	3	13	0.2484	21	0.2482	30	0.2479	50	0.2488	60	0.2483
0.3	1	12	0.2620	17	0.2740	22	0.2867	35	0.942	42	0.2944
	2	7	0.3632	14	0.3204	22	0.2866	37	0.2869	43	0.2912
	3	10	0.2949	15	0.2919	20	0.3002	34	0.2987	41	0.2978
0.35	1	9	0.3116	13	0.3132	16	0.3315	26	0.3385	31	0.3394
	2	6	0.3989	10	0.3589	17	0.3221	27	0.3318	32	0.3344
	3	8	0.3360	11	0.3403	15	0.3398	25	0.3440	29	0.3499

For Approach 3, we predict the future confidence interval in a same manner as Eq. (11.7). Since $\pi_L(X)$ and $\pi_U(X)$ are convex and concave functions in X around the reasonable values of α (e.g., 0.01, 0.05, or 0.10), respectively, Approach 3 produces a narrower future confidence interval prediction than Approach 1. Similar to the exact method, we can show that the arcsine interval prediction based on Approach 1 is a first-order approximation of Approach 3 using the Taylor expansion.

Some examples of sample size calculations and simulated confidence interval widths for the arcsine interval based on Approaches 1, 2, and 3 are shown in Table 11.3. Since Approach 3 has a shorter interval width than Approach 1, Approach 3 provides smaller sample sizes; however, Approach 3 intervals are closer to the user-desired confidence interval width without exceeding it. With the extreme values of p, Approach 2 largely does not achieve the desired confidence interval width, indicating that Approach 2 may not be appropriate for the sample size calculation. When p becomes close to 0.5 and a smaller confidence interval width is desirable, the differences between Approaches 1 and 3 decrease.

It is noteworthy that the sample sizes for the arcsine confidence interval for extreme values of p are smaller than those for the exact confidence interval. According to our Monte Carlo study (not shown in this chapter), with the extreme value of p (e.g., 0.05 and 0.10), both the arcsine interval and the exact confidence interval have much higher coverage probabilities than the target confidence level, consistent with the results of Pires and Amado (2008). Considering that the arcsine interval requires a smaller sample size, this result suggests that the arcsine interval may be a viable alternative to the exact confidence interval for the extreme value of p.

11.5 Applications

Following is an example demonstrating how the proposed approaches can work. For the past few years, the effectiveness of various stenting treatments in patients with acute stroke has been investigated through many small studies. Suppose that a new intracranial occluded artery stenting treatment for revascularization in acute ischemic stroke patients is investigated in a small pilot study. Such a study will not provide a definitive conclusion on efficacy of the treatment but rather focus on technical feasibility of the treatment before the treatment is tested in a large-scale randomized clinical trial. The primary endpoint of the study is the re-establishment of blood flow (recanalization) in occluded vessels. The recanalization is indicated by TIMI (Thrombolysis In Myocardial Infarction) grade 2 or 3 flow, i.e., delayed distal flow or distal flow without delay (TIMI Study Group 1985). Recently, a balloon-mounted stent treatment with similar patient groups reports an overall recanalization rate of 79% (Levy et al. 2006). Although the new study uses a different kind of stent treatment, investigators can reasonably assume that the study will have a similar recanalization rate to 79% or better as the study procedure has a slightly improved feature. Suppose that the desirable confidence interval width is 0.4 with the 95% confidence level for the exact confidence interval. The required sample sizes are 19, 21, 21, and 20 for Approaches 1, 2-1, 2-2 and 3. Because of the better properties discussed in Sect. 11.3, we recommend using Approach 3, of which sample size is about the middle of the three approaches.

The following example demonstrates a sensitivity analysis for different values of p and how the proper sample size can be decided. As briefly addressed at the end of Sect. 11.3, the choice of p impacts the prediction of confidence interval widths and subsequent sample size calculations. As described in the example above, also in our experiences, relevant information regarding study treatment commonly exists in the form of investigators' accumulated knowledge or in the published literature. For a study of a novel stenting treatment in patients with acute ischemic stroke, it may be of interest to investigate the rate of the symptomatic intracranial hemorrhage, an important safety endpoint of a stroke study, and one wishes to estimate the 95% confidence interval of the rate with the interval width, 0.3. The MERCI trial (Smith et al. 2005), a large clinical trial treating acute ischemic stroke patients with an embolectomy device, reports the intracranial hemorrhage rate, 7.8%. For a stenting treatment, the SARIS trial (Levy et al. 2006) reports the intracranial hemorrhage rate, 5.3%. Since no direct association between stenting and the adverse event is assumed in general, these numbers can be used for the prediction of the confidence interval width for the symptomatic intracranial hemorrhage rate. Based on Approach 3, the required sample sizes range from 15 to 17 for the rates from 5.3 to 7.8%. For a relevant sample size calculation, we choose the larger sample size as a conservative strategy. Based on the sample size 17, if the true rate is 5.3%, the predicted width is 0.267, a smaller width than the target width that is often more desirable.

11.6 Discussion

In this chapter, a few approaches were applied to predict future confidence intervals, and it was demonstrated that the prediction of the confidence interval can be widely different based on the approaches applied. The approach based on the expectation of boundaries of the confidence intervals (Approach 3) may be generally preferable in terms of adequate sample sizes and the desired confidence interval width. We also note that an application of Approach 3 to any confidence interval strategy is easily carried out since its computation is straightforward.

For normal approximation confidence intervals, while Approaches 1 and 2 can sometimes provide the same solutions (e.g., Wald's confidence interval), such a fact does not apply to the prediction of the arcsine interval used in this chapter. Such non-matching between Approaches 1 and 2 is generally true for the confidence interval strategy with corrections. The simulated results show that, for the extreme values of p, the arcsine interval with the correction may be comparable to the exact confidence interval in terms of the coverage rate. Also, note that the required sample sizes can be much different due to different confidence interval schemes (e.g., exact confidence interval in Table 11.2 vs. arcsine interval in Table 11.3). Thus, it is recommended that the same confidence interval strategy should be used at the planning stage of a study and after completing the study.

We briefly remark a potential Bayesian strategy in the prediction of the future confidence interval. In theory, if information regarding the rate of interest exists in various studies, this information can be aggregated in a form of the posterior distribution. Suppose that p has a known prior distribution $F(p)$ and the posterior distribution $F(p|\,\mathbf{x}^*)$ is based on the observations $\mathbf{x}^*$. In case of Approach 3, the boundaries for the future confidence interval can be predicted as

$$\int_{-\infty}^{\infty} E\left\{\pi_L(X)|p\right\} dF\left(p|\mathbf{x}^*\right), \quad \text{and} \quad \int_{-\infty}^{\infty} E\left\{\pi_U(X)|p\right\} dF\left(p|\mathbf{x}^*\right),$$

And, subsequently, the relevant sample size to achieve the target width can be calculated. Similar methods can be applied to Approaches 1 and 2. Commonly, the rate is considered to have a beta distribution (Stallard 1998), and the beta distribution with $\alpha = \beta = 1$ (equivalent to the uniform distribution on the support of (0, 1)) can be a reasonable choice of a prior distribution. A future research project would be to develop and investigate a method to properly incorporate information existing in various forms (e.g., studies with different sizes and correlation within a study) into the posterior distribution, together with the investigation of the impact of the Bayesian strategy on the actual interval width prediction in a small study.

We remark that the Clopper-Pearson exact interval may not be the most preferable confidence interval with the cases of extreme p and small n. If someone is willing to reduce the width of intervals while an actual confidence level is slightly compromised, other interval approaches can be more desirable. As we stated earlier, the proposed approaches are easily implemented for other methods to obtain the confidence interval. Consider an interval based on the log-likelihood ratio of the binomial distribution,

$$-2\left[x\log\left(p_0\right)+(n-x)\log\left(1-p_0\right)-\left\{x\log\left(\widehat{p}\right)+(n-x)\log\left(1-\widehat{p}\right)\right\}\right],$$

which is approximated to the χ_2^1 distribution in a large n. When x is 0 (or n), we can use 0.5 (or $n-0.5$). Given the knowledge of p, for Approach 1, we let $x=np$ and $\widehat{p}=p$. Then, p_0 satisfies a quantile of χ_2^1 distribution (e.g., 0.95-th quantile) will be obtained. For Approach 2, we let $p_0=p$ and $\widehat{p}=x/n$, then x satisfying a quantile of χ_2^1 distribution will be obtained. For Approach 3, x is replaced by the values of $1,\ldots,n$, which will produce n intervals. The expectation of low and upper boundaries will be obtained using the low and upper boundaries of those intervals. The likelihood method has a benefit that it can be extended to the Bayesian approach easily. For the Bayesian approach, the relevant posterior density will be considered to construct the likelihood ratio.

We finally comment that the concept proposed in this chapter can be extended to other statistics of interest, e.g., relative risk in two sample comparison. An implementation of the proposed approaches for various relevant confidence interval strategies warrants further investigations.

A.1 Appendix

The following R codes are to calculate the predicted width of confidence intervals proposed in Sects. 11.2 and 11.3. The parameters needed are n, p, and alpha for the sample size, true success rate, and 100(1-alpha)% confidence interval. The results of each function consist of predicted lower, upper bounds, and width.

R codes for Sect. 11.2

```
#####Approach 1######
approach1<-function(n,p,alpha){
x<-n*p
lc<-1/(1+(n-x+1)/(x*qf(0.025,2*x,2*(n-x+1))))
rc<-1/(1+(n-x)/((x+1)*qf(0.975,2*(x+1),2*(n-x))))
cat(round(lc,4),round(rc,4),round((rc-lc),4), fill=T)
}
#Usage
approach1(20,0.07,0.05)

#####Approach 2-1####
approach2.1<-function(n,p,alpha){
x<-seq(0,n,1)
probs<-pbinom(x,n,p)
L<-max(x[probs<=(alpha/2)],-0.5)+0.5
U<-min(min(x[probs>=(1-alpha/2)])+1-0.5,n)
cat(round(L/n,4),round(U/n,4), round((U/n-L/n),4), fill=T)
}
```

```
#Usage
approach2.1(20,0.07,0.05)

#####Approach 2-2####
  #Functions to solve Eq. (11.6)
incbetaL<-function(x,p,n,alpha){
duhaeyo<-0
ele1<-x+1
ele2<-n
for(i in ele1:ele2){
duhaeyo<-duhaeyo+factorial(ele2)/(factorial(i)*factorial
   (ele2-i))*
p^i*(1-p)^(ele2-i)
}
crit<-abs(duhaeyo-(1-alpha/2))
return(crit)
}
incbetaU<-function(x,p,n,alpha){
duhaeyo<-0
ele1<-x
ele2<-n
for(i in ele1:ele2){
duhaeyo<-duhaeyo+factorial(ele2)/(factorial(i)*factorial
   (ele2-i))*
p^i*(1-p)^(ele2-i)
}
crit<-abs(duhaeyo-(alpha/2))
return(crit)
}

approach2.2<-function(nn,pp,alphaa){
nval<-nn
pval<-pp
alphaval<-alphaa
x<-nval*pval
clow<-1/(1+(nval-x+1)/(x*qf((alphaval/2),2*x,2*(nval-x+1))))
cupp<-1/(1+(nval-x)/((x+1)*qf((1-alphaval/2),2*(x+1),
    2*(nval-x))))
lowval<-optimize(incbetaL, c(((clow-0.1)*nval),
    ((clow+0.1)*nval)),p=pval, n=nval, tol = 0.0001,
     alpha=alphaval)
uppval<-optimize(incbetaU, c(((cupp-0.1)*nval),
     ((cupp+0.1)*nval)),p=pval, n=nval, tol = 0.0001,
     alpha=alphaval)
lc<-max(lowval[[1]]/nval,0)
rc<-min(uppval[[1]]/nval,1)
cat(c(lc,rc,(rc-lc)))
}

#Usage
approach2.2(20,0.07,0.05)

####Approach 3#######
```

```
approach3<-function(n,p,alpha){
truep<-p
x<-seq(0,n,1)
probs<-dbinom(x,n,truep)
counts<-0
clc<-c()
crc<-c()
for(i in 1:(n+1)){
lc<-binom.test(x[i], n, conf.level = (1-alpha))$conf.int[1]
rc<-binom.test(x[i], n, conf.level = (1-alpha))$conf.int[2]
clc<-c(clc,lc*probs[i])
crc<-c(crc,rc*probs[i])
}
cat(round(sum(clc),4),round(sum(crc),4),round((sum(crc)-sum(clc)),
   4), fill=T)
}

#Usage
approach3(20,0.07,0.05)
```

R codes for Sect. 11.3

```
#Approach 1
approach1<-function(n,p,alpha){
c<-qnorm((1-alpha/2))
x<-n*p
lc<-max(sin(asin(sqrt((3/8+x)/(n+3/4)))-c/(2*sqrt(n)))^2,0)
rc<-min(sin(asin(sqrt((3/8+x)/(n+3/4)))+c/(2*sqrt(n)))^2,1)
cat(round(lc,4),round(rc,4),round((rc-lc),4))
}

#Approach 2
approach2<-function(n,p,alpha){
c<-qnorm((1-alpha/2))
lc<-max(((sin(asin(sqrt(p))-c/(2*sqrt(n)))^2)*(1+3/(4*n))
   -3/(8*n)),0)
rc<-min(((sin(asin(sqrt(p))+c/(2*sqrt(n)))^2)*(1+3/(4*n))
   -3/(8*n)),1)
cat(round(lc,4),round(rc,4),round((rc-lc),4))
}

#Approach 3
approach3<-function(n,p,alpha){
c<-qnorm((1-alpha/2))
truep<-p
x<-seq(0,n,1)
probs<-dbinom(x,n,truep)
counts<-0
clc<-c()
crc<-c()
for(i in 1:length(probs)){
lc<-max(sin(asin(sqrt((3/8+x[i])/(n+3/4)))-c/(2*sqrt(n)))^2,0)
rc<-min(sin(asin(sqrt((3/8+x[i])/(n+3/4)))+c/(2*sqrt(n)))^2,1)
```

```
clc<-c(clc,lc*probs[i])
crc<-c(crc,rc*probs[i])
}
crc<-c(crc,1*probs[(n+1)])
cat(round(sum(clc),4),round(sum(crc),4),round((sum(crc)
   -sum(clc)),4))
}
```

References

Agresti, A., & Coull, B. A. (1998). Approximate is better than "exact" for interval estimation of binomial proportions. *The American Statistician, 52*(2), 119–126.

Anscombe, F. J. (1948). The transformation of Poisson, binomial and negative-binomial data. *Biometrika, 35*(3/4), 246–254.

Blaker, H. (2000). Confidence curves and improved exact confidence intervals for discrete distributions. *Canadian Journal of Statistics, 28*(4), 783–798.

Blyth, C. R., & Still, H. A. (1983). Binomial confidence intervals. *Journal of the American Statistical Association, 78*(381), 108–116.

Bougnoux, P., Hajjaji, N., Ferrasson, M. N., Giraudeau, B., Couet, C., & Le Floch, O. (2009). Improving outcome of chemotherapy of metastatic breast cancer by docosahexaenoic acid: A phase II trial. *British Journal of Cancer, 101*(12), 1978–1985.

Breiman, L. (1992). *Probability*. Philadelphia, PA: SIAM.

Brown, L. D., Cai, T. T., & DasGupta, A. (2001). Interval estimation for a binomial proportion. *Statistical Science, 16*, 101–117.

Chang, M. N., & O'Brien, P. C. (1986). Confidence intervals following group sequential tests. *Controlled Clinical Trials, 7*(1), 18–26.

Clopper, C. J., & Pearson, E. S. (1934). The use of confidence or fiducial limits illustrated in the case of the binomial. *Biometrika, 26*(4), 404–413.

Freeman, M. F., & Tukey, J. W. (1950). Transformations related to the angular and the square root. *The Annals of Mathematical Statistics, 21*, 607–611.

Levy, E. I., Ecker, R. D., Horowitz, M. B., Gupta, R., Hanel, R. A., Sauvageau, E., et al. (2006). Stent-assisted intracranial recanalization for acute stroke: Early results. *Neurosurgery, 58*(3), 458–463.

Martin-Schild, S., Hallevi, H., Shaltoni, H., Barreto, A. D., Gonzales, N. R., Aronowski, J., et al. (2009). Combined neuroprotective modalities coupled with thrombolysis in acute ischemic stroke: A pilot study of caffeinol and mild hypothermia. *Journal of Stroke and Cerebrovascular Diseases, 18*(2), 86–96.

Newcombe, R. G., & Vollset, S. E. (1994). Confidence intervals for a binomial proportion. *Statistics in Medicine, 13*(12), 1283–1285.

Pires, A. M., & Amado, C. (2008). Interval estimators for a binomial proportion: Comparison of twenty methods. *REVSTAT–Statistical Journal, 6*(2), 165–197.

Rao, C. R. (1973). *Linear statistical inference and its applications*. New York: Wiley.

Rino, Y., Yukawa, N., Murakami, H., Wada, N., Yamada, R., Hayashi, T., et al. (2010). A phase II study of S-1 monotherapy as a first-line combination therapy of S-1 plus cisplatin as a second-line therapy, and weekly paclitaxel monotherapy as a third-line therapy in patients with advanced gastric carcinoma: A second report. *Clinical Medicine Insights: Oncology, 4*, CMO-S3920.

Sahai, H., & Ageel, M. I. (2000). *Analysis of variance: Fixed, random and mixed models*. Ann Arbor, MI: Sheridan Books.

Schiffer, F., Johnston, A. L., Ravichandran, C., Polcari, A., Teicher, M. H., Webb, R. H., et al. (2009). Psychological benefits 2 and 4 weeks after a single treatment with near infrared light to the forehead: A pilot study of 10 patients with major depression and anxiety. *Behavioral and Brain Functions, 5*(1), 46.

Smith, W. S., Sung, G., Starkman, S., Saver, J. L., Kidwell, C. S., Gobin, Y. P., et al. (2005). Safety and efficacy of mechanical embolectomy in acute ischemic stroke: Results of the MERCI trial. *Stroke, 36*(7), 1432–1438.

Sokalr, R. R., & Rohlf, F. J. B. (1981). *The principles and practice of statistics in biological research*. New York: WH Freeman.

Stallard, N. (1998). Sample size determination for phase II clinical trials based on Bayesian decision theory. *Biometrics, 54*, 279–294.

TIMI Study Group*. (1985). The thrombolysis in myocardial infarction (TIMI) trial: Phase I findings. *New England Journal of Medicine, 312*(14), 932–936.

Vollset, S. E. (1993). Confidence intervals for a binomial proportion. *Statistics in Medicine, 12*(9), 809–824.

Wang, W. (2014). An iterative construction of confidence intervals for a proportion. *Statistica Sinica, 24*, 1389–1410.

Chapter 12
Importance of Adjusting for Multi-stage Design When Analyzing Data from Complex Surveys

Trung Ha and Julia N. Soulakova

12.1 Introduction

12.1.1 Use of National Surveys in Behavioral Research

Data from national surveys are widely used to monitor the nation's health status, access to health care, improvements toward achieving national health objectives, and other important health-related goals. The Tobacco Use Supplement (TUS) to the Current Population Survey (CPS) data have been used to estimate trends in prevalence of smoking (U.S. Bureau of Labor Statistics and U.S. Census Bureau 2006; U.S. Department of Health and Human Services, Centers for Disease Control and Prevention, National Center for Health Statistics 2010; Fiore et al. 1989; Jemal et al. 2011; Soulakova et al. 2009), smoking initiation (Cummings and Shan 1995; Gilpin and Pierce 1997), and smoke-free workplaces (Shopland et al. 2001). Reliability of the TUS-CPS measures was demonstrated in several studies (Soulakova et al. 2012, 2015a, b; Soulakova and Crockett 2014). The National Health Interview Survey (NHIS) and Supplements (U.S. Department of Health and Human Services, Centers for Disease Control and Prevention, National Center for Health Statistics 2010) have been used to estimate the prevalence of chronic conditions and diseases including heart disease, hypertension, diabetes, and migraines (Blackwell et al. 2014; Burch et al. 2015; Stang and Osterhaus 1993), to address mental health (Allison et al. 1999; Blackwell et al. 2014), and to study health disparities among populations with diverse sexual orientations (Dahlhamer et al. 2014; Ward et al. 2014).

T. Ha · J. N. Soulakova (✉)
Burnett School of Biomedical Sciences, College of Medicine, University of Central Florida, Orlando, FL, USA
e-mail: Trung.Ha@ucf.edu; Julia.Soulakova@ucf.edu

Y. Zhao, D.-G. Chen (eds.), *New Frontiers of Biostatistics and Bioinformatics*, ICSA Book Series in Statistics, https://doi.org/10.1007/978-3-319-99389-8_12

Because national surveys commonly aim to gather a sample that is representative of the civilian noninstitutionalized US population, they use complex sampling (Centers for Disease Control and Prevention 2016; Parsons et al. 2014; U.S. Bureau of Labor Statistics and U.S. Census Bureau 2006). Thus, when performing analysis of data from these surveys, researchers should incorporate additional adjustments outlined in the survey's methodological guidelines. Specifically, the analytical approach for the 2006–2015 NHIS data is outlined in the Variance Estimation Guidelines; the Guidelines state that Taylor Linearization should be used when deriving the standard errors of the population estimates (Centers for Disease Control and Prevention 2016). Similarly, to adjust for the complex, multi-stage sampling in the TUS-CPS, the Balanced Repeated Replications (BRR) method should be used in data analysis (U.S. Bureau of Labor Statistics and U.S. Census Bureau 2006).

12.1.2 Variance Estimation Using BRR

For surveys utilizing complex designs, it is not sufficient to just incorporate the main weight in the analyses (Wolter 2007; Lohr 1999). The BRR method is one such common method that allows to correctly adjust for the design specifics. We briefly describe the BRR in the simplest case, where the population consists of L strata and each stratum has at least two Primary Sampling Units (PSUs). For simplicity, we consider the BRR method for sampling with replacement. While the BRR method can be used for sampling without replacement, the method is simpler in the former case (Wolter 2007).

First, we draw a stratified sample with 2 PSUs per stratum. As a result, we have a full sample containing L subsamples of two PSUs. Next, we draw one PSU from each subsample. The resulting set is termed a replicate or half sample. In general, we can draw as many as 2^L distinct replicates from the full sample. Because the maximum number of distinct replicates 2^L may be very large in practice, it is more convenient to consider only a subset of these replicates, e.g., a subset of k replicates. The number of replicates k should be chosen as the smallest integer, multiple of 4 such that $k \geq L$. To minimize computing time, however, one may choose the number of replicates to be less than L; this is illustrated in Sect. 12.1.3. The k replicates, i.e., balanced half samples, are chosen using Hadamard matrix (Wolter 2007, Chapter 3).

Then, the replicates and the full sample are used to compute $k + 1$ values of the statistic of interest. To avoid possible computing ambiguities, e.g., division by zero when computing a statistic based on a replicate, one can use Fay's method, also termed the modified half sample technique (Judkins 1990). In the final step, computed values are used to estimate variance via a scaled mean square difference. The formula for computing variance is depicted in formula (12.2) in Sect. 12.1.3.

The BRR and modified half sample technique have been extended and discussed for other types of complex sampling (Judkins 1990; Wolter 2007, Chapters 3 and 8). For example, in Sect. 12.1.3 we refer to a method termed Successive Difference Replication (SDR). The SDR was based on Successive Difference method

(Wolter 1984, 2007, Chapter 8) and initially was proposed for estimating variance of the estimated total when a systematic random sample is drawn from an ordered list (Fay and Train 1995). The SDR can be used when a stratum contains only one PSU and can also be applied to estimate variance of a general estimator (Ash 2014).

12.1.3 Application of BRR for the TUS-CPS Data Analysis

Since 2000 onwards, the CPS monthly sample size is about 60,000 households (U.S. Department of Commerce, U.S. Census Bureau 2015). These households (also termed housing units) are sampled using the following multi-stage sampling strategy (U.S. Bureau of Labor Statistics and U.S. Census Bureau 2006, Chapter 3). First, the PSUs are defined, where PSUs are usually a metropolitan area, a large county, or a group of smaller counties. The PSUs are grouped into 824 strata so that the strata are within state boundaries and are homogenous in terms of the labor force characteristics; there are 2025 PSUs. The strata are of two types: self-representing (SR) and non-self-representing (NSR). Each SR stratum consists of a single PSU, such as one of the 151 most populated metropolitan areas. Each NSR stratum consists of at least two PSUs. All 2000-based designs include 446 SR and 378 NSR strata. For each SR stratum, the PSU from the stratum is selected, and for each NSR stratum, only one PSU from the stratum is selected (using unequal probability sampling). As a result, there are 824 PSUs sampled in the first stage.

In the second stage, the Ultimate Sampling Units (USUs) are defined and selected from the sampled PSUs. The USUs represent a small group (usually four addresses) of housing units with similar demographic composition and geographic proximity. All housing units are surveyed within a sampled USU when the USU is small (15 housing units or less). The third stage, termed field subsampling, is implemented to select a subset of housing units when a USU is large (i.e., has more than 15 housing units identified for interview).

The sampled housing units are surveyed and a 4-8-4 rotation scheme is used (U.S. Bureau of Labor Statistics and U.S. Census Bureau 2006, Chapter 3). Specifically, each household is surveyed for 4 consecutive months, and 8 months later it is surveyed for 4 additional months. In any single month, 1/8 of the sample housing units is interviewed for the first time; another 1/8 is interviewed for the second time, and so on. The rotation chart was illustrated visually in Appendix 1 in Soulakova et al. (2009).

In research using TUS-CPS and other CPS Supplements, one needs to adjust for the multiple stages and other design characteristics, e.g., the number of sampled households differs across the Supplements. Specifically, one needs to incorporate the BRR with the main and replicate weights computed for the Supplement by the U.S. Census Bureau (U.S. Bureau of Labor Statistics and U.S. Census Bureau 2006, Chapter 14). The replicate weights are based on the main weight and correct for the difference in sampling the PSUs from the SR and NSR (Ash 2014; U.S. Bureau of Labor Statistics and U.S. Census Bureau 2006, Chapter 14). For the SR PSUs, the replicate weights are obtained using the SDR, while for the NSR PSUs, the weights are obtained using the modified half sample technique. For the 2000-based designs, there are 160 replicate weights, i.e., $k = 160$, that are available for public use and should be incorporated in all analyses (U.S. Bureau of Labor Statistics and U.S. Census Bureau 2006).

Let Y denote the parameter of interest, e.g., average age of single parents in the USA. Let n denote the total number of respondents in the sample, $\widehat{Y}$ denote the estimator of Y based on the sample and the main weight, and $\widehat{Y}_r$, $r = 1, 2, \ldots, R$, denote the estimator of Y based on the $r - th$ replicate weight. For example, in a case of estimating the mean, the following formulas should be used for computing $\widehat{Y}$ and $\widehat{Y}_r^{\prime}s$:

$$\widehat{Y} = \frac{1}{\sum_{i=1}^{n} W_i} \sum_{i=1}^{n} W_i Y_i \tag{12.1}$$

and

$$\widehat{Y}_r = \frac{1}{\sum_{i=1}^{n} W_{r,i}} \sum_{i=1}^{n} W_{r,i} Y_i, \quad r = 1, 2, \ldots, R,$$

where Y_i is the measurement for the $i - th$ respondent (e.g., age of the $i - th$ respondent in the sample), W_i is the main weight corresponding to the $i - th$ respondent, and $W_{r,\,i}$ is the value of the $r - th$ replicate weight corresponding to the $i - th$ respondent.

The estimated variance via BRR is given by

$$\text{var}\left(\widehat{Y}\right) = \frac{1}{R(1-\epsilon)^2} \sum_{r=1}^{R} \left(\widehat{Y}_r - \widehat{Y}\right)^2, \tag{12.2}$$

where $0 < \epsilon < 1$ is Fay's factor. For the 2000-based TUS-CPS data, $R = 160$ and $\epsilon = 0.5$. Thus, formula (12.2) reduces to

$$\text{var}_{TUS}\left(\widehat{Y}\right) = \frac{1}{40} \sum_{r=1}^{160} \left(\widehat{Y}_r - \widehat{Y}\right)^2. \tag{12.3}$$

We note that Fay's factor 0.5 is a default option in BRR-Fay variance estimation method in SAS® (SAS Institute Inc. 2016).

12.1.4 Three Analytical Methods

We discuss three methods of estimating several parameters and standard errors of estimators based on TUS-CPS data. Method I ignores any weighting and incorrectly treats the sample as if it is a simple random sample. Let $\widehat{Y}_I$ denote the mean estimated via method I, then

$$\widehat{Y}_I = \frac{1}{n} \sum_{i=1}^{n} Y_i \tag{12.4}$$

and

$$\mathrm{var}\left(\widehat{Y}_I\right) = \frac{1}{n(n-1)} \sum_{i=1}^{n} \left(Y_i - \widehat{Y}_I\right)^2. \tag{12.5}$$

Method II uses the main weights when computing the point estimates but ignores the replicate weights when estimating variance. Method III uses the main weight when computing the point estimate, and the main and replicate weights when computing the variance via formula (12.3). Let $\widehat{Y}_{II}$ and $\widehat{Y}_{III}$ denote the means estimated via methods II and III, respectively. Note that both point estimates are computed using formula (12.1) and thus, $\widehat{Y}_{II} = \widehat{Y}_{III}$.

Several methods can be used to estimate the variance of $\widehat{Y}_{II}$ while ignoring the replicate weights. Method II incorporates Taylor linearization. This is a default approach when one specifies the main weight but does not use any stratum/cluster option and does not specify variance method in the built-in procedures in SAS 9.4 *survey package* (SAS Institute Inc. 2016). Method II utilizes the following formula:

$$\mathrm{var}\left(\widehat{Y}_{II}\right) = \frac{n}{n-1} \sum_{i=1}^{n} \left(e_i - \overline{e}\right)^2, \tag{12.6}$$

where

$$e_i = \frac{W_i}{\sum_{i=1}^{n} W_i} \left(Y_i - \widehat{Y}_{II}\right)$$

and

$$\overline{e} = \frac{1}{n} \sum_{i=1}^{n} e_i.$$

For methods I, II, and III, standard errors can be computed, respectively, using

$$SE\left(\widehat{Y}_I\right) = \sqrt{\mathrm{var}\left(\widehat{Y}_I\right)},\ SE\left(\widehat{Y}_{II}\right) = \sqrt{\mathrm{var}\left(\widehat{Y}_{II}\right)} \quad \text{and} \\ SE\left(\widehat{Y}_{III}\right) = \sqrt{\mathrm{var}_{TUS}\left(\widehat{Y}_{III}\right)}, \tag{12.7}$$

where $\mathrm{var}\left(\widehat{Y}_I\right)$ is computed via formula (12.5), $\mathrm{var}\left(\widehat{Y}_{II}\right)$ is computed via formula (12.6), and $\mathrm{var}_{TUS}\left(\widehat{Y}_{III}\right)$ is computed via formula (12.3).

We illustrate three study claims:

Claim 1: All three methods could result in very similar point estimates.

Claim 2: Methods I and II/III could result in different point estimates.

Claim 3: Methods I, II, and III could result in discrepant standard errors and the pattern of these discrepancies can be different, e.g., method II could result in a smaller or larger standard error relative to method III.

Discrepancies in the results of the three methods are illustrated using 2014–2015 TUS-CPS data. While we used built-in procedures in SAS 9.4 *survey package* (SAS Institute Inc. 2016), we also performed additional computing to illustrate formulas presented in Sect. 12.1.3. We considered adult (18+ years old) single parents who lived with underage children (younger than 18 years old) in the USA. The measures of interest included parental smoking status (never smoker, former smoker, occasional smoker, and daily smoker), parental attitudes toward smoking bans, i.e., whether they support complete smoking bans in public places and cars, and smoking rules at home (a smoke-free home or not a smoke-free home).

The sample consisted of 6119 single parents and corresponded to the population of 9,223,391 single parents, i.e., $\sum_{i=1}^{6119} W_i = 9,223,391$. As depicted in Tables 12.1 and 12.2, the sample size slightly differed depending on the measure. For convenience of computing, we used the same population count of 9, 223, 391 for all measures when estimating the population total. About 10.4% of parents were between 18 and 24 years old, 66.3% of parents were between 25 and 44 years old, and 23.3% of parents were 45 years old or older. The parents were 16.6% male and 83.4% female, 46.0% non-Hispanic White, 28.3% non-Hispanic Black/African American, 20.1% Hispanic, and 5.6% other; 48.5% had a single child, 33.4% had two children, 12.1% had three children, and 6.0% had more than three children. We used the main weights when computing these sample summary statistics.

12.2 Examples

Claim 1. Table 12.1 illustrates that all three methods resulted in the same proportion of single parents who support complete smoking bans in outdoor children's areas (e.g., playgrounds): the proportion is 88.7% (when percentages are rounded to the tenths). The estimated value of $\widehat{Y}_I$ is computed via formula (12.4), where n = 6057 and the unweighted sample total is $\sum_{i=1}^{n} Y_i = 5371$. Thus, $\widehat{Y}_I = \frac{5371}{6057} = 88.7\%$. The estimated value of $\widehat{Y}_{II}$ $(\widehat{Y}_{III})$ is based on formula (12.1), where the weighted sample total is $\sum_{i=1}^{n} W_i Y_i = 8,098,238$ and the total population weight is $\sum_{i=1}^{n} W_i = 9,132,015$. Thus, $\widehat{Y}_{II} = \frac{8,098,238}{9,132,015} = 88.7\%$.

Table 12.1 also shows that the methods could result in very similar (if not equal upon rounding) prevalence estimates. For example, the differences did not exceed 0.2% for proportions of parents who support complete smoking bans in bars, cocktail lounges, and clubs, casinos, and cars when children are present. A similar discrepancy of 0.2% was observed in the estimated prevalence of occasional

Table 12.1 Proportion and total number of single parents supporting complete smoking bans in public places and cars

Single parents who support complete smoking bans in ...	Method I	Method II	Method III
Outdoor children's areas (n = 6057)			
Estimated proportion (SE)	88.7% (0.4072%)	88.7% (0.4743%)	88.7% (0.4459%)
95% CI for proportion	87.9–89.5%	87.7–89.6%	87.8–89.6%
Estimated total (SE)	8,181,148 (37,558)	8,181,148 (43,747)	8,181,148 (41,127)
95% CI for total	8,107,361–8,254,935	8,088,914–8,264,158	8,098,137–8,264,158
Bars, cocktail lounges, and clubs (n = 6001)			
Estimated proportion (SE)	52.6% (0.6446%)	52.7% (0.7657%)	52.7% (0.6774%)
95% CI for proportion	51.4–53.9%	51.2–54.2%	51.3–54.0%
Estimated total (SE)	4,851,504 (59,454)	4,860,727 (70,624)	4,860,727 (62,479)
95% CI for total	4,740,823–4,971,408	4,722,376–4,999,078	4,731,600–4,980,631
Casinos (n = 5983)			
Estimated proportion (SE)	51.2% (0.6463%)	51.0% (0.7677%)	51.0% (0.7527%)
95% CI for proportion	49.9–52.5%	49.5–52.5%	49.5–52.5%
Estimated total (SE)	4,722,376 (59,611)	4,703,929 (70,808)	4,703,929 (69,424)
95% CI for total	4,602,472–4,842,280	4,565,579–4,842,280	4,565,579–4,842,280
Car when children are present (n = 6027)			
Estimated proportion (SE)	94.2% (0.3005%)	94.3% (0.3467%)	94.3% (0.3022%)
95% CI for proportion	93.6–94.8%	93.6–95.0%	93.7–94.9%
Estimated total (SE)	8,688,434 (27,716)	8,697,658 (31,977)	8,697,658 (27,873)
95% CI for total	8,633,094–8,743,775	8,633,094–8,762,221	8,642,317–8,752,998
Car when other people are present (n = 6040)			
Estimated proportion (SE)	72.3% (0.5760%)	73.3% (0.6687%)	73.3% (0.6013%)
95% CI for proportion	71.2–73.4%	72.0–74.6%	72.1–74.5%
Estimated total (SE)	6,668,512 (53,127)	6,760,746 (61,676)	6,760,746 (55,460)
95% CI for total	6,567,054–6,769,969	6,640,842–6,880,650	6,650,065–6,871,426

Note: CI stands for "confidence interval"

Table 12.2 Proportion of smoke-free homes among single-parent households in the USA (n = 6119)

	Method I	Method II	Method III
Proportion (SE)	86.1% (0.4%)	86.4% (0.5%)	86.4% (0.6%)
95% confidence interval for proportion	85.2–87.0%	85.4–87.4%	85.3–87.5%

smokers (see Fig. 12.1). A discrepancy of 0.3% was observed in the estimated prevalence of smoke-free homes among single-parent households (see Table 12.2).

The above results could incorrectly suggest that the discrepancies are so minor that it is sufficient to incorporate the main weight in the analyses. However, the examples below illustrate that even small discrepancies in the proportions could be meaningful when the estimates are projected to the totals.

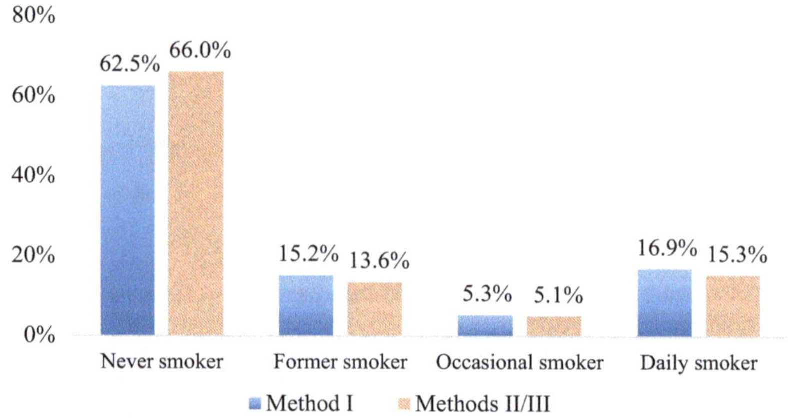

Fig. 12.1 Parental smoking status

Claim 2. Let us further discuss the results for smoke-free homes depicted in Table 12.2. While the difference in the proportions of smoke-free homes was only 0.3%, the estimated total numbers of smoke-free homes were quite different: they were 7,941,340 and 7,969,010 based on method I and methods II/III, respectively. The difference in the estimated totals was 27,670 households.

Figure 12.1 depicts the difference in the estimated prevalence of never smokers, former smokers, and daily smokers when methods I and II/III are used. The difference between the proportions (in absolute value) were relatively high: discrepancies ranged from 1.7 to 3.5%. The estimated total numbers of never smokers were 5,765,560 (method I) and 6,089,181 (methods II/III), the total numbers of former smokers were 1,403,330 (method I) and 1,250,618 (methods II/III), and the total number of daily smokers were 1,561,603 (method I) and 1,415,034 (methods II/III). The specific differences in the estimated totals for never smokers, former smokers, and daily smokers, respectively, were 323,621, 152,712, and 146,569.

Table 12.1 illustrates additional differences in the estimated total numbers of single parents who support complete smoking bans in bars, cocktail lounges, and clubs, casinos, and cars when children are present. The differences in estimated totals (in absolute value) ranged from 9223 to 92,234.

Claim 3. Table 12.1 depicts standard errors for estimated prevalence and totals computed via formulas (12.7). For example, standard errors for the estimated prevalence of parents who support complete smoking bans in cars when children are present ($n = 6027$) are based on the following formulas:

$$SE\left(\widehat{Y}_I\right) = \sqrt{\frac{1}{6027\,(6027-1)}327.9} = 0.3005\%,$$

$$SE\left(\widehat{Y}_{II}\right) = \sqrt{\frac{6027}{6027-1}1.1875 \times 10^{-5}} = 0.3467\%,$$

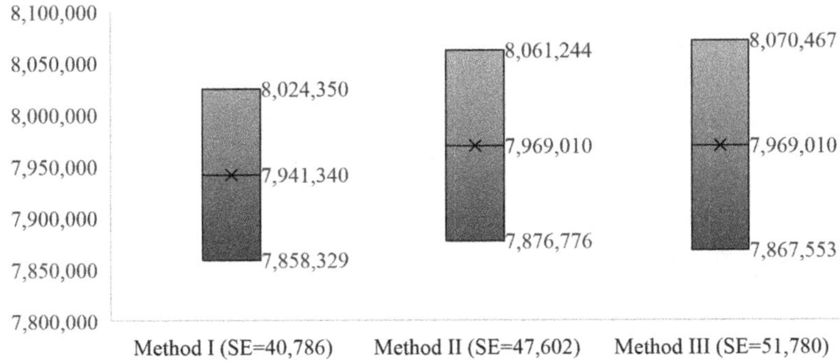

Fig. 12.2 Point estimates, statndard errors (SEs), and 95% confidence intervals for the total number of smoke-free homes among single-parent households in the USA

$$\text{and} \quad SE\left(\widehat{Y}_{III}\right) = \sqrt{\frac{1}{40} 3.6531 \times 10^{-4}} = 0.3022\%.$$

Table 12.1 illustrates that while the standard errors for estimated proportions were somewhat similar, the differences in the standard errors for estimated totals were quite pronounced. Note that method II led to larger standard errors relative to method III for all attitudinal measures depicted in Table 12.1. Method II corresponded to wider confidence intervals for the proportions and totals in comparison to method III (see Table 12.1). However, method II led to smaller standard errors relative to method III for the estimated proportion (see Table 12.2) and total number of smoke-free homes (see Fig. 12.2). Therefore, method II corresponded to narrower confidence intervals for the proportion (see Table 12.2) and total number of smoke-free homes in comparison to method III (see Fig. 12.2).

12.3 Discussion

The importance of incorporating the main weight in statistical analyses of survey data has been discussed in many survey sampling texts and publications (Hansen et al. 1953; Lohr 1999; Nassiuma 2001; Scheaffer et al. 2011). For example, in one study the distribution of annual salary was estimated using weighted and unweighted survey data; the corresponding histograms clearly showed a difference in the results (Lohr 2012).

While we have shown that incorporating the main weight is necessary when computing a point estimate, we have also found this is not sufficient when analyzing data from complex surveys such as TUS-CPS. Standard errors computed via methods I, II, and III could be very different, especially with respect to a total for a large population. In the latter case, the three methods could result in very different confidence intervals. For example, the standard error of the estimated total of smoke-free homes among single-parent households was largest for method III, smaller for method II, and smallest for method I. Thus, method I resulted in the narrowest confidence interval for the total number of smoke-free homes. However, it would be a mistake to claim that the result is valid because the method ignored the complex design features and thus, incorrectly underestimated the standard error. Because only method III incorporates the correct adjustments for the design specifics using the main and replicate weights, method III should be used in all analyses despite possibly giving less preferable results.

The study and considered examples have some limitations. First, the total count for the sample of single parents considered in the study was 9,223,391. This count was used when estimated totals (and standard errors) were computed based on estimated proportions (and standard errors). The count appears to be smaller than the total number of single-parent households with co-resident underage children reported by the U.S. Census Bureau in 2015: 10,432,000 (U.S. Census Bureau 2015). Thus, the presented population totals could be underestimated. An additional limitation is that method I resulted in smaller estimated variance relative to methods II and III in all considered examples. This, however, is not always the case.

We used simple built-in options in the survey package in SAS 9.4 (SAS Institute Inc. 2016), such as *proc surveyfreq* with BRR, main weight, and 160 replicate weights. Other major computing packages also provide built-in options for the variance estimation via BRR. From a computing standpoint, implementing correct adjustments is rather straightforward. We suggest following recommendations presented in the technical documentation for analysis of data from a national survey.

Acknowledgments We are thankful to James Holland, College of Medicine, University of Central Florida, for helping us improve the chapter.

Funding: Research reported in this publication was supported by the National Institute on Minority Health and Health Disparities of the National Institutes of Health under Award Number R01MD009718. The content is solely the responsibility of the authors and does not necessarily represent the official views of the National Institutes of Health.

References

Allison, D. B., Fontaine, K. R., Heo, M., Mentore, J. L., Cappelleri, J. C., Chandler, L. P., et al. (1999). The distribution of body mass index among individuals with and without schizophrenia. *The Journal of Clinical Psychiatry, 60*(4), 215–220.

Ash, S. (2014). Using successive difference replication for estimating variances. How to obtain more information. *Survey Methodology, 40*(1), 47–59.

Blackwell, D., Lucas, J., & Clarke, T. (2014). Summary health statistics for US adults: National health interview survey, 2012. *Vital and Health Statistics, 10*(260), 1–161.

Burch, R. C., Loder, S., Loder, E., & Smitherman, T. A. (2015). The prevalence and burden of migraine and severe headache in the United States: Updated statistics from government health surveillance studies. *Headache, 55*(1), 21–34.

Centers for Disease Control and Prevention. (2016). *Variance estimation guidance, NHIS 2016 (Adapted from NHIS Survey Description Documents)*. Retrieved from https://www.cdc.gov/nchs/data/nhis/2006var.pdf

Cummings, K. M., & Shan, D. (1995). Trends in smoking initiation among adolecents and young adults—United States, 1980-89. *Morbidity and Mortality Weekly Report, 44*(28), 521–525.

Dahlhamer, J., Galinsky, A., Joestl, S., & Ward, B. (2014). Sexual orientation in the 2013 national health interview survey: A quality assessment. *Vital and Health Statistics, 2*(169), 1–32.

Fay, R. E., & Train, G. (1995). Aspects of survey and model-based postcensal estimation of income and poverty characteristics for states and counties. In *Proceedings of the Section on Government Statistics, American Statistical Association, Alexandria, VA* (pp. 154–159).

Fiore, M. C., Novotny, T. E., Pierce, J. P., Hatziandreu, E. J., Patel, K. M., & Davis, R. M. (1989). Trends in Cigarette Smoking in the United States. The changing influence of gender and race. *JAMA, 261*(1), 49–55.

Gilpin, E. A., & Pierce, J. P. (1997). Trends in adolescent smoking initiation in the United States: Is tobacco marketing an influence? *Tobacco Control, 6*(2), 122–127.

Hansen, M., Hurwitz, W., & Madow, W. (1953). *Sample survey methods and theory, Vol. 1, Methods and applications*. New York: Wiley.

Jemal, A., Thun, M., Yu, X. Q., Hartman, A. M., Cokkinides, V., Center, M. M., et al. (2011). Changes in smoking prevalence among U.S. adults by state and region: Estimates from the Tobacco Use Supplement to the Current Population Survey, 1992-2007. *BMC Public Health, 11*, 512.

Judkins, D. (1990). Fay's method for variance estimation. *Journal of Official Statistics, 6*(3), 223–239.

Lohr, S. L. (1999). *Sampling: Design and analysis* (2nd ed.). Duxbury Press.

Lohr, S. L. (2012). *Using SAS® for the design, analysis, and visualization of complex surveys*. Retrieved from http://support.sas.com/resources/papers/proceedings12/343-2012.pdf

Nassiuma, D. K. (2001). *Survey sampling: Theory and methods*. Nairobi University Press.

Parsons, V., Moriarity, C., Jonas, K., Moore, T., Davis, K., & Tompkins, L. (2014). Design and estimation for the national health interview survey, 2006-2015. *Vital and Health Statistics, 2*(165), 1–53.

SAS Institute Inc. (2016). *SAS/STAT ® 14.2 user's guide*. Cary: SAS Institute Inc.

Scheaffer, R. L., Mendenhall III, W., Ott, R. L., & Gerow, K. G. (2011). *Elementary Survey Sampling. Zhurnal Eksperimental'noi i Teoreticheskoi Fiziki* (7th ed.). Brooks/Cole.

Shopland, D. R., Gerlach, K. K., Burns, D. M., Hartman, A. M., & Gibson, J. T. (2001). State-specific trends in smoke-free workplace policy coverage: The current population survey tobacco use supplement, 1993 to 1999. *Journal of Occupational and Environmental Medicine, 43*(8), 680–686.

Soulakova, J. N., & Crockett, L. J. (2014). Consistency and recanting of ever-smoking status reported by self and proxy respondents one year apart. *Journal of Addictive Behaviors, Therapy & Rehabilitation, 3*(4), 1000113. https://doi.org/10.4172/2324-9005.1000114.

Soulakova, J. N., Davis, W. W., Hartman, A., & Gibson, J. (2009). The impact of survey and response modes on current smoking prevalence estimates using TUS-CPS: 1992-2003. *Survey Research Methods, 3*(3), 123–137. Retrieved from http://www.ncbi.nlm.nih.gov/pubmed/21841957.

Soulakova, J. N., Hartman, A. M., Liu, B., Willis, G. B., & Augustine, S. (2012). Reliability of adult self-reported smoking history: Data from the tobacco use supplement to the current population survey 2002-2003 cohort. *Nicotine and Tobacco Research, 14*(8), 952–960. https://doi.org/10.1093/ntr/ntr313.

Soulakova, J. N., Bright, B. C., & Crockett, L. J. (2015a). Perception of time since smoking cessation: Time in memory can elapse faster. *Journal of Addictive Behaviors, Therapy & Rehabilitation, 4*(4). https://doi.org/10.4172/2324-9005.1000145.

Soulakova, J. N., Huang, H., & Crockett, L. J. (2015b). Racial/ethnic disparities in consistent reporting of smoking-related behaviors. *Journal of Addictive Behaviors, Therapy & Rehabilitation, 4*(4). https://doi.org/10.4172/2324-9005.1000147.

Stang, P., & Osterhaus, J. (1993). Impact of migraine in the United States: Data from the International Health Interview Survey. *Headache, 33*(1), 29–35.

U.S. Bureau of Labor Statistics & U.S. Census Bureau. (2006). *Design and methodology: Current population survey, Technical Paper 66*. Retrieved from https://www.census.gov/programs-surveys/cps/technical-documentation/complete.html

U.S. Census Bureau. (2015). *America's families and living arrangements: 2015*. Retrieved from https://www.census.gov/data/tables/2015/demo/families/cps-2015.html

U.S. Department of Commerce, U.S. Census Bureau. (2015). *Current population survey, May 2015: Tobacco use supplement*. Retrieved from https://www.census.gov/programs-surveys/cps/technical-documentation/complete.2015.html

U.S. Department of Health and Human Services, Centers for Disease Control and Prevention, National Center for Health Statistics. (2010). *The principal source of information on the health of the U.S. population*. Retrieved from https://www.cdc.gov/nchs/data/nhis/brochure2010January.pdf

Ward, B., Dahlhamer, J., Galinsky, A., & Joestl, S. (2014). Sexual orientation and health among US adults: National Health Interview Survey, 2013. *National Health Statistics Reports*, (77), 1–10.

Wolter, K. M. (1984). An investigation of some estimators of variance for systematic sampling. *Journal of the American Statistical Association, 79*(388), 781–790.

Wolter, K. M. (2007). *Introduction to variance estimation* (2nd ed.). Springer.

Chapter 13
Analysis of the High School Longitudinal Study to Evaluate Associations Among Mathematics Achievement, Mentorship and Student Participation in STEM Programs

Anarina L. Murillo, Hemant K. Tiwari, and Olivia Affuso

13.1 Introduction

Training the next generation of scientists is critical for the continued growth and success of technological and scientific developments in the United States. Many efforts have aimed to recruit, retain, and train individuals in the science, technology, engineering, and mathematics (STEM) fields. The National Science Foundation (NSF) (NSF 2012b) Math and Science Partnership (MSP) program as well as the NSF INCLUDES initiatives (NSF 2012a) have been put in place to improve K-12 STEM education and broaden the participation of diverse individuals in the STEM workforce. The American Statistical Association (ASA) has developed initiatives and programs to promote K-12 outreach as well as improve the recruitment and retention of students in statistics programs. However, according to the National Center for Education Statistics (NCES) (Chen 2013), only 28% of students surveyed during the 2003–2004 academic year pursued a bachelor's degree in a STEM field and 48% of those students left college or switched majors at the end of 6 years in Fall 2009.

To address these issues, several summer and year-long extracurricular programs have been created at the local, state, and national levels in order to recruit and retain STEM students. These programs expose students to opportunities in STEM fields, as well as prepare students for STEM programs through knowledge and skill development (King et al. 2017). It is well known that a strong background in mathematics and science is necessary for advanced training (e.g., baccalaureate

A. L. Murillo (✉) · H. K. Tiwari
Department of Biostatistics, University of Alabama at Birmingham, Birmingham, AL, USA
e-mail: amurillo@uab.edu; htiwari@uab.edu

O. Affuso
Department of Epidemiology, University of Alabama at Birmingham, Birmingham, AL, USA
e-mail: oaffuso@uab.edu

Y. Zhao, D.-G. Chen (eds.), *New Frontiers of Biostatistics and Bioinformatics*,
ICSA Book Series in Statistics, https://doi.org/10.1007/978-3-319-99389-8_13

and graduate school programs) in statistics, biostatistics, and bioinformatics. Data demonstrate that lower student retention rates are more prevalent among those with lower academic preparation and achievement in high school mathematics. More specifically, 41–46% of students with less than Algebra II/Trigonometry coursework completed and below a 2.5 High School GPA left STEM fields (Chen 2013). In contrast, only 12–14% of students with a High School GPA of 3.5 or higher and who took calculus in high school left STEM fields (Chen 2013). At the K-12 levels, mathematical literacy and comprehension set the foundation for statistics education and statistical literacy (BenZvi and Garfield 2008). This includes developing: a good understanding of the language of algebra and algebraic processes (e.g., proportional relationships and change; linear and nonlinear equations, inequalities, sequences and recursive relationships, etc.). While mathematics and statistics require different reasoning and intellectual skills (Kader and Perry 2006), a strong understanding of algebraic processes is an essential skill for statistical literacy and can be developed in both primary and secondary schools (Inter-university Consortium for Political and Social Research 2016). Thus, a strong foundation in the mathematical and statistical sciences is an essential skill for preparing students for STEM majors/careers.

Another benefit of these programs is creating the opportunity to foster relationships between students, peers, role models, and mentors. Mentorship is essential for career development and particularly important for recruiting and retaining underrepresented minorities (Griffin et al. 2010; Syed et al. 2012) who might be more susceptible to leaving STEM fields, such as women (Griffith 2010) and first generation college students, which is defined as students who are first members of their families to attend a college/university (Chen 2013). Additionally, prior studies showed that parental role modeling and mentoring may play a significant role in (Anderson and Minke 2007; Harackiewicz et al. 2012) shaping students interests in STEM majors/careers. In 2012–2013, STEM degrees conferred to underrepresented minorities were below average (16%) including non-Hispanic: Native Hawaiian/Pacific Islander (15%), American Indian/Alaska Native (14%), and Black/African-American (11%) students (Musu-Gillette et al. 2016). In the Hispanic population, only 14% of students graduated with a STEM degree (Musu-Gillette et al. 2016). Bachelor's degrees in STEM fields were awarded to a higher proportion of non-Hispanic Asian students (30%) in comparison with other students (Musu-Gillette et al. 2016). According to the NCES, the students who leave STEM programs at a higher rate include: underrepresented minorities, particularly non-Hispanic Black/African-American, Hispanic, and American Indian/Alaska Natives, and first-generation students (Chen 2013). Furthermore, a higher proportion of females (43%) were likely to leave STEM programs in comparison to males (Chen 2013). Hence, mentorship might be one approach to recruiting and retaining students in STEM majors/careers.

In this study, we investigated the effects of student participation in STEM activities, mentorship received from parents, teachers, and counselors on mathematics achievement and student enrollment plans in STEM majors/careers. The first aim was to evaluate the relationships among math achievement, student's participation in STEM activities, intent to pursue a STEM major/career, and mentorship received from parents, teachers, and counselors. The second aim was to evaluate the probability that students will pursue a STEM major/career given their math achievement, participation in STEM activities, and mentorship received. The overall goal was to assess these factors in order to inform potential STEM programs and policies aimed at high school students and underrepresented minorities to increase the STEM workforce. In particular, the results of this study may shed light on programs in the biostatistics, bioinformatics, statistics, and mathematics fields. This paper is organized as follows: methods are outlined in Sect. 13.2, results in Sect. 13.3, and discussion in Sect. 13.4.

13.2 Methods

13.2.1 Data and Sample

Data collected by The National Center for Education Statistics (NCES) as part of the High School Longitudinal Study (HSLS:09) in 2009–2013 were analyzed (Inter-university Consortium for Political and Social Research 2016). The HSLS:09 is a nationally representative longitudinal study that investigates 9th graders' paths from secondary to postsecondary transition plans including education and career choices. Baseline data were collected for Fall-term 9th graders in the 2009–2010 school year. The first follow-up took place in Spring of 2012. The same students were asked to complete the surveys, and dropouts and transfer students were also followed. In the Summer of 2013, data were collected for a postsecondary update, which only includes information for a subset of questions from the complete survey in order to track the cohort's postsecondary career/education plans. The second follow-up took place in 2016, a third follow-up is planned for 2021, and the final follow-up will take place in 2025. This present study only includes data collected from 2009 to 2013 since the 2016 data has not been released yet.

In the base year (2009), a sample of Fall-term 9th graders was randomly selected from more than 900 public and private high schools with both a 9th grade and a 11th grade class. The students, as well as the students' parents, principals, teachers in mathematics and science, and the school's main counselor completed surveys in order to evaluate the effects of social and educational factors on career paths into and out of STEM fields. Thus, the HSLS:09 is an ideal dataset for the goals of this present study and offers rich information on STEM participation at the high-school level and student's transition into future STEM career choices.

Participants were recruited for this complex survey design in a two-stage process. First, 1889 eligible schools were identified through stratified random sampling and school recruitment. Approximately 944 of eligible schools participated in HSLS:09, yielding a response rate of 55.5% (weighted) or 50.0% (unweighted). Second, students were randomly sampled from enrollment lists. There were 25,206 eligible selections, which is approximately 27 per school. Data were collected by computer-assisted telephone interviews (CATI), on-site questionnaires, telephone interviews, or through web-based surveys.

Sampling weights calculated by the HSLS:09 for the complex sampling design of the study were used to produce estimates for the target population with appropriate standard errors. Analytic weights were calculated at the school-level (e.g., for school administrator and counselor variables) and at the student-level. The student-level weights were provided for contextual data on science and math coursework at the school-level variables and for family and home contextual data for use with parent-level variables.

13.2.2 Study Variables

Data were collected at both the student-level and school-level. The students as well as students' parents, teachers (math and science only), school's principals and school's lead counselors were interviewed. Questions at the student-level included the following topics: career and school interests, STEM interests, coursework, attitudes and beliefs, social and cultural experiences, and exposure to STEM programs and majors/careers. At the parent-level, interview questions addressed: demographics, involvement and discussions with students career and postsecondary plans, as well as knowledge of options for college, career, and financing college education. At the teacher-level, respondents provided information on their own training and qualifications, quality of math and science curriculum at the school, and other topics pertaining to perceptions of parental involvement and attitudes about the school environment. At the administrator-level, information on school curriculum and career-related or transition programs were provided. Lastly, at the counselor-level, students' transition from high school to postsecondary education and/or careers transitions, coursework, advising, and availability of support were discussed.

In this analysis, we used publically available data collected between 2009 and 2013. The study variables included: demographic, mathematics achievement, mentoring, participation in STEM activities, and STEM interests collected in Fall 2009 and Spring 2012. Three additional questions on STEM major/career interests were obtained from Summer 2013. These variables are described in detail below. Primary outcome measures included: the student's mathematics achievement standardized theta score and student's intention to pursue a STEM career. Predictor variables included: student participation in STEM activities and mentoring.

Demographic and School Information Demographic variables collected at baseline used in this analysis included: sex, race/ethnicity, and school characteristics. Sex of the students, race/ethnicity, and school characteristics (type of school, urbanicity, and geographic region). Student's sex was recorded as either male or female. Student's self-reported race/ethnicity was categorized into seven dichotomous groups: (1) American Indian/Alaska Native, non-Hispanic; (2) Asian, non-Hispanic; (3) Black/African-American, non-Hispanic; (4) Hispanic, race specified or not specified; (5) More than one race, non-Hispanic; (6) Native Hawaiian/Pacific

Islander, non-Hispanic; and (7) White, non-Hispanic. The school environment (urbanicity) was recorded into four categories: (1) city, (2) suburb, (3) town, and (4) rural. The types of school included: (1) public or (2) private. The school geographic region was recorded into four dichotomous groups: (1) northeast, (2) midwest, (3) south, and (4) west.

Socioeconomic Status The socioeconomic status (SES) quintile variable was calculated by the NCES (Inter-university Consortium for Political and Social Research 2016) using parent/guardian's education, occupation, family income, and weighted using the estimated student weight collected at baseline and first follow-up (Spring 2012). This information was then used to categorize SES into three groups: (1) low SES ($\leq$20th percentile), (2) middle SES between the 20th and 80th percentile, and (3) high SES ($>$80th percentile).

Mathematics Achievement Mathematical comprehension was assessed using a 118-item test to measure algebraic reasoning which involved questions on algebraic content (e.g., linear equations, inequalities, proportional relationships) and algebraic processes (e.g., reasoning and problem solving). The mathematics standardized theta score is a rescaled estimate of the student's mathematics assessment score relative to the whole population (Fall 2009 9th graders) was used in this study. Math achievement scores obtained in Fall 2009 and Spring 2012 were used.

Intention to Pursue a STEM Career The student's intention to pursue a STEM career by age 30 was assessed using the coding scheme developed by the Occupational Information Network (O*NET) for the U.S. Department of Labor. Students selected which career they wanted by age 30, after which their responses were categorized as either desiring to pursue a STEM-related career or not a STEM-related occupation. Observations from Fall 2009 and Spring 2012 were used. Student's interests in STEM majors/careers obtained in the postsecondary update in Summer of 2013 were also analyzed.

Mentoring Mentoring variables were created using four specific survey questions including: (1) "*Why are you taking Fall 2009 math courses*"?, (2) "*Why are you taking Fall 2009 science courses*"?, (3) "*Why are you taking Spring 2012 math courses*"?, and (4) "*Why are you taking Spring 2012 science courses*"? Students were allowed to select one or more of the following responses: your parent(s) encouraged you to take it, a teacher encouraged you to take it, or a counselor suggested it. Three mentoring variables were created for parents, teachers, and school counselors. For each "yes" response to a mentoring-related question a score of "1" was recorded for the corresponding mentoring variable, and thus, each mentoring variable had a maximum value of 4.

STEM Activities The variables representing participation in STEM activities were created using two specific survey questions including: (1) "*Since 08–09, which activities did you participate in*"? (Fall 2009 question) and (2) "*Since Fall 2009, which activities did you participate*"? (Spring 2012 question). Students were allowed to select one or more of the following responses: math competition, math club, math camp, math summer program, science competition, science club, science program, or science camp. A score of "1" was recorded for each response selected to create one STEM variable for Fall 2009 (baseline) and for Spring 2012 (first follow-up). Each variable had a maximum value of 6.

13.2.3 Statistical Methods

Descriptive information including frequency (%), range (minimum and maximum values), mean ± SD are shown for all variables. Data were coded as missing by the NCES if at least one of the following criteria was met: (1) questions were not answered within the questionnaire, (2) sample member did not respond to the questionnaire, (3) questions were not answered because prior answered routed the respondent to another question, (4) participant didn't know, and for a few other reasons (e.g., item not administered because an abbreviated version of the questionnaire was not administered, unit nonresponse, etc.) detailed in the HSLS:09 User Guide (Inter-university Consortium for Political and Social Research 2016). All residuals were tested for normality from the regression models. Variance of residuals were also checked for equality from the regression models.

For the first aim, analysis of variance (ANOVA) tests was used to evaluate mean differences in mathematics achievement based on predictor variables (e.g., sex and race/ethnicity, student's engagement in STEM activities, intent to pursue a STEM career, and mentorship received from parents, teachers, and school counselors). Multiple regression models were used to predict math achievement scores in Fall 2009 and Spring 2012 based on these predictor variables. For the second aim, logistic regression models were used to evaluate the probability that students will pursue a STEM major/career given their math achievement, STEM activities, and mentorship received from parents, teachers, and school counselors. All statistical analyses were performed with statistical significance accepted when $P < 0.05$ and using sampling weights. The regression models were implemented using PROC SURVEYREG and PROC SURVEYLOGISTIC procedures in SAS 9.4 (SAS Institute 2015).

13.3 Results

13.3.1 Study Participants

The demographic and school characteristics are summarized in Table 13.1. A total of $n = 23{,}503$ students (50.95% male/49.05% female) were included. The study population was diverse including non-Hispanic: White (55.10%), Black/African-American (10.42%), and Asian (8.18%), as well as Hispanic (16.43%) participants.

Table 13.1 Demographic characteristics of the study sample (n = 23,503) are shown

Variable	Total, n (%) (n = 23,503)	Male, n (%) (n = 11,975)	Female, n (%) (n = 11,528)
Race/ethnicity			
American Indian/Alaska Native	181 (0.77)	101 (0.43)	80 (0.34)
Asian	1922 (8.18)	970 (4.13)	952 (4.05)
Black/African-American	2448 (10.42)	1276 (5.43)	1172 (4.99)
Hispanic	3862 (16.43)	1939 (8.25)	1923 (8.18)
Native Hawaiian/Pacific Islander	118 (0.50)	60 (0.26)	58 (0.25)
White	12,951 (55.10)	6594 (28.06)	6357 (27.05)
Other	2021 (8.60)	1035 (4.40)	986 (4.20)
SES			
Low SES	3262 (13.88)	1676 (7.13)	1586 (6.75)
Middle SES	14,815 (63.03)	7548 (32.12)	7267 (30.92)
High SES	5426 (23.09)	2751 (11.70)	2675 (11.38)
School type			
Public	19,273 (82.00)	9882 (42.05)	9391 (39.96)
Private	4230 (18.00)	2093 (8.91)	2137 (9.09)
School locale (urbanicity)			
City	6689 (28.46)	3369 (14.33)	3320 (14.13)
Suburb	8467 (36.03)	4307 (18.33)	4160 (17.70)
Town	2788 (11.86)	1418 (6.03)	1370 (5.83)
Rural	5559 (23.65)	2881 (12.26)	2678 (11.39)
School geographic region			
Northeast	3662 (15.58)	1794 (7.63)	1868 (7.95)
Midwest	6224 (26.48)	3215 (13.68)	3009 (12.80)
South	9587 (40.79)	4968 (21.14)	4619 (19.65)
West	4030 (17.15)	1998 (8.50)	4619 (19.65)
STEM major/career (Fall 2009)			
Yes	6490 (30.88)	2545 (12.11)	3945 (18,77)
No	14,528 (69.12)	8083 (38.46)	6445 (30.66)
Missing	2485	1347	1138
STEM major/career (Spring 2012)			
Yes	7098 (35.07)	2800 (13.83)	4298 (21.24)
No	13,142 (64.93)	7377 (36.45)	5765 (28.48)
Missing	3263	1798	1465
STEM job expected (Summer 2013)			
Yes	399 (4.72)	197 (2.33)	202 (2.39)
No	8052 (95.28)	4050 (47.92)	4002 (47.36)
Missing	15052	7728	7324

(continued)

Table 13.1 (continued)

Variable	Total, n (%) (n = 23,503)	Male, n (%) (n = 11,975)	Female, n (%) (n = 11,528)
STEM field expected (Summer 2013)			
Yes	2935 (25.54)	1920 (16.71)	1015 (8.83)
No	8558 (74.46)	3417 (29.73)	5141 (44.73)
Missing	12010	6638	5372
STEM current job (Summer 2013)			
Yes	245 (2.71)	113 (1.25)	132 (1.46)
No	8786 (97.29)	4378 (48.48)	4408 (48.81)
Missing	14,472	7484	6988

Less than 1% of the population were American Indian/Alaska Native (0.77%) and Native Hawaiian/Pacific Islander (0.50%). Approximately 14%, 63%, and 23% of the population were categorized as low, medium, and high SES, respectively.

Students represented both public (82%) and private (18%) schools. Approximately 52% of schools offered programs to encourage underrepresented students in STEM fields and 39% of schools had programs to inform parents about STEM higher education/careers. In Fall 2009 and Spring 2012, 69% and 65% of all students were not interested in pursuing a STEM major/career. In Summer 2013, approximately 95% of students expected a STEM major/career by November 2013 and 74% expected to pursue a STEM major/career field in November 2013. However, only 3% had a current job that was considered STEM-related.

13.3.2 Assessment of Student Mathematics Achievement

The mean(±SD) math achievement scores for the study sample was 51.11±10.08 in Fall 2009 and 51.50±10.15 in Spring 2012. No significant differences were found based on sex in Fall 2009 ($F(1, 21443) = 0.94$, $P = 0.3327$) nor Spring 2012 ($F(1, 21443) = 0.12$, $P = 0.7262$). However, math achievement scores did significantly vary by race in Fall 2009 $F(6, 21443) = 134.40$, $P < 0.001$) and Spring 2012 ($F(6, 20593) = 124.37$, $P < 0.001$). In Fall 2009 and Spring 2012, Asian students consistently had higher scores (58.07±10.62 in Fall 2009 and 58.68±10.58 in Spring 2012) followed by White, Native Hawaiian/Pacific Islander, Hispanic, and Black/African-American (see Table 13.2). Low scores were observed in American Indian/Alaska Native students (44.68±10.91 in Fall 2009 and 47.19±10.15 in Spring 2012), and was significantly lower in low SES individuals in comparison with medium and high SES student's in Fall 2009 ($F(3, 21443) = 654.68$, $P < 0.001$) and in Spring 2013 ($F(3, 20593) = 627.92$, $P < 0.001$).

Table 13.2 Summary of math achievement scores in Fall 2009 and Spring 2012 based on demographic, mentoring, STEM activities, and STEM interests

Variable	Fall 2009 *n*	Fall 2009 Mean (SD)	Fall 2009 *P*-value	Spring 2012 *n*	Spring 2012 Mean (SD)	Spring 2012 *P*-value
All	21,444	51.11 (10.08)		20,594	51.50 (10.15)	
Sex			0.332			0.776
Male$_{ref}$	10,887	51.08 (10.47)		10,384	51.59 (10.62)	
Female	10,557	51.15 (9.66)		10,210	51.41 (9.66)	
Race[a,b]			<0.0001			<0.0001
Amer. Ind./Alaska Native	163	44.68 (10.91)		142	47.19 (10.15)	
Asian	1672	58.07 (10.62)		1675	58.68 (10.58)	
Black/African-American	2219	46.40 (9.27)		2121	46.80 (8.80)	
Hispanic	3515	48.21 (9.38)		3271	48.64 (9.14)	
Native Haw./Pac. Islan.	1912	51.13 (9.48)		1756	51.45 (9.75)	
Other	110	49.24 (9.90)		97	50.52 (9.67)	
White$_{ref}$	11,853	51.97 (9.69)		11,532	52.20 (9.94)	
SES[a,b]			<0.0001			<0.0001
Low$_{ref}$	2862	46.39 (9.15)		3167	46.21 (9.03)	
Medium	13,512	49.94 (9.60)		12,066	50.29 (9.44)	
High	5070	56.90 (9.35)		5361	57.36 (9.66)	
Mentoring						
Teacher[a,b]			<0.0001			<0.0001
None	13,818	51.20 (9.68)		7607	51.70 (9.44)	
Once	1916	55.76 (9.41)		3334	54.55 (9.95)	
At least twice	845	58.09 (9.26)		3225	56.00 (9.90)	
Counselor[a]			<0.0001			0.728
None	14,460	51.89 (9.75)		7759	53.28 (9.84)	
Once	1390	52.59 (10.33)		2915	52.89 (10.11)	
At least twice	729	54.81 (10.17)		3507	53.81 (9.62)	
Parent[a,b]			<0.0001			<0.0001
None	13,410	50.90 (9.58)		12,958	49.62 (9.62)	
Once	1877	56.17 (9.39)		4047	52.85 (10.20)	
At least twice	1292	58.26 (9.12)		3589	56.77 (9.88)	
Activities[a,b]			<0.0001			<0.0001
0	18,935	50.43 (9.71)		16,142	50.51 (9.50)	
1	1432	55.91 (10.57)		1837	55.66 (10.37)	
2	510	59.98 (10.02)		861	58.51 (10.86)	
3	102	62.11 (10.60)		286	60.26 (12.64)	
4	62	64.30 (11.93)		191	62.40 (12.10)	
5	9	53.90 (15.08)		57	63.05 (13.5)	
6	26	51.77 (14.78)		67	52.19 (15.65)	

(continued)

Table 13.2 (continued)

Variable	Fall 2009			Spring 2012		
	n	Mean (SD)	P-value	n	Mean (SD)	P-value
STEM[a,b]			<0.0001			<0.0001
No	14,528	50.32 (10.07)		13,142	50.34 (10.08)	
Yes	6490	53.11 (9.75)		7098	53.84 (9.85)	

Mean (SD) with minimum and maximum values are shown. Statistically significant mean differences were determined by one-way ANOVA tests. The reference groups are denoted with subscript "*ref*"

[a]Bonferroni correction was used to correct for multiple testing (8 tests) for outcomes measured in Fall 2009 and in Spring 2012, significance was accepted when $P < 0.00625(0.05/8)$ in Fall 2009

[b]Bonferroni correction was used to correct for multiple testing (8 tests) for outcomes measured in Fall 2009 and in Spring 2012, significance was accepted when $P < 0.00625(0.05/8)$ in Spring 2012

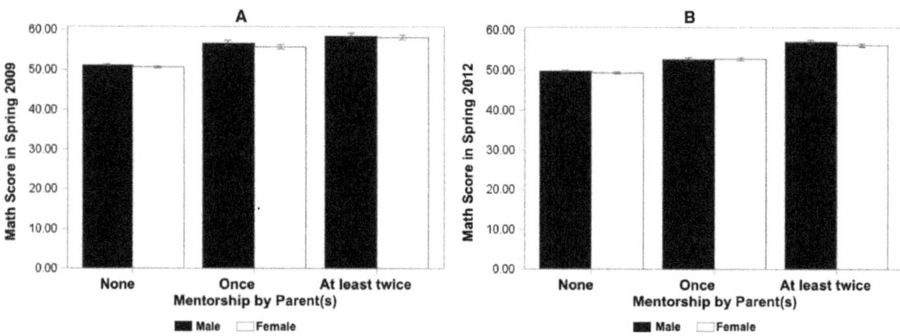

Fig. 13.1 Math achievement based on mentoring received from parents in Fall 2009 ((**a**) on the left) and in Spring 2012 ((**b**) on the right)

The largest differences in math achievement scores were found based on mentoring from parents in comparison with counselors and teachers in Fall 2009 and Spring 2012 (see Table 13.2 and Fig. 13.1). In Fall 2009, most students received no mentoring from teachers, counselors, and parents on science and math courses. However, mentoring received at least twice was greatest for parents in comparison with teachers and school counselors (see Table 13.2).

The students with no mentoring had a math achievement score of 51.20±9.68, which was significantly lower in comparison with students who received mentoring at least twice from teachers 58.09±9.26 ($F(1, 16578) = 446.29$, $P < 0.001$), school counselors 54.81±10.17 ($F(1, 16578) = 20.80$, $P < 0.001$), and parents 58.26±9.12 ($F(1, 16578) = 534.20$, $P < 0.001$). In contrast to Fall 2009, most students in Spring 2012 received mentoring from teachers (79.2%), counselors (82.8%), and parents (73.5%) on math and science courses at least twice since their

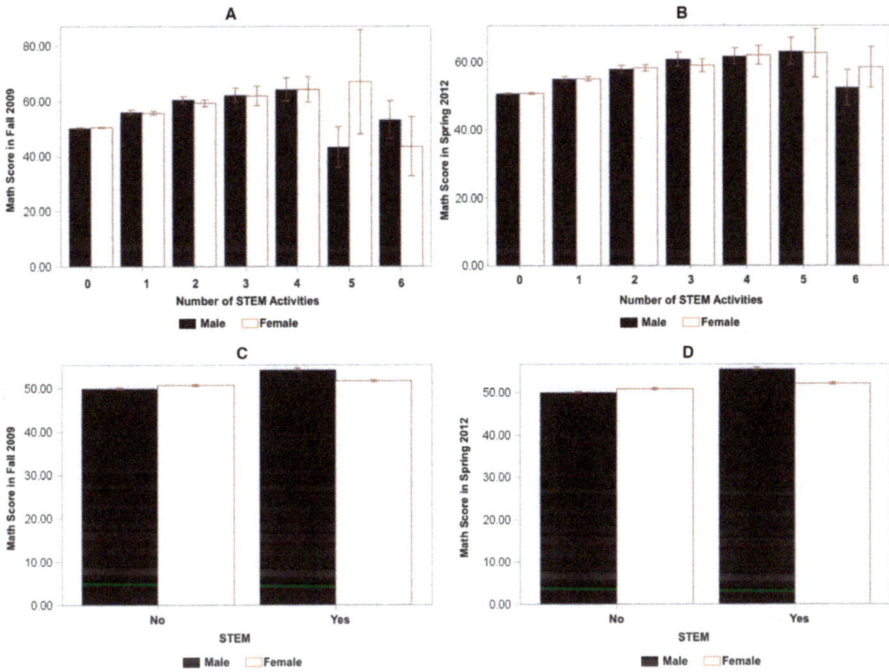

Fig. 13.2 Math achievement scores based on STEM activities are shown for Fall 2009 ((**a**) on top-left) and Spring 2012 ((**b**) on top-right), and also shown by STEM interests reported in Fall 2009 ((**c**) on bottom-left) and Spring 2012 ((**d**) on bottom-right)

freshman year. Math achievement scores were significantly higher for those that received mentoring from teachers ($F(1, 14165) = 127.83$, $P < 0.001$) or parents ($F(1, 20593) = 572.96$, $P < 0.001$). However, no significant differences in math achievement scores were found based on mentoring experience from counselors ($F(1, 14180) = 0.12$, $P = 0.7286$). Details of mentoring variables are shown in Table 13.10 in the Appendix.

Student participation in STEM activities remained low in Fall 2009 and Spring 2012 (see Fig. 13.2). Approximately 99% and 96% of students reported that they participated in two or less STEM activities in Fall 2009 and Spring 2012, respectively. Math achievement scores were significantly different based on participation in STEM activities in Fall 2009 ($F(1, 21075) = 143.70$, $P < 0.001$) and Spring 2012 ($F(1, 19440) = 95.82$, $P < 0.001$). Nearly 31% and 35% of students were interested in STEM majors/careers in Fall 2009 and Spring 2012, respectively (see Table 13.2). Math achievement scores were significantly different based on student's interest in STEM major/careers in Fall 2009 ($F(21017) = 120.94$, $P < 0.001$) and Spring 2012 ($F(1, 20239) = 160.76$, $P < 0.001$).

Table 13.3 Math achievement scores for Fall 2009 are presented in Model 1

Variable	β [95% CI]	P-value
Model 1		
Intercept	47.82 [47.04,48.60]	<0.001
Sex		
Female	−0.44 [−0.89,0.01]	0.057
Race/ethnicity		
American Indian/Alaska Native	−3.07 [−5.72,−0.42]	0.023
Asian	4.75 [3.81,5.69]	<0.001
Black/African-American	−4.64 [−5.50,−3.78]	<0.001
Hispanic	−1.45 [−2.14,−0.77]	<0.001
Native Hawaiian/Pacific Islander	−2.08 [−6.12,1.96]	0.312
Other	−1.62 [−2.42,−0.82]	<0.001
SES		
Medium	2.12 [1.40,2.85]	<0.001
High	6.82 [6.06,7.58]	<0.001
Mentoring		
Teacher	1.99 [1.58,2.40]	<0.001
Counselor	−0.65 [−1.14,−0.16]	0.011
Parent	1.96 [1.59,2.33]	<0.001
Activities	2.12 [1.63,2.61]	<0.001
STEM	1.53 [1.06,2.00]	<0.001

The reference groups for sex and race/ethnicity were male and White, respectively

Multiple regression models were used in this study to predict math achievement scores based on sex, race/ethnicity, mentoring, STEM activities, and STEM major/career interests. The first model predicted math achievement scores in Fall 2009 and the second model is used to predict math achievement scores in Spring 2012 (see Model 1 and 2 in Tables 13.3 and 13.4). In Model 1, significant predictors included SES, mentoring by teachers and parents, STEM activities, and STEM major/career interest ($R^2 = 0.17, F(8, 14104) = 149.04, P < 0.001$). In Model 2, significant predictors included sex, SES, mentoring, STEM activities and STEM major/career interest ($R^2 = 0.18, F(8, 13383) = 123.03, P < 0.001$). Race/ethnicity was significant for most groups but not all in Models 1 and 2. To check the normality assumption of the data, Q-Q plots and histograms, as well as descriptive statistics of the residuals were checked. Residuals for Model 1 (see Fig. 13.3a, b) had mean of −5.96, variance of 1.00, skewness of −0.18, and kurtosis of 0.08. Similarly, Model 2 (see Fig. 13.3c, d) has mean of −1.37, variance of 1.00, skewness of −0.18, and kurtosis of 0.01. Furthermore, given the large sample size of $N = 23{,}503$ and no heavy tails, we proceed with our analyses under the assumption that the data is approximately normally distributed.

Table 13.4 Math achievement scores for Spring 2012 are presented in Model 2

Variable	β [95% CI]	P-value
Model 2		
Intercept	47.99 [47.09,48.89]	<0.001
Sex		
Female	−0.77 [−1.26,−0.28]	0.002
Race		
American Indian/Alaska Native	−1.94 [−4.63,0.75]	0.158
Asian	4.32 [3.24,5.40]	<0.001
Black/African-American	−4.52 [−5.42,−3.62]	<0.001
Hispanic	−1.55 [−2.30,−0.81]	<0.001
Native Hawaiian/Pacific Islander	−1.18 [−5.10,2.74]	0.556
Other	−1.49 [−2.35,−0.63]	<0.001
SES		
Medium	2.69 [1.87,3.51]	<0.001
High	7.55 [6.67,8.43]	<0.001
Mentoring		
Teacher	1.20 [0.83,1.57]	<0.001
Counselor	−1.16 [−1.45,-0.87]	<0.001
Parent	1.11 [0.75,1.46]	<0.001
Activities	1.68 [1.25,2.11]	<0.001
STEM	1.96 [1.49,2.43]	<0.001

The reference groups for sex and race/ethnicity were male and White, respectively

13.3.3 Assessment of Student Enrollment in a STEM Major/Career

To assess student enrollment in STEM careers, participants were asked to indicate which job they expected to have at age 30. The percentage of students that were interested in STEM major/careers were 31% in Fall 2009 and 35% in Spring 2012. Logistic regression models were used to predict the probability of selecting a STEM major/career based on sex, race, SES, mentoring, participation in STEM activities, and math achievement scores (see Table 13.5 where Model 3 corresponds to Fall 2009 data and Table 13.6 where Model 4 corresponds to Spring 2012 data).

In Fall 2009, significant predictors were sex ($P < 0.001$), STEM activities ($P < 0.001$), math achievement ($P < 0.001$), and mentoring from parents ($P < 0.001$). However, mentoring by teachers ($P = 0.397$) and counselors ($P = 0.252$) were not significant (see Table 13.5). In Spring 2012, sex ($P < 0.001$), STEM activities ($P < 0.001$), and math achievement ($P < 0.001$) were significant (see Table 13.6). The c-statistic was 0.626 and 0.637 for models 3 and 4, respectively. The Hosmer-Lemeshow (H-L) test yielded a $\chi^2(8)$ of 86.79 ($P < 0.0001$), suggesting that the

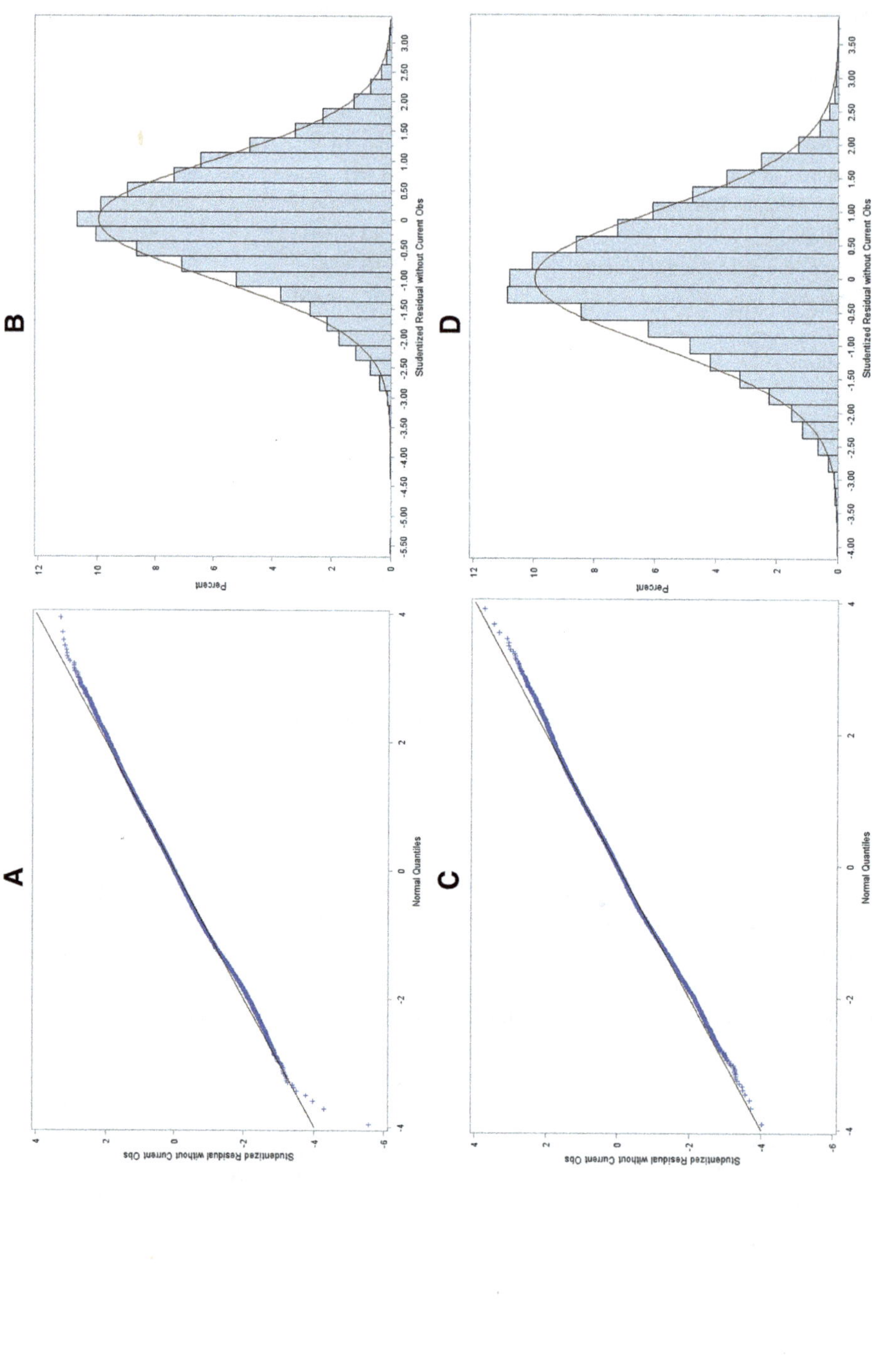

Fig. 13.3 Plots of residuals for Model 1 in (**a**) (Q-Q plot) and (**b**) (histogram) and for Model 2 in (**c**) (Q-Q plot) and (**d**) (histogram)

Table 13.5 Results for predicting expected STEM major/careers in Fall 2009 (Model 3)

Variable	OR [95% CI]	*P*-value
Model 3		
Sex		
Female	2.1 [1.9,2.3]	<0.001
Race		
American Indian/Alaska Native	0.9 [0.5,1.5]	0.467
Asian	1.4 [1.1,1.8]	0.020
Black/African-American	1.2 [0.9,1.5]	0.284
Hispanic	0.9 [0.8,1.0]	0.093
Native Hawaiian/Pacific Islander	0.9 [0.4,2.0]	0.745
Other	1.1 [0.9,1.4]	0.453
SES		
Medium	1.0 [0.8,1.3]	0.507
High	1.0 [0.8,1.2]	0.532
Mentoring		
Teacher	1.0 [0.9,1.2]	0.397
Counselor	1.1 [0.9,1.2]	0.252
Parent	1.2 [1.1,1.3]	<0.001
Activities	1.2 [1.1,1.4]	<0.001
Math	1.0 [1.0,1.0]	<0.001

The reference groups for sex and race/ethnicity were male and White, respectively

model was not fit to the data well. Similarly, the H-L test yielded a $\chi^2(8)$ of 102.83 ($P < 0.0001$), also suggesting that the model was not fit to the data well.

During the Summer of 2013 update, study participants were also asked to indicate which job they were expected to have in November 2013. Approximately 5% of students expected to have a STEM-related career by the end of the year. In Model 5, a logistic model was fit to the data to identify the significant predictors of STEM major/career by November 2013. Significant predictors included race ($P < 0.001$). Other variables such as sex, mentoring (in Fall 2009 and Spring 2012), and math achievement scores were not statistically significant (see results for Model 5 in Table 13.7). The c-statistic was 0.754 for model 5. The H-L test yielded a $\chi^2(8)$ of 6.27 ($P = 0.616$), suggesting that the model was fit to the data well.

In the Summer of 2013, 25% of students indicated that they were considering a STEM-related college major. Results of estimated model parameters for the probability that a student will enter a STEM-related major are shown in Table 13.8 ($\chi^2(13) = 45.62$, $P < 0.001$). Significant predictors ($P < 0.001$) included: sex, STEM activities (Spring 2012), and math achievement scores (Spring 2012). Other variables such as race/ethnicity and mentoring were not significant (see Table 13.8). The c-statistic was 0.626 for model 6. The H-L test yielded a $\chi^2(8)$ of 9.13 ($P = 0.331$), suggesting that the model was fit to the data well.

Table 13.6 Results for predicting expected STEM major/careers in Spring 2012 (Model 4)

Variable	OR [95% CI]	*P*-value
Model 4		
Sex		
Female	2.0 [1.8,2.3]	<0.001
Race		
American Indian/Alaska Native	1.4 [0.8,2.6]	0.268
Asian	0.9 [0.7,1.2]	0.382
Black/African-American	1.1 [0.9,1.4]	0.576
Hispanic	1.0 [0.8,1.2]	0.758
Native Hawaiian/Pacific Islander	1.0 [0.4,2.2]	0.823
Other	0.9 [0.8,1.2]	0.394
SES		
Medium	1.0 [0.8,1.2]	0.878
High	1.0 [0.8,1.2]	0.637
Mentoring		
Teacher	1.1 [0.9,1.2]	0.314
Counselor	1.0 [1.0,1.1]	0.314
Parent	1.1 [1.0,1.2]	0.041
Activities	1.2 [1.2,1.3]	<0.001
Math	1.0 [1.0,1.0]	<0.001

The reference groups for sex and race/ethnicity were male and White, respectively

Students also reported their current job in the Summer of 2013 at the time of the interview. Approximately 3% of students indicated that their current job was STEM-related (e.g., biological and biomedical sciences, engineering, etc.). Results of the model fitted to predict the probability that a student would have a STEM-related current job are shown in Table 13.9. Significant predictors included: all race groups except American Indian/Alaskan Native and individual in the category "other" ($P < 0.001$). Participation in STEM activities in Fall 2009 and mentoring by teachers in Spring 2012 were not significant. The c-statistic was 0.625 for model 7. The H-L test yielded a $\chi^2(8)$ of 14.30 ($P = 0.074$), suggesting that the model was fit to the data well.

13.4 Conclusions

The aim of this work was to assess the associations among mentorship (parents, teachers counselors), math achievement, student's intent to pursue STEM major/careers, and participation in STEM activities, and further, to investigate differences based on race/ethnicity, sex, and SES. Math achievement scores significantly varied in Fall 2009 and Spring 2012 based on race with the Asian race having

Table 13.7 Results for model fitted to predict the probability that the student expected to have a STEM career by November 2013

Variable	OR [95% CI]	*P*-value
Model 5		
Sex		
Female	1.0 [0.6,1.6]	0.959
Race		
American Indian/Alaska Native	−0.0 [0.0,0.0]	<0.001
Asian	2.4 [1.1,5.2]	<0.001
Black/African-American	1.5 [0.7,3.1]	<0.001
Hispanic	1.5 [0.8,2.8]	<0.001
Native Hawaiian/Pacific Islander	0.5 [0.2, 1.1]	<0.001
Other	0.5 [0.2, 1.1]	<0.001
SES		
Medium	0.8 [0.4,1.6]	0.262
High	1.2 [0.6,2.4]	0.304
Mentoring (Fall 2009)		
Teacher	0.9 [0.6,1.4]	0.701
Counselor	1.1 [0.8,1.7]	0.536
Parent	1.3 [0.9,1.9]	0.170
Activities (Fall 2009)	1.0 [0.7,1.5]	0.940
Math (Fall 2009)	1.0 [0.9,1.0]	0.356
Mentoring (Spring 2012)		
Teacher	0.9 [0.6,1.3]	0.561
Counselor	0.9 [0.6,1.2]	0.459
Parent	1.0 [0.7,1.5]	0.906
Activities (Spring 2012)	1.2 [0.9,1.5]	0.143
Math (Spring 2012)	1.0 [0.6,1.6]	0.136

The reference groups for sex and race/ethnicity were male and White, respectively. Observations were recorded in Fall 2009 and in Spring 2012

the highest scores (Fall 2009 and Spring 2012) and the American Indian/Alaska Native race (Fall 2009) or Black/African-American (Spring 2012) having the lowest scores, which is consistent with prior findings (Musu-Gillette et al. 2016). SES was significantly associated with math achievement scores in Fall 2009 and Spring 2012, where low SES had lowest scores and high SES had the highest scores in both Fall 2009 and Spring 2012. Math achievement scores were greater among students who received mentoring at least twice from teachers, school counselors, and parents in Fall 2009 and Spring 2012. Participation in STEM activities was significantly associated with math achievement scores in Fall 2009 and Spring 2012. Additionally, average math achievement scores were greater among students that were interested in STEM majors/careers. Among all the variables included in Models 1 (Fall 2009) and 2 (Spring 2012), significant predictors of math

Table 13.8 Results for model used to predict the probability that student will consider a STEM field for a future major/career

Variable	OR [95% CI]	P-value
Model 6		
Sex		
Female	0.3 [0.3,0.4]	<0.001
Race		
American Indian/Alaska Native	1.2 [0.3,3.9]	0.802
Asian	1.6 [1.1,2.5]	0.330
Black/African-American	1.1 [0.7,1.6]	0.289
Hispanic	0.2 [0.9,1.7]	0.561
Native Hawaiian/Pacific Islander	2.7 [0.9,7.8]	0.135
Other	1.1 [0.8,1.6]	0.372
SES		
Medium	1.2 [0.8,2.0]	0.910
High	1.5 [0.7,1.9]	0.741
Mentoring (Fall 2009)		
Teacher	1.2 [1.0,1.4]	0.032
Counselor	1.0 [0.9,1.1]	0.450
Parent	1.0 [0.9,1.1]	0.016
Activities (Fall 2009)	1.0 [0.9,1.2]	0.402
Math (Fall 2009)	1.0 [1.0,1.0]	0.281
Mentoring (Spring 2012)		
Teacher	1.0 [0.8,1.1]	0.708
Counselor	0.8 [0.7,0.9]	0.010
Parent	1.2 [1.0,1.4]	0.016
Activities (Spring 2012)	1.2 [1.1,1.4]	<0.001
Math (Spring 2012)	1.1 [1.0,1.1]	<0.001

The reference groups for sex and race/ethnicity were male and White, respectively. Observations were recorded in Fall 2009 and in Spring 2012

achievement scores were: sex, SES, mentoring (teachers, counselors, and parents), participation in STEM activities, and students interests in STEM major/careers. Race was significant for all groups except for Native Hawaiian/Pacific Islander (in Fall 2009 and Spring 2012) and American Indian/Alaska Native in Spring 2012.

Approximately 31% and 35% of students were interested in STEM major/careers in Fall 2009 and Spring 2012, respectively. The probability of students interest in STEM major/careers in Fall 2009 were significantly based on sex, the Asian race, mentoring by parents, participation in STEM activities and math achievement scores in Fall 2009. Similarly, the probability of students interest in STEM major/careers was significantly dependent on sex, mentoring by parents, participation in STEM activities, and math achievement scores in Spring 2012. However only 5% and 25% of students expected a major or career in STEM by November of 2013

Table 13.9 Results of model used to predict whether the current job will be in a STEM field

Variable	OR [95% CI]	*P*-value
Model 7		
Sex		
Female	0.8 [0.5,1.3]	0.415
Race		
American Indian/Alaska Native	1.3 [0.2,10.4]	0.088
Asian	2.6 [0.8,8.0]	<0.001
Black/African-American	1.4 [0.5,3.5]	<0.001
Hispanic	1.0 [0.4,2.3]	<0.001
Native Hawaiian/Pacific Islander	0.0 [0.0,0.0]	<0.001
Other	0.3 [0.1,1.5]	0.700
SES		
Medium	0.9 [0.4,2.2]	0.989
High	0.8 [0.3,2.0]	0.592
Mentoring (Fall 2009)		
Teacher	0.8 [0.4,1.3]	0.332
Counselor	1.0 [0.6,1.6]	0.943
Parent	1.1 [0.8,1.7]	0.474
Activities (Fall 2009)	1.5 [1.1,2.0]	0.003
Math (Fall 2009)	1.0 [1.0,1.0]	0.979
Mentoring (Spring 2012)		
Teacher	1.6 [1.1,2.4]	0.010
Counselor	0.9 [0.7,1.3]	0.600
Parent	0.7 [0.5,1.1]	0.140
Activities (Spring 2012)	0.9 [0.7,1.3]	0.735
Math (Spring 2012)	1.0 [1.0,1.1]2	0.411

The reference groups for sex and race/ethnicity were male and White, respectively. Observations were recorded in Fall 2009 and in Spring 2012

which was mainly based on race, math achievement, mentoring by teachers and parents, or participation in STEM activities. Interestingly, parents education levels and occupation (STEM or not STEM-related) were associated with fewer proportion of students interested in STEM major/careers. In other words, a greater proportion of students interested in STEM major/careers had one or more parent with either a high school education level only and/or were not in a STEM occupation.

Hence, parental mentoring was a significant predictor of math achievement scores and for the probability of students entering a STEM major/college, which is supported by prior studies (Anderson and Minke 2007; Harackiewicz et al. 2012). Participation in STEM activities was a significant indicator of math achievement scores, entering a STEM major/career, and student's future interests in STEM major/careers. Thus, our work suggests that programs aimed to recruit and retain STEM students may be effective if parents (Anderson and Minke 2007;

Harackiewicz et al. 2012) are involved in activities with these students. Math competition, summer programs, and other activities (King et al. 2017) may also play a significant role in exposing students to careers in biostatistics and bioinformatics, and furthermore, may help to develop their interests in these career areas earlier in life. Some limitations of this work include the statistical methods used. For instance, a multilevel model would have been ideal for this type of study design. Unfortunately, the cluster and strata data needed for a multilevel analysis were restricted and unavailable for this present study. Future work, may be to use a multilevel model as well as other school-level variables.

Acknowledgements This research was in part funded by the National Heart, Lung, and Blood Institute (NHLBI) of the National Institutes of Health (NIH) under grant number T32HL072757.

Appendix

Summary of observations used in this analysis are summarized in Table 13.10.

Table 13.10 Summary of variables used to create the mentoring and STEM activities variables

Variable	Total, *n* (%)	Male, *n* (%)	Female, *n* (%)
Since 08-09, the 9th grader participated in:			
Math competition	874 (4.15)	464 (2.20)	410 (1.95)
Math club	670 (3.18)	372 (1.77)	298 (1.41)
Math camp	123 (0.58)	67 (0.32)	56 (0.27)
Science competition	842 (4.00)	449 (2.13)	393 (1.86)
Science club	493 (2.34)	275 (1.30)	218 (1.03)
Science camp	205 (0.97)	109 (0.52)	96 (0.46)
Since Fall 2009, teenager participated in:			
Math competition	1102 (5.57)	614 (3.10)	488 (2.47)
Math club	895 (4.52)	464 (2.34)	431 (2.18)
Math Summer program	775 (3.92)	392 (1.98)	383 (1.94)
Science competition	1207 (6.10)	683 (3.45)	524 (2.65)
Science club	1253 (6.33)	601 (3.04)	652 (3.30)
Science program	742 (3.75)	365 (1.85)	377 (1.91)
9th grader is taking Fall 2009 math because:			
Teacher encouraged it	2504 (13.16)	1094 (5.75)	1410 (7.41)
Counselor suggested it	1784 (9.38)	878 (4.62)	906 (4.76)
Parent(s) encouraged it	2879 (15.14)	1300 (6.83)	1579 (8.30)

(continued)

Table 13.10 (continued)

Variable	Total, *n* (%)	Male, *n* (%)	Female, *n* (%)
9th grader is taking Spring 2012 math because:			
Teacher encouraged it	6075 (35.13)	2872 (16.61)	3203 (18.52)
Counselor suggested it	6010 (34.73)	2927 (16.91)	3083 (17.81)
Parent(s) encouraged it	5950 (34.43)	2875 (16.64)	3075 (17.79)
Family member encouraged it	2811 (16.29)	1435 (8.31)	1376 (7.97)
9th grader is taking Fall 2009 science because:			
Teacher encouraged it	1460 (8.33)	656 (3.74)	804 (4.59)
Counselor suggested it	1378 (7.86)	652 (3.72)	726 (4.14)
Parent(s) encouraged it	1942 (11.07)	843 (4.81)	1099 (6.27)
9th grader is taking Spring 2012 science because:			
Teacher encouraged it	5022 (32.24)	2328 (14.95)	2694 (17.30)
Counselor suggested it	5478 (35.14)	2564 (16.45)	2914 (18.69)
Parent(s) encouraged it	4708 (30.24)	2184 (14.03)	2524 (16.21)
Family member encouraged it	2467 (15.85)	1205 (7.74)	1262 (8.11)

References

Anderson, K. J., & Minke, K. M. (2007). Parent involvement in education: Toward an understanding of parents' decision making. *The Journal of Educational Research, 100*(5), 311–323.

BenZvi, D., & Garfield, J. (2008). Introducing the emerging discipline of statistics education. *School Science and Mathematics, 108*(8), 355–361.

Chen, X. *Stem Attrition: College Students' Paths into and Out of STEM Fields. Statistical Analysis Report. NCES 2014-001.* (2013). National Center for Education Statistics.

Griffin, K. A., Perez, D., Holmes, A. P., & Mayo, C. E. (2010). Investing in the future: The importance of faculty mentoring in the development of students of color in stem. *New Directions for Institutional Research, 2010*(148), 95–103.

Griffith, A. L. (2010). Persistence of women and minorities in stem field majors: Is it the school that matters? *Economics of Education Review, 29*(6), 911–922.

Harackiewicz, J. M., Rozek, C. S., Hulleman, C. S., & Hyde, J. S. (2012). Helping parents to motivate adolescents in mathematics and science: An experimental test of a utility-value intervention. *Psychological Science, 23*(8), 899–906.

Inter-university Consortium for Political and Social Research. (2016). High school longitudinal study, 2009–2013 [United States]. United States Department of Education. Institute of Education Sciences. National Center for Education Statistics.

Kader, G., & Perry, M. (2006). A framework for teaching statistics within the k-12 mathematics curriculum. In *Proceedings of the Seventh International Conference on Teaching Statistics.*

King, A. J., Fisher, A. M., Becich, M. J., & Boone, D. N. (2017). Computer science, biology and biomedical informatics academy: Outcomes from 5 years of immersing high-school students into informatics research. *Journal of Pathology Informatics, 8*, 2.

Musu-Gillette, L., Robinson, J., McFarland, J., KewalRamani, A., Zhang, A., & Wilkinson-Flicker, S. *Status and Trends in the Education of Racial and Ethnic Groups 2016. NCES 2016-007.* (2016). National Center for Education Statistics.

NSF. (2012a). Inclusion across the nation of communities of learners of underrepresented discoverers in engineering and science. https://www.nsf.gov/funding/pgm_summ.jsp?pims_id=505289

NSF. (2012b). Stem-c partnerships: Msp. https://nsf.gov/pubs/2012/nsf12518/nsf12518.htm?org=NSF

SAS Institute. (2015). Base sas 9.4 procedures guide.

Syed, M., Goza, B. K., Chemers, M. M., & Zurbriggen, E. L. (2012). Individual differences in preferences for matched-ethnic mentors among high-achieving ethnically diverse adolescents in stem. *Child Development, 83*(3), 896–910.

Chapter 14
Statistical Modeling for the Heart Disease Diagnosis via Multiple Imputation

Lian Li and Yichuan Zhao

14.1 Introduction

Missing data is a common challenge when analyzing data from the clinic. Missing data causes serious problems in statistical analysis, namely, reducing the power of a study, producing biased estimates, and even possibly leading to invalid conclusions. Incomplete datasets can occur via different means, such as mishandling of samples, low signal-to-noise ratio, measurement error, non-responses to questions, or aberrant value deletion. Based on the missing mechanism, Rubin (1976) defined the missing data as occurring in one of the three categories: (1) missing completely at random (MCAR): the probability of missingness for a variable is dependent on neither the known values nor the missing data; (2) missing at random (MAR): The probability of missingness for a variable may relate to the known values but not on the value of the missing data itself; (3) missing not at random (MNAR): The probability of missingness for a variable may depend on unobserved predictors or the missing value (Kang 2013).

The most common analytical procedure to deal with missing data is to exclude observations with any missing variable values from the analysis. Although this method of dealing with missing data is simple, a lot of information is lost with the decision to delete such data. Furthermore, if the missingness of data is not completely at random, excluding observations with missing values may cause a biased conclusion through ignoring the possible systematic difference between the

L. Li (✉)
Department of Epidemiology, Rollins School of Public Health, Emory University, Atlanta, GA, USA
e-mail: lli35@emory.edu

Y. Zhao
Department of Mathematics and Statistics, Georgia State University, Atlanta, GA, USA
e-mail: yichuan@gsu.edu

Y. Zhao, D.-G. Chen (eds.), *New Frontiers of Biostatistics and Bioinformatics*, ICSA Book Series in Statistics, https://doi.org/10.1007/978-3-319-99389-8_14

original dataset and the incomplete dataset. Single imputation is an alternative strategy for dealing with missing data. In this method, each missing value will be replaced by the variable mean which is obtained under the assumption of the data completeness. The uncertainty about the predictions of the missing data is not considered in this approach. Therefore, the estimated parameters drawing from this method may be biased (Graham 2009).

Multiple imputation (MI) is a more useful strategy for handling missing data. In multiple imputation, missing values are substituted with a set of derived possible values which contain the original variability and uncertainty of the actual values. Then a subset of complete data is used for standard analyses. Eventually, an inference is drawn from the combined results of a series of imputed datasets. The benefits of multiple imputation are that it not only restores the natural variability of the missing values, but also incorporates uncertainty due to the data missing. Therefore, the statistical inference of multiple imputation is widely accepted as a less biased and more valid result. Furthermore, multiple imputation is robust with regard to resisting violation of the normality assumptions and can produce appropriate results even with a small sample size or in the presence of a high rate of missing data (Sterne et al. 2009).

The goal of this chapter is to apply multiple imputation to a heart disease dataset and build a prediction model for heart disease diagnosis. The dataset comes from the UCI Machine Learning Repository website: http://archive.ics.uci.edu/ml/datasets/heart+-Disease. This dataset includes 920 observations and 14 variables. The outcome is the diagnosis of heart disease (1 = yes, 0 = no). In this dataset, ten variables have different extents of missing data. The most significant percentage of missingness is about 66%, and the variable with the second highest missed data rate is 53%. We will test if multiple imputation is appropriate to deal with such a large proportion of missing data.

The rest of the chapter is organized as follows. In Sect. 14.2, we first discuss the concept of two ways to impute missing data: a Markov chain Monte Carlo (MCMC) method and a fully conditional specification (FCS) method. Then, we outline the steps of model building with multiple imputation. In Sect. 14.3, we discuss some strategies, which combine with multiple imputation to build better statistical models. Finally, we make a conclusion in Sect. 14.4.

14.2 Data Analysis

14.2.1 Descriptive Analysis

This dataset includes 920 observations and 14 variables. "Age," "trestbps," "chol," "thalach," and "oldpeak" are continuous variables. "Slope" and "ca" are ordered variables. "Sex," "fbs," and "exang" are binary variables. "Restecg," "cp," and "thal" are nominal variables. The information about each variable in the dataset is presented in Table 14.1.

Table 14.1 Code book for the heart disease dataset

Variable information
age: age in years
sex: sex (1 = male; 0 = female)
cp: chest pain type
– Value 1: typical angina
– Value 2: atypical angina
– Value 3: non-anginal pain
– Value 4: asymptomatic
trestbps: resting blood pressure (in mm Hg on admission to the hospital)
chol: serum cholesterol in mg/dl
fbs: (fasting blood sugar >120 mg/dl) (1 = true; 0 = false)
restecg: resting electrocardiographic results
– Value 0: normal
– Value 1: having ST-T wave abnormality (T wave inversions and/or ST elevation or depression of >0.05 mV)
– Value 2: showing probable or definite left ventricular hypertrophy by Estes' criteria
thalach: maximum heart rate achieved
exang: exercise induced angina (1 = yes; 0 = no)
oldpeak = ST depression induced by exercise relative to rest
slope: the slope of the peak exercise ST segment
– Value 1: up sloping
– Value 2: flat
– Value 3: down sloping
ca: number of major vessels (0–3) colored by fluoroscopy
thal: 1 = normal; 2 = fixed defect; 3 = reversible defect
num: diagnosis of heart disease (angiographic disease status)
– Value 0: <50% diameter narrowing
– Value 1: >50% diameter narrowing

By exploring the dataset, some of the extrema can be easily found. For example, some of the patients have a value of "chol" (serum cholesterol in mg/dl) set at zero, which is impossible for the clinical test. The occurrence of zero values for this variable is probably caused by unstandardized data entry. Therefore, zero values are set as missing. After clearing the out-ranges, the continuous variables "age," "chol," "thalach," and "trestbps" resemble a bell-curve distribution without much skewness. Hence, those variables are kept as continuous variables in the dataset for imputation. Another continuous variable, "oldpeak," has about 45 of its values set to zero. To solve this problem, we generated a binary variable "oldpeak_c" basing on whether the value of "oldpeak" is higher than 0 or not. The proportion of "oldpeak_c" is different between the heart disease and non-heart disease groups. "Slope," "restecg," "thal," and "ca" are categorical variables with more than two levels. Some levels of variables are combined based on the results of the pair-wise chi-square test, and corresponding new binary variables are created. A Cochran-Armitage Trend Test did

Table 14.2 Descriptive statistics for categorical variables in dataset

Variable	Level	N = 920	%
HD	0	410	44.7
	1	510	55.3
Exang	0	528	57.4
	1	337	36.6
	Missing	55	6.0
Fbs	0	692	75.2
	1	138	15.0
	Missing	90	9.8
Sex	0	194	21.1
	1	726	78.9
slope_1	0	408	44.3
	1	203	22.1
	Missing	309	33.6
restecg_1	0	739	80.3
	1	179	19.5
	Missing	2	0.2
thal_3	0	238	25.9
	1	196	21.3
	Missing	486	52.8
ca_0	0	128	13.9
	1	181	19.7
	Missing	611	66.4
oldpeak_c	0	382	41.5
	1	476	51.7
	Missing	62	6.7

Table 14.3 Descriptive statistics for continuous variables in dataset

Variable	N	N Miss	Mean	Std Dev	Median	Maximum	Minimum
Age	920	0	53.51	9.42	54	77	28
Trestbps	860	60	132.29	18.54	130	200	80
Chol	718	202	246.83	58.53	239.5	603	85
Thalach	865	55	137.55	25.93	140	202	60

not find a trend for variable "cp." Therefore, "cp" was deleted from the dataset for imputation. Descriptive statistics for categorical variables are listed in Table 14.2, and continuous variable descriptive statistics are listed in Table 14.3.

14.2.2 Multiple Imputation

For multiple imputation, the variables in the dataset cannot be highly correlated. Otherwise, the EM algorithm (Expectation-Maximization) will not converge. Therefore, the correlation coefficient matrix must be checked first. Table 14.4 shows that

Table 14.4 Spearman correlation coefficient matrix for dataset

	Age	sex	Trestbps	chol	fbs	Thalach	exang	HD	oldpeak_c	slope_1	restecg_1	thal_3	ca_0
Age	1.00	0.06	0.26	0.10	0.23	−0.35	0.20	0.29	0.30	−0.12	0.14	−0.13	−0.38
Sex	0.06	1.00	0.02	−0.08	0.09	−0.18	0.18	0.31	0.07	−0.11	0.07	−0.38	−0.12
Trestbps	0.26	0.02	1.00	0.09	0.16	−0.09	0.15	0.11	0.12	−0.09	0.08	−0.10	−0.04
Chol	0.10	−0.08	0.09	1.00	0.05	−0.03	0.11	0.13	0.09	−0.04	−0.04	0.04	−0.14
Fbs	0.23	0.09	0.16	0.05	1.00	−0.05	0.03	0.14	0.09	−0.05	0.14	−0.12	−0.12
Thalach	−0.35	−0.18	−0.09	−0.03	−0.05	1.00	−0.38	−0.40	−0.17	0.42	−0.16	0.36	0.29
Exang	0.20	0.18	0.15	0.11	0.03	−0.38	1.00	0.46	0.39	−0.35	0.09	−0.33	−0.19
HD	0.29	0.31	0.11	0.13	0.14	−0.40	0.46	1.00	0.35	−0.38	0.11	−0.50	−0.48
oldpeak_c	0.30	0.07	0.12	0.09	0.09	−0.17	0.39	0.35	1.00	−0.37	0.02	−0.21	−0.17
slope_1	−0.12	−0.11	−0.09	−0.04	−0.05	0.42	−0.35	−0.38	−0.37	1.00	−0.08	0.32	0.12
restecg_1	0.14	0.07	0.08	−0.04	0.14	0.16	0.09	0.11	0.02	−0.08	1.00	−0.16	−0.05
thal_3	−0.13	−0.38	−0.10	0.04	−0.12	0.36	−0.33	−0.50	−0.21	0.32	−0.16	1.00	0.25
ca_0	−0.38	−0.12	−0.04	−0.14	−0.12	0.29	−0.19	−0.48	−0.17	0.12	−0.05	0.25	1.00

Table 14.5 Distribution of missing values in heart disease group and non-heart disease group

Heart disease	N Obs	Variable	N	N Miss	Miss percentage
0	410	Trestbps	410	20	4.87
		Chol	410	19	4.62
		Thalach	410	20	4.87
		Oldpeak_c	410	21	5.11
		Fbs	410	14	3.41
		Restecg_1	410	0	0.00
		Exang	410	20	4.87
		Slope_1	410	193	46.96
		Ca_0	410	246	59.85
		Thal_3	410	224	54.50
1	510	Trestbps	510	39	7.66
		Chol	510	11	2.16
		Thalach	510	35	6.88
		Oldpeak_c	510	41	8.06
		Fbs	510	76	14.93
		Restecg_1	510	2	0.39
		Exang	510	35	6.88
		Slope_1	510	116	22.79
		Ca_0	510	365	71.71
		Thal_3	510	262	51.47

there is no highly correlated pair of variables in the dataset. Thus, the multiple imputation dataset will keep all variables in the original dataset.

Multiple imputation assumes that the missing data follows the missing at random (MAR) pattern. However, sometimes this assumption is hard to verify in a real situation. Here, we only compare the percentage of the missing data in different outcome groups to ensure that the missingness is not related to heart disease status. Table 14.5 shows that there is no apparent pattern of missingness in different groups.

For datasets with arbitrary missing patterns, there are two methods that can be used to impute missing values: a Markov chain Monte Carlo (MCMC) method and a fully conditional specification (FCS) method.

The MCMC method assumes multivariate normality. The specific algorithm used in MCMC method is called the data augmentation (DA) algorithm developed by Tanner and Wong (1987). The algorithm imputes missing data by drawing the pseudo-random samples from a joint conditional distribution of Y_{mis} and θ given Y_{obs}: $P(Y_{mis},\theta | Y_{obs})$. The process of the MCMC method for imputing missing data is: Using some assumed numbers substitute the missing data Y_{mis} to get a complete data posterior distribution $P(\theta | Y_{mis}, Y_{obs})$, then simulate θ from the posterior distribution. Let $\theta^{(t)}$be the current simulated value of θ from the complete data posterior distribution, then ${Y_{mis}}^{(t)}$ can be drawn from the conditional predictive distribution ${Y_{mis}}^{(t+1)} \sim P(Y_{mis} | Y_{obs}, \theta^{(t)})$. Conditioning on ${Y_{mis}}^{(t+1)}$, the next simulated value of θ can be drawn from its complete data posterior distribution $\theta^{(t+1)} \sim P(\theta | Y_{obs}, {Y_{mis}}^{(t+1)})$. Repeat the above loop to yield a Markov chain $\{(\theta^{(t)}, {Y_{mis}}^{(t)}) : t = 1, 2, \ldots..\}$ which will reach its stationary distribution, the joint distribution of θ and Y_{mis} give Y_{obs}: $P(Y_{mis}, \theta | Y_{obs})$. For the details of MCMC method for multiple imputation, please refer to Zhang (2003) and the SAS support file for the MI procedure (Cary 2015).

The FCS method requires the existence of a joint distribution for all variables. In each imputation, FCS method includes two phases: preliminary filled-in phase and imputation phase. The procedure first replaces missing values with filled-in values for each variable in filled-in phase. That is, with P ordered variables $Y_1, Y_2, \ldots Y_p$, the missing values are filled in by using the sequence,

$$\begin{aligned}
\theta_1^{(0)} &\sim P\left(\theta_1 | Y_{1(obs)}\right) \\
Y_{1(*)}^{(0)} &\sim P\left(Y_1 | \theta_1^{(0)}\right) \\
Y_1^{(0)} &= \left(Y_{1(obs)}, Y_{1(*)}^{(0)}\right) \\
&\ldots \\
\theta_p^{(0)} &\sim P\left(\theta_p | Y_1^{(0)}, \ldots Y_{p-1}^{(0)}, Y_{p(obs)}\right) \\
Y_{p(*)}^{(0)} &\sim P\left(Y_p | \theta_p^{(0)}\right) \\
Y_p^{(0)} &= \left(Y_{p(obs)}, Y_{p(*)}^{(0)}\right)
\end{aligned}$$

Then, in the imputation phase, these filled-in values $Y_{p(*)}^{(0)}$ will be replaced by imputed values. For the details of FCS method for multiple imputation, please refer to (Van Buuren et al. 2006) and the SAS support file for the MI procedure (Cary 2015).

SAS 9.4 statistic software not only provides those two methods to impute the missing values but also has the MIANALYZE procedure to combine the results of the analyses of imputations and generates valid statistical inferences. The process of imputation in this chapter will be:

1. The missing data are imputed 5 times by MCMC method or FCS method to generate five complete datasets.
2. The imputed datasets are analyzed by using standard proc. logistic procedures.
3. Use "proc mianalyze" to combine the results from the imputed datasets for the inference.

14.2.3 Model Building

To avoid overfitting, the original dataset is randomly split 1:1 into a training dataset and a validation dataset. MI is applied to each dataset using both MCMC and FCS methods. Then, using the imputed training dataset, the model is built, and the imputed validation dataset is used to check the quality of the model.

Table 14.6 Stepwise selection of variables for model of original dataset

	MCMC-imputed dataset			FCS-imputed dataset			Original dataset		
Parameter	Estimate	Std Error	Pr > |t|	Estimate	Std Error	Pr > |t|	Estimate	Standard	Pr > ChiSq
Intercept	4.368	1.224	0.001	4.383	1.148	0.0002	6.728	1.848	0.0002
Thalach	−0.016	0.007	0.032	−0.017	0.007	0.015	−0.031	0.011	0.005
thal_3	−1.803	0.395	<0.0001	−1.735	0.388	<0.0001	−1.922	0.490	<0.0001
ca_0	−2.764	0.523	<0.0001	−2.620	0.505	<0.0001	−2.694	0.520	<0.0001
Exang	2.192	0.393	<0.0001	2.195	0.398	<0.0001	2.189	0.557	<0.0001

Table 14.7 Predicted results of the stepwise selection model

	MCMC-imputed dataset			FCS-imputed dataset			Original dataset		
Subset	False negative	False positive	Missing	False negative	False positive	Missing	False negative	False positive	Missing
Training	14.33	17.26	0	14.91	17.72	0	24.32	8.70	323
Valid	13.40	27.29	0	13.34	26.85	0	15.87	23.53	298

Because the outcome of heart disease (HD) is a binary variable, the logistic regression is a good model to predict the diagnosis of heart disease. Through stepwise selection, the best combination of independent variables is picked for the prediction model in the original training dataset. Variables "thalach," "thal_3," "ca_0," and "exang" are selected for the model by this method. Those variables are also used in MCMC and FCS imputation datasets to fit the regression. The estimated parameters and significant test *p*-values are listed in Table 14.6. Table 14.7 represents the predicted results.

The variables in the model are selected by the stepwise method based on the original dataset, which has a good deal of missingness. If we were to use the imputed dataset that has missing information replaced through multiple imputation, the model might be different. We used the MCMC imputed dataset and a backward selection method to get the best combination of variables, which are "slope_1," "thal_3," "ca_0," "exang," and "sex" (Table 14.8). The prediction accuracy after backward selection of variables and either MCMC- or FCS-imputation methods is shown in Table 14.9.

In the logistic regression modeling, outliers can produce extremely large residuals and may affect the results of the analysis and lead to incorrect inferences (Sarkar et al. 2011). Through different types of diagnostic plots, we found some outliers and deleted them from the training dataset (Fig. 14.1). The prediction accuracy of the backward selection model without outliers is presented in Table 14.10.

Some researchers have reported that stratifying the dataset before imputation may produce better results (Von Hippel 2009). Therefore, we also used this strategy on the heart disease dataset. The best model from the stratified imputation dataset was slightly different from the earlier model building on imputation (Table 14.11). But,

Table 14.8 Backward selection of variables into model by imputed dataset

	MCMC-imputed dataset		FCS-imputed dataset				Original dataset		
Parameter	Estimate	Std Error	Pr > |t|	Estimate	Std Error	Pr > |t|	Estimate	Standard	Pr > ChiSq
Intercept	1.952	0.618	0.002	1.787	0.583	0.002	1.607	0.667	0.016
slope_1	−1.352	0.376	0.000	−1.381	0.392	0.001	−0.781	0.485	0.107
thal_3	−1.419	0.447	0.002	−1.350	0.430	0.002	−1.442	0.521	0.006
ca_0	−2.987	0.527	<0.0001	−2.843	0.520	<0.0001	−2.901	0.530	<0.0001
Exang	2.206	0.410	<0.0001	2.235	0.412	<0.0001	2.447	0.564	<0.0001
Sex	0.932	0.417	0.026	0.977	0.404	0.016	0.879	0.548	0.109

Table 14.9 Predicted results by backward selection model

	MCMC-imputed dataset			FCS-imputed dataset			Original dataset		
Subset	False negative	False positive	Missing	False negative	False Positive	Missing	False negative	False positive	Missing
Training	12.63	17.41	0	14.16	18.33	0	20.27	11.96	323
Valid	10.60	24.85	0	11.69	24.54	0	15.87	25.00	298

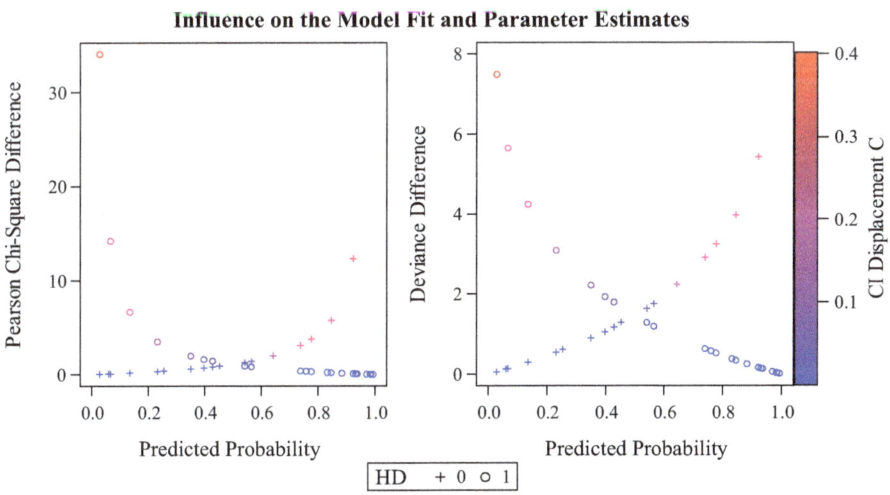

Fig. 14.1 Finding outliers in the backward selection model

the prediction accuracy with stratification was not better than direct imputation in the validation dataset (Table 14.12).

Based on the prediction accuracy, "slope_1," "thal_3," "ca_0," "exang," and "sex" are selected to build a final model. For the odds ratio estimates (Table 14.13), we find that the up sloping of the peak exercise ST segment, and that there is a reversible defect of "thal," such that zero for the number of major vessels colored by fluoroscopy has a negative effect on odds of heart disease. Exercise

Table 14.10 Predicted results by backward selection model without outliers

	MCMC-imputed dataset			FCS-imputed dataset			Original dataset		
Subset	False negative	False positive	Missing	False negative	False positive	Missing	False negative	False positive	Missing
Training	5.96	8.95	0	8.60	10.48	0	9.23	5.81	323
Valid	10.34	21.87	0	9.83	20.31	0	13.11	23.88	298

Table 14.11 Comparison of models for stratified imputation dataset and direct imputation dataset

Stratified imputation by MCMC				Direct imputation by MCMC			
Parameter	Estimate	Std Error	Pr > |t|	Parameter	Estimate	Std Error	Pr > |t|
Intercept	0.7618	0.4901	0.1217	Intercept	1.9522	0.6177	0.0018
slope_1	−1.6521	0.3641	<0.0001	slope_1	−1.3516	0.3762	0.0004
ca_0	−2.9005	0.6144	<0.0001	thal_3	−1.4189	0.447	0.0017
Exang	2.5774	0.4195	<0.0001	ca_0	−2.9869	0.5267	<0.0001
Sex	1.4872	0.3815	0.0001	Exang	2.2061	0.410	<0.0001
				Sex	0.9320	0.4174	0.0259

Table 14.12 Predicted results by model for stratified imputation dataset and direct imputation dataset

Stratified imputation_MCMC				Direct impution_MCMC			
Subset	False negative	False positive	Missing	Subset	False negative	False positive	Missing
Training	12.33	17.79	0	Training	12.63	17.41	0
Valid	16.19	22.12	0	Valid	10.60	24.85	0

Table 14.13 Odds ratio estimates for final model

	Estimates by MCMC imputed			Estimates by FCS imputed			Estimates by original		
Effect	Point estimate	95% Wald confidence limits		Point estimate	95% Wald confidence limits		Point estimate	95% Wald confidence limits	
slope_1	0.26	0.12	0.54	0.25	0.12	0.54	0.46	0.18	1.19
thal_3	0.24	0.10	0.58	0.26	0.11	0.60	0.24	0.09	0.66
ca_0	0.05	0.02	0.14	0.06	0.02	0.16	0.06	0.02	0.16
Exang	9.08	4.06	20.32	9.35	4.16	21.01	11.55	3.83	34.85
Sex	2.54	1.12	5.76	2.64	1.20	5.83	2.41	0.82	7.06

induced angina and being a male have positive impacts on odds of heart disease. The final model for MCMC-imputed dataset will be: logit(probability of heart disease) $= 1.877 - 1.362 \times \text{slope\_1} - 1.398 \times \text{thal\_3} - 2.991 \times \text{ca\_0} + 2.234 \times$ Exang $+ 0.922 \times$ Sex.

14.3 Discussion

The heart disease dataset examined in this chapter has a high missing rate. Excluding observations with any missing variable values from the analysis would remove about 67.5% of patient information, which would then not be used for prediction. Through MI, we can use all the information provided by the dataset and give a predicted value for each missing piece of data. Comparing the models built on different datasets, the MCMC dataset and FCS dataset generated coefficients, which have a smaller standard error than the same datasets without MI. Therefore, the models based on imputed data are more precise than the model of the original dataset. The predicted results generated by imputation datasets demonstrate a lower false negative rate and higher false positive rate for training and validation datasets than for the original dataset (Table 14.8). In addition, if one takes the false penalty into account, the predicted results of using the imputed dataset will be much better than using the original dataset because the penalty for a false negative is five times that of a false positive.

Comparing the coefficients and predicted results generated by MCMC dataset and FCS dataset, there is no significant difference between those two methods. The real-time of the procedure MCMC MI is shorter than the real-time of the procedure FCS MI. This difference may be due to the FCS method's including an additional preliminary filled-in phase.

Using the dataset without outliers to fit the model further significantly improved a prediction accuracy in the training dataset. However, the accuracy of the model fit on the validation dataset did not change much (Table 14.10). This method can only detect outliers by fitting a model for the original dataset that does not have complete information. For the imputation data, each imputed dataset can generate different outliers. A better way to combine the imputed datasets and discover outliers in the imputation dataset needs further investigation.

The method that stratifies the dataset before imputation separately imputes data for patients and non-patients, which may reduce the bias in the imputation. However, we did not get a better result from the stratified method for this heart disease dataset. The possible reason could be the high missing rate and a difference in missing rates for patients and non-patients. After stratifying, some variables have an even higher missing rate for the patient group or non-patient group. Hence, the imputation loses its precision in comparison to the non-stratified imputation.

14.4 Conclusion

Multiple imputation is a good strategy for dealing with missing values in the above heart disease dataset. Using multiple imputation, we maximize the information provided by the dataset. The coefficients in the model built on the imputed dataset have a more precise confidence interval and a better prediction accuracy than the model established on the original dataset. Furthermore, even with a high missing rate, as occurs in this dataset, multiple imputation is seen to be robust and to produce appropriate results.

Acknowledgements The authors are grateful to the two reviewers for their helpful comments, which improved the manuscript significantly. The authors would like to thank Lisa Elon for invaluable advice and Dr. Eric Dammer for critical reading of the manuscript.

References

Cary, N. (2015). *SAS/STAT® 14.1 User's Guide*. Cary, NC: SAS Institute Inc.

Graham, J. W. (2009). Missing data analysis: Making it work in the real world. *Annual Review of Psychology, 60*, 549–576.

Kang, H. (2013). The prevention and handling of the missing data. *Korean Journal of Anesthesiology, 64*, 402–406.

Rubin, D. B. (1976). Inference and missing data. *Biometrika, 63*, 581–592.

Sarkar, S. K., Midi, H., & Rana, S. (2011). Detection of outliers and influential observations in binary logistic regression: An empirical study. *Journal of Applied Sciences, 11*(1), 26–35.

Sterne, J. A., White, I. R., Carlin, J. B., Spratt, M., Royston, P., Kenward, M. G., Wood, A. M., & Carpenter, J. R. (2009). Multiple imputation for missing data in epidemiological and clinical research: Potential and pitfalls. *BMJ, 338*, b2393.

Tanner, M. A. & Wong W. H. (1987). Source: *Journal of the American Statistical Association, 82*(398), 528–540.

Van Buuren, S., Brand, J. P. L., Groothuis-Oudshoorn, C. G. M., & Rubin, D. B. (2006). Fully conditional specification in multivariate imputation. *Journal of Statistical Computation and Simulation, 76*, 1049–1064.

Von Hippel, P. T. (2009). How to impute interactions, squares, and other transformed variables. *Sociological Methodology, 39*, 265–291.

Zhang, P. (2003). Multiple imputation: Theory and method. *International Statistical Review, 71*, 581–592.

Part IV
High-Dimensional Gene Expression Data Analysis

Chapter 15
Learning Gene Regulatory Networks with High-Dimensional Heterogeneous Data

Bochao Jia and Faming Liang

15.1 Introduction

The emergence of high-throughput technologies has made it feasible to measure the activities of thousands of genes simultaneously, which provides scientists with a major opportunity to infer gene regulatory networks. Accurate inference of gene regulatory networks is pivotal to gaining a systematic understanding of the molecular mechanism, to shedding light on the mechanism of diseases that occur when cellular processes are dysregulated, and to identifying potential therapeutic targets for the diseases. Given the high dimensionality and complexity of high-throughput data, inference of gene regulatory networks largely relies on the advance of statistical modeling and computation. The Gaussian graphical model is a promising tool to achieve this challenge.

The Gaussian graphical model uses a network to depict conditional independence relationships for a large set of Gaussian random variables, where the absence of an edge between two variables indicates independence of the two variables conditioned on all other variables. In the literature, a variety of methods have been proposed to learn Gaussian graphical networks. To name a few, they include covariance selection (Dempster 1972), nodewise regression (Meinshausen and Bühlmann 2006), graphical Lasso (Yuan and Lin 2007; Friedman et al. 2008), adaptive graphical Lasso (Fan et al. 2009), projected covariance matrix method (Fan et al. 2015), and ψ-learning (Liang et al. 2015). In general, these methods assume that the data are homogeneous, i.e., all samples are drawn from a single Gaussian

B. Jia (✉)
Eli Lilly and Company, Lilly Corporate Center, IN, USA
e-mail: jbc409@ufl.edu

F. Liang
Department of Statistics, Purdue University, West Lafayette, IN, USA
e-mail: fmliang@purdue.edu

Y. Zhao, D.-G. Chen (eds.), *New Frontiers of Biostatistics and Bioinformatics*, ICSA Book Series in Statistics, https://doi.org/10.1007/978-3-319-99389-8_15

distribution. However, in practice, we have often the data that are heterogeneous, i.e., the samples are drawn from a mixture Gaussian distribution, while a single Gaussian graphical network still needs to be learned for all the samples in a fashion of data integration. Here are some examples:

1. **Data with hidden biological/clinical subtypes.** It is known that complex diseases such as cancer can have significant heterogeneity in response to treatments, and this heterogeneity is often reflected in gene expression. For example, the gene expression patterns can vary with subtypes of the cancer. Since for many types of cancers, the definition of subtypes is still unclear and the number of samples from each subtype can be very small, it is impractical to construct an individual gene regulatory network for each subtype. In this case, we might still be interested in constructing a single gene regulatory network for the heterogeneous data in a fashion of data integration. Such an integrated gene regulatory network can facilitate us to identify fundamental patterns common to the development and progression of the disease.
2. **Data with hidden confounding factors.** In real-world applications, the gene expression data may contain some systematic differences caused by known or unknown confounding factors, such as study cohorts, sample collection, and experimental batches. Due to the limited number of samples from each level of the confounding factors, we also prefer to learn a single gene regulatory network for the heterogeneous data in a fashion of data integration. Moreover, for many problems, the confounding factors can be unknown.

In this paper, we develop a mixture model method to learn Gaussian graphical networks for heterogeneous data with hidden clusters. The new method is developed based on the imputation-consistency (IC) algorithm proposed by Liang et al. (2018) and the ψ-learning algorithm proposed by Liang et al. (2015). The IC algorithm is a general algorithm for dealing with high-dimensional missing data problems. Like the EM algorithm (Dempster et al. 1977), the IC algorithm works in an iterative manner, iterating between an I-step and a C-step. The I-step is to impute the missing data conditioned on the observed data and the current estimate of parameters, and the C-step is to find a "consistent" estimator for the minimizer of a Kullback–Leibler divergence defined on the pseudo-complete data. For high-dimensional problems, the "consistent" estimate can be found with sparsity constraints or screened data. Refer to Fan and Lv (2008) and Fan and Song (2010) for variable screening methods. Under quite general conditions, Liang et al. (2018) showed that the average of the "consistent" estimators across iterations is consistent to the true parameters. The ψ-learning algorithm is originally designed for learning Gaussian graphical models for homogeneous data. The proposed method can be viewed as a combination of the IC algorithm and the ψ-learning algorithm, which simultaneously clusters samples to different groups and learn an integrated network across all the groups. When applying the IC algorithm to cluster samples, their cluster membership is treated as missing data.

We note that the proposed mixture model method is different from the methods for joint estimation of multiple Gaussian graphical models, such as fused Lasso (Danaher et al. 2014) and Bayesian nodewise regression (Lin et al. 2017). For the latter methods, the samples' cluster membership is known a priori and the goal is to learn an individual network for each cluster of samples. In contrast, the proposed method works for the case that the cluster membership is unknown and the goal is to learn an integrated network across all hidden groups. The proposed method is also different from the methods proposed by Ruan et al. (2011) and Lee et al. (2018). For the former, the goal is to learn an individual network for each cluster of samples, although it assumes that the cluster membership is unknown. The latter is to first group samples to different clusters using an eigen-analysis-based approach and then apply the ψ-learning algorithm to learn the network structure. Since the method did not account for the uncertainty of sample clustering, it often performs less well.

The rest part of this paper is organized as follows. In Sect. 15.2, we describe the proposed method. In Sect. 15.3, we illustrate the performance of the proposed method using simulated examples. In Sect. 15.4, we apply the proposed method to learn a gene regulatory network for breast cancer with a heterogeneous gene expression dataset. In Sect. 15.5, we conclude the paper with a brief discussion.

15.2 Mixture Gaussian Graphical Models

15.2.1 Algorithms for Homogeneous Data

To have a better description for the proposed method, we first give a brief review for the existing Gaussian graphical model algorithms for homogeneous data.

Let $\boldsymbol{V} = \{\boldsymbol{X}_1, \ldots, \boldsymbol{X}_p\}$ denote a set of p Gaussian random variables, where $\boldsymbol{X}_i = \{X_{i1}, \ldots X_{in}\}$ denotes n observations of variable i. In the context of gene regulatory networks, X_{ij} refers to the expression level of gene i measured in experiment j. Let $\boldsymbol{X}^{(j)} = (X_{1j}, \ldots, X_{pj})^T$ denote the expression levels of all p genes measured in experiment j, which is assumed to follow a Gaussian distribution $N_p(\boldsymbol{\mu}, \boldsymbol{\Sigma})$ with the mean vector $\boldsymbol{\mu}$ and covariance matrix $\boldsymbol{\Sigma}$. Let $\boldsymbol{E} = (e_{ij})$ denote the adjacency matrix, where $e_{ij} = 1$ if the edge is present and 0 otherwise. The adjacency matrix specifies the structure of the Gaussian graphical network. Let $\rho_{ij|V\setminus\{i,j\}}$ denote the partial correlation coefficient of variable i and variable j conditioned on all other variables. Let $\boldsymbol{C} = (C_{ij}) = \boldsymbol{\Sigma}^{-1}$ denote the concentration matrix, also known as the precision matrix. Let $\beta_i^{(j)}$'s denote the coefficients of the regressions

$$\boldsymbol{X}_j = \beta_i^{(j)}\boldsymbol{X}_i + \sum_{r\in V\setminus\{i,j\}} \beta_r^{(j)}\boldsymbol{X}_r + \epsilon^{(j)}, \quad j = 1, 2, \ldots, p, \tag{15.1}$$

where $\epsilon^{(j)}$ is a zero-mean Gaussian random vector. Since $\rho_{ij|V\{i,j\}}$ can be expressed as $\rho_{ij|V\setminus\{i,j\}} = -C_{ij}/C_{ii}C_{jj}$ and $\beta_i^{(j)}$'s can be expressed as $\beta_i^{(j)} = -C_{ji}/C_{jj}$ and $\beta_j^{(i)} = -C_{ji}/C_{ii}$, the following relationship holds:

$$e_{ij} = e_{ji} = 1 \Leftrightarrow \rho_{ij|V\setminus\{i,j\}} \neq 0 \Leftrightarrow C_{ij} \neq 0 \Leftrightarrow \beta_i^{(j)} \neq 0 \text{ and } \beta_j^{(i)} \neq 0. \quad (15.2)$$

Based on the relation between partial correlation coefficients and the concentration matrix, Dempster (1972) proposed the covariance selection method, which identifies the edges of the Gaussian graphical network by identifying the nonzero elements of the concentration matrix. However, this method cannot be applied to the problems with $p > n$, where the sample covariance matrix is nonsingular and thus the concentration matrix cannot be calculated. To tackle this difficulty, Yuan and Lin (2007) proposed to estimate the concentration matrix with l_1-regularization. Soon, this method was accelerated by Friedman et al. (2008) using the coordinate descent algorithm in a similar way to Lasso regression (Tibshirani 1996), which leads to the so-called graphical Lasso algorithm. Based on the relation between partial correlation coefficients and regression coefficients, Meinshausen and Bühlmann (2006) proposed the nodewise regression method, which is to learn Gaussian graphical networks by identifying nonzero regression coefficients of the regressions given in (15.1) with a sparsity constraint.

Alternative to estimating the concentration matrix and regression coefficients, the ψ-learning algorithm (Liang et al. 2015) is to provide an equivalent measure for the partial correlation coefficient in the sense that

$$\psi_{ij} = 0 \Longleftrightarrow \rho_{ij|V\setminus\{i,j\}} = 0, \quad (15.3)$$

where ψ_{ij} is the partial correlation coefficient of variable i and variable j conditioned on a subset of $V \setminus \{i, j\}$ and the subset is obtained via correlation screening. Since the ψ-learning algorithm is used as a component of the proposed mixture model method for learning Gaussian graphical models with grouped samples, the details of the algorithm are given below.

Algorithm 1 (ψ-learning)

(a) (Correlation screening) Determine the reduced neighborhood for each variable $\boldsymbol{X}_i$.

  (i) Conduct a multiple hypothesis test to identify the pairs of variables for which the empirical correlation coefficient is significantly different from zero. This step results in a so-called empirical correlation network.

  (ii) For each variable X_i, identify its neighborhood in the empirical correlation network, and reduce the size of the neighborhood to $O(n/\log(n))$ by removing the variables having lower correlation (in absolute value) with X_i. This step results in a so-called reduced correlation network.

(b) (ψ-calculation) For each pair of variables i and j, identify a subset of nodes S_{ij} based on the reduced correlation network resulted in step (a) and calculate $\psi_{ij} = \rho_{ij|S_{ij}}$, where $\rho_{ij|S_{ij}}$ denotes the partial correlation coefficient of X_i and X_j conditioned on the variables $\{X_k : k \in S_{ij}\}$. In this paper, we set $S_{ij} = S_i \setminus \{j\}$ if $|S_i \setminus \{j\}| \leq |S_j \setminus \{i\}|$ and $S_j \setminus \{j\}$ otherwise, where S_i denotes the neighborhood of node i in the reduced correlation network, and $|\cdot|$ denotes the cardinality of a set.
(c) (ψ-screening) Conduct a multiple hypothesis test to identify the pairs of vertices for which ψ_{ij} is significantly different from zero, and set the corresponding element of the adjacency matrix to 1.

The multiple hypothesis tests involved in the algorithm can be done using the empirical Bayesian method developed in Liang and Zhang (2008), which allows for the general dependence between test statistics. Other multiple hypothesis testing procedures that account for the dependence between test statistics, e.g., the two-stage procedure of Benjamini et al. (2006), can also be applied here. The correlation screening step involves two procedures, (1) multiple hypothesis test and (2) sure independence screening, to control the neighborhood size for each variable. The two procedures seem redundant, but actually they are not. Indeed the multiple hypothesis test is able to identify the pairs of independent variables, but the size of each neighborhood cannot be guaranteed to be less than $O(n/\log(n))$ as established in Liang et al. (2015). We have tried to use the sure independence screening procedure only, which results in the same neighborhood size $O(n/\log(n))$ for each variable. However, in this case, the enlarged neighborhood may contain some variables that are independent of the central one, and thus the power of the followed ψ-screening test will be reduced.

The ψ-learning algorithm consists of two free parameters, namely α_1 and α_2, which refer to the significance levels used in correlation screening and ψ-screening, respectively. Following the suggestion of Liang et al. (2015), we specify their values in terms of q-values (Storey 2002); setting $\alpha_1 = 0.2$ and $\alpha_2 = 0.05$ or 0.1 in all computations. In particular, we set $\alpha_2 = 0.05$ for the simulated examples and $\alpha_2 = 0.1$ for the real data example. A large value of α_2 avoids to lose more potential interactions between different genes.

Under mild conditions, e.g., the joint Gaussian distribution of $\boldsymbol{X}_1, \ldots, \boldsymbol{X}_p$ satisfies the faithfulness condition, Liang et al. (2015) showed that the ψ-partial correlation coefficient is equivalent to the true partial correlation coefficient in determining the structure of Gaussian graphical models in the sense of (15.3). Compared to other Gaussian graphical model algorithms, the ψ-learning algorithm has a significant advantage that it has reduced the computation of partial correlation coefficients from a high-dimensional problem to a low dimensional problem via correlation screening and thus can be used for very high-dimensional problems. As shown in Liang et al. (2015), the ψ-learning algorithm is consistent; the resulting network will converge to the true one in probability as the sample size becomes large. The ψ-learning algorithm tends to produce better numerical performance and cost less CPU time than the existing algorithms, such as gLasso and nodewise regression, especially when p is large.

15.2.2 The Mixture Gaussian Graphical Model Method

Let $\mathscr{X} = \{X^{(1)}, \ldots, X^{(n)}\}$ denote a set of n independent samples which are drawn from a mixture Gaussian distribution with M components, where the sample size n can be much smaller than the dimension p. Suppose that M is known. Later, we will describe a Bayesian information criterion (BIC) to determine the value of M. The log-likelihood function of the samples is given by

$$\ell(\mathscr{X}|\Theta) = \sum_{k=1}^{M} \log\left(\pi_k \phi(X_i|\mu_k, \Sigma_k)\right), \tag{15.4}$$

where $\Theta = \{(\pi_k, \mu_k, \Sigma_k) : k = 1, \ldots, M\}$ denotes the collection of unknown parameters, π_k's are mixture proportions, μ_k's are mean vectors, and Σ_k's are covariance matrices of the M Gaussian components, respectively; and $\phi(\cdot|\mu_k, \Sigma_k)$ denotes the density function of the multivariate Gaussian distribution. Let τ_i denote the indicator variable for the component/cluster membership of sample i, for $i = 1, 2, \ldots, n$. That is, $p(\tau_i = k) = \pi_k$ and $X_i|\tau_i = K \sim N(\boldsymbol{\mu}_k, \Sigma_k)$ for $k = 1, \ldots, M$ and $i = 1, 2, \ldots, n$. Henceforth, we will use cluster to denote the group of samples assigned to a component of the mixture Gaussian graphical model. Cluster and component are also used exchangeably in this paper.

If the sample size n is greater than p, then the parameters Θ can be estimated using the EM algorithm as described in what follows. Let $\pi_k^{(t)}$, $\mu_k^{(t)}$ and $\Sigma_k^{(t)}$ denote, respectively, the estimates of π_k, μ_k, and Σ_k obtained at iteration t. Let $\Theta_k^{(t)} = (\pi_k^{(t)}, \mu_k^{(t)}, \Sigma_k^{(t)})$. The E-step calculates the conditional expectation of τ_i given X_i and the current estimate of Θ, i.e.,

$$\gamma_{ik}^{(t)} = P(\tau_i = k|X_i; \Theta^{(t)}) = \frac{\pi_k^{(t)} \phi(X_i|\mu_k^{(t)}, \Sigma_k^{(t)})}{\sum_{l=1}^{M} \pi_l^{(t)} \phi(X_i|\mu_l^{(t)}, \Sigma_l^{(t)})}. \tag{15.5}$$

which leads to the so-called Q-function,

$$Q(\Theta, \Theta^{(t)}) = \sum_{k=1}^{M} \left[\sum_{i=1}^{n} log(\phi(X_i|\mu_k^{(t)}, \Sigma_k^{(t)}))\gamma_{ik}^{(t)} \right] = \sum_{k=1}^{M} Q_k(\Theta, \Theta^{(t)}). \tag{15.6}$$

The M-step updates $\Theta^{(t)}$ by maximizing the Q-function, which can be done by maximizing Q_k with respect to $\Theta_k = (\pi_k, \mu_k, \Sigma_k)$ for each k. For each value of k, $\Theta_k^{(t)}$ can be updated by setting

$$\begin{aligned}
\pi_k^{(t+1)} &= \frac{1}{n}\sum_{i=1}^{n} \gamma_{ik}^{(t)}, \\
\mu_k^{(t+1)} &= \frac{\sum_{i=1}^{n} \gamma_{ik}^{(t)} X_i}{\sum_{i=1}^{n} \gamma_{ik}^{(t)}}, \\
\Sigma_k^{(t)} &= \sum_{i=1}^{n} \left(\frac{\gamma_{ik}^{(t)}}{\sum_{j=1}^{n} \gamma_{jk}^{(t)}} (X_i - \mu_k^{(t+1)})(X_i - \mu_k^{(t+1)})' \right).
\end{aligned} \tag{15.7}$$

However, this algorithm does not work when $n < p$, as $\Sigma_k^{(t)}$'s will be singular in this case.

When $n < p$, to avoid the issues caused by the singularity of $\Sigma_k^{(t)}$'s, we propose the following algorithm. For the proposed algorithm, we assume that all components of the mixture Gaussian graphical model share a common adjacency matrix, although their covariance and precision matrices can be different from each other. The new algorithm consists of two stages. The first stage is to apply the imputation-consistency (IC) algorithm to generate a series of estimates for the common adjacency matrices, and the second stage is to average the estimates to get a stable estimate for the common adjacency matrix. Note that, as can be seen below, the IC algorithm generates a Markov chain.

To learn the common adjacency matrix at each iteration, a ψ-integration procedure is needed, which is to integrate the adjacency matrices learned for each component into one adjacency matrix. This procedure can be described as follows. Let $\psi_{kij}^{(t)}$ denote the ψ-partial correlation coefficient calculated for the k-th cluster at iteration t, which can be transformed to a z-score via Fisher's transformation:

$$Z_{kij}^{(t)} = \frac{\sqrt{n_k^{(t)} - |S_{kij}^{(t)}| - 3}}{2} \log\left[\frac{1+\hat{\psi}_{kij}^{(t)}}{1-\hat{\psi}_{kij}^{(t)}}\right], \quad i, j = 1, \ldots p, k = 1, \ldots M. \tag{15.8}$$

where $|S_{kij}^{(t)}|$ denotes the conditioning set used in calculating $\psi_{kij}^{(t)}$, and $n_k^{(t)}$ is the number of samples assigned to cluster k at iteration t. For convenience, we call the z-score a ψ-score. The ψ-scores from different clusters can be combined using Stouffer's meta-analysis method (Stouffer et al. 1949) by setting

$$Z_{ij}^{(t)} = \frac{\sum_{k=1}^{M} \omega_k^{(t)} z_{kij}^{(t)}}{\sqrt{\sum_{k=1}^{M} (\omega_k^{(t)})^2}}, \quad i, j = 1, \ldots p, \tag{15.9}$$

where $\omega_k^{(t)}$ is a nonnegative weight assigned to cluster k at iteration t. In this paper, we set $\omega_k^{(t)} = n_k^{(t)}/n$. Stouffer's method falls into the class of Fisher's combined probability tests used for combining the results from several independent tests bearing upon the same overall hypothesis. For mixture GGMs, since for each

t and each pair of nodes (i, j), $z_{kij}^{(t)}$'s are mutually independent and thus $Z_{ij}^{(t)}$ defined in (15.9) approximately follows a standard normal distribution under the null hypothesis $H_0 : e_{ij} = 0$. Then a multiple hypothesis test can be conducted on $Z_{ij}^{(t)}$'s to identify the pairs of nodes for which $Z_{ij}^{(t)}$ is differentially distributed from the standard normal $N(0, 1)$, and the adjacency matrix common to all components of the mixture model can be determined thereby. In this paper, we adopted the multiple hypothesis testing procedure of Liang and Zhang (2008) to conduct the test. This testing procedure allows general dependence between test statistics. Alternative to the meta-analysis method, some regularization approaches, such as fused graphical Lasso and group graphical Lasso (Danaher et al. 2014), can also be applied here for estimating the common adjacency matrix, as long as the resulting estimator is consistent following from the theory of the IC algorithm (Liang et al. 2018).

Given the ψ-integration procedure, the first stage of the proposed method can be summarized as follows: It starts with an initial estimate $\Theta^{(0)} = \{(\pi_k^{(0)}, \mu_k^{(0)}, \Sigma_k^{(0)}) : k = 1, \ldots, M\}$, and then iterates between the following steps:

Algorithm 2 (IC Estimation for Mixture Gaussian Graphical Models)

(a) (imputation) Impute the indicator variable $\tau_i^{(t+1)}$ by drawing from the multinomial distribution as defined in (15.5) for each $i = 1, 2, \ldots, n$.

(b) (consistency) Based on the imputed values of $\tau_i^{(t+1)}$'s, update the estimate $\Theta^{(t)}$ by

- (i) setting $n_k^{(t+1)} = \sum_{i=1}^n I(\tau_i^{(t+1)} = k)$, $\pi_k^{(t+1)} = n_k^{(t+1)}/n$, and $\mu_k^{(t+1)} = \sum_{j \in \{i: \tau_i^{(t+1)} = k\}} \boldsymbol{X}_j / n_k^{(t+1)}$;
- (ii) applying the ψ-learning algorithm to learn an adjacency matrix for each cluster of the samples.
- (iii) applying the ψ-integration procedure to integrate the adjacency matrices learned in step (ii) into one.
- (iv) applying the algorithm given in Hastie et al. (2009, p. 634) to recover the covariance matrices for each cluster, given the common adjacency matrix learned in step (iii).

Let $\boldsymbol{\tau}^{(t)} = \{\tau_1^{(t)}, \ldots, \tau_n^{(t)}\}$ for $t = 1, 2, \ldots$. As in the stochastic EM algorithm (Celeux and Govaert 1995; Nielsen 2000), the sequence $\Theta^{(0)} \to \boldsymbol{\tau}^{(1)} \to \Theta^{(1)} \to \ldots \to \boldsymbol{\tau}^{(t)} \to \Theta^{(t)} \to \cdots$ forms two interleaved Markov chains. Intuitively, the Markov chain $\{\Theta^{(t)}\}$ will converge to a stationary distribution whose mean is close to the true parameter value Θ and variance reflects the variation of $\boldsymbol{\tau}^{(t)}$ introduced in imputation. This intuition has been justified rigorously in Liang et al. (2018) under quite general conditions. Following from Theorems 3 and 4 of Liang et al. (2018), the Markov chain $\{\Theta^{(t)}\}$ is almost surely ergodic if the sample size n is sufficiently large, and the average of $\Theta^{(t)}$'s along with iterations forms a consistent estimate of Θ when t is sufficiently large. In this paper, the adjacency matrices are averaged in the following way, which corresponds to the second stage of the proposed method. Define

$$Z_{ij} = \sum_{t=t_0+1}^{T} Z_{ij}^{(t)}/(T - t_0), \quad i, j = 1, 2, \ldots, p,$$

where t_0 denotes the number of burn-in iterations of the IC algorithm, and then the final estimate of the adjacency matrix can be obtained by conducting another multiple hypothesis test for Z_{ij}'s. As before, under the null hypothesis H_0: $e_{ij} = 0$, Z_{ij} follows the standard normal distribution.

Thus far, we have treated the number of clusters M as known. In practice, M can be determined using an information criterion, such as AIC or BIC. Following Ruan et al. (2011), we define the degree of freedom for a model with M components as

$$df(M) = M\left[p + \sum_{i \leqslant j} \hat{e}_{ij}\right], \tag{15.10}$$

where p represents the dimension of the mean vector, and $\hat{e}_{ij}$ denotes the (i, j)-th element of the estimated common adjacency matrix. Although we have assumed that the mixture Gaussian graphical model has a common adjacency matrix for all components, it can have a completely different concentration matrix for each component. Hence, for each component, we count each nonzero entry of the concentration matrix as a different parameter. The BIC score is then given by

$$BIC(M) = -2\ell(\mathscr{X}|\hat{\Theta}(M)) + \log(n)df(M), \tag{15.11}$$

where $\ell(\mathscr{X}|\hat{\Theta}(M))$ is the log-likelihood function given by Eq. (15.4), and M can be determined by minimizing $BIC(M)$.

In (15.10), we did not count for the parameters $\pi_1, \ldots, \pi_{M-1}$. This is due to two reasons. First, the problem is considered under the high-dimensional scenario where p is allowed to be greater than and grow with n. However, M is considered as fixed or to grow at a lower order of $\log(n)$. Therefore, including $M - 1$ or not in (15.10) will not affect much the performance of the criterion when n becomes large. Second, we ignore $M - 1$ in (15.10) to make the definition of the BIC score (15.11) consistent with the one used in Ruan et al. (2011), which facilitates comparisons.

15.3 Simulation Studies

We compare the performance of the proposed method with some methods developed for homogeneous data such as gLasso (Friedman et al. 2008), nodewise regression (Meinshausen and Bühlmann 2006), and ψ-learning (Liang et al. 2015), as well as the EM-regularization method developed by Ruan et al. (2011) for mixture Gaussian graphical models. As aforementioned, the method by Ruan et al. (2011) is different from the proposed one, as whose goal is to estimate an individual Gaussian graphical network for each cluster. Moreover, since Ruan et al. (2011) applied the gLasso algorithm to learn an individual Gaussian graphical network for each cluster, it will be very hard to integrate those networks into a common one.

15.3.1 Example 1

We began with the case where the number of clusters M of the mixture model is known and the components are different in means. For this simulation study, we fix $M = 3$ and the total number of samples $n = 300$, and varied the dimension p between 100 and 200. We set the component means as $\boldsymbol{\mu}_1 = 0$, $\boldsymbol{\mu}_2 = m\mathbf{1}_p$, and $\boldsymbol{\mu}_3 = -m\mathbf{1}_p$, where $\mathbf{1}_p$ denotes a p-dimensional vector of ones. We let all the three components share the same precision matrix C:

$$C_{ij} = \begin{cases} 0.5, & \text{if} |j-i| = 1, i = 2, \ldots, (p-1), \\ 0.25, & \text{if} |j-i| = 2, i = 3, \ldots, (p-2), \\ 1, & \text{if} i = j, i = 1, \ldots, p, \\ 0, & \text{otherwise}, \end{cases} \tag{15.12}$$

and generated 100 samples from each component of the mixture model. The samples from different components are combined and shuffled. Three different values of m are considered, including $m = 0$, 0.3, and 0.5. Under each setting of m and p, 50 independent datasets were generated.

The proposed method was applied to this heterogeneous dataset. To initialize π_k's and $\boldsymbol{\mu}_k$'s, we randomly grouped the samples into three clusters and calculated their respective proportions and means. To initialize the covariance matrices, we first applied the ψ-learning algorithm to the whole dataset to obtain a common adjacency matrix, and then applied the algorithm by Hastie et al. (2009, p. 634) to estimate the covariance matrix for each cluster with the common adjacency matrix. The IC algorithm converges very fast, usually in about 10 iterations. For this example, the algorithm was run for 20 iterations for each dataset.

To access the performance of the proposed method, we calculated the precision and recall for the estimate of the adjacency matrix obtained with each dataset. Estimating the adjacency matrix can be viewed as a set of binary decision problems with each corresponding to one potential edge. For a set of binary decision problems, the precision and recall are defined by

$$\text{precision} = \frac{TP}{TP+FP}, \qquad \text{recall} = \frac{TP}{TP+FN},$$

where TP, FP, and FN denote true positives, false positives, and false negatives, respectively, and they are defined via a binary decision table (see Table 15.1). In general, the method producing a larger area under the precision–recall curve is considered as a better method. The area under the precision–recall curve

Table 15.1 Outcomes of binary decision

	$A_{ij} = 1$	$A_{ij} = 0$
$\hat{A}_{ij} = 1$	True positive (TP)	False positive (FP)
$\hat{A}_{ij} = 1$	False negative (FN)	True negative (TN)

is often denoted by AUC (area under curve) in the literature. Unlike *receiver operating characteristic (ROC)* curves, which are always increasing monotonically, the precision–recall curves are not monotonic; that is, an increase in recall does not always lead to a reduction in precision. Refer to Aggarwal (2018, pp. 228–230) for more discussions on this issue. Figure 15.1 shows the averaged precision–recall curves produced by the proposed method, where each curve was obtained by averaging over those from 50 different datasets.

For comparison, we have applied other methods, including the EM-regularization method of Ruan et al. (2011), ψ-learning, gLasso, and nodewise regression, to this example. As shown in Fig. 15.1, under both scenarios with $m = 0$ (representing homogeneous data) and $m \neq 0$, the proposed method outperforms all others. Moreover, when the value of m increases, the performance of the ψ- learning, gLasso, and nodewise regression methods deteriorates as they are designed for homogeneous data. The EM-regularization method is robust to the value of m, and tends to outperform gLasso, nodewise regression, and ψ-learning when m becomes large. Note that the EM-regularization method produced a different network for each cluster and thus three precision–recall curves in total. Figure 15.1 showed only the best one, i.e., the curve with the largest value of AUC. Table 15.2 summarizes the areas under the precision–recall curves produced by different methods, where the average areas (over 50 datasets) were reported with the standard deviations given in the corresponding parentheses. The comparison indicates that the proposed method significantly outperforms other methods for the heterogeneous data. For the homogeneous data (i.e., $m = 0$), the proposed method performs as well as the ψ-learning method, while significantly outperforms others. This indicates generality of the proposed method, which can be applied to homogeneous data without significant harms. For the EM-regularization method, its poor performance for the homogeneous data may be due to two reasons. Firstly, the gLasso procedure employed there tends to perform less well than the ψ-learning and nodewise regression methods as shown in Fig. 15.1a, b. Secondly, the EM-regularization method produced three different networks, which are not allowed to be integrated under its current procedure. For the purpose of comparison, we reported only the result for the network with the largest AUC area. However, this "best" network may still be worse than the properly integrated one for the homogeneous data.

In addition to underlying networks, we are interested in parameter estimation and cluster identification for the mixture Gaussian graphical model. To access the accuracy of parameter estimation, we adopt the criteria used by Ruan et al. (2011), which include the averaged spectral norm defined by

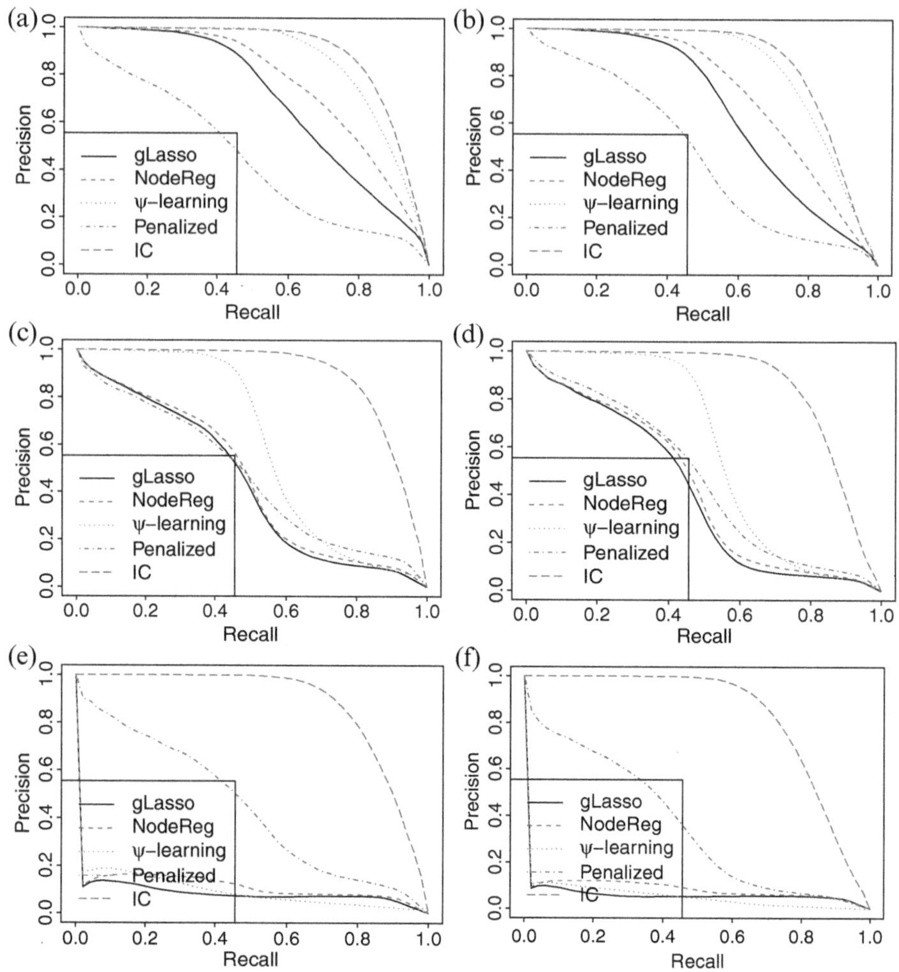

Fig. 15.1 Comparison of different methods for recovering underling networks for heterogeneous data with different cluster means: "gLasso" refers to the graphical Lasso method, "NodeReg" refers to the nodewise regression method, "Penalized" refers to the EM-regularization method of Ruan et al. (2011), ψ-learning refers to the ψ-learning methods, and "IC" refers to the proposed method. The plots (**a**) and (**b**) represent the scenario of homogeneous data. (**a**) $m = 0$ and $p = 100$, (**b**) $m = 0$ and $p = 200$, (**c**) $m = 0.3$ and $p = 100$, (**d**) $m = 0.3$ and $p = 200$, (**e**) $m = 0.5$ and $p = 100$, (**f**) $m = 0.5$ and $p = 200$

$$SL = \frac{1}{M} \sum_{k=1}^{M} \| \hat{\Sigma}_k^{-1} - \Sigma_k^{-1} \|, \tag{15.13}$$

where $\|A\|$ is the largest singular value of matrix A; the averaged Frobenius norm defined by

Table 15.2 Average AUCs produced by different methods for the heterogeneous data with different cluster means

	m	gLasso	NodeReg	ψ-learning	Penalized	IC
$p = 100$	0	0.696(0.002)	0.765(0.003)	0.859(0.003)	0.662(0.003)	0.888(0.004)
	0.3	0.437(0.002)	0.453(0.003)	0.602(0.003)	0.634(0.003)	0.892(0.004)
	0.5	0.084(0.001)	0.112(0.001)	0.095(0.003)	0.459(0.020)	0.876(0.004)
$p = 200$	0	0.658(0.002)	0.731(0.002)	0.834(0.002)	0.654(0.002)	0.855(0.004)
	0.3	0.402(0.002)	0.421(0.002)	0.585(0.002)	0.597(0.008)	0.857(0.004)
	0.5	0.059(0.001)	0.084(0.001)	0.051(0.002)	0.439(0.015)	0.829(0.004)

$$FL = \frac{1}{M}\sum_{k=1}^{M} \|\hat{\Sigma}_k^{-1} - \Sigma_k^{-1}\|_F \tag{15.14}$$

$$= \frac{1}{M}\sum_{k=1}^{M} \sqrt{\sum_{i,j} (\hat{\Sigma}_k^{-1}(i,j) - \Sigma_k^{-1}(i,j))^2}, \tag{15.15}$$

and the averaged Kullback–Leibler (KL) loss defined by

$$KL = \frac{1}{M}\sum_{k=1}^{M} KL(\Sigma_k, \hat{\Sigma}_k), \tag{15.16}$$

where

$$KL(\Sigma, \hat{\Sigma}) = tr(\Sigma\hat{\Sigma}^{-1}) - log|\Sigma\hat{\Sigma}^{-1}| - p. \tag{15.17}$$

To assess the accuracy of cluster identification, we calculated the averaged false and negative selection rates over different clusters. Let s_k denote the index set of observations for cluster k, and let $\hat{s}_k$ denote its estimate. Define

$$fsr = \frac{1}{M}\sum_{k=1}^{M} \frac{|\hat{s}_k \backslash s_k|}{|\hat{s}_k|}, \qquad nsr = \frac{1}{M}\sum_{k=1}^{M} \frac{|s_k \backslash \hat{s}_k|}{|s_k|} \tag{15.18}$$

where $|\cdot|$ denotes the set cardinality. The smaller the values of fsr and nsr are, the better the performance of the method is. The comparison was summarized in Table 15.3 where, for each setting of m and p, each method was evaluated based on 50 datasets with the averaged evaluation results reported. The numbers in the parentheses represent the standard deviations of the corresponding averages. The comparison indicates that the proposed method significantly outperforms the other methods in both parameter estimation and cluster identification.

Table 15.3 Comparison of different methods in parameter estimation and cluster identification for the heterogeneous data with different cluster means

		m	SL	FL	KL	fsr	nsr
$p = 100$	Penalized	0	3.642(0.015)	22.309(0.018)	149.701(1.217)	–	–
		0.3	3.618(0.008)	22.231(0.004)	149.058(0.569)	0.453(0.004)	0.393(0.005)
		0.5	3.488(0.027)	22.414(0.056)	160.736(1.540)	0.014(0.002)	0.016(0.002)
	IC	0	3.261(0.045)	11.222(0.072)	24.619(0.281)	–	–
		0.3	2.984(0.037)	10.508(0.084)	21.439(0.251)	0.008(0.001)	0.008(0.001)
		0.5	3.025(0.035)	10.635(0.081)	21.701(0.276)	0(0)	0(0)
$p = 200$	Penalized	0	3.644(0.009)	31.480(0.007)	296.131(1.483)	–	–
		0.3	3.578(0.022)	31.529(0.056)	304.948(4.098)	0.512(0.004)	0.533(0.010)
		0.5	3.143(0.033)	32.712(0.124)	388.672(5.041)	0.015(0.002)	0.021(0.007)
	IC	0	3.437(0.042)	16.102(0.107)	51.161(0.413)	–	–
		0.3	3.350(0.042)	15.800(0.039)	49.069(0.311)	0(0)	0(0)
		0.5	2.732(0.036)	16.312(0.010)	50.177(0.246)	0(0)	0(0)

15.3.2 Example 2

To make the problem harder, we consider the model for which each component has a different mean vector as well as a different concentration matrix, although the adjacency matrix is still the same for all components. As for Example 1, we fix $M = 3$ and the total sample size $n = 300$, varied the dimension p between 100 and 200, and set the cluster mean vectors as $\boldsymbol{\mu}_1 = 0$, $\boldsymbol{\mu}_2 = m\mathbf{1}_p$, and $\boldsymbol{\mu}_3 = -m\mathbf{1}_p$, where $\mathbf{1}_p$ denotes a p-dimensional vector of ones. The common pattern of the concentration matrix is given by

$$C_{ij}^{(k)} = \begin{cases} c_k, & \text{if } |j-i| = 1, i = 2, \ldots, (p-1), \\ c_k/2, & \text{if } |j-i| = 2, i = 3, \ldots, (p-2), \\ 1, & \text{if} i = j, i = 1, \ldots, p, \\ 0, & \text{otherwise}, \end{cases} \tag{15.19}$$

for $k = 1, 2, 3$. We set $c_1 = 0.6$, $c_2 = 0.5$, and $c_3 = 0.4$ for the three components, respectively. From each component, we generated 100 samples. Three different values of m are considered, which are 0, 0.3, and 0.5. Under each setting of m and p, 50 independent datasets were generated.

Figure 15.2 shows the precision–recall curves produced gLasso, nodewise regression, ψ-learning, EM-regularization, and the proposed method. It indicates that the proposed method outperforms others. The two plots in the first row of Fig. 15.2 compare the performance of different methods when $m = 0$, which represents a very difficult scenario that each cluster is only slightly different in precision matrices and thus the samples will be extremely difficult to be clustered. However, the proposed method still outperform others under this scenario.

Table 15.4 compares the areas under the precision–recall curves produced by different methods, and Table 15.5 compares the performance of different methods in parameter estimation and cluster identification. For each setting of m and p, each method was evaluated based on 50 datasets the averaged evaluation results reported. The numbers in the parentheses of the two tables represent the standard deviations of the corresponding averages. The comparison indicates that the proposed method outperforms others in both parameter estimation and cluster identification.

15.3.3 Identification of Cluster Numbers

When the number of clusters M is unknown, we propose to determine its value according to the BIC criterion give in (15.11). In what follows, we illustrated the performance of the proposed method under this scenario using some simulated examples. We considered the cases with $M = 2$ and 3 and $p = 100$ and 200. For each combination of (M, p), we simulated 100 samples from each cluster with

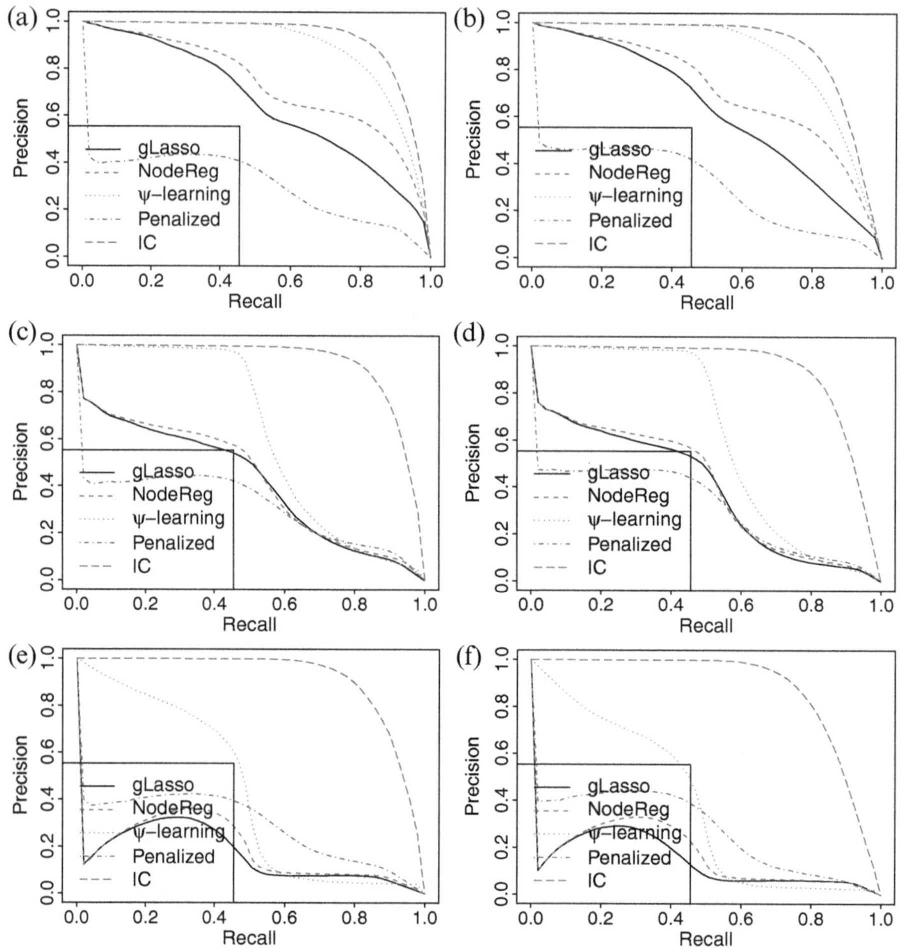

Fig. 15.2 Comparison of different methods for recovering underlying networks for heterogeneous data with different cluster means as well as different cluster precision matrices: "gLasso" refers to the graphical Lasso method, "NodeReg" refers to the nodewise regression method, ψ-learning refers to the ψ-learning method, "Penalized" refers to the EM-regularization method, and "IC" refers to the proposed method. (**a**) $m = 0$ and $p = 100$, (**b**) $m = 0$ and $p = 200$, (**c**) $m = 0.3$ and $p = 100$, (**d**) $m = 0.3$ and $p = 200$, (**e**) $m = 0.5$ and $p = 100$, (**f**) $m = 0.5$ and $p = 200$

Table 15.4 Comparison of average AUCs produced by different methods for the heterogeneous data with different cluster means as well as different cluster precision matrices

	m	gLasso	NodeReg	ψ-learning	Penalized	IC
	0	0.653(0.002)	0.732(0.002)	0.888(0.003)	0.595(0.003)	0.927(0.003)
$p = 100$	0.3	0.416(0.003)	0.429(0.003)	0.624(0.002)	0.571(0.003)	0.926(0.003)
	0.5	0.162(0.001)	0.184(0.001)	0.434(0.005)	0.460(0.003)	0.914(0.003)
	0	0.625(0.002)	0.711(0.002)	0.858(0.001)	0.573(0.002)	0.896(0.002)
$p = 200$	0.3	0.388(0.002)	0.401(0.003)	0.615(0.002)	0.555(0.002)	0.898(0.003)
	0.5	0.136(0.001)	0.161(0.001)	0.380(0.004)	0.358(0.018)	0.878(0.003)

Table 15.5 Comparison of different methods in parameter estimation and cluster identification for the heterogeneous data with different cluster means as well as different cluster precision matrices

		m	SL	FL	KL	fsr	nsr
$p = 100$	Penalized	0	3.745(0.019)	22.952(0.033)	258.397(3.042)	0.633(0.121)	0.651(0.106)
		0.3	3.723(0.022)	22.902(0.038)	257.245(3.395)	0.174(0.009)	0.196(0.013)
		0.5	3.749(0.019)	22.973(0.032)	257.954(2.943)	0.159(0.035)	0.177(0.100)
	IC	0	3.453(0.049)	11.782(0.069)	42.363(0.369)	0.103(0.017)	0.094(0.011)
		0.3	3.387(0.042)	11.572(0.060)	41.161(0.284)	0.010(0.003)	0.010(0.003)
		0.5	3.292(0.041)	11.521(0.069)	42.098(0.419)	0(0)	0(0)
$p = 200$	Penalized	0	3.797(0.002)	32.411(0.004)	505.035(3.343)	0.597(0.231)	0.678(0.429)
		0.3	3.740(0.020)	32.336(0.045)	510.729(4.079)	0.479(0.104)	0.345(0.085)
		0.5	3.752(0.012)	32.333(0.034)	508.792(1.768)	0.267(0.009)	0.108(0.005)
	IC	0	3.405(0.014)	17.045(0.080)	92.140(0.709)	0.212(0.010)	0.209(0.009)
		0.3	3.533(0.043)	16.989(0.075)	91.503(0.746)	0.005(0.001)	0.004(0.001)
		0.5	3.573(0.041)	17.194(0.090)	94.059(0.701)	0(0)	0(0)

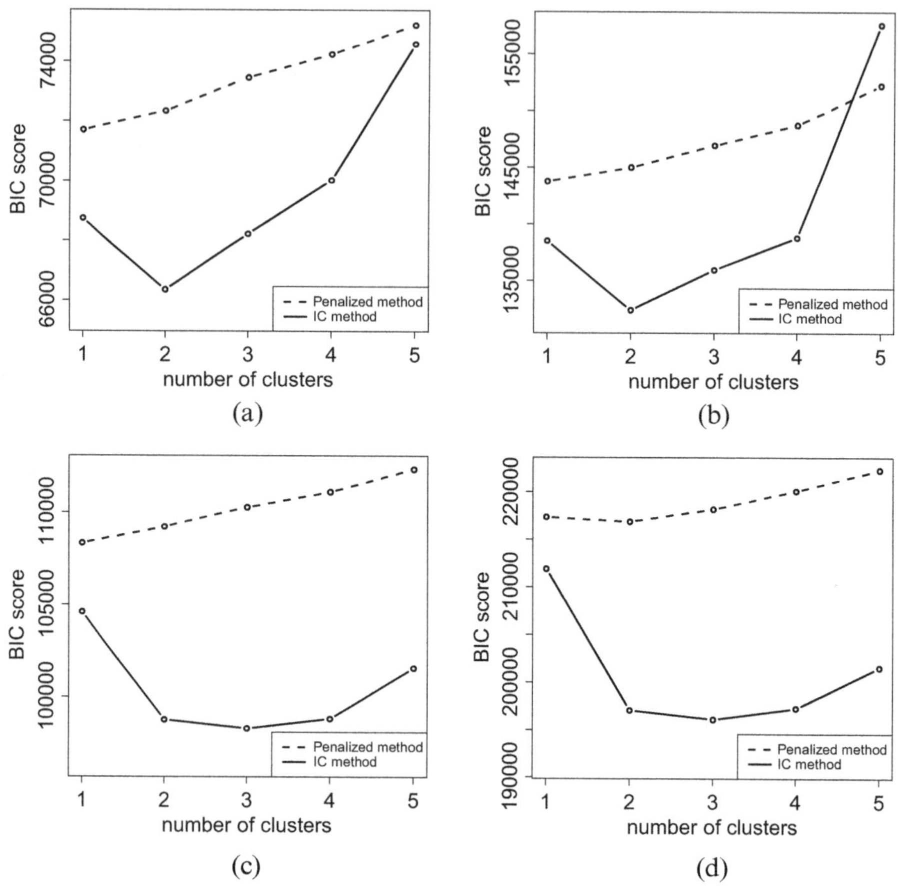

Fig. 15.3 BIC scores produced by the EM-regularization (Penalized) method and the proposed method (IC) for different settings of (M, p). (**a**) $M = 2$ and $p = 100$, (**b**) $M = 2$ and $p = 200$, (**c**) $M = 3$ and $p = 100$, (**d**) $M = 3$ and $p = 200$

the same precision matrix as defined in (15.12). For the cluster means, we set $\boldsymbol{\mu}_1 = 0.5\mathbf{1}_p$ and $\boldsymbol{\mu}_2 = -0.5\mathbf{1}_p$ for $M = 2$, and set $\boldsymbol{\mu}_1 = 0$, $\boldsymbol{\mu}_2 = 0.5\mathbf{1}_p$, and $\boldsymbol{\mu}_3 = -0.5\mathbf{1}_p$ for $M = 3$.

Figure 15.3 compares the performance of the EM-regularization method and the proposed method in identification of cluster numbers. It indicates that for the simulated example, the proposed method was able to correctly identify the true value of M according to the BIC criterion, while the EM-regularization method could not.

15.4 A Real Data Example

Breast cancer is one of the most prevalent types of cancer which can be classified into four molecular subtypes, namely luminal A, basal-like, HER2-enriched, and luminal B, based on their tumor expression profiles (Haque et al. 2012). In this study, we aim to construct a single gene regulatory network across the four subtypes to discover the overall gene regulation mechanism in breast cancer. The gene expression data for breast cancer are available at The Cancer Genome Atlas (TCGA), which contains 768 patients and 20,502 genes. For each patient, some clinical information such as survival time, age, gender, and tumor stages are also available, but the cancer subtypes are unknown. Since the data might be heterogeneous given the existence of breast cancer subtypes, the proposed method can be applied here. For this study, we are interested in learning a gene regulatory network related to the survival time of patients. For this reason, we first applied a marginal screening method to select the survival time-related genes. For each gene, we calculated its p-value using the marginal Cox regression after adjusting the effects of age, gender, and tumor stages, and then selected 592 genes according to a multiple hypothesis test at a false discovery rate (FDR) level of 0.05. We used the empirical Bayes method of Liang and Zhang (2008) to conduct the test.

To determine the number of components for the mixture model, we calculated BIC scores for $M = 1, 2, \ldots, 5$ with the results shown in Fig. 15.4. According to the BIC scores, we set $M = 3$. The resulting three clusters consist of 338, 191, and 238 patients, respectively. Figure 15.5a shows the Kaplan–Meier curves of the three clusters. A log-rank test for the three curves produced a p-value of

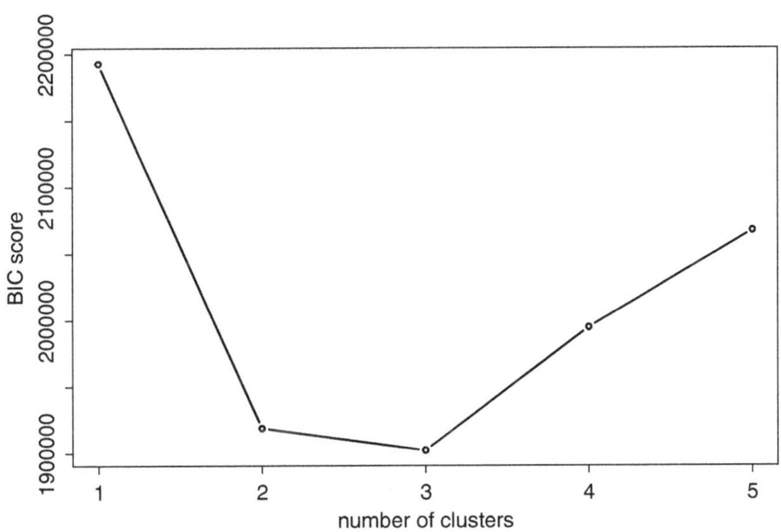

Fig. 15.4 BIC scores produced by the proposed method for breast cancer data

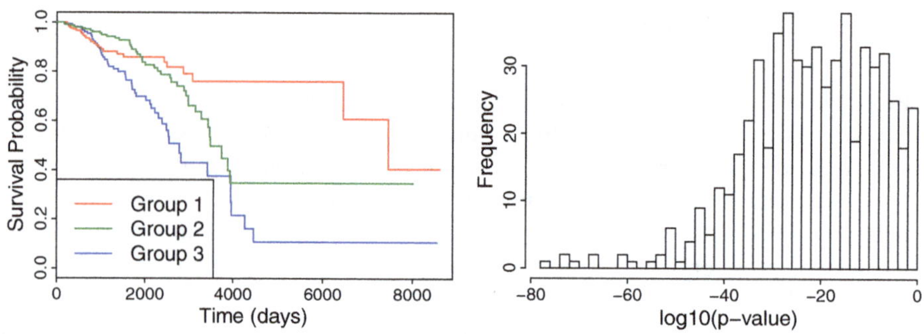

Fig. 15.5 The left panel shows the Kaplan–Meier curves for three different patient groups, and the right panel shows the histogram of the logarithm of p-values obtained for each gene in the ANOVA test

3.89×10^{-5}, which indicate that the patients in different clusters have different survival probabilities. Further, for each gene, we conducted an ANOVA test for its mean expression level across the three clusters. The resulting p-values are shown in Fig. 15.5b, where most p-values are very close to 0. This implies that the three clusters have different means and thus the data are heterogeneous. We note that the clustering results produced by the proposed method are biologically meaningful, which is likely due to the existence of hidden subtypes of breast cancer. As stated in Haque et al. (2012), women with luminal A tumors had the longest survival time, women with HER2-enriched and luminal B tumors had a much shorter survival time, and women with basal-like tumors had an intermediate survival time with deaths occurring earlier than those with luminal A tumors.

With M being set to 3, the proposed method produced a gene regulatory network (shown in Fig. 15.6), from which some hub genes can be identified. The hub genes refer to those with high connectivity in the gene regulatory network, and they tend to play important roles in gene regulation. To provide a stable way to identify hub genes, we consider a cross-validation like method. We divide the dataset into five subsets equally and then run the proposed method for five times, each applying to four of the five subsets only. In each run, we identified ten hub genes according to their connectivity. The results were summarized in Table 15.6, where the genes were ranked by their frequencies being selected as the hub gene among the 5 runs. The results indicate that the performance of the proposed method is quite stable: quite a few genes are frequently selected as the hub gene in different runs.

Our findings of hub genes are pretty consistent with the existing knowledge. Among the top 10 hub genes, 8 of them have been verified in the literature to be related with breast cancer. For example, LHFPL3, the gene has the most connectivities in the networks, is characteristic of primary glioblastoma which are important processes for cancer development and progression (Milinkovic et al. 2013). The gene SEPP1 is significantly associated with breast cancer risk among

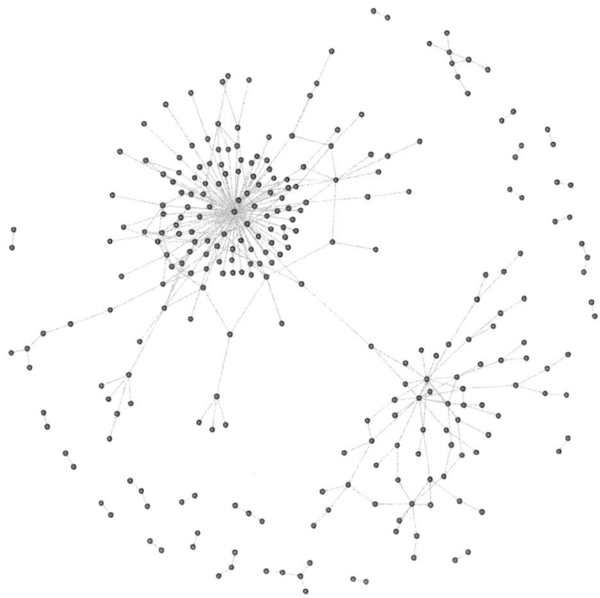

Fig. 15.6 The gene regulatory network constructed by the proposed method for breast cancer

Table 15.6 Top ten hub genes identified by the proposed method, where "Freq" denotes the number of times that the gene was selected as a hub gene in the five subset runs, "Links" denotes the average number of edges connected to the gene in the five networks with its standard deviation given in the parentheses, and the superscript * indicates that this gene has been verified in the literature to be related with breast cancer

Rank	Gene	Freq	Links	Rank	Gene	Freq	Links
1	LHFPL3*	4	49.2(9.6)	6	KRT12	3	13.4(5.1)
2	SEPP1*	4	8.4(1.4)	7	FXYD1*	2	5.4(2.3)
3	MYH11	4	8.6(1.4)	8	SCARA5*	2	6.4(1.6)
4	F13A1*	3	12.2(3.8)	9	CLEC3B*	2	7.8(2.8)
5	MAMDC2*	3	5.4(1.0)	10	LRRC70*	2	5.8(1.7)

women (Mohammaddoust et al. 2018). The gene F13A1 is known as a thrombotic factor that plays a major role in tumor formation (Ahmadi et al. 2016). In the cancer coexpression network developed by Meng et al. (2016), they found that MAMDC2 plays a key role in the development of breast invasive ductal carcinoma. Our results also reveal some new findings, such as the gene MYH11. Li et al. (2016) reported that MYH11 plays a role in tumor formation by disturbing stem cell differentiation or affecting cellular energy balance and has been identified as a driver gene in human colorectal cancer, although few researches identify its function in breast cancer.

15.5 Discussion

In this paper, we have proposed a new method for constructing gene regulatory networks for heterogeneous data, which is able to simultaneously cluster samples to difference groups and learn an integrated network across the groups. The proposed method was illustrated using some simulated examples and a real-world gene expression data example. The numerical results indicate that the proposed method significantly outperforms the existing ones, such as graphical Lasso, nodewise regression, ψ-learning, and EM-regularization. For the real-world gene expression data example, we conducted a detailed post-clustering analysis, which indicates the heterogeneity of the data and justifies the importance of the proposed method for real problems.

In addition to microarray gene expression data, the proposed method can be applied to next generation sequencing (NGS) data based on the transformations developed in Jia et al. (2017). To learn gene regulatory networks from NGS data, which are often assumed to follow a Poisson or negative binomial distribution, Jia et al. (2017) developed a random effect model-based transformation to continuize the NGS data. Further, the continuized data can be transformed to Gaussian using the nonparanormal transformation (Liu et al. 2012), and the proposed method can be applied then. We expect that the proposed method can also find wide applications in other scientific fields.

Acknowledgements The authors thank the book editor Dr. Yichuan Zhao and two referees for their constructive comments which have led to significant improvement of this paper. Liang's research was supported in part by the grants DMS-1612924 and DMS/NIGMS R01-GM117597.

References

Aggarwal, C. C. (2018). *Machine learning for text*. New York: Springer.

Ahmadi, M., Nasiri, M., & Ebrahimi, A. (2016). Thrombosis-related factors FV and F13A1 mutations in uterine myomas. *Zahedan Journal of Research in Medical Sciences, 18*(10), e4836.

Benjamini, Y., Krieger, A. M., & Yekutieli, D. (2006). Adaptive linear step-up procedures that control the false discovery rate. *Biometrika, 93*(3), 491–507.

Celeux, G., & Govaert, G. (1995). Gaussian parsimonious clustering models. *Pattern Recognition, 28*(5), 781–793.

Danaher, P., Wang, P., & Witten, D. M. (2014). The joint graphical lasso for inverse covariance estimation across multiple classes. *Journal of the Royal Statistical Society, Series B, 76*(2), 373–397.

Dempster, A. P. (1972). Covariance selection. *Biometrics, 28*, 157–175.

Dempster, A. P., Laird, N. M., & Rubin, D. B. (1977). Maximum likelihood from incomplete data via the EM algorithm. *Journal of the Royal Statistical Society. Series B, 39*, 1–38.

Fan, J., Feng, Y., & Wu, Y. (2009). Network exploration via the adaptive LASSO and SCAD penalties. *The Annals of Applied Statistics, 3*(2), 521.

Fan, J., Feng, Y., & Xia, L. (2015). A projection based conditional dependence measure with applications to high-dimensional undirected graphical models. ArXiv preprint arXiv:1501.01617.

Fan, J., & Lv, J. (2008). Sure independence screening for ultrahigh dimensional feature space. *Journal of the Royal Statistical Society, Series B, 70*, 849–911.

Fan, J., & Song, R. (2010). Sure independence screening in generalized linear model with NP-dimensionality. *Annals of Statistics, 38*, 3567–3604.

Friedman, J., Hastie, T., & Tibshirani, R. (2008). Sparse inverse covariance estimation with the graphical lasso. *Biostatistics, 9*, 432–441.

Haque, R., Ahmed, S. A., Inzhakova, G., Shi, J., Avila, C., Polikoff, J., et al. (2012). Impact of breast cancer subtypes and treatment on survival: An analysis spanning two decades. *Cancer Epidemiology and Prevention Biomarkers, 21*(10), 1848–1855.

Hastie, T., Tibshirani, R., & Friedman, J. (2009). *The elements of statistical learning* (2nd ed., 763 pp.). Berlin, Springer.

Jia, B., Xu, S., Xiao, G., Lamba, V., & Liang, F. (2017). Learning gene regulatory networks from next generation sequencing data. *Biometrics, 73*, 1221–1230.

Lee, S., Liang, F., Cai, L., & Xiao, G. (2018). A two-stage approach of gene network analysis for high-dimensional heterogeneous data. *Biostatistics, 19*(2), 216–232.

Li, Y., Tang, X. Q., Bai, Z., & Dai, X. (2016). Exploring the intrinsic differences among breast tumor subtypes defined using immunohistochemistry markers based on the decision tree. *Scientific Reports, 6*, 35773.

Liang, F., Jia, B., Xue, J., Li, Q., & Luo, Y. (2018). An imputation-consistency algorithm for high-dimensional missing data problems and beyond. ArXiv preprint arXiv:1802.02251.

Liang, F., Song, Q., & Qiu, P. (2015). An equivalent measure of partial correlation coefficients for high dimensional gaussian graphical models. *Journal of the American Statistical Association, 110*, 1248–1265.

Liang, F., & Zhang, J. (2008). Estimating the false discovery rate using the stochastic approximation algorithm. *Biometrika, 95*, 961–977.

Lin, Z., Wang, T., Yang, C., & Zhao, H. (2017). On joint estimation of Gaussian graphical models for spatial and temporal data. *Biometrics, 73*(3), 769–779.

Liu, H., Han, F., Yuan, M., Lafferty, J., & Wasserman, L. (2012). High-dimensional semiparametric Gaussian copula graphical models. *The Annals of Statistics, 40*(4), 2293–2326.

Meinshausen, N., & Bühlmann, P. (2006). High-dimensional graphs and variable selection with the Lasso. *Annals of Statistics, 34*, 1436–1462.

Meng, L., Xu, Y., Xu, C., & Zhang, W. (2016). Biomarker discovery to improve prediction of breast cancer survival: Using gene expression profiling, meta-analysis, and tissue validation. *OncoTargets and Therapy, 9*, 6177.

Milinkovic, V., Bankovic, J., Rakic, M., Stankovic, T., Skender-Gazibara, M., Ruzdijic, S., et al. (2013). Identification of novel genetic alterations in samples of malignant glioma patients. *PLoS One, 8*(12), e82108.

Mohammaddoust, S., Salehi, Z., & Saeidi Saedi, H. (2018). SEPP1 and SEP15 gene polymorphisms and susceptibility to breast cancer. *British Journal of Biomedical Science, 75*, 36–39.

Nielsen, S. F. (2000). The stochastic EM algorithm: Estimation and asymptotic results. *Bernoulli, 6*(3), 457–489.

Ruan, L., Yuan, M., & Zou, H. (2011). Regularized parameter estimation in high-dimensional gaussian mixture models. *Neural computation, 23*(6), 1605–1622.

Storey, J. D. (2002). A direct approach to false discovery rates. *Journal of the Royal Statistical Society: Series B (Statistical Methodology), 64*(3), 479–498.

Stouffer, S. A., Suchman, E. A., DeVinney, L. C., Star, S. A., & Williams, R. M., Jr. (1949). *The American soldier, Vol. 1: Adjustment during army life*. Princeton, NJ: Princeton University Press.

Tibshirani, R. (1996). Regression analysis and selection via the Lasso. *Journal of the Royal Statistical Society, Series B, 58*, 267–288.

Yuan, M., & Lin, Y. (2007). Model selection and estimation in the Gaussian graphical model. *Biometrika, 94*, 19–35.

Chapter 16
Performance Evaluation of Normalization Approaches for Metagenomic Compositional Data on Differential Abundance Analysis

Ruofei Du, Lingling An, and Zhide Fang

16.1 Introduction

Classical microbiological research requires microbial culture, by which the studied microbes reproduce in culture medium (Handelsman 2004). However, since a community of microbes (i.e., microbiome) is usually not able to survive under the predetermined laboratory condition, our understanding of microbes at aggregate level had been much hindered (National Research Council 2007). The scenario started to change since mid-1980s when a different approach was innovated (Woese 1987), in which microbiome samples are obtained from the site in situ; DNA contents are extracted and sequenced; sequence alignment is subsequently performed, and then followed by computational or statistical analysis (Wooley et al. 2010). The related study is named metagenomics, especially boosted by the rapid advancement of DNA sequencing technologies in the past decade (Metzker 2010; Bragg and Tyson 2014).

Having the entire genomic DNA or particular DNA contents (e.g., 16S rDNA) sequenced, metagenomic datasets can be classified as whole-genome sequence

R. Du
Biostatistics Shared Resource, University of New Mexico Comprehensive Cancer Center, Albuquerque, NM, USA
e-mail: RDu@salud.unm.edu

L. An
Department of Agricultural and Biosystems Engineering, University of Arizona, Tucson, AZ, USA

Interdisciplinary Program in Statistics, University of Arizona, Tucson, AZ, USA

Z. Fang (✉)
Biostatistics Program, School of Public Health, Louisiana State University Health Sciences Center, New Orleans, LA, USA
e-mail: zfang@lsuhsc.edu

Y. Zhao, D.-G. Chen (eds.), *New Frontiers of Biostatistics and Bioinformatics*, ICSA Book Series in Statistics, https://doi.org/10.1007/978-3-319-99389-8_16

(WGS) data or marker-gene survey data. They are together termed as metagenomic sequence data, or metagenomic count data in this chapter. The obtained sequence reads can be aligned against a database for taxonomic analysis (e.g., RDP database (Cole et al. 2013)) or functional analysis (e.g., COGs (Tatusov et al. 2003), eggNOGs (Powell et al. 2014) databases). The number of reads aligned to a feature, either a taxonomic unit or a functional family, indicates the abundance level of the feature in a sample. It is often of primary interest to identify the features of which the abundance levels differ between conditions, for example, to find the microbial species more abundantly appeared in a diseased human gut than in a healthy gut (Shreiner et al. 2015). This comparative study is named differential abundance analysis. However, due to the fact that the total amount of DNA undergone sequencing, conventionally referred as to library size, may differ substantially as observed, normalization of library size is inescapable before the differential abundance analysis is performed. Otherwise, a differentially abundant feature may be claimed because of uneven library sizes instead of the difference in the abundance of study interest.

Various normalization methods have also been developed for RNA-Seq data analysis (Dillies et al. 2013). As both metagenomic sequence data and RNA-Seq data share a common structure: the count of reads aligned to a feature (e.g., a gene for RNA-Seq data), there have been suggestions proposed to treat metagenomic sequence data as another variant of RNA-Seq data and simply apply the existing normalization methods for RNA-Seq data to metagenomics data analysis (Fernandes et al. 2014; Anders et al. 2013). Towards differential abundance analysis, McMurdie and Holmes (2014) classified the existing normalization methods widely used for metagenomic count data into three groups: (1) Model/None, in which a parametric model is employed to normalize the data or no normalization is applied in some cases, includes the Upper Quartile (UQ) (Bullard et al. 2010), Relative Log Expression (RLE) (Anders and Huber 2010), Trimmed Mean of M-value (TMM) (Robinson and Oshlack 2010), and Cumulative Sum Scaling (CSS) (Paulson et al. 2013); (2) Rarefied (McMurdie and Holmes 2013), in which samples with library size being less than a specified value will be discarded and the remaining samples will be subsampled such that all library sizes equal to the specified value (detailed later); (3) Proportion, in which raw counts are divided by total library size, is named as Total Sum Scaling (TSS) in this chapter. The UQ normalization shares the same spirit with CSS method, so we do not evaluate UQ method. The basic conclusion McMurdie and Holmes drew from their study is that "both proportions and rarefied counts result in a high rate of false positives in tests for species that are differentially abundant across sample classes" and they suggest that it is fine to use the normalization methods from Model/None group, "In particular, an analysis that models counts with the Negative Binomial—as implemented in DESeq2 or in edgeR with RLE normalization—was able to accurately and specifically detect differential abundance over the full range of effect sizes, replicate numbers, and library sizes that we simulated" (McMurdie and Holmes 2014).

There is increasing evidence that many metagenomic count data may be regarded as samples from the microbial ecosystems, and the count of reads to a feature indicates the relative abundance (i.e., compositional proportion) of the feature in the ecosystem (Tsilimigras and Fodor 2016; Gloor et al. 2016). Mandal et al. (2015) provided an excellent example explaining the difference between the comparison of abundance across specimens, and that across microbial ecosystems. We summarize that the former is about absolute abundance, while the latter is about relative abundance. Weiss et al. (2017) explicitly pointed out that the metagenomic data from 16S rDNA amplicon sequencing possess the compositional data characteristics, and studied six normalization methods combined with different test approaches for differential abundance analysis. The simulation studies conducted in their paper utilized Multinomial, Dirichlet-multinomial, and Gamma-Poisson distributions. However, as indicated in the same paper, both Multinomial and Dirichlet-multinomial distributions may not be appropriate for metagenomic compositional data as these distributions imply a negative correlation between any pair of the features, while Gamma-Poisson distribution does not impose the simplex (i.e., the relative abundances sum to 1). Adequate simulation criteria are strongly needed for drawing correct conclusions about the performance of normalization methods on metagenomic compositional data.

In this chapter, we adopt a metagenomic dataset to show the ineffectiveness of some normalization methods, list the details of conducting simulation based on the characteristics learned from the dataset, and demonstrate the impact of normalization methods on the differential abundance analysis. We advocate, in order to avoid ineffective normalization, case-by-case simulation should be conducted according to the dataset to be analyzed. We are drawing attention to the research community and calling for normalization methods specially designed for metagenomics compositional data.

16.2 Motivating Example

The NIH Human Microbiome Project (HMP) (https://hmpdacc.org/hmp/ (Peterson et al. 2009)) provides the 16S rDNA sequencing output and the processed datasets, collected from different sites of healthy human bodies. We downloaded the saliva and stool sample data (170 saliva samples vs. 191 stool samples) from http://www.hmpdacc.org/HMQCP/ (last visited on February 28, 2018). The sequencing reads were processed by the bioinformatics tool Quantitative Insights Into Microbial Ecology (QIIME, (Caporaso et al. 2010)). For each taxonomic unit, the coefficients of variation (CV: the ratio of the sample standard deviation over the sample mean) of the counts can be calculated for the saliva and the stool samples, respectively. As the CV is an indication of the level of standardized variation between the samples for a feature, it is expected that after appropriate normalization the CV values from all the features under the same condition will decrease in general since the variation due to unequal library sizes should have been reduced. A subsampled dataset is obtained using the steps: randomly selecting the same number (i.e., 361) of samples from the HMP saliva and stool dataset with replacement, and then removing the

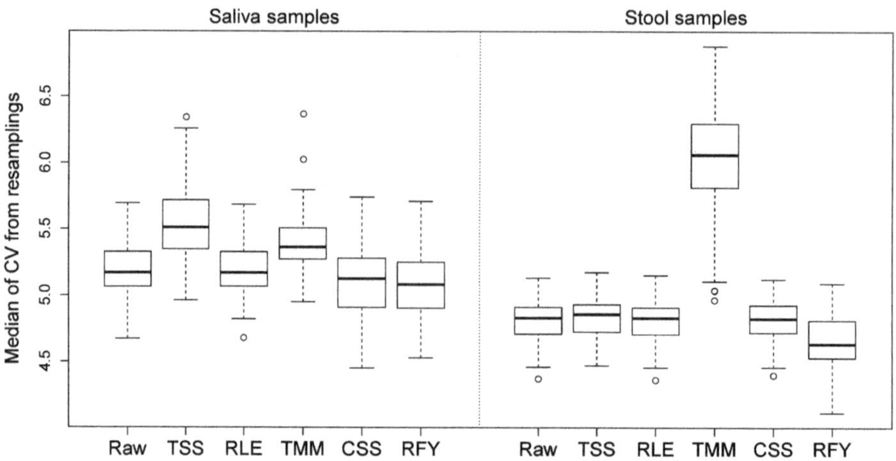

Fig. 16.1 Boxplots of the median values of Coefficients of Variation of the counts in the non-normalized subsampled datasets (Raw), and the normalized subsampled datasets by five different methods from the HMP saliva and stool dataset

duplicated ones. The resampling process repeated one hundred times. Figure 16.1 shows the boxplots of the median CV values of the non-normalized subsampled datasets (Raw), and the normalized subsampled datasets by five different methods. We can see instead of reducing the CV, the TMM normalization has noticeably increased CV values in both saliva and stool samples. This may imply that the TMM normalization is ineffective for the data intended for differential abundance analysis between saliva and stool microbiota. The TSS normalization results in higher CV values for the datasets subsampled from the saliva samples as well. However, it is worth noting that reduced CV itself does not sufficiently mean a good normalization because overreducing sample variation could lead to additional false positives. That is, we cannot conclude that RFY is superior than the other normalizations for this dataset either. This CV analysis on the HMP saliva and stool dataset shows a striking example, which motivated us to investigate how the existing normalization methods perform with metagenomic compositional datasets.

16.3 Data Notation and Methods

A metagenomic dataset can be organized as shown in Table 16.1. A column contains the sequence counts for all the features in a sample; a row lists the counts for a feature across all the samples. For example, y_{ij} denotes the count for feature i from sample j.

With these notations, the steps and the formula of the normalization methods studied in this chapter are briefly introduced as follows.

Table 16.1 Format of a metagenomic dataset of two conditions

	Condition 1			Condition 2		
	Sample 1	...	Sample n_1	Sample $n_1 + 1$	...	Sample $n_1 + n_2$
Feature 1	y_{11}	...	y_{1n_1}	y_{1,n_1+1}	...	y_{1,n_1+n_2}
Feature 2	y_{21}	...	y_{2n_1}	y_{2,n_1+1}	...	y_{2,n_1+n_2}
⋮	⋮	⋮	⋮	⋮	⋮	⋮
Feature m	y_{m1}	...	y_{mn_1}	y_{m,n_1+1}	...	y_{m,n_1+n_2}

TSS (White et al. 2009): The total sum of the counts in a sample serves as the estimate of the library size of the sample. A TSS normalized count is calculated as

$$\tilde{y}_{ij}^{TSS} = \frac{y_{ij}}{\sum_i y_{ij}} N^{TSS},$$

where N^{TSS} is an appropriately chosen normalization constant.

RLE (Anders and Huber 2010): The geometric mean of the counts to a feature from all the samples is first calculated. The ratio of a raw count over the geometric mean to the same feature is then computed. The scale factor of a sample is obtained as the median of the ratios for the sample. A RLE normalized count can be calculated as

$$\tilde{y}_{ij}^{RLE} = y_{ij}/\text{median}_i \left\{ \frac{y_{ij}}{\left(\prod_j y_{ij}\right)^{\frac{1}{n_1+n_2}}} \right\}.$$

TMM (Robinson and Oshlack 2010): The ratio of two observed relative abundances for a feature in two samples is considered to be an estimate of the scale factor between the two samples. The log2 of the ratio is named M value; and the log2 of the geometric mean of the observed relative abundances is called A value. This name convention follows the M and A values given originally in the M-A plot (Yang et al. 2002). That is, for feature i from samples j, l,

$$M_{i(jl)} = \log_2 \frac{y_{ij}/\sum_i y_{ij}}{y_{il}/\sum_i y_{il}}; \quad A_{i(jl)} = \frac{1}{2} \log_2 \left(\frac{y_{ij}}{\sum_i y_{ij}} \frac{y_{il}}{\sum_i y_{il}} \right).$$

The features with specified upper or lower percent of M (default 30%) or A (default 5%) values are trimmed out. The weighted sum of the M values can be used to derive the scale factor,

$$\log_2 \left(SF_{jl}^{TMM} \right) = \frac{\sum_{i \in m_{jl}^{TMM}} \left(w_{i(jl)} M_{i(jl)} \right)}{\sum_{i \in m_{jl}^{TMM}} \left(w_{i(jl)} \right)},$$

where SF_{jl}^{TMM} denotes the scale factor of sample j relative to sample l by TMM method, and m_{jl}^{TMM} denotes the remaining features after the trimming step for the two samples. The weight $w_{i(jl)}$ is computed by,

$$w_{i(jl)} = \frac{\sum_i y_{ij} - y_{ij}}{y_{ij} \sum_i y_{ij}} + \frac{\sum_i y_{il} - y_{il}}{y_{il} \sum_i y_{il}}.$$

After appropriate steps, a TMM normalized count can also be expressed as the quotient of y_{ij} and some attainable value.

CSS (Paulson et al. 2013): For a sample, CSS is defined as the sum of counts that are less than or equal to a percentile, determined by the data. This cumulative sum excludes the raw counts from features that are preferentially amplified, and thus is considered to be relatively invariant across the samples. Using this sum as the scale factor, a CSS normalized count can be calculated as

$$\tilde{y}_{ij}^{CSS} = \frac{y_{ij}}{\sum_{i \in m^{CSS}} (y_{ij})} N^{CSS},$$

where N^{CSS} is an appropriately chosen normalization constant, and m^{CSS} denotes the features included in the cumulative summation for the sample.

RFY (McMurdie and Holmes 2013): Rarefying normalization starts with selection of a library size, N^{RFY}. Then any sample, with library size less than N^{RFY}, is considered defective and discarded. For any remaining sample, the features are resampled using their counts as sampling weights. The resampled dataset, or the normalized samples, share the same library size. In this chapter, we use the same criterion as that in McMurdie and Holmes (2014) to set the 15th percentile of total sums of the counts of raw samples as the N^{RFY}. Note that, RFY does not provide an estimate of scale factor of a sample as other normalizations do. In this sense, TSS, RLE, TMM, and CSS are called scaling normalizations, but RFY is not.

16.4 Simulation Study

16.4.1 Parameters and Data Characteristics

Mandal et al. (2015) has made a remarkable comment for metagenomic compositional data analysis: "It is critical to understand what the observed data represent and what statistical parameters are being tested." As discussed in Introduction, in our opinion, the answer to the comment is: metagenomic compositional data should be deemed as samples from the microbial ecosystems, and the read counts to the features should be used as the indication of the relative abundances (i.e., compositional proportions) of the features in the ecosystems. For a statistical test, the relative abundance is the underlying parameter to be compared between conditions. The relative abundance of feature i for condition k is denoted by $p_i^{(k)}$, subject to the simplex, i.e., $\sum_{i=1}^{m} p_i^{(k)} = 1$.

Through more than a decade of metagenomics research, it has been recognized that metagenomic data possess at least three outstanding characteristics: (1) a great proportion of the features have a sparse count, meaning that the data contain an inflated proportion of zero counts (Paulson et al. 2013; Sohn et al. 2015); (2) the data suffer from the under-sampling issue, that is, more features are found from sample with larger library size, in other words, zero counts could also be associated to library size (Srinivas et al. 2013); (3) the counts are usually overdispersed (McMurdie and Holmes 2014).

16.4.2 Data Simulation

Data simulation encompasses two consecutive steps: learning of real dataset on the characteristics outlined above, and statistical simulation using the parameters learned. To emphasize, both the learning of real dataset and statistical simulation are carried out for each condition separately.

Learning of real dataset. The expectation of y_{ij} is expressed as $\mu_{ij} = \mu_j p_i^{(k)}$, where μ_j is the expectation of the sum of the counts in sample j and is named sample scale here. An estimate of μ_{ij} can be obtained by,

$$\widehat{\mu}_{ij} = \widehat{\mu}_j \cdot \widehat{p}_i^{(k)} = \sum_i y_{ij} \cdot \frac{\sum_{j\in(k)} y_{ij}}{\sum_{i,j\in(k)} y_{ij}},$$

where $j \in (k)$ represents the samples from condition k only. Note that, as an estimate of count, $\widehat{\mu}_{ij}$ is rounded to the nearest integer.

The observed counts, with the same estimated expectation, of all the samples under the same condition, are put together to fit a Negative Binomial (NB) distribution. There is a fitted size parameter of NB distribution from each of the grouped raw counts. This size parameter indicates the level of overdispersion of the counts, which is detailed in Appendix. We will use the average of the fitted size values for the simulation.

After the NB fitting, for the group of observed counts that share the same estimated expectation, the probability of zero can be calculated using the fitted NB distribution. If the observed proportion of zeros is greater than this probability, their difference is recorded as the estimated probability of inflated zero counts for that expectation.

The samples (or columns in Table 16.1) under the same condition are sorted according to the values of $\widehat{\mu}_j$ (i.e., $\sum_i y_{ij}$) from the least to the greatest. Then, for a feature (or a row in Table 16.1), the cumulative sums of the counts from sample 1 to another sample are calculated, i.e., $\sum_{j=1}^{J} y_{ij}$, $J = 1, \ldots n_c$, where n_c is the sample size under that condition. Thus, for a feature, we use the maximum of the $\widehat{\mu}_j$'s, over the samples (or columns) with the cumulative sums ≤ 3, to estimate the boundary library size of the under-sampling.

Simulation steps. Simulation is carried out for each of the conditions separately as well. First, the $\widehat{\mu}_j$'s from the real dataset are used to build an empirical distribution from which random numbers can be generated and serve as the sample scales (μ_j^{sim}'s) for the simulation. Second, the expectation of count is obtained following $\mu_{ij}^{sim} = \mu_j^{sim} \cdot \widehat{p}_i^{(k)}$. The simulated count ($y_{ij}^{sim}$) is randomly selected from either a zero point, or a random number from the NB distribution with the learned parameter values. Third, in simulated sample *j*, if the estimated boundary size of under-sampling for a feature is greater than $\sum_i y_{ij}^{sim}$, the corresponding count is replaced by zero. R codes for learning of a real dataset and subsequently data simulation are available at a Github webpage https://github.com/rdu2017/Normalization-Evaluation.

16.4.3 Normalization Performance

The purpose of normalization is to adjust all the samples to the same scale for differential abundance analysis. Although it is conventional to say that normalization is for library size, it is essentially the sample scale that needs to be normalized. After normalization, the counts for a feature in different samples under the same condition are assumed to have the same expectation. The expectations are compared between conditions to draw the conclusion for the analysis. Thereupon, the sample scale, the sum of expectations of counts in the sample, needs to be normalized among all the samples. In turn, the relative abundance is compared.

Using the HMP saliva and stool sample data as template, we generated 100 simulated datasets. The four methods (TSS, RLE, TMM, and CSS) were applied for estimation of the sample scales in the normalization. Since the RFY approach does not perform normalization through estimating sample scale, it is not included here. The Pearson correlation coefficient between the estimated sample scales and the true values is calculated to show how well a normalization works. The estimate is better when the coefficient is closer to one. Figure 16.2 displays the boxplots of the coefficients from the 100 simulated datasets. Among these four methods, TMM appears uncompetitive. Both TSS and CSS perform better than RLE, while the median of TSS (0.625) is slightly higher than that of CSS (0.61) but with two times larger standard deviation (0.08 vs. 0.04).

16.4.4 Impact of Normalization on Differential Abundance Analysis

To be able to set true/false differentially abundant features explicitly, we take only the simulated data from one condition, i.e., the stool metagenome. A simulated

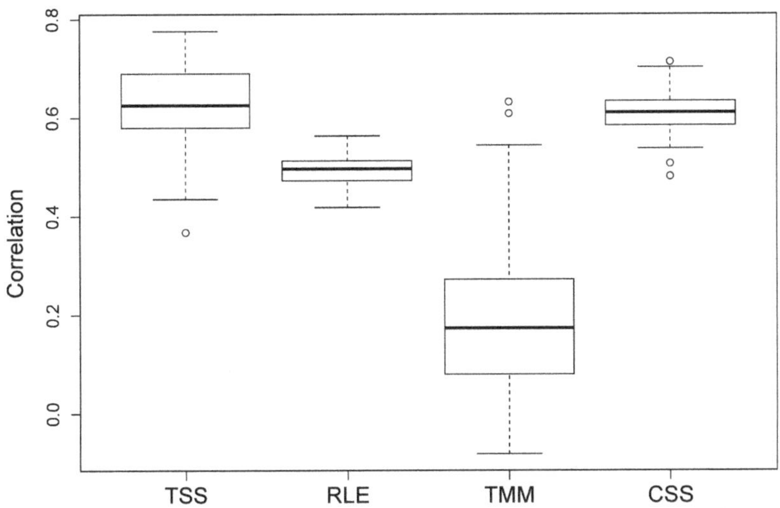

Fig. 16.2 Boxplots of Pearson correlation coefficients of sample scales estimated after different normalization methods and the true values, using the 100 simulated datasets

dataset, containing 191 samples, is randomly partitioned into two smaller datasets with 96 and 95 samples in each. Meanwhile, we intend to keep the compositional characteristics of the data. The quartiles of $p_i^{(k)}$'s are calculated. In the dataset that contains 96 samples, the features (i.e., rows) from the third and fourth quartiles are randomly swapped with the features from the first and second quartiles. By so doing, the two partitioned datasets share 50% true and 50% false differentially abundant features with the compositional structure still maintained. A two-sided T test is first performed to compare the normalized counts for each feature, and followed by the Benjamini-Hochberg procedure (Benjamini and Hochberg 1995) for false discovery rate controlling at 0.05 among all the tests. Figure 16.3a shows the boxplots of the observed false discovery rates (FDR) in the analysis output after a normalization procedure. It is noticeable that TMM normalization has much higher FDR, with 46% tests showing FDR greater than 0.05. RFY normalization performs the second worst regarding FDR controlling, with 12% tests having FDR greater than 0.05. As indicated in Fig. 16.3b, the true positive rates (TPR) associated with TMM and non-normalization are lower than TPR associated with the other normalizations. It is clear that an ineffective normalization (TMM here) will discourage the differential analysis in error rate controlling or statistical power, or both.

Our focus is to examine how a normalization impacts subsequent differential analysis. However, it should be pointed out that differential abundance analysis itself is influenced by both normalization and the statistical approach used for analysis. Figure 16.4 shows FDR and TPR on the same shuffled datasets above but analyzed using NB regression approach. It seems NB approach has better TPR rate but worse FDR controlling compared to T test. Nonetheless, TMM still shows ineffectiveness in the NB approach.

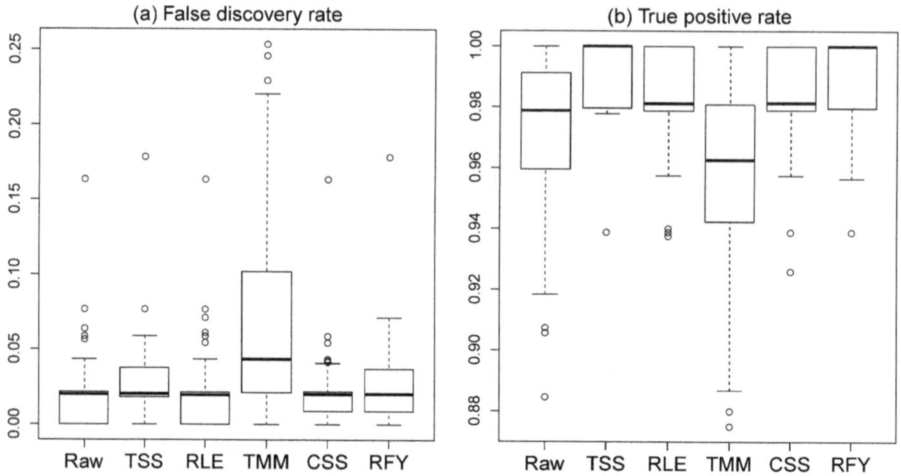

Fig. 16.3 Impact of normalization on differential abundance analysis in both FDR and TPR, by two-sided T test with 100 datasets shuffled from stool metagenome dataset. (**a**) False discovery rate. (**b**) True positive rate

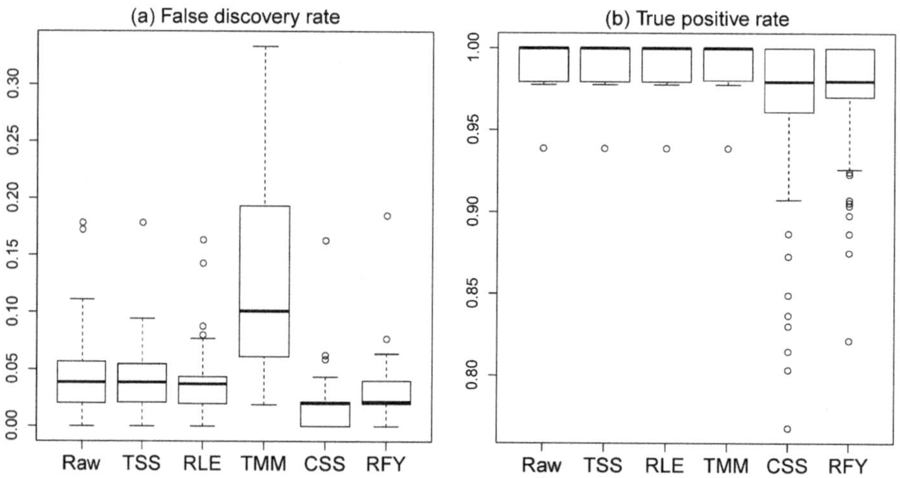

Fig. 16.4 Impact of normalization on differential abundance analysis in both FDR and TPR, by Negative Binomial regression with 100 datasets shuffled from stool metagenome dataset. (**a**) False discovery rate. (**b**) True positive rate

16.5 Discussion

16.5.1 TMM and RLE with Metagenomic Compositional Dataset

For gene expression studies, there is a widely used assumption that the majority of genes do not express differentially between conditions. Many of RNA-Seq normalization methods were developed based on this assumption, including TMM and RLE. The "non-differential" in the assumption is implemented as non-differential absolute abundance after normalization. Subsequent differential analysis is also to compare the normalized counts between conditions, instead of comparing the relative abundances as it is for compositional data. In Appendix, we use hypothetical datasets to explain why TMM and RLE normalizations may not work well with metagenomic compositional dataset. We would then like to suggest using RNA-Seq normalization with caution for metagenomic compositional data analysis.

16.5.2 Simulation Benchmark

Metagenomic studies have been frustrated by lack of good simulation benchmarks (Johnson et al. 2014). Meanwhile, contrary conclusions have been seen from the simulation studies conducted with different criteria (McMurdie and Holmes 2014; Weiss et al. 2017; Costea et al. 2014; Paulson et al. 2014). In our vision, the practice needs improvement from at least two aspects. First, the idea that a simulation study should be designed to apply for overall situations may not be realistic. Instead, a case-by-case simulation practice should be encouraged, based on the real dataset to analyze. Second, in terms of metagenomic compositional data, all the important data characteristics should be included when designing a simulation. Using a convenient statistical distribution is not a good strategy because it may not be capable to reflect the complex in a real dataset.

We suggest that a simulation be carried out for each condition independently for metagenomic compositional data. The distribution of library size, the relative abundance, the overdispersion parameter, the probability of zero count from a zero mass state, and the boundary library size in terms of under-sampling are learned from a real metagenomic dataset. Hopefully, the simulation approach we provide in this chapter can serve as a good basis for building up simulation benchmarks in the research community of metagenomic data analysis.

16.5.3 Novel Normalization Methods Are Needed

As observed, TMM method should be avoided for analysis of the HMP saliva and stool dataset. In Appendix, we also provide a figure showing that the RLE normalization does not work well for the mouse stool metagenomic dataset, which has been used as the benchmark dataset in the chapter where CSS was introduced (Paulson et al. 2013). From our experience, no matter with real or simulated data, in most situations the CSS does not identify the data-driven percentile, up to which the raw counts will be summed, and then the default value 50th percentile is used. It is questionable to us whether there commonly exists a claimed percentile so that the raw counts are distributed differently lower or greater than it (see Supplementary Figure 1 in Paulson et al. 2013). In addition, there is no specific consideration of the compositional characteristics in the development of CSS. Conceptually, TSS may be fine for compositional data normalization as it uses a count divided by the total sum of the counts of a sample, as an estimate of the relative abundance. However, as many previous studies have shown, TSS is unreliable against the overdispersed counts, under-sampling issue, and aberrant counts in many situations. In a word, novel normalizations, specifically designed for metagenomic compositional data, are highly in demand. Developing novel normalization methods is our future research topics.

Acknowledgements The authors are grateful to two anonymous reviewers for their careful reading of the manuscript and their comments and suggestions. ZF's research is supported by grant U54 GM104940 from the National Institute of General Medical Sciences of the National Institutes of Health, which funds the Louisiana Clinical and Translational Science Center of Pennington Biomedical Research Center. LA's research is partially supported by National Science Foundation [DMS-1222592] and United States Department of Agriculture [Hatch project, ARZT-1360830-H22-138]. RD's research was supported in part by the UNM Comprehensive Cancer Center, a recipient of NCI Cancer Support Grant 2 P30 CA118100-11 (PI: Cheryl L. Willman, MD).

A.1 Appendix

A.1.1 Supplementary Data Distribution

Negative Binomial Distribution. A NB distribution is defined as,

$$P\left(X=x\right)=\frac{\Gamma\left(x+r\right)}{\Gamma\left(\mathrm{r}\right)\mathrm{x}!}p^{r}\left(1-p\right)^{x},$$

where r and p are two parameters, and r is called size parameter. The mean of the NB distribution is,

$$\mu=\frac{r\left(1-p\right)}{p},$$

and the variance is,

$$V=\frac{r\left(1-p\right)}{p^{2}}=\mu+\frac{1}{r}\mu^{2}.$$

Thus, r indicates the level of overdispersion in the counts.

Raw counts

	Condition 1		Condition 2	
Feature 1	10	10	200	200
Feature 2	10	10	80	80
Feature 3	10	10	80	80
Feature 4	10	10	10	10
Feature 5	10	10	10	10
Feature 6	100	100	10	10
Total	150	150	390	390

→

Relative abundance ratio compared to 1st sample

	Condition 1		Condition 2	
Feature 1	1	1	7.69	7.69
Feature 2	1	1	3.08	3.08
Feature 3	1	1	3.08	3.08
Feature 4	1	1	0.38	0.38
Feature 5	1	1	0.38	0.38
Feature 6	1	1	0.038	0.038

→

Scale factor relative to 1st sample

Condition 1		Condition 2	
1	1	1.73	1.73

Fig. 16.5 TMM normalization on a hypothetical dataset

A.1.2 Supplementary Illustration of TMM and RLE with Compositional Dataset

For gene expression studies, there is a widely used assumption that the majority of genes do not express differentially between conditions. Many of RNA-Seq normalization methods were developed based on this assumption, including TMM and RLE. The "non-differential" in the assumption is implemented as non-differential absolute abundance after normalization. Subsequent differential analysis is also to compare the normalized counts between conditions, instead of comparing the relative abundances as it is for compositional data.

Focusing on the essence of a normalization procedure, the hypothetical datasets are made of the expectations of counts. For TMM approach, the logarithm function and the weighted sum are not applied since those are designed for reducing the effect of count variation. In Fig. 16.5, the relative abundance ratio, compared to the first sample, is first calculated from the raw counts, i.e., $\frac{y_{ij}/\sum_i y_{ij}}{y_{i1}/\sum_i y_{i1}}$. The trimmed mean of the ratios for each sample, after trimming the largest and smallest values, is used as the scale factor. The true scale factor is 2.6 (390/150), but the output from TMM is 1.73. Figure 16.5 shows a very likely situation for metagenomic compositional data, in which the relative abundances vary largely between conditions. TMM may not work well for such data since it merely relies on the assumption that after normalization most of features should share the same absolute abundance.

For RLE normalization, the geometric mean of the counts to each feature from all the samples is first calculated, see Fig. 16.6. Next, the ratio of a raw count over the mean count for the same feature is computed. The scale factor for a sample is obtained as the median of the ratios for the sample. For this hypothetic dataset, RLE approach does not suggest any normalization adjustment since all the scale factors equal to 1; however, the true library sizes are very different (e.g., 210 vs. 310).

Raw counts

	Condition 1		Condition 2	
Feature 1	10	10	200	200
Feature 2	20	20	20	20
Feature 3	20	20	20	20
Feature 4	30	30	30	30
Feature 5	30	30	30	30
Feature 6	100	100	10	10
Total	210	210	310	310

→

Average count

	Average
Feature 1	44.7
Feature 2	20
Feature 3	20
Feature 4	30
Feature 5	30
Feature 6	31.6

→

Ratio: raw count / average count

	Condition 1		Condition 2	
Feature 1	0.22	0.22	4.47	4.47
Feature 2	1	1	1	1
Feature 3	1	1	1	1
Feature 4	1	1	1	1
Feature 5	1	1	1	1
Feature 6	3.16	3.16	0.32	0.32

→

Scale factor of each sample

Condition 1		Condition 2	
1	1	1	1

Fig. 16.6 RLE normalization steps on a hypothetical dataset

In Fig. 16.6, it is clear that the scale factor is determined by the absolute count of Feature 2, 3, 4, or 5, instead of the relative abundance of one of those features. A subsequently comparative analysis would reach the conclusion that there is no differential abundance for Feature 2, 3, 4, or 5 between the conditions. However, the relative abundances of the features have altered, for Feature 4 it is 14% and 10% under the two conditions, respectively.

A.1.3 Supplementary Example

Mouse stool metagenomic data. Fresh or frozen adult human fecal microbial communities were transplanted into guts of germ-free C57BL/6J mice. Here, germ-free environment is referred to as mice gut that does not previously expose to microbes. Following the transplanting, 12 recipient mice were fed with a standard low-fat, plant polysaccharide-rich diet for 4 weeks; after that, six mice were switched to take high-fat/high-sugar Western diet for another 6 weeks. Amplification and pyrosequencing of V2 region of 16S rRNA genes were performed periodically to record the changes of microbial community structure of fecal samples of the mice (Turnbaugh et al. 2009). There are 85 samples under condition one (associated to low-fat diet fed mice), and 54 samples under condition two (associated to Western diet fed mice). The bioinformatic tool RDP (Wang et al. 2007) was used to generate the count data, which is featured at species level.Together, there are 52 genera

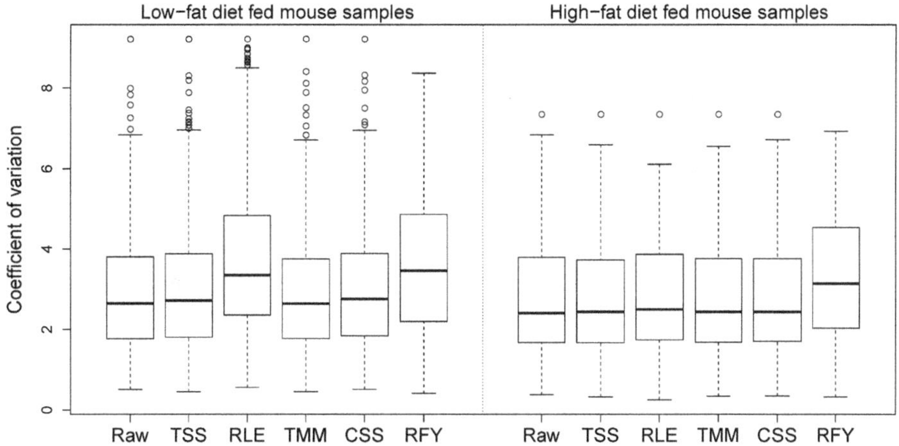

Fig. 16.7 Boxplots of coefficients of variation of counts in the raw data, and the normalized data for the mouse stool metagenomic data

shown under both conditions, and the data is considered to represent low complex metagenomic data. Figure 16.7 demonstrates that RLE and RFY should not be recommended for normalization of the metagenomic data.

References

Anders, S., & Huber, W. (2010). Differential expression analysis for sequence count data. *Genome Biology, 11*(10), R106.

Anders, S., et al. (2013). Count-based differential expression analysis of RNA sequencing data using R and Bioconductor. *Nature Protocols, 8*(9), 1765–1786.

Benjamini, Y., & Hochberg, Y. (1995). Controlling the false discovery rate: A practical and powerful approach to multiple testing. *Journal of the Royal Statistical Society. Series B (Methodological)*, 289–300.

Bragg, L., & Tyson, G. W. (2014). Metagenomics using next-generation sequencing. *Environmental Microbiology: Methods and Protocols, 1096*, 183–201.

Bullard, J. H., et al. (2010). Evaluation of statistical methods for normalization and differential expression in mRNA-Seq experiments. *BMC Bioinformatics, 11*(1), 94.

Caporaso, J. G., et al. (2010). QIIME allows analysis of high-throughput community sequencing data. *Nature Methods, 7*(5), 335–336.

Cole, J. R., et al. (2013). Ribosomal Database Project: Data and tools for high throughput rRNA analysis. *Nucleic Acids Research, 42*(D1), D633–D642.

Costea, P. I., et al. (2014). A fair comparison. *Nature Methods, 11*(4), 359.

Dillies, M.-A., et al. (2013). A comprehensive evaluation of normalization methods for Illumina high-throughput RNA sequencing data analysis. *Briefings in Bioinformatics, 14*(6), 671–683.

Fernandes, A. D., et al. (2014). Unifying the analysis of high-throughput sequencing datasets: Characterizing RNA-seq, 16S rRNA gene sequencing and selective growth experiments by compositional data analysis. *Microbiome, 2*(1), 15.

Gloor, G. B., et al. (2016). It's all relative: Analyzing microbiome data as compositions. *Annals of Epidemiology, 26*(5), 322–329.

Handelsman, J. (2004). Metagenomics: Application of genomics to uncultured microorganisms. *Microbiology and Molecular Biology Reviews, 68*(4), 669–685.

Johnson, S., et al. (2014). A better sequence-read simulator program for metagenomics. *BMC Bioinformatics, 15*(9), S14.

Mandal, S., et al. (2015). Analysis of composition of microbiomes: A novel method for studying microbial composition. *Microbial Ecology in Health and Disease, 26*(1), 27663.

McMurdie, P. J., & Holmes, S. (2013). phyloseq: an R package for reproducible interactive analysis and graphics of microbiome census data. *PloS One, 8*(4), e61217.

McMurdie, P. J., & Holmes, S. (2014). Waste not, want not: Why rarefying microbiome data is inadmissible. *PLoS Computational Biology, 10*(4), e1003531.

Metzker, M. L. (2010). Sequencing technologies—The next generation. *Nature Reviews Genetics, 11*(1), 31–46.

National Research Council. (2007). *The new science of metagenomics: Revealing the secrets of our microbial planet*. Washington, DC: National Academies Press.

Paulson, J. N., et al. (2013). Differential abundance analysis for microbial marker-gene surveys. *Nature Methods, 10*(12), 1200–1202.

Paulson, J. N., Bravo, H. C., & Pop, M. (2014). Reply to: "A fair comparison". *Nature methods, 11*(4), 359–360.

Peterson, J., et al. (2009). The NIH human microbiome project. *Genome Research, 19*(12), 2317–2323.

Powell, S., et al. (2014). eggNOG v4. 0: Nested orthology inference across 3686 organisms. *Nucleic Acids Research, 42*(D1), D231–D239.

Robinson, M. D., & Oshlack, A. (2010). A scaling normalization method for differential expression analysis of RNA-seq data. *Genome Biology, 11*(3), R25.

Shreiner, A. B., Kao, J. Y., & Young, V. B. (2015). The gut microbiome in health and in disease. *Current Opinion in Gastroenterology, 31*(1), 69.

Sohn, M. B., Du, R., & An, L. (2015). A robust approach for identifying differentially abundant features in metagenomic samples. *Bioinformatics, 31*(14), 2269–2275.

Srinivas, G., et al. (2013). Genome-wide mapping of gene–microbiota interactions in susceptibility to autoimmune skin blistering. *Nature Communications, 4*, 2462.

Tatusov, R. L., et al. (2003). The COG database: An updated version includes eukaryotes. *BMC Bioinformatics, 4*(1), 1.

Tsilimigras, M. C., & Fodor, A. A. (2016). Compositional data analysis of the microbiome: Fundamentals, tools, and challenges. *Annals of Epidemiology, 26*(5), 330–335.

Turnbaugh, P. J., et al. (2009). The effect of diet on the human gut microbiome: A metagenomic analysis in humanized gnotobiotic mice. *Science Translational Medicine, 1*(6), 6ra14.

Wang, Q., et al. (2007). Naive Bayesian classifier for rapid assignment of rRNA sequences into the new bacterial taxonomy. *Applied and Environmental Microbiology, 73*(16), 5261–5267.

Weiss, S., et al. (2017). Normalization and microbial differential abundance strategies depend upon data characteristics. *Microbiome, 5*(1), 27.

White, J. R., Nagarajan, N., & Pop, M. (2009). Statistical methods for detecting differentially abundant features in clinical metagenomic samples. *PLoS Computational Biology, 5*(4), e1000352.

Woese, C. R. (1987). Bacterial evolution. *Microbiological Reviews, 51*(2), 221.

Wooley, J. C., Godzik, A., & Friedberg, I. (2010). A primer on metagenomics. *PLoS Computational Biology, 6*(2), e1000667.

Yang, Y. H., et al. (2002). Normalization for cDNA microarray data: A robust composite method addressing single and multiple slide systematic variation. *Nucleic Acids Research, 30*(4), e15.

Chapter 17
Identification of Pathway-Modulating Genes Using the Biomedical Literature Mining

Zhenning Yu, Jin Hyun Nam, Daniel Couch, Andrew Lawson, and Dongjun Chung

17.1 Introduction

In system biology, each gene is not considered as an independent player but instead they are studied in the context of a complex network among them. As a result, a biological pathway is considered as the *de facto* functional unit and hence, the accurate and effective identification of novel pathways is of great interest. Here, a biological pathway is often defined as a set of genes that share and constitute a common biological function. Various experimental approaches such as RNA-seq and ChIP-seq are often employed to study functions of genes and pathways. However, they are still limited in the sense that each of these experiments focuses only on one aspect of biology while two genes can be related through various biological functions and in multiple layers. Biomedical literature, especially those available in the *PubMed* database (https://www.ncbi.nlm.nih.gov/pubmed/), is considered as a valuable resource to overcome this limitation because relationship among genes is comprehensively characterized in the biomedical literature. However, effective utilization of biomedical literature to study relationship among genes still remains challenging because most abstracts have information for only a single gene and as a result, it is not straightforward to infer the relationship among genes from the biomedical literature (Qin et al. 2014).

In order to address this issue, we recently developed a framework of literature mining and its Bayesian analysis focusing on indirect relationship among genes mediated by gene ontology (GO) terms (Chung et al. 2017). This framework does not suffer from the fact that most abstracts have information for only a single

Z. Yu (✉) · J. H. Nam · D. Couch · A. Lawson · D. Chung
Department of Public Health Sciences, Medical University of South Carolina, Charleston, SC, USA
e-mail: yuz@musc.edu; namj@musc.edu; couchd@musc.edu; lawsonab@musc.edu; chungd@musc.edu

Y. Zhao, D.-G. Chen (eds.), *New Frontiers of Biostatistics and Bioinformatics*, ICSA Book Series in Statistics, https://doi.org/10.1007/978-3-319-99389-8_17

gene because two genes in two different abstracts can still be linked through GO terms shared between these two abstracts. Moreover, this approach also facilitates easy interpretation of novel findings because it identifies GO terms relevant to each gene without any additional downstream analysis, which will significantly reduce burden of biologists. In this approach, we first implemented a text mining of PubMed literature to recognize gene and GO term entities. Then, we quantified the association between a gene and a GO term using a hypergeometric test, which also takes into account how much each gene and each biological function has been studied in the literature. Finally, after assembling the hypergeometric test p-values as a matrix of GO terms $\times$ genes, a novel Bayesian bi-clustering approach, namely *bayesGO*, was applied to simultaneously identify gene clusters and GO term clusters and to figure out the association between each gene and each GO term. This Bayesian model also allows computationally efficient posterior inference based on the Metropolis-Hastings within Gibbs sampler and the poor man's reversible jump Markov chain Monte Carlo approaches. We validated this approach using the experimental validation data and an application to studies of pathway-modulating genes in yeast. In order to further facilitate easy application of this approach, we developed a web interface `"GAIL"` for the PubMed literature mining and an `R` package `"bayesGO"` for identifying pathways and facilitating their interpretation. In this chapter, we provide a step-by-step guideline showing how to use these software to investigate the relationship among genes with the PubMed literature mining data obtained using human gene entities and GO terms.

17.2 Methods

The overall workflow of the proposed GAIL-bayesGO analysis is shown in Fig. 17.1. Specifically, users can query genes and GO terms of interest using the web interface `GAIL` (Sect. 17.2.2) and investigate relationships among genes, along with associated GO terms, in the PubMed literature (Sect. 17.2.1). Using this web interface, users can download the corresponding hypergeometric p-value matrix indicating association between genes and GO terms in the literature. This matrix can be used as input of the `R` package `bayesGO` (Sect. 17.2.2). By taking this matrix as input, `bayesGO` allows users to identify pathway-modulating genes and GO terms enriched for these gene using a simple interface (Sect. 17.2.3).

17.2.1 Text Mining of Biomedical Literature

We use a text mining approach to find associations between genes and GO terms in PubMed abstracts. The first step involves cooccurrence-based name entity recognitions for genes and GO terms. Cooccurrence-based approaches have been used in many biomedical text mining studies (Frijters et al. 2008; Jenssen et al.

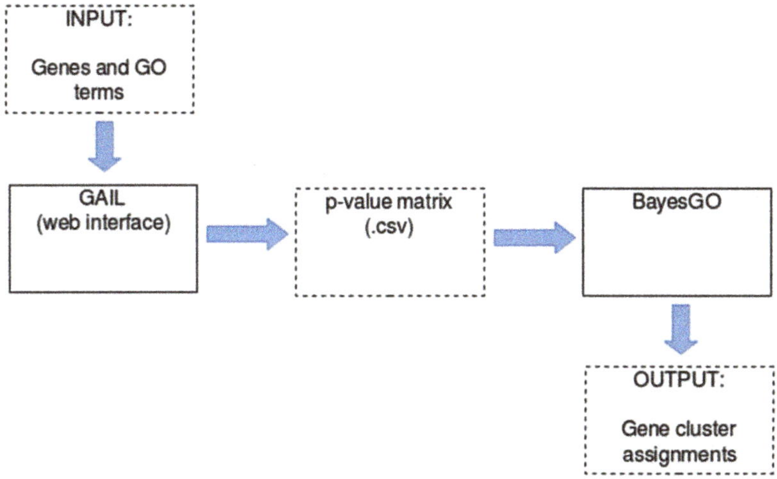

Fig. 17.1 Overall workflow of the GAIL-bayesGO analysis framework

2001). It simply looks for the concepts that appear in the same texts, and in our study, if a gene and a GO term appear in the same abstract text, it assumes a relationship between the gene and the GO term. Notably, the same gene and the GO term can be expressed in various ways by different authors. For example, "*CARTPT*," "*CART*," "*cocaine and amphetamine regulated transcript*" all refer to the same cocaine-related gene. To address the variability issues, we built dictionaries for variants of genes and GO terms through integrating multiple data sources. Specifically, we used genes with approved gene symbol in the *HUGO Gene Nomenclature Committee (HGNC)* (https://www.genenames.org/) as our keys and mapped alternative symbols and names from *Ensembl* (https://www.ensembl.org/index.html) and *GenBank* (https://www.ncbi.nlm.nih.gov/genbank/) to them. As such, gene and GO dictionaries were constructed based on their identifiers, names, and synonyms. Though similar dictionaries were introduced in other studies (Liu et al. 2005; Mitsumori et al. 2005), none of them consider the invalid name entity problems. Invalid entities can occur due to common words that are used in other concepts or can be some short-length words, both of which cause ambiguous meanings. For example, the GO identifier "*GO:0007612*" corresponds to the name "learning" and if we search this term against abstracts, we retrieve many abstracts discussing irrelevant topics such as "machine learning" and "medical learning" rather than this actual GO term. To solve this problem, we used methods proposed by Koike and Takagi (2004). Specifically, we used *WordNet*, a lexical database for general English, to filter out genes and GO entities corresponding to common words. Finally, *NCBI EFetch* was used to search abstracts associated with genes and GO terms by querying the PubMed database and then associations between genes and GO terms were built if they appear together in the same abstract.

17.2.2 Database and Web Interface for Biomedical Literature Mining

Using the *Django* web framework, we developed `GAIL` (Gene-gene Association Inference based on Literature mining), a web interface to facilitate analysis and visualization of the data obtained from the literature mining. We use the graph database *Neo4j* (https://neo4j.com/) as its backend since it is very efficient at storing and retrieving graph-structured data (e.g., gene-GO-abstract relationships). The interface contains a *Query* page (Fig. 17.2) wherein the user inputs a list of genes and GO terms of interest. For the genes, valid forms of input include gene IDs from *Entrez*, *Ensembl*, and *HUGO*, as well as the *HUGO*-approved official gene symbol. GO terms can be inputted either by their GO ID or by their name.

After submitting their query, the user is redirected to an *Association Network* page, providing a graph-based visualization of the results (Fig. 17.3). Currently, the *D3.js* (https://d3js.org/) JavaScript library is being used for rendering the network. The nodes in the graph represent genes and the edges represent the association between genes, determined by hypergeometric p-values obtained via literature mining between genes and GO terms. Clicking on an edge provides additional information on the relationship, including a partial correlation coefficient between the genes and showing the GO terms shared by the two. Along with all of this information, the *Association Network* page allows the user to download the data retrieved by the interface. Among others, users can download a matrix of hypergeometric p-values for genes and GO terms in the comma-separated values (CSV) file format. This file can be taken as input for the *bayesGO* software that will be described in detail in the following sections. The interface and database are currently hosted at http://chunglab.io/GAIL.

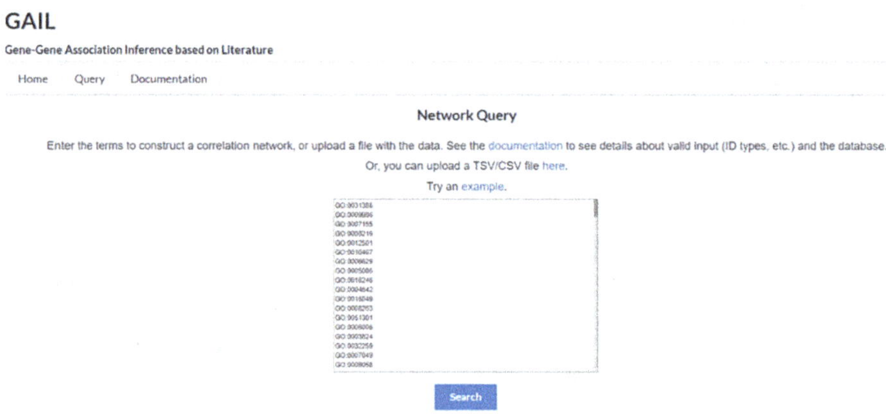

Fig. 17.2 Web interface "GAIL" for biomedical literature mining: the "Query" page

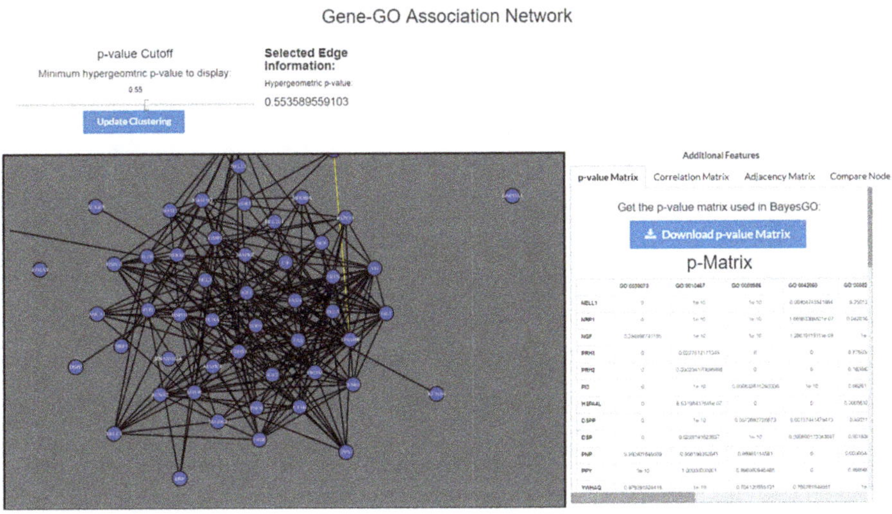

Fig. 17.3 Web interface "GAIL" for biomedical literature mining: the "Association Network" page

17.2.3 bayesGO: Bayesian Hierarchical Model to Identify Pathway-Modulating Genes

Next, we identify pathway-modulating genes and facilitate functional interpretation of these genes using *bayesGO*, the Bayesian hierarchical model proposed in Chung et al. (2017). Here we provide a brief review of bayesGO. In this approach, we first calculate a hypergeometric test p-value for each pair of a gene and a GO term using the literature mining data obtained as described in the previous section. We denote the hypergeometric test p-value for i-th gene and t-th GO term as Y_{ti} for $i = 1, \cdots, G$ and $t = 1, \cdots, T$. Note that this p-value reflects degree of association between a gene and a GO term. Then, we take a probit transformation of these p-values to facilitate data visualization and modeling. We denote the probit-transformed p-value as $Z_{ti} = \Phi^{-1}(Y_{ti})$, where $\Phi(\cdot)$ is the cumulative standard normal distribution function.

Given this dataset, bayesGO identifies gene clusters and GO term clusters as follows. Here, it is assumed that G genes constitute unobserved K gene clusters and T GO terms constitute V unobserved GO term clusters. This assumption is based on the rationale that genes in the same cluster are considered to be related to similar biological functions and that there is also strong correlation among GO terms. Note that here we allow that a gene can belong to only a single gene cluster and similarly, a GO term can belong to only a GO term cluster. We denote the membership of i-th gene to a gene cluster as M_i and the membership of t-th GO term to a GO term cluster as L_t, where $M_i \in \{1, \cdots, K\}$ and $L_t \in \{1, \cdots, V\}$. Finally, in order to reflect the fact that each gene cluster can be described with a set of related GO terms, we introduce a latent binary indicator for the enrichment of t-th GO term for i-th gene, denoted as E_{ti}, where $E_{ti} = 1$ if t-th GO term is enriched for i-th gene and $E_{ti} = 0$ otherwise.

The main distribution hierarchy of the Bayesian model can be described as follows:

$$\begin{aligned}
(Z_{ti}|E_{ti} = 1, \mu_{i1}, \tau_{i1}) &\sim N(\mu_{i1}, 1/\tau_{i1}),\\
(Z_{ti}|E_{ti} = 0, \mu_{i0}, \tau_{i0}) &\sim N(\mu_{i0}, 1/\tau_{i0}),\\
(E_{ti}|\boldsymbol{\Theta}, L_t, M_i) &\sim Bernoulli(\theta_{L_t M_i}),\\
(M_i|\boldsymbol{\alpha}) &\sim Categorical\,(\alpha_1, \cdots, \alpha_K)\,,\\
(L_t|\boldsymbol{\beta}) &\sim Categorical\,(\beta_1, \cdots, \beta_V)\,,
\end{aligned}$$

for $t = 1, \cdots, T$ and $i = 1, \cdots, G$. In other words, we model the enrichment status of a GO term for a gene cluster (E_{ti}) using a Bernoulli mass function by taking into account their clustering structure (M_i and L_t). Then, conditional on this enrichment status, the emission distribution of probit-transformed p-values (Z_{ti}) is modeled using a mixture of Gaussian densities. Finally, we consider semi-conjugate priors for the conditional emission distributions and a conjugate prior for the enrichment status.

In the description above, K and V are assumed to be known in advance. However, in practice, it is usually not easy to know or determine these values *a priori*. Based on this rationale, we implement a data-driven selection method to determine the optimal values of K and V using the poor man's reversible jump Markov chain Monte Carlo approach. Let us denote the maximum possible number of gene clusters as K_{max} while letting K be the effective number of gene clusters considered above, i.e., $K \leq K_{max}$. Then, we generate the cluster index for i-th gene (M_i) as follows:

$$\begin{aligned}
(\alpha_1^*, \cdots, \alpha_{K_{max}}^*|\alpha_0) &\sim Dirichlet(\alpha_0, \cdots, \alpha_0),\\
(\phi_k|\eta) &\sim Bernoulli(\eta),\\
\alpha_k &= \frac{\phi_k \alpha_k^*}{\sum_{k'=1}^{K_{max}} \phi_{k'}\alpha_{k'}^*},\\
(M_i|\boldsymbol{\alpha}) &\sim Categorical\left(\alpha_1, \cdots, \alpha_{K_{max}}\right),
\end{aligned}$$

for $i = 1, \cdots, G$ and $k = 1, \cdots, K_{max}$. Here α_k^* can be interpreted as the relative proportion of genes belonging to the k-th cluster while ϕ_k indicates whether the k-th cluster participates in the model or not. Then, the final value for proportion of genes belonging to the k-th cluster (α_k) is calculated using only the clusters that participate in the model, i.e., those such that $\phi_k = 1$. Similarly, we can determine the cluster index for t-th GO term (L_t) and the effective number of GO term cluster (V) which is less than or equal to the maximum possible number of GO term clusters (V_{max}). We used weakly informative priors for remaining hyperpriors. We implemented sensitivity analyses to evaluate potential impacts of hyperpriors on gene and GO term clustering and confirmed that bayesGO is relatively robust against misspecification of priors, especially for gene and GO term clusters with high confidence. We also implemented convergence diagnostics for various simulation studies and real data analyses and found that the proposed algorithm converges relatively quickly for most cases. Please check Chung et al. (2017) for more details about the hyperprior settings, the sensitivity analysis results, and the convergence diagnostics results.

In practice, it is relatively easy to set the values of K_{max} and V_{max} because K_{max} and V_{max} affect only upper bounds for the numbers of gene clusters and GO term clusters and as a result, it suffices to set them large enough so that $K \leq K_{max}$ and $V \leq V_{max}$. Based on our experience of analyzing real datasets, we recommend to set $K_{max} = 0.1G$ and $V_{max} = 0.1T$. Moreover, it is also straightforward to check whether our setting of K_{max} or V_{max} is appropriate (i.e., large enough) by monitoring K and V values across MCMC iterations. Please check Chung et al. (2017) for more details about the guidelines for parameter settings and interpretation. We implemented this approach as an `R` package `bayesGO`, which is currently publicly available in its GitHub webpage (https://dongjunchung.github.io/bayesGO/).

17.3 Results

17.3.1 Summary and Preprocessing of Literature Mining Results

The gene dictionary contains 39,820 genes with their HGNC approved symbol as keys and other synonyms as associated values. The GO dictionary contains 16,386 terms related to *homo sapiens*, with their official GO identifiers as keys and their names and other synonyms as values. From the literature mining, a total number of 8,453,254 and 5,599,412 abstracts were found to be related to at least one gene and GO term, respectively. After the co-occurrence analysis, a final total of 1,138,344 associations between genes and GO terms were stored in the database. Based on this literature mining result, in the section, we focus on a small subset of this data for purpose of illustration. Specifically, we first considered top 200 genes with the smallest numbers of missing cells. Then, we selected GO terms with less than 5% missing for these genes, which gave us 95 GO terms. Finally, we removed genes with an average p-value that is larger than or equal to 0.8 for these 95 GO terms in order to exclude the set of genes that will clearly not be benefited from this analysis. This preprocessing step resulted in a matrix of 95 GO terms and 77 genes, where only 86 cells were missing among the 7315 cells in this matrix ($= 77 \times 95$), i.e., less than 2% cells are missing. This dataset will be used for the analysis described below.

17.3.2 bayesGO Analysis

The R package `bayesGO` can be installed from its GitHub webpage (https://dongjunchung.github.io/bayesGO/) using the `R` package `devtools`.

```
R> library(devtools)
R> install_github("dongjunchung/bayesGO")
```

The small subset of literature mining data described above is included as an example data in this R package. This example dataset can be loaded using the following command line and it can be found as an object named `pmat`, which is a matrix of 95 rows (GO terms) and 77 columns (genes).

```
R> library(bayesGO)
R> data(pmat)
```

The bayesGO model can be fitted using the following command line. The maximum possible numbers of gene and GO term clusters can be specified using the arguments `Kmax` and `Vmax`, respectively. Here, both `Kmax` and `Vmax` are set to 10. When we ran this command line using a single core CPU at 2GHz, it took about 294 min to analyze this `pmat` matrix, which contains 95 GO terms and 77 genes. This assumed the MCMC updates of two chains, each of 30,000 iterations. In practice, a smaller number of MCMC iterations can also be considered as we set 30,000 iterations conservatively. In general, the computation time increases as a function of numbers of both genes and GO terms. We are currently working on improving the computation efficiency, especially using multi-core parallel computing approaches.

```
R> fit.bayesGO <- bayesGO( pmat, Vmax=10, Kmax=10 )
```

Simply typing the resulting object name (here `fit.bayesGO`) gives the summary of model fit, as shown below. Specifically, while we allowed at most 10 clusters for each of gene and GO term clusters, we ended up finding 7 gene clusters and 8 GO term clusters.

```
R> fit.bayesGO
Summary: Bayesian ontology fingerprint analysis
results
(class: BayesGO)
--------------------------------------------------
Model settings:
Number of genes to be analyzed: 77
Number of GO terms to be analyzed: 95
Maximum possible number of gene clusters: 10
Maximum possible number of GO term clusters: 10
--------------------------------------------------
Data analysis results:
Number of identified gene clusters: 7
Number of identified GO term clusters: 8
Association between GO terms (rows) and genes
(columns):
    0.00 0.83 0.08 1.00 0.00 0.00 0.56
    0.01 0.03 0.07 0.00 0.00 0.61 0.58
    0.09 0.00 0.02 0.67 0.03 0.82 0.58
    0.05 0.02 0.02 0.04 0.01 0.24 0.03
    0.45 0.04 0.08 0.28 0.54 0.59 0.19
    0.06 0.03 0.01 0.03 0.57 0.83 0.01
    0.71 0.31 0.02 1.00 0.21 0.73 1.00
    0.03 0.38 0.02 0.70 0.10 0.21 0.10
--------------------------------------------------
```

This summary also gives a matrix of enrichment of each GO term cluster for each gene cluster (the last part), which is also provided in Table 17.1. In this association matrix, the number in each cell indicates the proportion that the enrichment is observed for a pair of a gene and a GO term across the MCMC iterations, averaged over a block of the gene cluster and the GO term cluster. Hence, each cell has a value between zero and one and the value close to one means that the GO term cluster might be important to explain the function of the gene cluster. The enrichment matrix of each GO term for each gene (i.e., not at the level of clusters, but each element) can also be visualized using the function `plot()`, as depicted in Fig. 17.4.

```
R> plot( fit.bayesGO )
```

Table 17.1 Enrichment of each GO term cluster for each gene cluster

GO term	Gene cluster						
cluster	1	2	3	4	5	6	7
1	0.00	0.83	0.08	1.00	0.00	0.00	0.56
2	0.01	0.03	0.07	0.00	0.00	0.61	0.58
3	0.09	0.00	0.02	0.67	0.03	0.82	0.58
4	0.05	0.02	0.02	0.04	0.01	0.24	0.03
5	0.45	0.04	0.08	0.28	0.54	0.59	0.19
6	0.06	0.03	0.01	0.03	0.57	0.83	0.01
7	0.71	0.31	0.02	1.00	0.21	0.73	1.00
8	0.03	0.38	0.02	0.70	0.10	0.21	0.10

The number in each cell indicates the proportion that the enrichment is observed for a pair of a gene and a GO term across the MCMC iterations, averaged over a block of the gene cluster and the GO term cluster

Finally, members of each of the gene and GO term clusters can be checked using the following command line. Specifically, the function `predict()` returns a list object with three elements: (1) a list object for gene clusters (M_i), where each element provides the information about members of each gene cluster and the corresponding assignment probabilities (Tables 17.4, 17.6, 17.7, and 17.8); (2) a list object for GO term clusters (L_t), where each element provides the information about members of each GO term cluster and the corresponding assignment probabilities (Tables 17.2, 17.3, and 17.5); and (3) an association matrix (E_{ti}; Table 17.1). In Tables 17.2, 17.3, 17.4, 17.5, 17.6, 17.7, and 17.8, the assignment probabilities indicate the degree of confidence for the membership of each gene or GO term to the corresponding cluster, where assignment probabilities closer to 1 indicate higher degree of assignment confidence. For example, the assignment probability of 0.90 means that a gene or a GO term is assigned to a certain cluster in 90% of times across the MCMC iterations. In practice, assignment probabilities larger than 0.9 can be considered to be assignments with high confidence.

```
R> predict( fit.bayesGO )
```

Tables 17.2 and 17.3 show members of each of 6 among the 8 GO term clusters identified by bayesGO. GO term clusters 2, 3, 5, 6, 7, and 8 essentially represent programmed cell death, kinase activity, growth factor receptor binding, immune response, cell growth, and receptor binding, respectively. The GO term cluster 4 turns out to be a “garage collector,” a cluster of GO terms that are not enriched for any of gene clusters (Table 17.1), and hence is not presented in these tables. As expected, this GO term cluster is a mixed bag of GO terms representing different

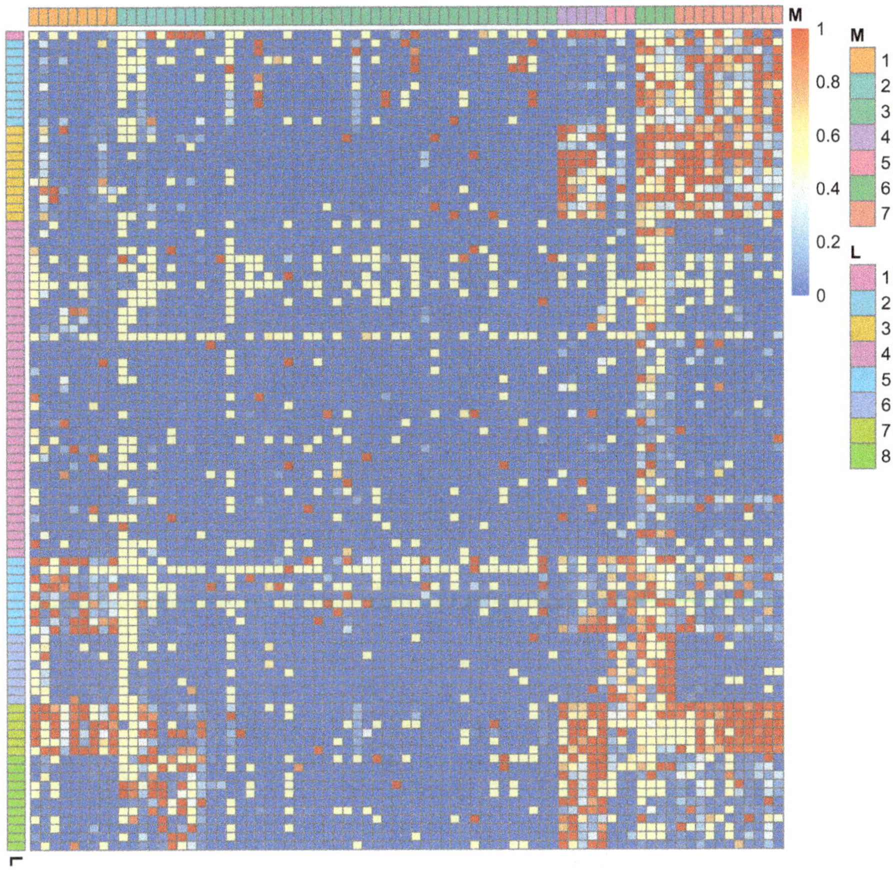

Fig. 17.4 Enrichment heatmap of each GO term (row) for each gene (column), where each cell indicates the proportion that the enrichment is observed for a pair of a gene and a GO term across the MCMC iterations. The row (L) and column sidebars (M) indicate the GO term and gene cluster indices, respectively

biological functions. The GO term cluster 1 is also not presented here either because it has only a single GO term, i.e., singleton. Readers who are interested in these GO term clusters can still find their members in Appendix (Table 17.5).

Next, Table 17.4 shows 5 among the 7 gene clusters identified by bayesGO. Similar to the case of GO term clustering, gene clusters 2 and 3 turn out to be "garbage collectors," for which none of GO term clusters are enriched (Table 17.1), and hence are not presented in this table. We note that the genes belonging to these gene clusters are not necessarily "garbage" but instead it simply means that we do not have GO terms that are associated with these genes in our data matrix. Hence, a larger set of GO terms needs to be considered for more meaningful analysis of these genes. Similarly, the gene cluster 5 is also not presented here either to avoid over-interpretation because in this gene cluster, assignment probabilities are not

Table 17.2 Members of each GO term cluster, along with their assignment probabilities and descriptions (GO term clusters 2, 3, and 5)

GO term cluster 2		
GO term name	Assignment probability	Description
GO:0006914	0.96	Autophagy
GO:0016236	0.96	Macroautophagy
GO:0007050	0.96	Cell cycle arrest
GO:0007049	0.96	Cell cycle
GO:0051726	0.82	Regulation of cell cycle
GO:0006309	0.80	Apoptotic DNA fragmentation
GO:0008219	0.80	Cell death
GO:0031966	0.76	Mitochondrial membrane
GO:0006915	0.70	Apoptotic process
GO:0012501	0.60	Programmed cell death
GO term cluster 3		
GO term name	Assignment probability	Description
GO:0004691	1.00	cAMP-dependent protein kinase activity
GO:0016301	0.99	Kinase activity
GO:0004697	0.99	Protein kinase C activity
GO:0016477	0.99	Cell migration
GO:0016246	0.97	RNA interference
GO:0005952	0.95	cAMP-dependent protein kinase complex
GO:0005790	0.93	Smooth endoplasmic reticulum
GO:0033673	0.88	Negative regulation of kinase activity
GO:0048870	0.62	Cell motility
GO:2000144	0.58	Positive regulation of DNA-templated transcription, initiation
GO:0004722	0.33	Protein serine/threonine phosphatase activity
GO term cluster 5		
GO Term name	Assignment probability	Description
GO:0005160	0.91	Transforming growth factor beta receptor binding
GO:0005104	0.88	Fibroblast growth factor receptor binding
GO:0006412	0.86	Translation
GO:0071897	0.85	DNA biosynthetic process
GO:0005161	0.75	Platelet-derived growth factor receptor binding
GO:0003677	0.51	DNA binding
GO:0010467	0.49	Gene expression
GO:0006955	0.49	Immune response
GO:0009058	0.36	Biosynthetic process

Table 17.3 Members of each GO term cluster, along with their assignment probabilities and descriptions (GO term clusters 6, 7, and 8)

GO term cluster 6		
GO term name	Assignment probability	Description
GO:0045087	0.56	Innate immune response
GO:0030246	0.56	Carbohydrate binding
GO:0006954	0.56	Inflammatory response
GO:0005777	0.56	Peroxisome
GO:0001775	0.56	Cell activation
GO:0004298	0.56	Threonine-type endopeptidase activity
GO:0004601	0.55	Peroxidase activity
GO:0031386	0.50	Protein tag
GO term cluster 7		
GO term name	Assignment probability	Description
GO:0016310	1.00	Phosphorylation
GO:0008283	1.00	Cell proliferation
GO:0007165	1.00	Signal transduction
GO:0005154	1.00	Epidermal growth factor receptor binding
GO:0016049	0.99	Cell growth
GO:0004707	0.99	MAP kinase activity
GO term cluster 8		
GO term name	Assignment probability	Description
GO:0005102	0.96	Receptor binding
GO:0031012	0.95	Extracellular matrix
GO:0006897	0.94	Endocytosis
GO:0070085	0.93	Glycosylation
GO:0043235	0.91	Receptor complex
GO:0016021	0.88	Integral component of membrane
GO:0005006	0.81	Epidermal growth factor-activated receptor activity
GO:0005783	0.80	Endoplasmic reticulum
GO:0044214	0.65	Spanning component of plasma membrane
GO:0009986	0.54	Cell surface
GO:0007155	0.48	Cell adhesion

Table 17.4 Members of each gene cluster, along with their assignment probabilities (gene clusters 1, 4, 6, and 7)

Gene cluster 1	
Gene name	Assignment probability
IGF1	1.00
LIF	0.99
EGR1	0.99
FOS	0.99
PTPN11	0.98
IGF2	0.96
MAP2	0.43
PNO1	0.41
SOAT1	0.40
Gene cluster 4	
Gene name	Assignment probability
NRP1	1.00
EPHA8	1.00
ACKR3	0.99
EPHA3	0.99
HSPG2	0.58
Gene cluster 6	
Gene name	Assignment probability
RUNX1	0.96
PCNA	0.96
NFKBIA	0.45
NFKB1	0.44
Gene cluster 7	
Gene name	Assignment probability
MARK2	0.94
MAPK3	0.92
FOXM1	0.78
YWHAQ	0.76
PI3	0.74
MAPK8	0.73
MSMP	0.63
MCL1	0.63
CFDP1	0.62
PROK1	0.54
MTOR	0.53

high enough for any of its gene members. Readers who are interested in these gene clusters can still find the relevant information in Appendix (Tables 17.6, 17.7, and 17.8).

The gene cluster 1 includes genes such as *IGF1* (insulin-like growth factor 1), *IGF2*, *EGR1* (early growth response 1), and *FOS* (Fos proto-oncogene, AP-1 transcription factor subunit), all of which are known to be involved in cell growth and proliferation. This is consistent with the fact that the GO term clusters 5 (growth factor) and 7 (cell proliferation and growth) are enriched for the gene cluster 1. The gene cluster 4 includes genes such as *NRP1* (neuropilin-1; a membrane-bound coreceptor to a tyrosine kinase receptor), *EPHA8* (ephrin type-A receptor 8), *EPHA3*, and *ACKR3* (atypical chemokine receptor 3). These genes are associated with various receptor functions and this is consistent with the GO term cluster 8 (receptor) being enriched for this gene cluster. The gene cluster 6 includes genes such as *RUNX1* (runt-related transcription factor 1), *NFKB1* (nuclear factor NF-kappa-B p105 subunit), and *NFKBIA* (NFKB inhibitor alpha). These genes are involved in various immune and inflammatory responses and this is consistent with the GO term cluster 6 (immune response) being enriched for this cluster. Finally, the gene cluster 7 includes genes such as *MAPK2* (mitogen-activated protein kinase 2), *MAPK3*, *MAPK8*, *FOXM1* (forkhead box protein M1), and *MTOR* (mechanistic target of rapamycin). These genes are involved in cell proliferation, cell differentiation, cell cycle, apoptosis, and autophagy, among other functions. We note that the GO term clusters 2 (autophagy, cell cycle, apoptosis), 3 (kinase activity), and 7 (cell proliferation) are enriched for these gene clusters.

17.4 Conclusion

In this chapter, we described approaches for the text mining of biomedical literature and the statistical analysis of this literature mining data. In particular, we described the web interface and the R package `bayesGO` that allow easy application of these approaches for various biological applications. We illustrated that these approaches can provide a principled way of utilizing the biomedical literature to investigate pathway-modulating genes and facilitating interpretation of these novel genes, which can be useful for the investigation of various biological problems. Currently, we are actively improving the user interface and implementing additional features for both the web interface and the R package `bayesGO`, which will further enhance user experiences in the future.

Acknowledgements This work was supported by the NIH/NIGMS grant (R01 GM122078) and the NIH/NCI grant (R21 CA209848).

Appendix

See Tables 17.5 for GO term clusters and Tables 17.6–17.8 for gene clusters omitted in the main text.

Table 17.5 Members of each GO term cluster, along with their assignment probabilities and descriptions (GO term clusters 1 and 4)

GO term cluster 1		
GO term name	Assignment probability	Description
GO:0005886	0.51	Plasma membrane
GO term cluster 4		
GO term name	Assignment probability	Description
GO:0042060	0.99	Wound healing
GO:0009056	0.99	Catabolic process
GO:0001503	0.99	Ossification
GO:0043657	0.99	Host cell
GO:0016032	0.99	Viral process
GO:0005604	0.99	Basement membrane
GO:0045098	0.99	Type III intermediate filament
GO:0008081	0.99	Phosphoric diester hydrolase activity
GO:0006281	0.99	DNA repair
GO:0042493	0.99	Response to drug
GO:0046323	0.99	Glucose import
GO:0030163	0.99	Protein catabolic process
GO:0006629	0.98	Lipid metabolic process
GO:0006260	0.98	DNA replication
GO:0010468	0.98	Regulation of gene expression
GO:0006119	0.98	Oxidative phosphorylation
GO:0051301	0.98	Cell division
GO:0007268	0.98	Chemical synaptic transmission
GO:0043005	0.97	Neuron projection
GO:0009792	0.97	Embryo development ending in birth or egg hatching
GO:0016458	0.97	Gene silencing
GO:0006006	0.96	Glucose metabolic process
GO:0043234	0.96	Protein complex
GO:0048468	0.94	Cell development
GO:0005524	0.94	ATP binding
GO:0032259	0.93	Methylation
GO:0009790	0.93	Embryo development
GO:0007585	0.91	Respiratory gaseous exchange
GO:0043687	0.90	Post-translational protein modification

(continued)

Table 17.5 (continued)

GO term cluster 4		
GO term name	Assignment probability	Description
GO:0051641	0.90	Cellular localization
GO:0030073	0.89	Insulin secretion
GO:0030154	0.89	Cell differentiation
GO:0005131	0.87	Growth hormone receptor binding
GO:0016791	0.86	Phosphatase activity
GO:0019835	0.64	Cytolysis
GO:0019787	0.53	Ubiquitin-like protein transferase activity
GO:0045155	0.53	Electron transporter, transferring electrons from CoQH2-cytochrome c reductase complex and cytochrome c oxidase complex activity
GO:0004842	0.52	Ubiquitin-protein transferase activity
GO:0003824	0.36	Catalytic activity

Table 17.6 Members of the gene cluster 2, along with their assignment probabilities

Gene cluster 2	
Gene name	Assignment probability
CD177	1.00
GNPDA1	0.99
INSR	0.96
CD53	0.94
CAV1	0.88
DNAH8	0.84
CDH17	0.81
ABCA1	0.78
FAS	0.36

Table 17.7 Members of the gene cluster 3, along with their assignment probabilities

Gene cluster 3	
Gene name	Assignment probability
HK1	0.94
PNP	0.93
SOD1	0.93
REG1A	0.93
CFP	0.92
CXCR4	0.92
DSP	0.92
ABCB1	0.91
HSPA4L	0.88
ARTN	0.87
XRCC1	0.87
RIMS2	0.86
B3GAT1	0.84
RMDN2	0.84
RB1	0.83
MAPK14	0.82
RAC1	0.81
NGF	0.81
HDAC9	0.81
HSPD1	0.80
CTSD	0.80
ZNF629	0.80
ANXA5	0.79
SNCA	0.78
CTSB	0.76
PRH2	0.76
PROS1	0.75
RUNX2	0.75
BCL2L1	0.74
IFNB1	0.73
NR5A1	0.72
TH	0.71
CNBP	0.71
TIMP1	0.69
RNASEH2A	0.67
BCL2	0.64

Table 17.8 Members of the gene cluster 5, along with their assignment probabilities

Gene cluster 5	
Gene name	Assignment probability
IL1B	0.64
RELA	0.48
IL1A	0.48

References

Chung, D., Lawson, A., & Zheng, W. J. (2017). A statistical framework for biomedical literature mining. *Statistics in Medicine, 36*(22), 3461–3474.

Frijters, R., Heupers, B., van Beek, P., Bouwhuis, M., van Schaik, R., de Vlieg, J., et al. (2008). Copub: A literature-based keyword enrichment tool for microarray data analysis. *Nucleic Acids Research, 36*(suppl_2), W406–W410.

Jenssen, T. K., Lægreid, A., Komorowski, J., & Hovig, E. (2001). A literature network of human genes for high-throughput analysis of gene expression. *Nature Genetics, 28*(1), 21–28.

Koike, A., & Takagi, T. (2004). Gene/protein/family name recognition in biomedical literature. In *Proceedings of BioLink 2004 Workshop: Linking Biological Literature, Ontologies and Databases: Tools for Users* (Vol. 42, p. 56).

Liu, H., Hu, Z. Z., Zhang, J., & Wu, C. (2005). Biothesaurus: A web-based thesaurus of protein and gene names. *Bioinformatics, 22*(1), 103–105.

Mitsumori, T., Fation, S., Murata, M., Doi, K., & Doi, H. (2005). Gene/protein name recognition based on support vector machine using dictionary as features. *BMC Bioinformatics, 6*(1), S8.

Qin, T., Matmati, N., Tsoi, L. C., Mohanty, B. K., Gao, N., Tang, J., et al. (2014). Finding pathway-modulating genes from a novel Ontology Fingerprint-derived gene network. *Nucleic Acids Research, 42*(18), e138–e138.

Chapter 18
Discriminant Analysis and Normalization Methods for Next-Generation Sequencing Data

Yan Zhou, Junhui Wang, Yichuan Zhao, and Tiejun Tong

18.1 Introduction

Next-generation sequencing data are getting more popular in biological and medical researches for the increased specificity and sensitivity of gene expression (Mardis 2008; Wang et al. 2009; Morozova et al. 2009). They aim to analyze much low cost and less noisy data as well as to enable certain applications that are not achievable by microarray data.

In medical and biological studies, it is a fundamental issue to discriminate which type of diseases a new patient or sample belongs to. With the reduced cost in sequencing, more and more researchers or practitioners have adopted next-generation sequencing data to diagnose diseases (Lorenz et al. 2014). For discriminant analysis of microarray data, the discriminant methods have been well developed in the past years. To name a few, they include the diagonal linear discriminant analysis, the diagonal quadratic discriminant analysis in Dudoit et al. (2002), the bias-corrected rules for discriminant analysis in Huang et al. (2010) and

Y. Zhou
College of Mathematics and Statistics, Institute of Statistical Sciences, Shenzhen University, Shenzhen, China
e-mail: zhouy1016@szu.edu.cn

J. Wang
School of Data Science, City University of Hong Kong, Kowloon, Hong Kong
e-mail: j.h.wang@cityu.edu.hk

Y. Zhao
Department of Mathematics and Statistics, Georgia State University, Atlanta, GA, USA
e-mail: yichuan@gsu.edu

T. Tong (✉)
Department of Mathematics, Hong Kong Baptist University, Kowloon, Hong Kong
e-mail: tongt@hkbu.edu.hk

Y. Zhao, D.-G. Chen (eds.), *New Frontiers of Biostatistics and Bioinformatics*,
ICSA Book Series in Statistics, https://doi.org/10.1007/978-3-319-99389-8_18

the bias-corrected geometric diagonalization method for regularized discriminant analysis in Zhou et al. (2017b).

To describe the next-generation sequencing technology and it difference from microarray data, we note that next generation sequencing (Shendure and Ji 2008) is a revolutionary technology for the modern biomedical and biological research. The increasing popularity of next-generation sequencing has indirectly triggered a competition among several companies, which aims to manufacture a sequencing platform for high-quality sequences with longer read and more throughput at low cost.

There are many notable papers in the area of transcriptome or gene analysis using next-generation sequencing technology, such as yeast sequencing in Nagalakshmi et al. (2008), human sequencing in Cloonan et al. (2008), Morin et al. (2008), mouse sequencing in Mortazavi et al. (2008), and so forth. There are millions of short reads from the transcript population of interest in next-generation sequencing technology and these reads are mapped to the reference genome, thus next-generation sequencing produces counts and offers a better way to detect novel transcripts. That is, a count number is measured for the expression level of each gene in next-generation sequencing data. We note that there are many methods for analyzing next-generation sequencing data in Anders and Huber (2010), Birchler and Kavi (2008), The Cancer Genome Atlas Research Network (2014), Dillies et al. (2013).

Discriminant analysis is to predict the category of a new observation with the features from the training data. First, the training data are divided into a number of categories with the dependent variables. One main objective of discriminant analysis is to develop discriminant score functions that will discriminate between the categories of the dependent variables in a perfect way. Researchers can examine whether significant features exist among the groups, i.e., among the predictor variables. It also evaluates the accuracy of the classification.

Unlike microarray data that follow a Gaussian distribution, RNA-seq data follow a discrete distribution such as a Poisson or negative binomial distribution (Bullard et al. 2010; Robinson and Smyth 2008; Robinson et al. 2010; Love et al. 2014; Lin et al. 2014). As a result, the existing methods for discriminating microarray data may not perform well or may not even be applicable for next-generation sequencing data. In this chapter, we introduce a few newly developed discriminant analysis methods and normalization methods for next-generation sequencing data.

The rest of the chapter is organized as follows. In Sect. 18.2, we briefly introduce the discriminant analysis methods for continuous microarray data. In Sect. 18.3, we present in detail the three discriminant analysis methods for next-generation sequencing data. In Sect. 18.4, we introduce some normalization methods for next-generation sequencing data. In Sect. 18.5, simulation studies are carried out to evaluate the performance of the introduced methods. We further illustrate their practical usefulness by analyzing two next-generation sequencing datasets in Sect. 18.6. Finally, we conclude the chapter in Sect. 18.7 with some future work.

18.2 Discriminant Analysis for Microarray Data

We first introduce the discriminant analysis methods for microarray data. With proper normalization, we assume that the samples in each class are randomly drawn from a G-dimensional multivariate normal distribution with mean vector $\boldsymbol{\mu}_k = (\mu_{k1}, \ldots, \mu_{kG})^T$ and covariance matrix Σ_k, where $k = 1, \ldots, K$ with K being the total number of classes, and T represents the transpose. To be specific, there are n_k independent and identically distributed (i.i.d.) random vectors in the kth class such that

$$\boldsymbol{x}_{k,1}, \ldots, \boldsymbol{x}_{k,n_k} \overset{\text{i.i.d.}}{\sim} \text{MVN}(\boldsymbol{\mu}_k, \Sigma_k). \tag{18.1}$$

Let $n = \sum_{k=1}^{K} n_k$ be the total sample size of all classes. Given a new observation, $\boldsymbol{x}^*$, the main goal of discriminant analysis is to assign the class label that the new observation belongs to.

18.2.1 *Linear Discriminant Analysis*

Let π_k be the prior probability of observing a sample from the kth class such that $\sum_{k=1}^{K} \pi_k = 1$. By the Bayes rule, the posterior probability that the new observation, $\boldsymbol{x}^*$, belongs to the kth class is

$$P(y^* = k|\boldsymbol{x}^*) = \frac{f_k(\boldsymbol{x}^*)\pi_k}{\sum_{k=1}^{K} f_k(\boldsymbol{x}^*)\pi_k}, \tag{18.2}$$

where y^* represents the class label of $\boldsymbol{x}^*$, and f_k is the probability density function of the sample in class k. We then select the value of k that maximizes the posterior probability $P(y^* = k|\boldsymbol{x}^*)$ as the assigned label of the new observation. The linear discriminant analysis (LDA) (Wald and Kronmal 1977) assumes that the covariance matrices are equal for all classes, i.e. $\Sigma_k = \Sigma$. With formulas (18.1) and (18.2), the linear discriminant scores are given as

$$d_k^L(\boldsymbol{x}^*) = (\boldsymbol{x}^* - \boldsymbol{\mu}_k)^T \Sigma^{-1} (\boldsymbol{x}^* - \boldsymbol{\mu}_k) - 2 \ln \pi_k. \tag{18.3}$$

Note that $\boldsymbol{\mu}_k$, Σ, and π_k in (18.3) are unknown and they need to be estimated from the sample data. One common procedure for estimating these parameters are as follows:

(a) Estimate $\boldsymbol{\mu}_k$ with the sample mean in each class, $\bar{\boldsymbol{x}}_k = \sum_{i=1}^{n_k} x_{k,i}/n_k$.
(b) Estimate Σ with the pooled sample covariance matrix $S_{\text{pool}} = \sum_{k=1}^{K} (n_k - 1) S_k/(n - K)$, where $S_k = \sum_{i=1}^{n_k} (\boldsymbol{x}_{k,i} - \bar{\boldsymbol{x}}_k)(\boldsymbol{x}_{k,i} - \bar{\boldsymbol{x}}_k)^T/(n_k - 1)$ are the respective sample covariance matrices.

(c) Estimate π_k with $\hat{\pi}_k = n_k/n$.

With the above estimates, we have the sample version of the linear discriminant scores as

$$\hat{d}_k^L(\boldsymbol{x}^*) = (\boldsymbol{x}^* - \bar{\boldsymbol{x}}_k)^T S_{\text{pool}}^{-1}(\boldsymbol{x}^* - \bar{\boldsymbol{x}}_k) - 2\ln\hat{\pi}_k. \tag{18.4}$$

Given that LDA assumes equal covariance matrices for all classes, it may not be realistic in practice. Taking into account the discrepancy among the covariance matrices, one may consider the quadratic discriminant analysis (QDA) with the discriminant scores as

$$d_k^Q(\boldsymbol{x}^*) = (\boldsymbol{x}^* - \boldsymbol{\mu}_k)^T \Sigma_k^{-1}(\boldsymbol{x}^* - \boldsymbol{\mu}_k) + \ln|\Sigma_k| - 2\ln\pi_k. \tag{18.5}$$

Accordingly, the sample version of the quadratic discriminant scores is given as

$$\hat{d}_k^Q(\boldsymbol{x}^*) = (\boldsymbol{x}^* - \bar{\boldsymbol{x}}_k)^T S_k^{-1}(\boldsymbol{x}^* - \bar{\boldsymbol{x}}_k) + \ln|S_k| - 2\ln\hat{\pi}_k. \tag{18.6}$$

18.2.2 Diagonal Linear Discriminant Analysis

For microarray data, the number of genes (or features) is often larger than the sample size. Under the "large G small n" scenario, LDA and QDA may not be applicable due to the singularity of the sample covariance matrices. To overcome this problem, a simple yet efficient approach in the literature is to apply the diagonalization methods to LDA and QDA for discrimination (Dudoit et al. 2002).

Let $D_k = \text{diag}(s_{k1}^2, \ldots, s_{kG}^2)$ be the diagonal matrix of the sample covariance matrix S_k, where s_{kg} is the sample variance of gene g. Let also $D_{\text{pool}} = \text{diag}(s_1^2, \ldots, s_G^2)$ be the diagonal matrix of the pooled sample covariance matrix S_{pool}. Since a diagonal matrix without any zero entries is often invertible, by replacing S_{pool} by D_{pool} in (18.4), we have the diagonal linear discriminant analysis (DLDA) with the discriminant scores as

$$\hat{d}_k^{DL}(\boldsymbol{x}^*) = \sum_{g=1}^{G}(x_g^* - \bar{x}_{kg})^2/s_g^2 - 2\ln\hat{\pi}_k. \tag{18.7}$$

Similarly, we can define the diagonal quadratic discriminant analysis (DQDA) with the discriminant scores as

$$\hat{d}_k^{DQ}(\boldsymbol{x}^*) = \sum_{g=1}^{G}(x_g^* - \bar{x}_{kg})^2/s_{kg}^2 + \sum_{g=1}^{G}\ln s_{kg}^2 - 2\ln\hat{\pi}_k. \tag{18.8}$$

Even though DLDA and DQDA are widely applicable, they are not the optimal rules, and improvements over them are available in the recent literature. For instance, Huang et al. (2010) proposed two bias-corrected rules for DLDA and DQDA, and Zhou et al. (2017b) developed a diagonalization method for the regularized discriminant analysis (RDA) with the same spirit as DLDA and DQDA. Other directions for extending DLDA and DQDA are also available, e.g. in Friedman (1989), Hastie et al. (1995, 1994), Hastie and Tibshirani (1996), Clemmensen et al. (2011), Grosenick et al. (2008), Leng (2008), Mai et al. (2012).

18.3 Discriminant Analysis for Next-Generation Sequencing Data

Next-generation sequencing data are entirely different from microarray data, and the discriminant analysis methods for microarray data cannot be directly applied to next-generation sequencing data. In this section, we introduce some newly developed discriminant methods for analyzing next-generation sequencing data. They include the Poisson linear discriminant analysis in Witten (2011), the zero-inflated Poisson logistic discriminant analysis in Zhou et al. (2018), and the negative binomial linear discriminant analysis in Dong et al. (2016).

18.3.1 Poisson Linear Discriminant Analysis

For next-generation sequencing data, Witten (2011) proposed a Poisson linear discriminant analysis (PLDA) by assuming that the data follow a Poisson distribution. Let X_{ig} be the number of reads mapped to gene g in sample i, where $i = 1, \ldots, n$ and $g = 1, \ldots, G$. We assume that

$$X_{ig} \sim \text{Poisson}(l_i \lambda_g), \tag{18.9}$$

where l_i is the size factor that scales the gene counts for the ith sample, and λ_g is the total number of reads for the gth gene. Let K be the number of classes and n be the total number of observations drawn from all K classes. Then the class-specific model for next-generation sequencing data is

$$(X_{ig} | y_i = k) \sim \text{Poisson}(l_i \lambda_g d_{kg}), \tag{18.10}$$

where d_{kg} allows the differential expression between different classes for the gth gene (Witten 2011). Let $\boldsymbol{x}^* = (x_1^*, \ldots, x_G^*)^T$ be a new observation with size factor l^* and class label y^*.

By the Bayes rule, we have

$$P(y^* = k|\boldsymbol{x}^*) \propto f_k(\boldsymbol{x}^*)\pi_k, \tag{18.11}$$

where f_k is the probability mass function associated with the kth class, and π_k is the prior probability that one sample is drawn from the kth class. This leads to the discriminant scores of PLDA as

$$\ln P(y^* = k|\boldsymbol{x}^*) = \sum_{g=1}^{G} x_g^* \ln d_{kg} - \sum_{g=1}^{G} l^* \lambda_g d_{kg} + \ln \pi_k + C, \tag{18.12}$$

where C is a constant independent of k. For the estimation of the unknown parameters, one may refer to Witten (2011). Finally, we choose k that maximizes the discriminant score as the class label of the new observation.

18.3.2 Zero-Inflated Poisson Logistic Discriminant Analysis

In practice, however, there may have excess zeros in next-generation sequencing data, especially for small RNA or microRNA. For example, the cervical cancer dataset in Witten et al. (2010), Witten (2011) contains about 47.6% zeros, and the liver and kidney dataset in Marioni et al. (2008) contains about 45.5% zeros, among all numerical values. In such cases, the zero-inflated distributions ought to be considered for modeling next-generation sequencing data.

Let X_{ki_kg} be the number of reads mapped to gene g in sample i_k of the kth class, where $k = 1, \ldots, K$, $i_k = 1, \ldots, n_k$ and $g = 1, \ldots, G$. Let n_k be the sample size in class k and $n = \sum_{k=1}^{K} n_k$ be the total sample size of all classes. The zero-inflated Poisson distribution is given as

$$X_{ki_kg} \sim \begin{cases} \delta_{\{0\}} & p_{ki_kg} \\ \text{Poisson}(\mu_{ki_kg}) & (1 - p_{ki_kg}), \end{cases} \tag{18.13}$$

where $\delta_{\{0\}}$ is the zero distribution, μ_{ki_kg} and p_{ki_kg} are the expected value and the probability of $\delta_{\{0\}}$ for gene g in sample i_k in class k, respectively. We further assume $\mu_{ki_kg} = l_{i_k}\lambda_g d_{kg}$ is the same as in Sect. 18.3.1.

Following the logistic models in Ridout et al. (1998) and Mouatassim and Ezzahid (2012), one may also consider a logistic relation between the probability of zeros and the mean of the genes with the sequencing depth as

$$\ln\{\frac{P(X_{ki_kg} = 0)}{1 - P(X_{ki_kg} = 0)}\} = \alpha + \beta_1(\frac{N_{ki_k}}{N_{1i_1}}) + \beta_2\mu_{ki_kg}, \tag{18.14}$$

where N_{ki_k} is the total sequencing depth of sample i_k in class k, and α, β_1, and β_2 are the intercept and coefficients of N_{ki_k}/N_{1i_1} and $\mu_{ki_k g}$, respectively.

Given the new observation $\boldsymbol{x}^*$, Zhou et al. (2018) proposed the zero-inflated Poisson logistic discriminant analysis (ZIPLDA) with the discriminant scores as

$$\begin{aligned}\ln P(y^* = k|\boldsymbol{x}^*) = &\sum_{g=1}^{G} I_{(x_g^*=0)} \ln\left(p_{kg}^* + (1 - p_{kg}^*)e^{(-d_{kg}l^*\lambda_g)}\right)\\ &- \sum_{g=1}^{G} I_{(x_g^*>0)} d_{kg}l^*\lambda_g + \sum_{g=1}^{G} I_{(x_g^*>0)} \ln(1 - p_{kg}^*)\\ &+ \sum_{g=1}^{G} I_{(x_g^*>0)} x_g^* \ln(d_{kg}) + \ln \pi_k + C,\end{aligned} \tag{18.15}$$

where C is a constant independent of k. For the estimation of the unknown parameters, one may refer to Zhou et al. (2018). Finally, we choose k that maximizes the discriminant score as the class label of the new observation. When $p_{kg}^* \to 0$, by formula (18.15) it yields that

$$\ln\left(p_{kg}^* + (1 - p_{kg}^*)e^{(-d_{kg}l^*\lambda_g)}\right) \to d_{kg}l^*\lambda_g.$$

That is, the discriminant scores of ZIPLDA will reduce to the discriminant scores of PLDA in (18.12) when there are no excess zeros.

18.3.3 Negative Binomial Linear Discriminant Analysis

For genes with adequate sequencing depth, the Poisson or zero-inflated Poisson distribution may not provide a good modeling due to the overdispersion issue in the data. This section introduces the negative binomial linear discriminant analysis (NBLDA) in Dong et al. (2016) for next-generation sequencing data.

Let X_{ig} be the number of reads mapped to gene g in sample i, where $i = 1, \ldots, n$ and $g = 1, \ldots, G$. By assuming a negative binomial distribution for the data, we have

$$X_{ig} \sim \text{NB}(\mu_{ig}, \phi_g), \quad \mu_{ig} = l_i\lambda_g, \tag{18.16}$$

where l_i and λ_g are the same as in Sect. 18.3.1, and $\phi_g \geq 0$ is the dispersion parameter. We have $E(X_{ig}) = \mu_{ig}$ and $\text{Var}(X_{ig}) = \mu_{ig} + \mu_{ig}^2\phi_g$. Further, the class-specific model for the data is

$$(X_{ig}|y_i = k) \sim \text{NB}(\mu_{ig}d_{kg}, \phi_g), \tag{18.17}$$

where d_{kg} is defined the same as in Sect. 18.3.1, and $y_i = k \in \{1, \ldots, K\}$ is the class label of sample i.

Then for the new observation, $\boldsymbol{x}^*$, by the Bayes rule and the probability mass function of $X_{ig} = x_{ig}$ in (18.17), the discriminant scores of NBLDA are given as

$$\ln P(y^* = k|\boldsymbol{x}^*) = \sum_{g=1}^{G} x_g^* \left[\ln d_{kg} - \ln(1 + l^* \lambda_g d_{kg} \phi_g)\right] - \sum_{g=1}^{G} \phi_g^{-1} \ln(1 + l^* \lambda_g d_{kg} \phi_g) + \ln \pi_k + C, \quad (18.18)$$

where C is a constant independent of k. For the estimation of the unknown parameters, one may refer to Dong et al. (2016). Finally, we assign the new observation $\boldsymbol{x}^*$ to class k that maximizes the quantity (18.18). It is noteworthy that NBLDA will be equivalent to PLDA when there is no dispersion in the data, i.e. when $\phi_g = 0$ for all genes.

18.4 Normalization Methods for Next-Generation Sequencing Data

It is well known that normalization is an important step for pre-processing the gene expression microarray data. Accordingly, it is equally important to perform normalization for next-generation sequencing data. In this section, we review several methods for RNA-seq data normalization, including those for same species and different species.

18.4.1 Normalization for Same Species

We introduce two normalization methods for RNA-seq data with same species: a scale normalization method in Robinson and Oshlack (2010) and a hypothesis testing based normalization method in Zhou et al. (2017a). For ease of notation, we assume that the true expression level and the observed count of gene g in library k are μ_{gk} and Y_{gk}, respectively, where $k = 1, 2$ and $g = 1, \ldots, G$. Let also L_g be the length of gene g, and N_k be the total number of reads in library k. By formulating the expected value of the count in a sample by the product of the true expression level and the gene length, the expected value of Y_{gk} is given as

$$E[Y_{gk}] = \frac{\mu_{gk} L_g}{S_k} N_k, \quad (18.19)$$

where $S_k = \sum_{g=1}^{G} \mu_{gk} L_g$ is referred to as the total RNA-seq expression of sample k.

18.4.1.1 The Trimmed Mean of M-Values Normalization Method

A trimmed mean of M-values normalization method (TMM) based on model (18.19) was proposed in Robinson and Oshlack (2010). We first introduce the gene log-fold-changes and the absolute expression levels (MA). Note that the unknown S_k in model (18.19) may not be estimated directly from the data. As an alternative, however, we can estimate $f_{kr} = S_k/S_r$ which is the relative RNA production of two samples. The log-fold-changes for sample k relative to sample r for gene g are defined as

$$M_{gk}^{r} = \log_2 \frac{Y_{gk} N_r}{Y_{gr} N_k},$$

and the absolute expression levels are defined as

$$A_{gk}^{r} = \frac{1}{2} \log_2 (\frac{Y_{gk}}{N_k} \times \frac{Y_{gr}}{N_r}).$$

The TMM method takes the average value after removing the upper and lower 5% of the log-fold-changes M_{gk}^{r} and the absolute expression levels A_{gk}^{r}. The trim method means removing the different expression genes which may effect the normalization. In order to balance the data, the TMM method takes a weighted mean of M_{gk}^{r} by the inverse of the approximate asymptotic variances in Casella and Berger (2002). After trimming, there are $G*$ valid M_{gk}^{r} and A_{gk}^{r} values. Consequently, a global scaling factor $\text{TMM}_k^{(r)}$ for sample k relative to sample r can be calculated as follows:

$$\log_2(\text{TMM}_k^{(r)}) = \frac{\sum_{g \in G*} \frac{M_{gk}^{r}}{w_{gk}^{r}}}{\sum_{g \in G*} \frac{1}{w_{gk}^{r}}},$$

where

$$w_{gk}^{r} = \frac{N_k - Y_{gk}}{N_k Y_{gk}} + \frac{N_r - Y_{gr}}{N_r Y_{gr}}, \quad Y_{gk} > 0,\ Y_{gr} > 0.$$

Note that the genes with $Y_{gk} = 0$ or $Y_{gr} = 0$ are excluded from the computation of the scaling factor $\text{TMM}_k^{(r)}$. The main reason is that the log-fold-changes cannot be calculated in the case of $Y_{gk} = 0$ or $Y_{gr} = 0$. Finally, we combine the test method in Robinson and Oshlack (2010) and the TMM normalization factor to detect DE genes.

18.4.1.2 A Hypothesis Testing Based Normalization Scaling Factor Method

For normalization of RNA-seq data with same species, a number of other methods are also available in the literature, e.g. in Bolstad et al. (2003), Bullard et al. (2010), Anders and Huber (2010), Huang et al. (2014), Huang and Girimurugan (2018), Huang and Yu (2016), Huang (2016). In this section, we introduce another efficient method called the hypothesis testing based normalization (HTN) method, which was proposed in Zhou et al. (2017a) that utilizes the available knowledge of housekeeping genes to reduce the bias of normalization. The main objective is to test which genes are differentially expressed in two samples or libraries. For gene g, we test

$$\mathrm{H}_{0g} : \mu_{g1} = \mu_{g2} \quad \text{vs} \quad \mathrm{H}_{1g} : \mu_{g1} \neq \mu_{g2} \quad \text{for all} \quad g. \tag{18.20}$$

Under the assumptions that the read counts for each gene follow a Poisson distribution, by model (18.19) the above hypothesis is equivalent to

$$\mathrm{H}_{0g} : \lambda_{g1} = c\frac{N_1}{N_2}\lambda_{g2} \quad \text{vs} \quad \mathrm{H}_{1g} : \lambda_{g1} \neq c\frac{N_1}{N_2}\lambda_{g2} \quad \text{for all} \quad g, \tag{18.21}$$

where $c = S_2/S_1$ is the scaling factor of sample 2 relative to sample 1.

Based on the above hypothesis testing framework, we are able to calculate the p-value for each gene g. Specifically, conditioning on $Y_{g1} + Y_{g2} = n_g$, the probability of Y_{g1} equals y_{g1} is given as

$$P(Y_{g1} = y_{g1} \mid Y_{g1} + Y_{g2} = n_g) = \frac{n_g!}{y_{g1}!(n_g - y_{g1})!} p_0^{y_{g1}} (1 - p_0)^{n_g - y_{g1}}. \tag{18.22}$$

From the above equation, we can see that Y_{g1} follows a binomial distribution, where $p_0 = \lambda_{g1}/(\lambda_{g1} + \lambda_{g2}) = (cN_1/N_2)/(1 + cN_1/N_2)$.

By the conditional distribution of Y_{g1} in (18.22), the p-values are

$$\begin{aligned} p_g(c) &= P(|Y_{g1} - n_g p_0| \geq |y_{g1} - n_g p_0| \big| n_g) \\ &= P\left(\left| (1 + c\frac{N_1}{N_2})Y_{g1} - c\frac{N_1}{N_2}n_g \right| \geq \left| (1 + c\frac{N_1}{N_2})y_{g1} - c\frac{N_1}{N_2}n_g \right| \Big| n_g \right), \end{aligned} \tag{18.23}$$

where y_{g1} and $y_{g2} = n_g - y_{g1}$ are the observed values of Y_{g1} and Y_{g2}, respectively.

Given the true value of scaling factor c, we can calculate the p-values for all genes and test which genes are differentially expressed. The HTN method is proposed to find the optimal scaling factor by utilizing the stability of housekeeping genes. Since housekeeping genes are assumed to be non-DE genes, its p-values follow a uniform distribution on (0, 1) when the true value of c is given. Then for the significance level α, the false positive rate of those genes is supposed to be around the nominal level. In real data, we can get a set H of housekeeping genes in priori from the published studies or based on certain biological information, e.g., the GO terms of the genes (Chen et al. 2014). Finally, we find the optimal c value by the following criterion:

$$c_{\mathrm{opt}} = \underset{c>0}{\operatorname{argmin}} \left| \frac{1}{m} \sum_{g \in H} I(p_g(c) < \alpha \mid H_0, c) - \alpha \right|, \tag{18.24}$$

where m is the number of housekeeping genes in set H. Theoretically, the choice of α has no effect on the c_{opt} value. Simulations also show that c_{opt} keeps stable for varying α. For this, we suggest to find the c_{opt} value with a grid search method.

18.4.2 Normalization for Different Species

One significant difference between the same and different species is the numbers and lengths of genes. For RNA-seq data with same species, the numbers and lengths of genes are equal to each other. However, those with different species may have different gene numbers and different gene lengths. The normalization methods for same species cannot be applied to different species directly. In the section, we introduce a scale based normalization (SCBN) method for RNA-seq data with different species, which was recently proposed in Zhou et al. (submitted).

Let G_o be the set of one-to-one orthologous genes that are to be tested for differential expression, which is a subset of the complete gene set from two species. Let $Y_{g_i s}$ be the random variable that represents the count of reads mapped to the orthologous gene g_i in species s, and $y_{g_i s}$ be the real observation, where $g_i \in G_o$ and $s \in \{1, 2\}$.

We consider the hypothesis testing to detect differential expressions of each orthologous gene g_i between two species as follows:

$$H_0^{g_i} : \mu_{g_i 1} = \mu_{g_i 2} \quad \text{vs} \quad H_1^{g_i} : \mu_{g_i 1} \neq \mu_{g_i 2}, \tag{18.25}$$

where $\mu_{g_i s}$ is the true expression level for orthologous gene g_i in specie s.

Let $c = S_2/S_1$ be the scaling factor for normalization. Based on model (18.19), we assume that the reads mapped to the orthologous genes follow a Poisson distribution so that $Y_{g_i s} \sim \text{Poisson}(\lambda_{g_i s})$, where $\lambda_{g_i s} = E(Y_{g_i s})$. The above hypothesis is then equivalent to

$$H_0^{g_i} : \lambda_{g_i 1} = \frac{L_{g_i 1}}{L_{g_i 2}} \frac{N_1}{N_2} c \lambda_{g_i 2} \quad \text{vs} \quad H_1^{g_i} : \lambda_{g_i 1} \neq \frac{L_{g_i 1}}{L_{g_i 2}} \frac{N_1}{N_2} c \lambda_{g_i 2}.$$

The difference between the above hypothesis and that for the same species is that the length $L_{g_i s}$ may vary for different s.

Next, we calculate the p-value for each orthologous gene. Similar to the same species, conditioning on $Y_{g_i 1} + Y_{g_i 2} = n_{g_i}$, the random variable $Y_{g_i 1}$ follows a binomial distribution, with parameters as follows:

$$Y_{g_i 1} \big| Y_{g_i 1} + Y_{g_i 2} = n_{g_i} \sim \text{Binomial}(n_{g_i}, p_0^{g_i}), \tag{18.26}$$

where

$$p_0^{g_i} = \frac{\lambda_{g_i 1}}{\lambda_{g_i 1} + \lambda_{g_i 2}} = \frac{c L_{g_i 1} N_1}{L_{g_i 2} N_2 + c L_{g_i 1} N_1}$$

is the probability of success under the null hypothesis of (18.25). By formula (18.26), we get the p-value of the test for different species for a given scaling factor c, that is

$$p_{g_i}(c) = P(|(1 + \frac{L_{g_i 1}}{L_{g_i 2}} \frac{N_1}{N_2} c) Y_{g_i 1} - \frac{L_{g_i 1}}{L_{g_i 2}} \frac{N_1}{N_2} c n_{g_i}| \geq$$
$$|(1 + \frac{L_{g_i 1}}{L_{g_i 2}} \frac{N_1}{N_2} c) y_{g_i 1} - \frac{L_{g_i 1}}{L_{g_i 2}} \frac{N_1}{N_2} c n_{g_i}| | n_{g_i}). \tag{18.27}$$

The purpose of the SCBN method is to find the optimal scaling factor c and then detect the DE genes for two species. By the reported studies or by certain biological measures (Brawand et al. 2011; Chen et al. 2014), we may know some non-DE genes in priori. Assume that we know in priori a set H of conserved orthologous genes, which are considered as non-DE genes for its stability between species. Then for the significance level α and the scaling factor c, the value of $\sum_{g_i \in H} (1/m) I(p_{g_i}(c) < \alpha | H_0; c)$ for conserved orthologous genes and the nominal level at α should be close to each other. This hence suggests to search for the optimal scaling factor by the following criterion:

$$c_{\text{opt}} = \underset{c>0}{\text{argmin}} \left| \sum_{g_i \in H} \frac{1}{m} I(p_{g_i}(c) < \alpha | H_0; c) - \alpha \right|, \tag{18.28}$$

where m be the number of genes in the set H. As in Sect. 8.1, the optimal c value can be derived by the grid search method.

18.5 Simulation Studies

In this section, we assess the performance of the classification methods via a number of simulation studies. We consider a total of five classification methods, including

PLDA in Witten (2011), ZIPLDA in Zhou et al. (2018), NBLDA in Dong et al. (2016), the support vector machines (SVM) classifier in Meyer et al. (2014), and the k nearest neighbors (kNN) classifier in Ripley (1996). As described in Tan et al. (2014), SVM and kNN can be applied to discrete data without modification when $p > n$. In our experiments, we use the R packages "PoiClaClu" for PLDA and "e1071" for SVM. We also consider the number of nearest neighbors as $1, 3$, or 5 for kNN.

18.5.1 Simulation Design

We first generate the data from the negative binomial distribution:

$$X_{ki_kg} \sim \text{NB}(l_{i_k}\lambda_g d_{kg}, \phi), \tag{18.29}$$

and then set $X_{ki_kg} = 0$ with probability p_{ki_kg}, which is related to $d_{kg}l_{i_k}\lambda_g$ and the sequence depth. Note that X_{ki_kg} follows a negative binomial distribution when $p_{ki_kg} = 0$, and it follows a Poisson distribution when $p_{ki_kg} = 0$ and $\phi = 0$. In each simulation study, we compare the misclassification rates by varying only one parameter and fixing all others.

We consider the binary classification with $K = 2$. The parameters l_{i_k}, λ_g, and d_{kg} are set as the same as those in Witten (2011). Specifically, the size factors l_{i_k} are from the uniform distribution on [0.2, 2.2], the λ_g values are from the exponential distribution with expectation 25, and the $\ln d_{kg}$ values are from $N(0, \sigma^2)$. In each experiment, we generate n (the summation of all classes) samples as the training set and generate another n samples as the test set.

First, we generate the data from the Poisson distribution with $p_{ki_kg} = 0$ and $\phi = 0$. In Study 1, we fix $\sigma = 0.2$ and consider the case that the number of features $p = 100$ or 1000, 20% or 40% of which are differentially expressed between the two classes. Then we compare the misclassification rates of all methods with different sample sizes, $n = 8$, 16, 24, 40 and 64, for two classes. In Study 2, we investigate the performance of the methods when the proportions of differentially expressed genes are 0.2, 0.4, 0.6, 0.8, and 1.0 with fixed sample size $n = 8$ or 20. In this study, we also set $\sigma = 0.2$ and $p = 100$ or 1000.

Second, we generate the data from the zero-inflated Poisson distribution. In Study 3, we fix $\phi = 0.001$ and $\sigma = 0.2$ and consider the case that the number of features $p = 100$ or 1000, 20% or 40% of which are differentially expressed between the two classes. We set p_{ki_kg} as a random variable following a uniform distribution on [0,1] for each sample. Then we compare the misclassification rates of all methods with different sample sizes, $n = 8$, 16, 24, 40, and 64, for two classes. In Study 4, we investigate the performance of the methods when the proportions of differentially expressed genes are 0.2, 0.4, 0.6, 0.8, and 1.0 with fixed sample size $n = 8$ or 20. In this study, we also set $\phi = 0.001$, $\sigma = 0.2$, and $p = 100$ or 1000.

Third, we generate the data from the negative binomial distribution with $p_{ki_kg} = 0$. In Study 5, we fix $\phi = 1$ and $\sigma = 0.2$ and consider the case that the number of features $p = 100$ or 1000, 40% or 80% of which are differentially expressed (DE) between the two classes. Then, we compare the misclassification rates of all methods with different sample sizes, $n = 16$, 24, 40, 64, and 80, for two classes. In Study 6, we investigate the performance of the methods when the proportions of differentially expressed genes are 0.2, 0.4, 0.6, 0.8, and 1.0 with fixed sample size $n = 40$ or 80. In this study, we set $\phi = 1$, $\sigma = 0.2$ and $p = 100$ or 1000.

18.5.2 Simulation Results

For each simulated data, we use the misclassification rate for evaluation, which is computed by repeating the simulation 1000 times and taking an average over all the simulations. Here, we use the TMM method to normalize the data before classification. We report the misclassification rates along with various parameters in Figs. 18.1 and 18.2.

Studies 1 and 2 investigate the effect of different sample sizes and the proportions of differentially expressed genes for the binary classification when the data are from the Poisson distribution. Figure 8 of the supplement in Zhou et al. (2018) showed that the misclassification rates of all methods have decreased with an increasing number of sample sizes. ZIPLDA and PLDA perform significantly better than the other methods in all settings, especially for small number of genes. ZIPLDA performs nearly the same as PLDA. Figure 9 of the supplement in Zhou et al. (2018) showed that the misclassification rates of all methods are the same tend as Study 1. ZIPLDA and PLDA are better than the other methods in Study 2.

Studies 3 and 4 investigate the effect of different sample sizes and the proportions of differentially expressed genes for the binary classification when the data are from zero-inflated Poisson distribution. Figure 2 in Zhou et al. (2018) showed that the misclassification rates of all methods have decreased with an increasing number of sample sizes. ZIPLDA performs significantly better than the other methods in all settings, especially for small number of genes. Figure 3 in Zhou et al. (2018) showed that the misclassification rates of all methods are decreased with an increasing number of differentially expressed genes. ZIPLDA shows its superiority over the other methods in Study 4.

Studies 5 and 6 investigate the effect of different sample sizes and the proportions of differentially expressed genes for the binary classification when the data are from the negative binomial distribution. Figure 18.1 showed that the misclassification rates of NBLDA are totally better than the other methods except small sample size. ZIPLDA performs nearly the same as PLDA. Figure 18.2 showed that the misclassification rates of all methods are decreased with an increasing number of differentially expressed genes. NBLDA is better than the other methods in Study 6.

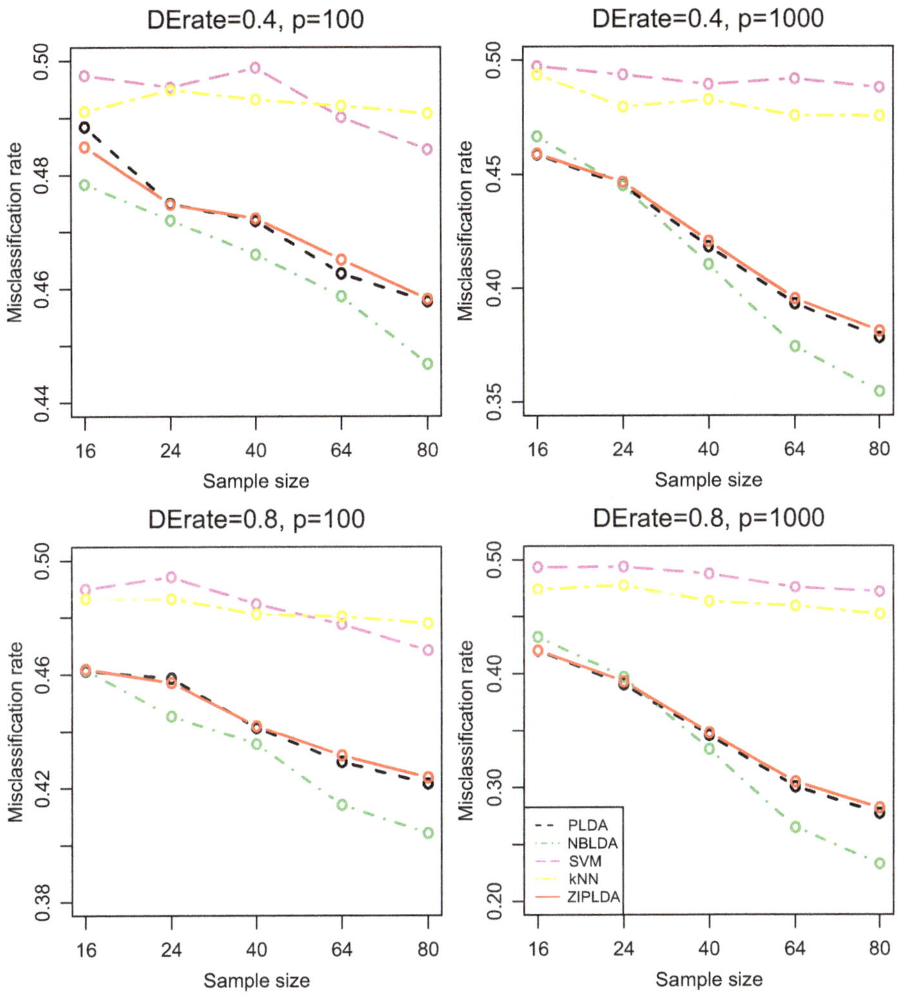

Fig. 18.1 The misclassification rates of all methods with different sample sizes for three classes (Study 5). Here, $\sigma = 0.2$ and the data are drawn from negative binomial distribution for all plots. The left panels have 100 features with different DE rates. The right panels have 1000 features with different DErates

18.6 Real Data Analysis

We apply the five methods to analyze two RNA-seq datasets: the cervical cancer dataset in Witten et al. (2010) and the Caucasian race dataset in Wang et al. (2008).

The first dataset is a microRNA-seq dataset from the Gene Expression Omnibus (GEO) (https://www.ncbi.nlm.nih.gov/geo/query/acc.cgi?acc=GSE79017) with access number GSE79017, which is also available in Wolenski et al. (2017).

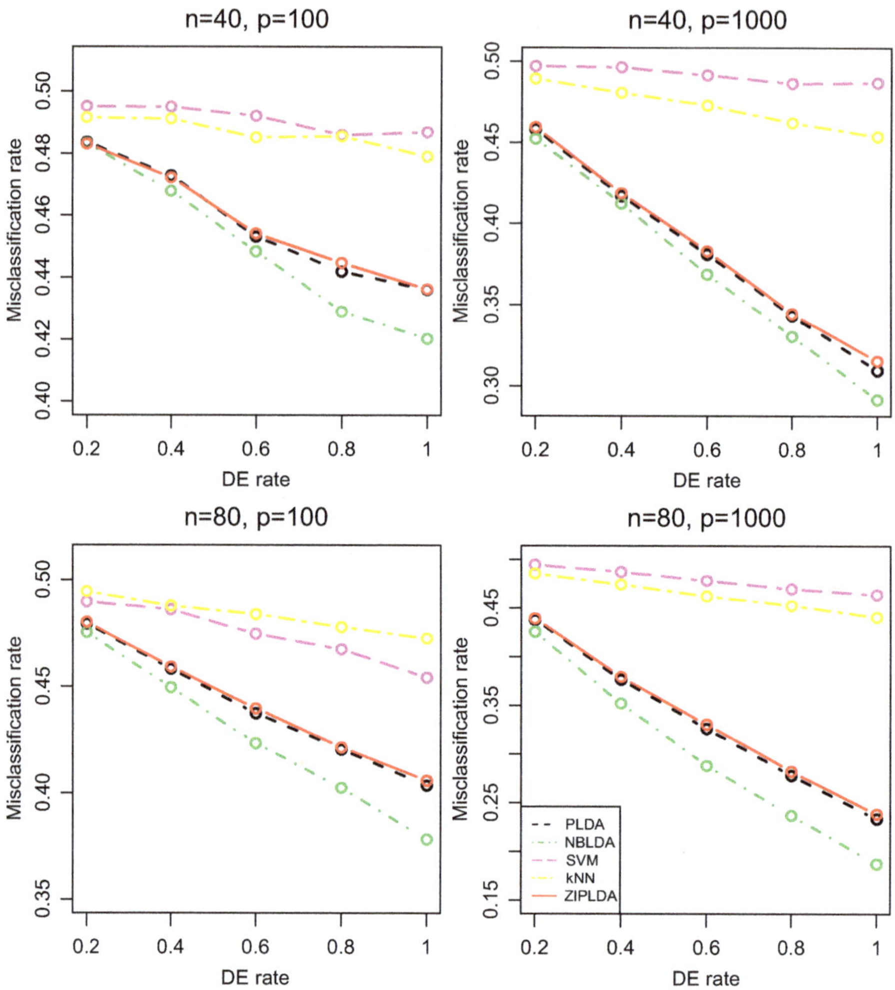

Fig. 18.2 The misclassification rates of all methods with different DE rates for three classes (Study 6). Here, $\sigma = 0.2$ and the data are drawn from Poison distribution for all plots. The left panels have 100 features with different sample sizes. The right panels have 1000 features with different sample sizes

MicroRNA is a type of small RNAs and the length is from 18 to 30 nucleotides. MicroRNA plays important regulatory roles in diverse biological processes (Birchler and Kavi 2008; Stefani and Slack 2008). There are three classes in the dataset, including 12 samples from liver, 18 samples from urine, and 18 samples from plasma.

The second dataset is a Caucasian race dataset from ReCount (http://bowtie-bio.sourceforge.net/recount/), an online resource consisting of RNA-seq gene count datasets built from the raw data. The dataset is also released in Wang et al. (2008).

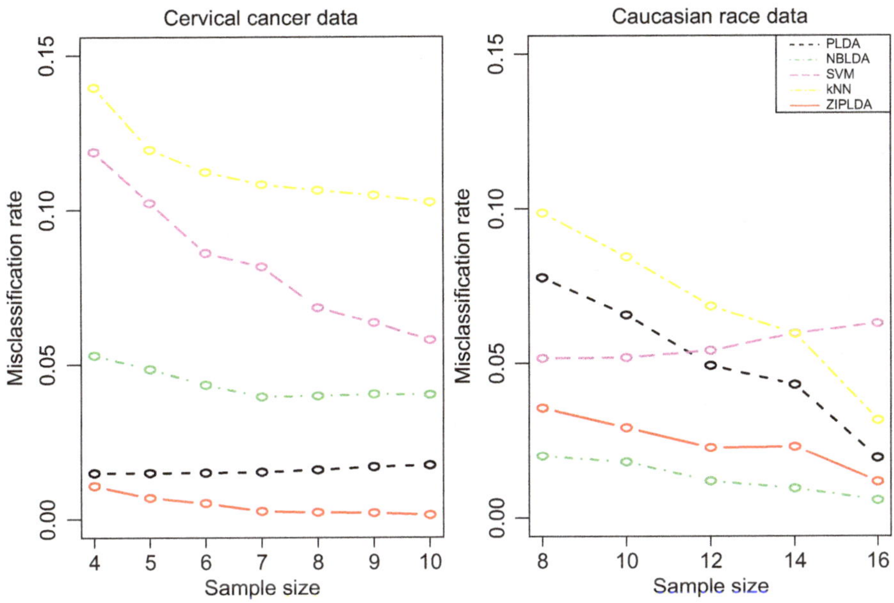

Fig. 18.3 The misclassification rates of all methods for the cervical cancer dataset and the Caucasian race dataset

There are 11 Caucasian race and 9 non-Caucasian race samples with measurements on 52,580 transcripts.

To compare the classification methods, we randomly draw some of the samples from each class to build the training set, and set the rest samples as the test set. We also apply the TMM method to normalize the data before classification. We repeat 1000 times and calculate the average misclassification rates for each method. For the cervical cancer dataset (see the left panel of Fig. 18.3), it is evident that the performance of ZIPLDA is better than those of the other methods for different sample sizes. For the Caucasian rate dataset (see the right panel of Fig. 18.3), however, NBLDA outperforms all other methods.

18.7 Discussion

Discrimination of different disease types for next-generation sequencing data is of great importance in medical research, such as disease diagnosis and drug discovery. In this chapter, we introduce three discriminant analysis methods and also three normalization methods for next-generation sequencing data. Simulations and two real data examples are examined to compare these methods.

Unlike microarray data, we note that the current methods are still far less satisfactory for classification and normalization of next-generation sequencing data. Many problems remain to be solved, such as the very high overdispersion in RNA-seq data, e.g., when the dispersion is larger than 5. In such situations, the existing discriminant methods may not provide the optimal performance in practice. As a future work, a mixture distribution with a point mass at zero and a negative binomial distribution can be considered for analyzing RNA-seq data.

Acknowledgements The authors thank the editor and two referees for their helpful comments that have led to some significant improvements of this chapter. Yan Zhou's research was supported by the National Natural Science Foundation of China (Grant No. 11701385), National Statistical Research Project (Grant No. 2017LY56), the Doctor Start Fund of Guangdong Province (Grant No. 2016A030310062), and the National Social Science Foundation of China (Grant No. 15CTJ008). Junhui Wang's research was supported by HK RGC grants GRF-11302615 and GRF-11331016. Yichuan Zhao's research was partially supported by the NSF Grant DMS-1406163 and NSA Grant H98230-12-1-0209. Tiejun Tong's research was supported by the Health and Medical Research Fund (Grant No. 04150476) and the National Natural Science Foundation of China (Grant No. 11671338).

References

Anders, S., & Huber, W. (2010). Differential expression analysis for sequence count data. *Genome Biology, 11*, R106.

Birchler, J. A., & Kavi, H. H. (2008). Slicing and dicing for small RNAs. *Science, 320*, 1023–1024.

Bolstad, B. M., Irizarry, R. A., Astrand M., & Speed, T. P. (2003). A comparison of normalization methods for high density oligonucleotide array data based on variance and bias. *Bioinformatics, 19*, 185–193.

Brawand, D., Soumillon, M., Necsulea, A., Julien, P., Csardi, G., Harrigan, P., et al. (2011). The evolution of gene expression levels in mammalian organs. *Nature, 478*, 343–348.

Bullard, J. H., Purdom, E., Hansen, K. D., & Dudoit, S. (2010). Evaluation of statistical methods for normalization and differential expression in mRNA-Seq experiments. *BMC Bioinformatics, 11*, 94.

Casella, G., & Berger, R. L. (2002). *Statistical inference*. Pacific Grove, CA: Duxbury.

Chen, C. M., Lu, Y. L., Sio, C. P., Wu, G. C., Tzou, W. S., & Pai, T. W. (2014). Gene ontology based housekeeping gene selection for RNA-seq normalization. *Methods, 67*, 354–363.

Clemmensen, L., Hastie, T., Witten, D., & Ersbfill, B. (2011). Sparse discriminant analysis. *Technometrics, 53*, 406–413.

Cloonan N., Forrest A. R., Kolle G., Gardiner B. B., Faulkner G. J., Brown M. K., et al. (2008). Stem cell transcriptome profiling via massive-scale mRNA sequencing. *Nature Methods, 5*, 613–619.

Dillies, M. A., Rau, A., Aubert, J., Hennequet-Antier, C., Jeanmougin, M., Servant, N., et al. (2013). A comprehensive evaluation of normalization methods for Illumina high-throughput RNA sequencing data analysis. *Briefings in Bioinformatics, 14*, 671–683.

Dong, K., Zhao, H., Tong, T., & Wan, X. (2016). NBLDA: Negative binomial linear discriminant analysis for RNA-Seq data. *BMC Bioinformatics, 17*, 369.

Dudoit, S., Fridlyand, J., & Speed, T. P. (2002). Comparison of discrimination methods for the classification of tumors using gene expression data. *Journal of the American Statistical Association, 97*, 77–87.

Friedman, J. H. (1989). Regularized discriminant analysis. *Journal of the American Statistical Association, 84*, 165–175.

Grosenick, L., Greer, S., & Knutson, B. (2008). Interpretable classifiers for FMRI improve prediction of purchases. *IEEE Transactions on Neural Systems and Rehabilitation Engineering, 16*, 539–548.

Hastie, T., Buja, A., & Tibshirani, R. (1995). Penalized discriminant analysis. *The Annals of Statistics, 23*, 73–102.

Hastie, T., Tibshirani, R., & Buja, A. (1994). Flexible discriminant analysis by optimal scoring. *Journal of the American Statistical Association, 89*, 1255–1270.

Hastie, T., & Tibshirani, R. (1996). Discriminant analysis by Gaussian mixtures. *Journal of the Royal Statistical Society. Series B (Methodological), 58*, 155–176.

Huang, H. H. (2016). Ensemble method of k-mer and natural vector for the phylogenetic analysis of multiple-segmented viruses. *Journal of Theoretical Biology, 398*, 136–144.

Huang, H. H., & Girimurugan, S. B. (2018). A novel real-time genome comparison method using discrete wavelet transform. *Journal of Computational Biology, 25*(4), 406–416.

Huang, H. H., & Yu, C. (2016). Clustering DNA sequences using the out-place measure with reduced n-gram. *Journal of Theoretical Biology, 406*, 61–72.

Huang, H. H., Yu, C., Hernandez, T., Zheng, H., Yau, S. C., He, R.L., et al. (2014). Global comparison of multiple-segmented viruses in 12-dimensional genome space. *Molecular Phylogenetics and Evolution, 81*, 29–36.

Huang, S., Tong, T., & Zhao, H. (2010). Bias-corrected diagonal discriminant rules for high-dimensional classification. *Biometrics, 66*, 1096–1106.

Leng, C. (2008). Sparse optimal scoring for multiclass cancer diagnosis and biomarker detection using microarray data. *Computational Biology and Chemistry, 32*, 417–425.

Lin, B., Zhang, L., & Chen, X. (2014). LFCseq: A nonparametric approach for differential expression analysis of RNA-seq data. *BMC Genomics, 15*, S7.

Lorenz, D. J., Gill, R. S., Mitra, R., & Datta, S. (2014). Using RNA-seq data to detect differentially expressed genes. In S. Datta & D. Nettleton (Eds.), *Statistical analysis of next generation sequencing data* (pp. 25–49). New York: Springer.

Love, M. I., Huber, W., & Anders, S. (2014). Moderated estimation of fold change and dispersion for RNA-seq data with DESeq2. *Genome Biology, 15*, 550.

Mai, Q., Zou, H., & Yuan, M. (2012). A direct approach to sparse discriminant analysis in ultra-high dimensions. *Biometrika, 99*, 29–42.

Mardis, E. R. (2008). Next-generation DNA sequencing methods. *Annual Review of Genomics and Human Genetics, 9*, 387–402.

Marioni, J. C., Mason, C. E., Mane, S. M., Stephens, M., & Gilad, Y. (2008). RNA-seq: An assessment of technical reproducibility and comparison with gene expression arrays. *Genome Research, 18*, 1509–1517.

Meyer, O., Bischl, B., & Weihs, C. (2014). Support vector machines on large data sets: simple parallel approaches. In M. Spiliopoulou, L. Schmidt-Thieme, & R. Janning (Eds.), Data analysis, machine learning and knowledge discovery. Studies in Classification, Data Analysis, and Knowledge Organization (pp. 87–95). Cham: Springer.

Morin, R. D., O'Connor, M. D., Griffith, M., Kuchenbauer, F., Delaney A., Prabhu A. L., et al. (2008). Application of massively parallel sequencing to micro RNA profiling and discovery in human embryonic stem cells. *Genome Research, 18*, 610–621.

Morozova, O., Hirst, M., & Marra, M. A. (2009). Applications of new sequencing technologies for transcriptome analysis. *Annual Review of Genomics and Human Genetics, 10*, 135–151.

Mortazavi, A., Williams, B. A., McCue, K., Schaeffer, L., & Wold, B. (2008). Mapping and quantifying mammalian transcriptomes by RNA-seq. *Nature Methods, 5*, 621–628.

Mouatassim, Y., & Ezzahid, E. H. (2012). Poisson regression and Zero-inflated Poisson regression: Application to private health insurance data. *European Actuarial Journal, 2*, 187–204.

Nagalakshmi, U., Wang, Z., Waern, K., Shou, C., Raha, D., Gerstein, M., et al. (2008). The transcriptional landscape of the yeast genome defined by RNA sequencing. *Science, 320*, 1344–1349.

Ridout, M., Demetrio, C. G. B., & Hinde, J. (1998). Models for count data with many zeros. In *International biometric conference*, Cape Town.

Ripley, B. D. (1996). *Pattern recognition and neural networks*. New York: Cambridge.

Robinson, M. D., McCarthy, D. J., & Smyth, G. K. (2010). edgeR: A bioconductor package for differential expression analysis of digital gene expression data. *Bioinformatics, 26*, 139–140.

Robinson, M. D., & Oshlack, A. (2010). A scaling normalization method for differential expression analysis of RNA-seq data. *Genome Biology, 11*, R25.

Robinson, M. D., & Smyth, G. K. (2008). Small-sample estimation of negative binomial dispersion, with applications to SAGE data. *Biostatistics, 9*, 321–332.

Shendure, J., & Ji, H. (2008). Next-generation DNA sequencing. *Nature Biotechnology, 26*, 1135–1145.

Stefani, G., & Slack, F. J. (2008). Small non-coding RNAs in animal development. *Nature Reviews Molecular Cell Biology, 9*, 219–230.

Tan, K. M., Petersen, A., & Witten, D. M. (2014). Classification of RNA-seq data. In *Statistical analysis of next generation sequencing data* (pp. 219–246). New York: Springer.

The Cancer Genome Atlas Research Network (2014). Comprehensive molecular characterization of gastric adenocarcinoma. *Nature, 513*, 202–209.

Wald, P. W., & Kronmal, R. A. (1977). Discriminant functions when covariances are unequal and sample sizes are moderate. *Biometrics, 33*, 479–484.

Wang, E. T., Sandberg, R., Luo, S. J., Khrebtukova, I., Zhang, L., Mayr, C., et al. (2008). Alternative isoform regulation in human tissue transcriptomes. *Nature, 456*, 470–476.

Wang, Z., Gerstein, M., & Snyder, M. (2009). RNA-Seq: A revolutionary tool for transcriptomics. *Nature Reviews Genetics, 10*, 57–63.

Witten, D. M. (2011). Classification and clustering of sequencing data using a Poisson model. *The Annals of Applied Statistics, 5*, 2493–2518.

Witten, D. M., Tibshirani, R., Gu, S. G., Fire, A., & Lui, W. (2010). Ultra-high throughput sequencing-based small RNA discovery and discrete statistical biomarker analysis in a collection of cervical tumours and matched controls. *BMC Biology, 8*, 58.

Wolenski, F. S., Shah, P., Sano, T., Shinozawa, T., Bernard, H., Gallacher, M. J., et al. (2017). Identification of microRNA biomarker candidates in urine and plasma from rats with kidney or liver damage. *Journal of Applied Toxicology, 37*, 278–286.

Zhou, Y., Wan, X., Zhang, B. X., & Tong, T. (2018). Classifying next-generation sequencing data using a zero-inated Poisson model. *Bioinformatics, 34*(8), 1329–1335.

Zhou, Y., Wang, G., Zhang, J., & Li, H. (2017). A hypothesis testing based method for normalization and differential expression analysis of RNA-Seq data. *PLoS One, 12*, e0169594.

Zhou, Y., Zhang, B., Li, G., Tong, T., & Wan, X. (2017). GD-RDA: A new regularized discriminant analysis for high dimensional data. *Journal of Computational Biology, 24*, 1099–1111.

Zhou, Y., Zhu, J. D., Tong, T., Wang, J. H., Lin, B. Q., & Zhang, J. (submitted). A statistical normalization method and differential expression analysis for RNA-seq data between different species.

Part V
Survival Analysis

Chapter 19
On the Landmark Survival Model for Dynamic Prediction of Event Occurrence Using Longitudinal Data

Yayuan Zhu, Liang Li, and Xuelin Huang

19.1 Introduction

Predicting the risk of adverse clinical events is important to both medical research and clinical practice. When the time of the occurrence of the adverse event differs among patients, the prediction is often carried out through survival regression models such as the Cox proportional hazard model, with predictors being the baseline variables and the outcome being the time from baseline to the event occurrence (Steyerberg 2009). In many longitudinal cohort studies and electronic health record databases, the prognostic information is collected longitudinally over a series of clinical visits, until the occurrence of the adverse clinical event. Building a prediction model with only the prognostic variables at baseline may be suboptimal because it does not fully utilize the large amount of longitudinal information collected at the follow-up visits. Since these follow-up visits are temporally closer to the occurrence of the event of interest, the data at these visits may have stronger association with the risk of the event. From the perspective of clinical practice, the patient and physician may want to review the disease progress and update the prognosis at each clinical visit when new data become available. Developing a prediction model with baseline predictors and applying it at the follow-up visits is generally not appropriate because that model fails to adjust for the changing at-risk population, i.e., the population of patients who have not experienced the adverse event at the time of the follow-up visit.

Our research is motivated by a study of chronic myelogenous leukemia (CML) with a focus on the early detection of disease progression (Quintás-Cardama et al. 2014). CML is a myeloproliferative disorder of blood stem cells (Sawyers 1999).

Y. Zhu · L. Li (✉) · X. Huang
Department of Biostatistics, MD Anderson Cancer Center, Houston, TX, USA
e-mail: LLi15@mdanderson.org

Y. Zhao, D.-G. Chen (eds.), *New Frontiers of Biostatistics and Bioinformatics*, ICSA Book Series in Statistics, https://doi.org/10.1007/978-3-319-99389-8_19

The causative molecular defect is the BCR-ABL protein (Faderl et al. 1999). BCR-ABL fusion gene has been found in up to 95% of the patients who were diagnosed with CML (Yan et al. 2017). An increase of the BCR-ABL transcript levels is often noted before clinical symptoms of progression in clinical practice. Therefore, BCR-ABL can be viewed as an important biomarker to predict the time to progression (Quintás-Cardama et al. 2014). Imatinib has been used as the first generation Tyrosine kinase inhibitor (TKI) to treat patients with CML by inhibiting the expression of BCR-ABL. In spite of the significant efficacy, resistance to this agent has been widely considered as a notable clinical issue (Gorre and Sawyers 2002). Dasatinib, as the second generation TKI, is indicated for the treatment of patients with CML who are resistant to or intolerant of imatinib (Wong 2009) and has been found to significantly improve the patient outcome (Hochhaus et al. 2007). The study that we considered was a randomized trial that used dasatinib to treat patients with chronic phase CML after failure of imatinib therapy and compared different dose schedules of dasatinib. Our analysis set is composed of 618 patients who are resistant to imatinib. Follow-up is defined as the time between the start and the end of certain dose schedule, and our interest is in the prediction of time to progression under this dose schedule. In this data set, the follow up is up to 6.5 years with a median of 2.5 years, and patients have on average 8–9 measurements of BCR-ABL per person before disease progression occurs or dose schedule ends. Figure 19.1 shows the Kaplan-Meier estimate of the marginal progression-free survival and the longitudinal trajectories of log(BCR-ABL) for 20 randomly selected patients. We normalized the biomarker BCR-ABL to be between 0 and 100. Left panel in Fig. 19.1 indicates that progression rate is low among the patients who are treated with dasatinib; only up to 25% of the patients had progression while using dasatinib, and most of progressions occurred within the first 3 years. Right panel in Fig. 19.1 displays highly diverse patterns of BCR-ABL trajectories on individual level. In fact, patients were scheduled to be followed every 3 months in the first year, every 6 months in the second year, and annually thereafter, though the actual visit times vary among different subjects. Later, we will use this CML data set as an illustration of dynamic prediction to predict the risk of disease progression using BCR-ABL as a time-varying predictor.

Dynamic prediction methodology has been studied in the statistical literature to generate subject-specific prediction of the probability of the event using longitudinal data at any time during follow-up in the aforementioned context (Rizopoulos 2011; van Houwelingen and Putter 2011). Let $i = 1, 2, \ldots, n$ indicate the n subjects in the dataset from which the prediction model is to be developed. For the i-th subject, T_i denotes the time when the event of interest occurs and C_i denotes the censoring time. We observe $Y_i = \min(T_i, C_i)$ and the censoring indicator $\delta_i = 1\{T_i \le C_i\}$. Let $\boldsymbol{X}_i(s)$ denote the vector of the predictors at time s. The predictors $\boldsymbol{X}_i(s)$ may include proper numerical summaries of the longitudinal history of subject i from baseline to the time of prediction s, including time-invariant covariates and both internal and external time-dependent covariates (Kalbfleisch and Prentice 2002). The predictor $\boldsymbol{X}_i(s)$ is defined on $[0, T_i)$ but can only be observed at the measurement times $\{t_{ij}; j = 1, 2, \ldots, n_i\}$, where $0 = t_{i1} < \ldots < t_{in_i} < Y_i$.

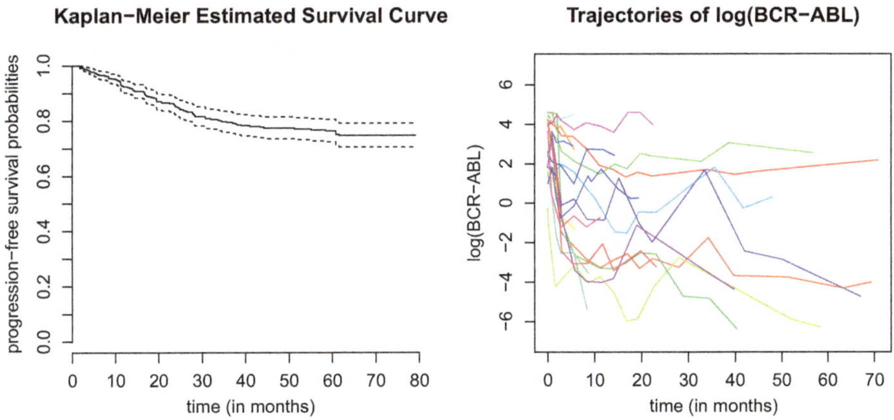

Fig. 19.1 Kaplan-Meier curve for marginal progression-free survival (left) and the longitudinal trajectories of log(BCR-ABL) for 20 randomly selected patients (right). Each color represents the trajectory for one individual

Denote the longitudinal data by $\boldsymbol{X}_{ij} = \boldsymbol{X}_i(t_{ij})$. The data from the n subjects, consisting of $\{Y_i, \delta_i, \boldsymbol{X}_{ij}, t_{ij} | i = 1, 2, \ldots, n; j = 1, 2, \ldots, n_i\}$, are independent and identically distributed from the population to which the prediction model is to be applied. The measurement times are assumed to be non-informative in the sense that $\{t_{ij}\}$ are independent of $\boldsymbol{X}_i(.)$ and T_i. The censoring time C_i is assumed to be independent of T_i, $\boldsymbol{X}_i(.)$ and $\{t_{ij}\}$. The goal of dynamic prediction is to estimate the predicted probability

$$\text{pr}(T \in (s, s+\tau] | T > s, \boldsymbol{X}(s)) \tag{19.1}$$

for any new subject from the same at-risk population (i.e., did not experience the event) at the time of prediction s. The prediction is based on $\boldsymbol{X}(s)$ and the prediction horizon is τ. For example, (19.1) may be the subject's probability of disease occurrence in the next τ years, given that this subject survived s years since baseline without the disease, and conditional on $\boldsymbol{X}(s)$, the average biomarker test results in the last 12 months prior to s.

There are generally two approaches to dynamic prediction: an approach based on the joint model of longitudinal and time-to-event data (Blanche et al. 2015; Rizopoulos 2011; Rizopoulos et al. 2014; Taylor et al. 2013) and an approach based on the landmark or "partly conditional" model of survival (Huang et al. 2005; van Houwelingen 2007; van Houwelingen and Putter 2011; Zheng and Heagerty 2005). Each approach has advantages and disadvantages (Li et al. 2017). Generally speaking, the landmark approach is computationally much simpler than joint modeling and may be the only viable option when there are multiple time-varying predictor variables, predictors with diverse nonlinear longitudinal trajectories, or categorical time-varying predictors (e.g., hospitalization episodes and medication use). Therefore, the landmark approach has broader scope of application. Recent empirical research suggests that the two approaches have comparable prediction accuracy for scenarios to which both are applicable (Maziarz et al. 2016).

This paper studies the following landmark Cox model (Li et al. 2017) and its extensions. Let $\mathcal{R}(s)$ be the population of subjects at risk at time s, i.e., subjects with $T > s$. The landmark Cox model specifies that, for a subject in $\mathcal{R}(s)$, the conditional distribution of the residual survival time $T(s) = T - s$ given predictors $\boldsymbol{X}(s)$ follows a Cox model. Its survival function, evaluated at time $s + u$ $(u \geq 0)$ and conditional on the covariates at time s, is given by

$$\text{pr}(T(s) > u|\boldsymbol{X}(s)) = \exp\left[-\exp\{\boldsymbol{X}(s)^T\boldsymbol{\beta}(s)\}\int_0^u h_0(v, s)dv\right] \quad (u \geq 0). \quad (19.2)$$

Here s is called the landmark time, which is the time that has elapsed since the baseline. It can also be viewed as the time of prediction when the model is used to estimate the predicted probabilities. A prediction at landmark time s can only be made for subjects in $\mathcal{R}(s)$. The time u denotes the time elapsed since s. For a given s, model (19.2) is a Cox model with time-independent covariates in the sense that $\boldsymbol{X}(s)$ does not vary with u. But $\boldsymbol{X}(s)$ may vary with the landmark time s. All the parameters of this model, including the baseline hazard function $h_0(u; s)$ and the log hazard ratio $\boldsymbol{\beta}(s)$, may vary with the landmark time s. Both the univariate function $\boldsymbol{\beta}(s)$ and the non-negative bivariate function $h_0(u; s)$ are assumed to take smooth but otherwise unrestricted shapes. This is a general class of landmark dynamic prediction models. The partly conditional survival model (Maziarz et al. 2016; Zheng and Heagerty 2005) is a special case of (19.2) with the baseline hazard function not dependent on s. The super Cox models of van Houwelingen and Putter (2008, 2011) involve some additional assumptions on the shape of $\boldsymbol{\beta}(s)$ and $h_0(u, s)$. Nonparametric estimation of this model was studied by Li et al. (2017). With model (19.2), the predicted probability (19.1) is estimated by

$$1 - \exp\{-H_0(\tau; s)\theta_i(s)\},$$

where $H_0(\tau; s) = \int_0^\tau h_0(u; s)\text{d}u$ is the baseline cumulative hazard function evaluated at the prediction horizon τ, and $\theta_i(s) = \exp\{\boldsymbol{X}_i(s)^T\boldsymbol{\beta}(s)\}$ is the exponentiated linear predictor.

Here is an illustration of the aforementioned landmark dynamic prediction model in the context of the CML example. The research interest is to predict the risk of disease progression at a selected landmark time s among the at-risk patients, using all the available prognostic information collected up to that time. To illustrate, we select two landmark times, $s = 6$ and $s = 18$ months, to predict the risk of progression in 1 year (i.e., prediction horizon τ is 12 months). In addition to the time-varying biomarker, BCR-ABL, we also include gender, age at baseline ($\geq$60), and dose schedule (four categories) as subject-specific predictors in the prediction model (19.2). We plot the predicted conditional survival curves $\exp\{-H_0(\tau; s)\theta_i(s)\}$ and the observed trajectories of log(BCR-ABL) for two randomly selected patients at each s in Fig. 19.2. Prediction performance is

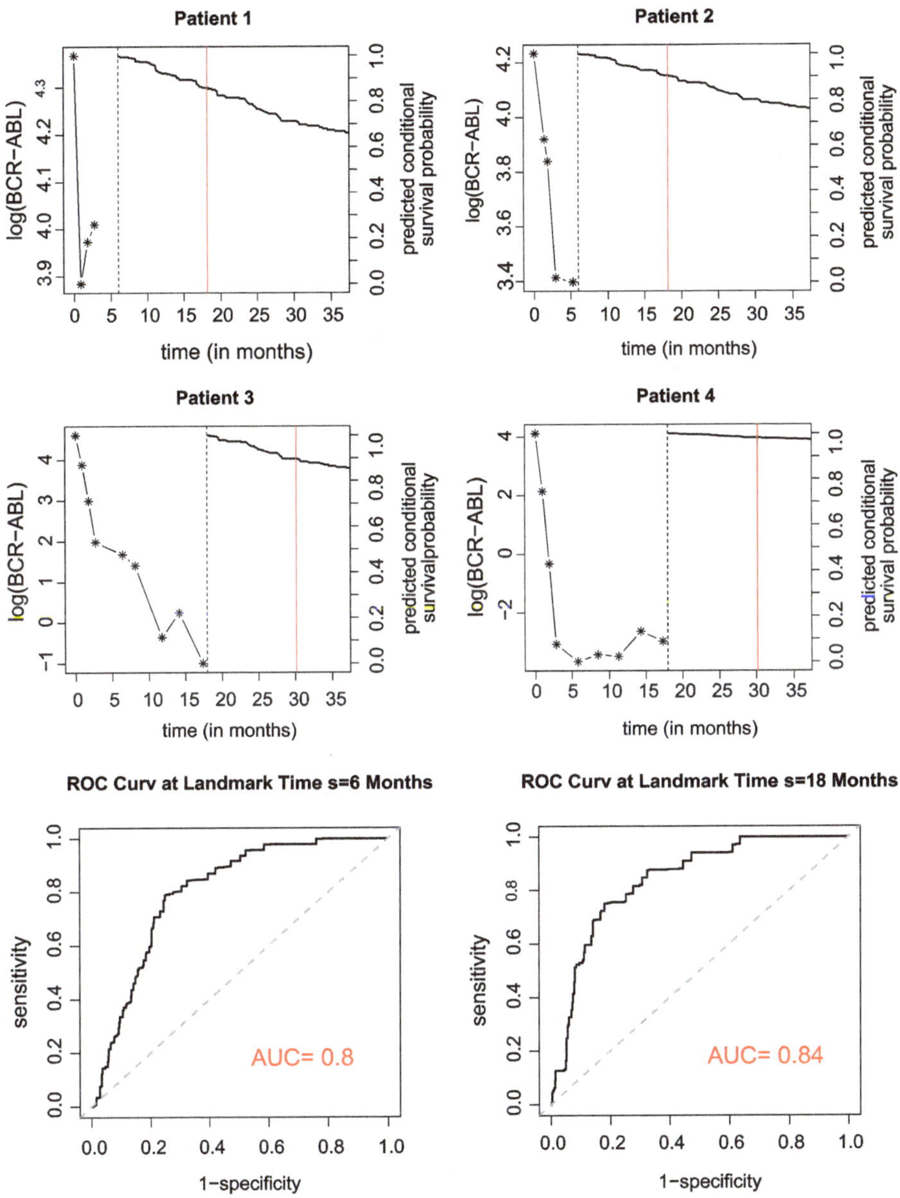

Fig. 19.2 Trajectories of log(BCR-ABL) along with the predicted survival probability curves for four patients; black dashed vertical line denotes the landmark time, red solid vertical line denotes where the prediction is evaluated (prediction horizon is 12 months). The two plots at the bottom are the time-dependent receiver operating characteristic (ROC) curves for prediction at landmark time $s = 6$ months and $s = 18$ months

evaluated by a time-dependent receiver operating characteristic (ROC) curve and the area under the ROC (i.e., AUC). The time-dependent ROC curves and the AUCs are estimated based on a nonparametric method proposed by Li et al. (2018), which has been implemented in the R package *tdROC*. Figure 19.2 shows that the AUC of the prediction at $s = 6$ and $s = 18$ months is, respectively, 0.80 and 0.84. Please note that the results above serve as an illustration of the use of a landmark dynamic prediction model. We have focused on the simple case where separate Cox models are fitted at two selected landmark times. More sophisticated estimation methodology has been developed for more complicated situations where it is necessary to estimate $\beta(s)$ as a smooth function with irregularly spaced clinical visit times (e.g., Li et al. (2017)), but that is not related to the main focus of this article and will not be discussed here.

Model (19.2) implies infinitely many Cox models, each defined at a different landmark time s and linking $T(s)$ and $\boldsymbol{X}(s)$. A fundamental difficulty in the current research of landmark dynamic prediction models is that it is unclear whether a joint probability distribution of $\{T, C, \boldsymbol{X}(s)\}$ exists that satisfies model (19.2) and, if so, how to simulate datasets from such a distribution for the purpose of simulation studies. This difficulty is widely recognized (Li et al. 2017; Maziarz et al. 2016; van Houwelingen and Putter 2011; Zheng and Heagerty 2005). As a result, statistical properties of the landmark model sometimes have to be studied in simulations in which the datasets are simulated from the shared parameter model (Huang et al. 2005; Maziarz et al. 2016). In such situations, the landmark model operates under misspecification, which complicates the interpretation of its numerical performance. This problem also makes it difficult to study the asymptotic property of the estimation procedure theoretically. Zheng and Heagerty (2005) presented a method for data simulation but it works only with constant $\boldsymbol{\beta}(s)$ functions, a single longitudinal biomarker variable on equally spaced measurement times, and a positive stable distribution assumption on the biomarker. Due to this unsolved problem, the landmark model is sometimes viewed as a working model or an algorithm (Maziarz et al. 2016) instead of a comprehensive probability model (van Houwelingen and Putter 2011).

In this paper, we show that there exists a joint distribution of longitudinal and survival data that satisfies the landmark Cox model (19.2) without additional restrictions other than the model assumptions themselves. We provide an algorithm to generate data from this joint distribution. The work in this paper may facilitate further research on the landmark survival model in both the theoretical and empirical fronts. In Sect. 19.2, we derive the joint distribution and propose a data generating algorithm. We further demonstrate that the landmark Cox model can be generalized to a landmark linear transformation model that includes both the Cox model and the accelerated failure time model as special cases. This extension gives greater flexibility to the landmark dynamic prediction models than the models presently in the literature, but the proposed joint distribution and data generating algorithm can still be extended to this situation (Sect. 19.3). We present a simulation to demonstrate the proposed methodology in Sect. 19.4.

19.2 Joint Distribution of the Longitudinal and Time-to-Event Data for the Landmark Cox Model

According to model (19.2), the Cox model at landmark time s can be expressed as

$$G_0(T(s); s) = -\boldsymbol{X}(s)^T \boldsymbol{\beta}(s) + \epsilon(s) = -\log\theta(s) + \epsilon(s) ,$$

where $G_0(t; s) = \log H_0(t; s)$ and $\epsilon(s)$ has a standard extreme value distribution with density function $f(x) = \exp(x - e^x)$, $x \in (-\infty, \infty)$ (Cheng et al. 1995). We first consider the situation where $\theta(s)$ among the subjects in $\mathcal{R}(s)$ follows a gamma distribution with shape parameter $\alpha(s)$ and rate parameter $\eta(s)$. Under this situation, the marginal survival function of $T(s)$ can be calculated as

$$\mathrm{pr}(T(s) > u) = \left[\frac{\eta(s)}{H_0(u; s) + \eta(s)}\right]^{\alpha(s)} , \quad \forall\, u \geq 0 . \tag{19.3}$$

The conditional distribution of $\theta(s)$ given $W(s) = G_0(T(s); s)$ is

$$\begin{aligned} f(\theta(s)|W(s)) &\propto f(W(s)|\theta(s))f(\theta(s)) \\ &\propto \theta(s)^{(\alpha(s)+1)-1} \exp\left\{-(\eta(s) + e^{W(s)})\theta(s)\right\} \end{aligned} \tag{19.4}$$

This is a gamma distribution with shape parameter $\alpha(s) + 1$ and rate parameter $\eta(s) + e^{W(s)} = \eta(s) + H_0(T(s); s)$.

A key observation of this paper is that the following equality holds:

$$\mathrm{pr}(T(s) > u) = \mathrm{pr}(T > s + u|T > s) , \quad \forall\, u \geq 0 ,\ s \geq 0. \tag{19.5}$$

The left-hand side of this equality is the survival probability of the residual survival time among subjects in the risk set $\mathcal{R}(s)$. The conditional probability on the right-hand side is specified for all the subjects in the population. The probabilities on both sides of the equality are marginal probabilities involving survival time T only, i.e., not conditional on $\boldsymbol{X}(.)$.

Both (19.3) and (19.5) imply that

$$\left[\frac{H_0(u; s) + \eta(s)}{\eta(s)}\right]^{\alpha(s)} = \left[\frac{H_0(s + u; 0) + \eta(0)}{H_0(s; 0) + \eta(0)}\right]^{\alpha(0)} .$$

Hence, given $\alpha(s)$, $\eta(s)$ and $H_0(u; 0)$, $H_0(u; s)$ is uniquely determined by

$$H_0(u; s) = \eta(s)\left\{\left[\frac{H_0(s + u; 0) + \eta(0)}{H_0(s; 0) + \eta(0)}\right]^{\alpha(0)/\alpha(s)} - 1\right\} . \tag{19.6}$$

It can be shown that $H_0(u; s)$ is a monotone increasing function in u at any fixed s and equals 0 when $u = 0$. Therefore, $H_0(u; s)$ as given by this equation is a proper cumulative hazard function.

The two parameter functions $\alpha(s)$ and $\eta(s)$ characterize the distribution of $\theta(s)$, the exponentiated linear predictor, in the time-varying risk set $\mathcal{R}(s)$. The univariate function $H_0(t; 0)$ is the baseline cumulative hazard function of the Cox model at baseline (i.e., $s = 0$). Equation (19.6) shows that these three functions uniquely determine the baseline cumulative hazard function of all the subsequent Cox models (i.e., $\forall s > 0$).

Based on the result above, we propose the following algorithm to generate data from the joint distribution of the longitudinal data $\boldsymbol{X}(s)$ and survival data $\{T, C\}$ that satisfies the landmark Cox model (19.2).

1. Prespecify a time grid for the longitudinal measurement times of all subjects, denoted by $s_1 = 0 < s_2 <, \ldots, < s_K$.
2. Prespecify $\alpha(s)$, $\eta(s)$, $\beta(s)$, $H_0(t; 0)$, and calculate $H_0(u; s)$ using Eq. (19.6).
3. For subject $i = 1, 2, \ldots, n$,

   (a) Simulate T_i from its marginal distribution. Since $T_i = T_i(0)$, the survival function of this distribution is given by (19.3), with $s = 0$. Let K_i be the number of longitudinal measurement times that fall within $[0, T_i)$. Calculate $T_i(s) = T_i - s$ with $s = s_1, s_2, \ldots, s_{K_i}$.
   (b) We generate $\{\theta_i(s); s = s_1, \ldots, s_{K_i}\}$ from a K_i-dimensional Gaussian copula distribution with correlation parameter ρ. Based on (19.4), the marginal distribution of $\theta_i(s_j)$ $(j = 1, 2, \ldots, K_i)$ in the copula is a gamma distribution with shape parameter $\alpha(s_j) + 1$ and rate parameter $\eta(s_j) + H_0(T_i(s_j); s_j)$.
   (c) When there is a single covariate in $\boldsymbol{X}(s)$, calculate $X_i(s) = \log \theta_i(s)/\beta(s)$ for $s = s_1, s_2, \ldots, s_{K_i}$. When there are M covariates $(M > 1)$, write $\log \theta_i(s) = \sum_{m=1}^{M} \beta_m(s) X_{mi}(s)$. Since $\theta_i(s)$ is generated in Step 3(b), and all the $\beta_m(.)$ functions are pre-specified, any time-invariant or time-dependent longitudinal covariate processes $X_{mi}(s)$ $(m = 1, 2, \ldots, M)$ that satisfy the linear constraint above are sufficient.

4. Generate a random censoring time C_i for each subject and censor both T_i and the longitudinal data process.

Remark 19.1 The algorithm above uses a common grid of longitudinal measurement times for all the subjects. In applications where it is desirable to allow for irregularly spaced measurement times, two approaches can be used. The first one is to use a very dense grid of time points and large ρ, generate all the longitudinal data, and then randomly "knock out" some data. The second approach is to randomly generate different longitudinal measurement times for each subject in Step 1, and parameterize the copula so that the correlation among adjacent $\theta(s)$ is adaptive to their time gap. The autocorrelation structure in Generalized Estimating Equations analysis can be used for this purpose (Diggle et al. 2002).

Remark 19.2 In Step 3(b), when there is an s_j where $\beta(s_j) = 0$, there is no association between $X_i(s_j)$ and $T_i(s_j)$ and $X_i(s_j)$ cannot be generated by dividing $\log\theta_i(s_j)$ with $\beta(s_j)$. In such a situation, any distribution for $X_i(s_j)$ can be used. However, it would be more plausible to use a distribution that is similar to the distribution of $X(s)$ at the adjacent measurement times and incorporate some correlation among them.

In the derivation above, we show that the joint distribution of longitudinal and survival data that satisfies the landmark Cox model (19.2) exists, and provide an algorithm to generate data from such a joint distribution. The only additional assumption, other than the assumptions of the model (19.2) itself, is that $\theta(s)$ has a gamma distribution. This assumption is used here for illustration. In the following, we show that the joint distribution and data generating algorithm above can be extended to situations where $\theta(s)$ is any distribution.

Let the density function of $\theta(s)$ in the risk set $\mathcal{R}(s)$ be denoted by $f(\theta(s); s)$. Then

$$\begin{aligned}\text{pr}(T(s) > u) &= \int_0^\infty \exp\{-H_0(u; s)\theta(s)\} f(\theta(s); s)\mathrm{d}\theta(s)\,, \quad \forall\, u \geq 0\\ &= E_{\theta(s)}\left[\exp\{-H_0(u; s)\theta(s)\}\right]\,,\end{aligned}$$

where $E_{\theta(s)}(.)$ denotes the expectation with respect to the distribution of $\theta(s)$ among subjects in the risk set $\mathcal{R}(s)$. In this situation, equality (19.5) implies that

$$E_{\theta(s)}\left[e^{-H_0(u;s)\theta(s)}\right] = \frac{E_{\theta(0)}\left[e^{-H_0(s+u;0)\theta(0)}\right]}{E_{\theta(0)}\left[e^{-H_0(s;0)\theta(0)}\right]}\,. \tag{19.7}$$

For data generation, the density function $f(\theta(s); s)$ is pre-specified, which is analogous to the pre-specification of $\alpha(s)$ and $\eta(s)$ above. $H_0(t; 0)$ is also pre-specified. For given values of s and u, we can solve (19.7) for $H_0(u; s)$ numerically in Step 2 of the data generating algorithm. Moreover, it can be easily seen that $H_0(u; s)$ is an increasing function with respect to u and $H_0(0; s) = 0$ for any fixed s. Hence it is a proper cumulative hazard function. A required regularity condition in Eq. (19.7) is that all the expectations must exist. This is satisfied in practical situations where $\theta(s)$ has a bounded support.

Step 3(b) of the data generating algorithm above can be modified for any marginal distribution of $\theta(s)$, with the following conditional distribution of $\theta_i(s)$ given $W_i(s) = G_0(T_i(s); s)$:

$$f(\theta_i(s)|W_i(s)) = \frac{f(W_i(s)|\theta_i(s))f(\theta_i(s); s)}{\int_0^\infty f(W_i(s)|\theta_i(s))f(\theta_i(s); s)\mathrm{d}\theta_i(s)}\,. \tag{19.8}$$

19.3 Extension to the Landmark Linear Transformation Model

To our knowledge, the landmark dynamic prediction models reported in the current literature are all based on the Cox model. In this section, we extend model (19.2) to a more general landmark linear transformation model and show that the joint distribution and data generation results in Sect. 19.2 still apply. Following Cheng et al. (1995), linear transformation models can be written as

$$g\{S(t)\} = \gamma(t) + \boldsymbol{X}^T\boldsymbol{\beta}\,, \tag{19.9}$$

where $\gamma(t)$ is an unspecified increasing function, which maps the positive real line onto the whole real line, $\boldsymbol{\beta}$ is a vector of unknown regression coefficients, and $\boldsymbol{X}$ is a vector of time-independent covariates. The link function $g(x)$ is a known decreasing function that maps from (0, 1) to the real line. It links the survival function of the failure times given covariates and the linear predictor. When $g(x) = \log\{-\log(x)\}$ and $\gamma(t)$ is the log of the baseline cumulative hazard function, (19.9) produces the Cox model. When $g(x) = -\log\{x/(1-x)\}$, it produces the proportional odds model. The accelerated failure time (AFT) model has the following form: $\log(T) = \boldsymbol{X}^T\boldsymbol{\beta} + \sigma\xi$, where ξ is a random disturbance term with a standard location-scale distribution, and $\sigma > 0$ is the scale parameter. The AFT model is also a special case of the linear transformation model, as it can be expressed as

$$S_\xi^{-1}\{S(t)\} = \log(t)/\sigma - \boldsymbol{X}^T\boldsymbol{\beta}/\sigma \doteq \gamma(t) + \boldsymbol{X}^T\boldsymbol{\beta}^\dagger\,,$$

where $S_\xi(x) = \mathrm{pr}(\xi > x)$ is the survival function of ξ, $g(x) = S_\xi^{-1}(x)$, $\gamma(t) = \log(t)/\sigma$, and $\boldsymbol{\beta}^\dagger = -\boldsymbol{\beta}/\sigma$.

The landmark linear transformation model assumes that at each landmark time s, $T(s)$ given $\boldsymbol{X}(s)$ follows the model

$$g\{S(u;s)\} = \gamma(u;s) + \log\theta(s)\,,$$

where $\theta(s) = \exp\{\boldsymbol{X}^T(s)\boldsymbol{\beta}(s)\}$, $S(u;s)$ is the survival function of $T(s)$ given $\theta(s)$, and $\gamma(u;s)$ is an unspecified bivariate function that increases in u (defined for $T(s)$). The $\gamma(u;s)$ function resembles $G_0(u;s)$ in the landmark Cox model of Sect. 19.2. The marginal survival distribution of T is

$$\mathrm{pr}(T > t) = E_{\theta(0)}\left[g^{-1}\{\gamma(t;0) + \log\theta(0)\}\right].$$

The data generating algorithm is similar to that in Sect. 19.2. Equation (19.7) becomes

$$E_{\theta(s)}\{g^{-1}[\gamma(u;s)+\log\theta(s)]\} = \frac{E_{\theta(0)}\{g^{-1}[\gamma(s+u;0)+\log\theta(0)]\}}{E_{\theta(0)}\{g^{-1}[\gamma(s;0)+\log\theta(0)]\}}.$$

We can use this equation to solve for $\gamma(u;s)$ at any s and u, with pre-specified $\gamma(t;0)$ and $f(\theta(s);s)$. The conditional distribution of $\theta(s)$ given $W(s) = \gamma(T(s);s)$ is obtained from Eq. (19.8).

19.4 Simulation

We illustrate the proposed data generating algorithm with the following simulation. The landmark times are pre-specified at $s = 0, 2, 4, \cdots, 10$. The marginal distribution of $\theta(s)$ is a gamma distribution with $\alpha(s) = 1 + 0.15\,s$ and $\eta(s) = 1.2 - 0.02\,s$. The cumulative hazard function at $s = 0$ is from a Weibull distribution with $H_0(u;0) = (\lambda u)^{\kappa}$, $\lambda = 0.15$ and $\kappa = 3$. The administrative end of follow-up is 15.

Given $\alpha(0)$, $\eta(0)$, and $H_0(u;0)$, following the algorithm described in Sect. 19.2, we first generate the event time T at $s = 0$ by (19.3). Based on (19.4), $\theta(s)$ is simulated from the gamma distribution with shape parameter $\alpha(s) + 1$ and rate parameter $\eta(s) + H_0(T(s);s)$, where $T(s) = T - s$. We use an exchangeable correlation structure with $\rho = 0$ or 0.6 in the Gaussian copula to impose serial correlation on $\theta(s)$. We set the regression coefficient as a function of the landmark time s, $\beta(s) = 0.3 - 0.015\,s$. The time-varying biomarker is obtained as $X_i(s) = \log\theta_i(s)/\beta(s)$. In addition, random drop-out times are generated from a Weibull distribution to keep the censoring rate between 20% and 25%. Two sample sizes $n = 100$ and $n = 500$ are examined in this simulation study. For each sample size, $N = 1000$ datasets are simulated.

We fit a separate Cox model at each landmark time on the at-risk subjects, and compare the estimated $\beta(s)$ with its true value at that landmark time. The results are summarized in Table 19.1. It can be seen that at each landmark time, the estimated regression coefficient from the Cox model has negligible bias and the coverage probability is close to the nominal level of 95%. We also conduct a 0.05-level z-test by function *cox.zph* in R to check the proportional hazards assumption in model (19.2) at each landmark time. From Table 19.1, we see that the acceptance rate over 1000 simulations is around 95%. Trajectories of $X(s)$ for 20 randomly selected subjects for $\rho = 0$ or 0.6 are plotted in Fig. 19.3, which shows that a higher correlation parameter ρ leads to visually smoother trajectories. Figure 19.3 also shows that the estimated baseline cumulative hazard function at each landmark time is close to the true curve in the analysis of a single simulated dataset with $n = 500$. The average results over all 1000 simulated datasets completely overlap with the true curves (plot omitted).

Table 19.1 Bias, empirical standard error (ESE), coverage probability (CP) of 95% confidence intervals for estimating $\beta(s)$ and the acceptance rate (AR) of the z-test for the proportional hazards assumption

Landmark	**n=100**	$\rho = 0$				$\rho = 0.6$			
time (s)	True	Bias	ESE	CP	AR	Bias	ESE	CP	AR
0	0.30	0.006	0.044	0.955	0.944	0.007	0.044	0.957	0.957
2	0.27	0.006	0.044	0.956	0.948	0.006	0.043	0.950	0.950
4	0.24	0.006	0.047	0.952	0.943	0.005	0.047	0.958	0.937
6	0.21	0.007	0.055	0.947	0.958	0.006	0.056	0.960	0.955
8	0.18	0.009	0.075	0.950	0.954	0.010	0.076	0.937	0.955
10	0.15	0.024	0.114	0.963	0.955	0.115	2.996	0.949	0.958
Landmark	**n=500**	$\rho = 0$				$\rho = 0.6$			
time (s)	True	Bias	ESE	CP	AR	Bias	ESE	CP	AR
0	0.30	0.002	0.020	0.941	0.945	0.001	0.020	0.939	0.945
2	0.27	0.001	0.020	0.945	0.947	0.001	0.020	0.935	0.949
4	0.24	0.002	0.020	0.944	0.952	0.002	0.020	0.950	0.948
6	0.21	0.002	0.023	0.952	0.946	0.002	0.023	0.951	0.956
8	0.18	0.002	0.028	0.962	0.952	0.002	0.028	0.953	0.950
10	0.15	0.003	0.036	0.943	0.949	0.003	0.036	0.954	0.951

Results are obtained at landmark times $s = 0, 2, \cdots, 10$, with a sample size of $n = 100$ or $n = 500$ and $N = 1000$ Monte Carlo replicates

19.5 Discussion

This paper presents some new results that are of fundamental importance to the landmark survival model for dynamic prediction. First, we extend the conventional Cox model based landmark model to a more general class of landmark linear transformation model. Second, we show that models in this class no longer need to be viewed as "working models": joint distributions of the longitudinal and survival data exist that satisfies the model assumptions at infinitely many landmark times. Third, we propose an algorithm to simulate data from such distributions. This work facilitates the future development of the landmark dynamic prediction models in both the theoretical and empirical fronts. The R code for the proposed algorithm is available upon request.

Our results do not imply that the landmark models in this paper satisfy the consistency condition of Jewell and Nielsen (1993). In the context of model (19.2), the consistency condition requires

$$E\left[h_0(0, s_2)\exp\left\{\boldsymbol{X}(s_2)^T\boldsymbol{\beta}(s_2)\right\}|\boldsymbol{X}(s_1)\right] = h_0(s_2, s_1)\exp\left\{\boldsymbol{X}(s_1)^T\boldsymbol{\beta}(s_1)\right\}$$

for any landmark times $s_1 < s_2$. It remains an open question whether the landmark survival models, in general or in certain practically meaningful special cases, are consistent according to that definition. In survival prediction, the estimand of

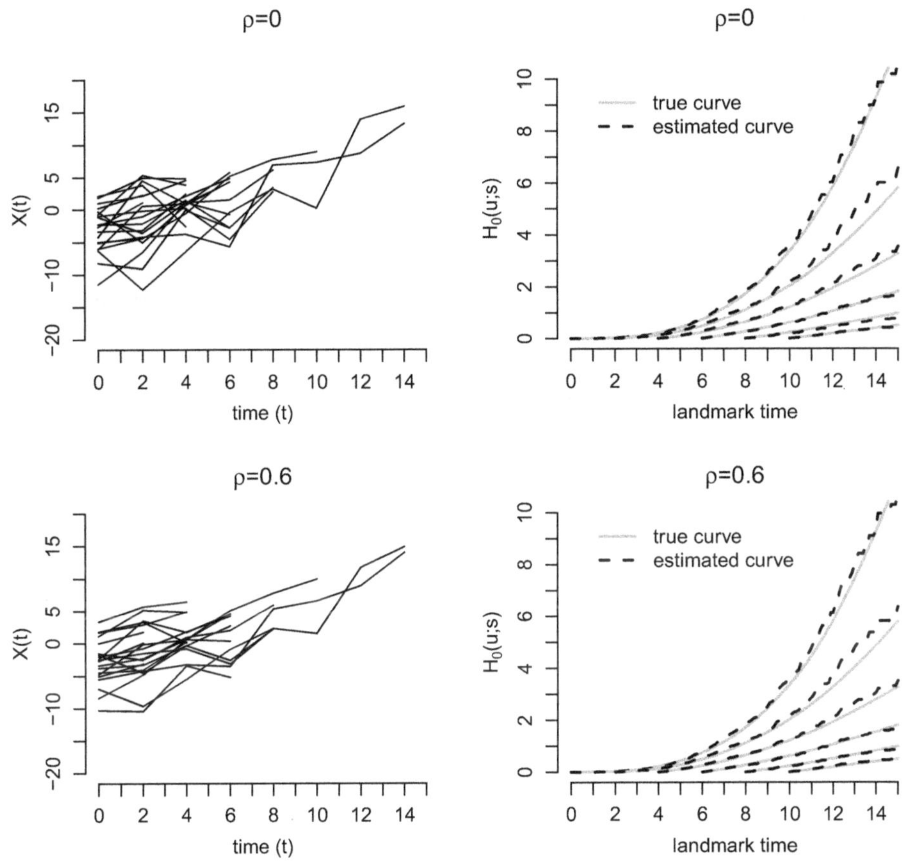

Fig. 19.3 Trajectories of $X(s)$ for 20 subjects (left) and the estimated $H_0(u; s)$ from one simulated dataset (right). $\rho = 0$ (top) or 0.6 (bottom). In this dataset, the number of at-risk subjects at landmark times 0, 2, 4, 6, 8, 10 are 500, 479, 393, 261, 159, 95

interest is usually the survival probability at a prediction horizon, not the hazard function at a future time point. For this reason, our paper studies a different question from that in Jewell and Nielsen (1993). We study whether the pair of residual life time $T(s)$ and time-varying predictor $\boldsymbol{X}(s)$ can satisfy the Cox model (or linear transformation model) at all landmark times simultaneously. If this is not the case, then the landmark model must work under misspecification in some landmark times, causing biased prediction. Our result shows that the landmark survival model is an appropriate probability model that can be used to obtain unbiased prediction at all landmark times.

Acknowledgements The authors gratefully acknowledge the financial support for this research by the National Institutes of Health (grant 5P30CA016672 and 5U01DK103225) and MD Anderson Cancer Center.

References

Blanche, P., Proust-Lima, C., Loubère, L., Berr, C., Dartigues, J. F., & Jacqmin-Gadda, H. (2015). Quantifying and comparing dynamic predictive accuracy of joint models for longitudinal marker and time-to-event in presence of censoring and competing risks. *Biometrics, 71*(1), 102–113.

Cheng, S. C., Wei, L. J., & Ying, Z. (1995). Analysis of transformation models with censored data. *Biometrika, 82*(4), 835–845.

Diggle, P. J., Heagerty, P., Liang, K.-Y., & Zeger, S. L. (2002). *Analysis of longitudinal data* (2nd ed.). Oxford: Oxford University Press.

Faderl, S., Talpaz, M., Estrov, Z., O'Brien, S., Kurzrock, R., & Kantarjian, H. M. (1999). The biology of chronic myeloid leukemia. *New England Journal of Medicine, 341*(3), 164–172.

Gorre, M. E., & Sawyers, C. L. (2002). Molecular mechanisms of resistance to STI571 in chronic myeloid leukemia. *Current Opinion in Hematology, 9*(4), 303–307.

Hochhaus, A., Kantarjian, H. M., Baccarani, M., Lipton, J. H., Apperley, J. F., Druker, B. J., et al. (2007). Dasatinib induces notable hematologic and cytogenetic responses in chronic-phase chronic myeloid leukemia after failure of imatinib therapy. *Blood, 109*(6), 2303–2309.

Huang, X., Yan, F., Ning, J., Feng, Z., Choi, S. & Cortes, J. (2005). A two-stage approach for dynamic prediction of time-to-event distributions. *Statistics in Medicine, 35*(13), 2167–2182.

Jewell, N. P., & Nielsen, J. P. (1993). A framework for consistent prediction rules based on markers. *Biometrika, 80*(1), 153–164.

Kalbfleisch, J. D., & Prentice, R. L. (2002). *The Statistical Analysis of Failure Time Data* (2nd ed.). Hoboken, NJ: Wiley.

Li, L., Greene, T., & Hu, B. (2018). A simple method to estimate the time-dependent receiver operating characteristic curve and the area under the curve with right censored data. *Stat Methods Med Res. 27*(8), 2264–2278.

Li, L., Luo, S., Hu, B., & Greene, T. (2017). Dynamic prediction of renal failure using longitudinal biomarkers in a cohort study of chronic kidney disease. *Stat Biosci. 9*(2), 357–378.

Maziarz, M., Heagerty, P., Cai, T., & Zheng, Y. (2016). On longitudinal prediction with time-to-event outcome: Comparison of modeling options. *Biometrics* Epub ahead of print, https://doi.org/10.1111/biom.12562.

Quintás-Cardama, A., Choi, S., Kantarjian, H., Jabbour, E., Huang, X., & Cortes, J. (2014). Predicting outcomes in patients with chronic myeloid leukemia at any time during tyrosine kinase inhibitor therapy. *Clinical Lymphoma Myeloma and Leukemia, 14*(4), 327–334.

Rizopoulos, D. (2011). Dynamic predictions and prospective accuracy in joint models for longitudinal and time-to-event data. *Biometrics, 67*(3), 819–829.

Rizopoulos, D., Hatfield, L. A., Carlin, B. P., & Takkenberg, J. J. M. (2014). Combining dynamic predictions from joint models for longitudinal and time-to-event data using Bayesian model averaging. *Journal of the American Statistical Association, 109*(508), 1385–1397.

Sawyers, C. L. (1999). Chronic myeloid leukemia. *New England Journal of Medicine, 340*(17), 1330–1340.

Steyerberg, E. W. (2009). *Clinical prediction models: A practical approach to development, validation, and updating*. New York: Springer.

Taylor, J. M., Park, Y., Ankerst, D. P., Proust-Lima, C., Williams, S., Kestin, L., et al. (2013). Real-time individual predictions of prostate cancer recurrence using joint models. *Biometrics, 69*(1), 206–213.

van Houwelingen, H. C. (2007). Dynamic prediction by landmarking in event history analysis. *Scandinavian Journal of Statistics, 34*(1), 70–85.

van Houwelingen, H. C., & Putter H. (2008). Dynamic predicting by landmarking as an alternative for multi-state modeling: an application to acute lymphoid leukemia data. *Lifetime Data Analysis, 14*(4), 447–463.

van Houwelingen, H., & Putter, H. (2011). *Dynamic prediction in clinical survival analysis*. Boca Raton, FL: Chapman & Hall/CRC.

Wong, S. F. (2009). New dosing schedules of dasatinib for CML and adverse event management. *Journal of Hematology & Oncology, 2*(1), 10.

Yan, F., Lin, X., & Huang, X. (2017). Dynamic prediction of disease progression for leukemia patients by functional principal component analysis of longitudinal expression levels of an oncogene. *The Annals of Applied Statistics, 11*(3), 1649–1670.

Zheng, Y. Y., & Heagerty, P. J. (2005). Partly conditional survival models for longitudinal data. *Biometrics, 61*(2), 379–391.

Chapter 20
Nonparametric Estimation of a Cumulative Hazard Function with Right Truncated Data

Xu Zhang, Yong Jiang, Yichuan Zhao, and Haci Akcin

20.1 Introduction

A truncated sample contains realizations of random variables (L, T) subject to the constraint $L \leq T$. Two types of truncation coexist in a truncated sample: T is left truncated by L and L is right truncated by T. In many real-world truncated samples T is the failure time and of study interest while L is the study entrance time. Consequently, left truncation is also known as late entrance (Kaplan and Meier 1958). Right truncation may occur due to retrospective sampling. The AIDS incubation time described in the following example is right truncated.

Records of AIDS cases have been collected at the Centers for Disease Control and Prevention (CDC). If infection of HIV virus was due to blood transfusion, the precise infection date could be traced back. For a particular study closing date, patients developing AIDS after the closing date would not be included in the sample. Therefore, the AIDS incubation time is right truncated by the time between the infection date and the closing date. Kalbfleisch and Lawless (1989) provided one AIDS data set that included 295 AIDS cases infected by blood transfusion by July

X. Zhang (✉)
Center for Clinical and Translational Sciences, University of Texas Health Science Center, Houston, TX, USA
e-mail: Xu.Zhang@uth.tmc.edu

Y. Jiang
MetLife Inc., Whippany, NJ, USA

Y. Zhao
Department of Mathematics and Statistics, Georgia State University, Atlanta, GA, USA
e-mail: yichuan@gsu.edu

H. Akcin
Department of Risk Management and Insurance, Georgia State University, Atlanta, GA, USA
e-mail: hakcin1@gsu.edu

Y. Zhao, D.-G. Chen (eds.), *New Frontiers of Biostatistics and Bioinformatics*, ICSA Book Series in Statistics, https://doi.org/10.1007/978-3-319-99389-8_20

1, 1986 and reported to CDC by January 1, 1987. It has been noted that the right truncation nature of incubation time should be properly addressed. This AIDS data set were analyzed by Lui et al. (1986), Medley et al. (1987), Kalbfleisch and Lawless (1989) among others.

In a truncated sample, distribution functions of T and L are routinely estimated by the truncated version Kaplan-Meier estimators (Kaplan and Meier 1958; Woodroofe 1985). Asymptotic properties of the truncated version Kaplan-Meier estimators have been studied by Woodroofe (1985), Wang et al. (1986), Chao and Lo (1988), Keiding and Gill (1990), assuming quasi-independence between T and L (Tsai 1990). Bilker and Wang (1996) and Chi et al. (2007) studied two-sample comparison of the distribution functions for right truncated data. Bilker and Wang (1996) extended the Mann-Whitney test to truncated samples, assuming parameterized truncation mechanism. Chi et al. (2007) developed a test to compare the integrated weighted differences between two survival functions.

For a truncated sample, variable transformation is often employed for right truncation. Let τ be a large constant. The transformed variable $\tau - L$ is left truncated by $\tau - T$. Because of this feature, some survival quantities of the variable subject to right truncation are defined on the reversed time axis and the existing inferences for left truncation are applicable to these reverse-time quantities. Much research work has been done on the reverse-time hazard. Lagakos et al. (1988) proposed a weighted log-rank test for equivalent reverse-time hazards throughout the entire study period. Kalbfleisch and Lawless (1991), as well as Gross and Huber-Carol (1992), studied the Cox model on the reverse-time hazard.

It has been noted that the reverse-time hazard is very different from the forward-time hazard function (Lagakos et al. 1988) and there is no natural interpretation associated with a reverse-time hazard (Finkelstein et al. 1993). Several statisticians started to focus on the hazard function for the right truncated data. Finkelstein et al. (1993) studied the proportional hazards regression model and constructed a score test to assess effects of covariates. Using the inverse probability weighting technique (Wang 1989), Shen (2010) proposed a class of semiparametric tests to compare the weighted integrated hazard functions given known parametric distribution of truncation variable. Shen proved asymptotic properties of the proposed test statistic and suggested the resampling method to estimate the asymptotic variance of the test statistic because semiparametric weighting causes complexity in the composition of the asymptotic variance. Here we focus on the setting of random truncation that the distribution of the truncation variable is unspecified. We develop a family of weighted tests and derived the asymptotic distribution of the test statistic.

The remainder of this book chapter is organized as follows. Section 20.2 describes the technical background for the reverse-time hazard. Section 20.3 centers on the nonparametric inference for the cumulative hazard function, as well as a one-sample log-rank test. Section 20.4 introduces a group of weighted log-rank tests for the two-sample context. Results of the simulation studies are presented in Sect. 20.5. In Sect. 20.6, the AIDS blood transfusion data set is analyzed to illustrate the two-sample tests. Concluding remarks are given in Sect. 20.7.

20.2 Nonparametric Inference for the Reverse-Time Hazard Function

A truncated sample contains realizations of the random variables (L, T) with the constraint $L \leq T$. The sample can be described as $\{L_i^0, T_i^0\}, i = 1, \cdots, n$, and $L_i^0 \leq T_i^0$. Right truncation occurs if the variable L is of study interest and the variable T is the truncation variable. Suppose that L and T are positive random variables with distribution functions G and F, and satisfy the condition of quasi-independence (Tsai 1990).

We define (a_k, b_k) be the inner support of a distribution function $K(t)$, where $a_k = \inf\{z > 0 : K(z) > 0\}, b_k = \sup\{z > 0 : K(z) < 1\}$. Consequently, (a_G, b_G) and (a_F, b_F) are, respectively, the interior supports of G and F. G and F are estimable only if $a_G < b_F$. Identifiability is a challenging issue for a truncated sample. Practically, one can choose $a^* = \min(L_1^0, \cdots, L_n^0), b^* = \max(T_1^0, \cdots, T_n^0)$, and then the conditional distribution functions $F^*(t) = P(T \leq t | T \geq a^*)$ and $G^*(t) = P(L \leq t | L \leq b^*)$ are identifiable (Wang 1989; Klein and Moeschberger 2003). In general, the study interest on L has to be restricted to the quantities associated with the conditional distribution function G^*. For the purpose of simplicity we assume $a_G = 0$ and $b_G < b^*$, so that G^* would agree with the unconditional distribution function G.

Let $\lambda(t)$ and $\Lambda(t)$ be, respectively, the hazard and cumulative hazard functions of L, with the definitions

$$\lambda(t) = \lim_{\Delta t \to 0} \frac{P(t \leq L < t + \Delta t | L \geq t)}{\Delta t}$$

and

$$\Lambda(t) = \int_0^t \lambda(s) ds = \int_0^t \frac{dG(s)}{P(L \geq s)}. \tag{20.1}$$

For a truncated sample, variable transformation is often employed for right truncation. Let τ be the largest observed time of the sample. The transformed variable $L^* = \tau - L$ is left truncated by $\tau - T$. The hazard function of L^*, which is measured on the reversed time axis, is called as the reverse-time hazard or "retro-hazard." Let $\lambda^*(t)$ denote the reverse-time hazard function and its definition can be found in Lagakos et al. (1988),

$$\lambda^*(t) dt = P(t - dt < L \leq t | L \leq t).$$

The cumulative reverse-time hazard is defined as

$$\Lambda^*(t) = -\int_t^\tau d\Lambda^*(s) ds = \int_t^\tau \lambda^*(s) ds = \int_t^\tau \frac{dG(s)}{P(L \leq s)}. \tag{20.2}$$

Note that the negative sign in the second term reflects monotone decreasing in $\Lambda^*(t)$.

The Nelson-Aalen estimator is a routine estimator of a cumulative hazard function (Nelson 1969; Aalen 1978). For a truncated sample, define the counting processes, $N_i(t) = I(L_i^0 \leq t), Y_i(t) = I(L_i^0 \leq t \leq T_i^0)$ and let $\bar{N}(t) = \sum_{i=1}^n N_i(t), \bar{Y}(t) = \sum_{i=1}^n Y_i(t)$. The Nelson-Aalen estimator of the reverse-time cumulative hazard is given by

$$\widehat{\Lambda}^*(t) = \sum_{i=1}^n \int_t^\tau \frac{dN_i(s)}{\bar{Y}(s)}. \tag{20.3}$$

Weak convergence of $\sqrt{n}\{\widehat{\Lambda}^*(t) - \Lambda^*(t)\}$ was discussed by Keiding and Gill (1990), based on the standard result of a martingale. Here, we wish to provide some details about the martingale associated with the reverse-time hazard. Consider the counting process $N_i^L(t) = I(L_i^0 \geq t)$. It increases from 0 to 1 when moving backwards along the time axis from the origin τ. $\int_\tau^t Y_i(s)d\Lambda^*(s)$ is the compensator of the counting process. The martingale is yielded when we subtract the compensator from the counting process,

$$M_i(t) = N_i^L(t) - \int_\tau^t Y_i(s)d\Lambda^*(s).$$

The estimator $\widehat{\Lambda}^*(t)$ can be alternatively expressed as $\sum_{i=1}^n \int_\tau^t \{dN_i^L(u)/\bar{Y}(u)\}$. It follows that $\widehat{\Lambda}^*(t) - \Lambda^*(t) = \sum_{i=1}^n \int_\tau^t \{dM_i(u)/\bar{Y}(u)\}$. Based on the martingale central limit theorem, it can be proved that $n^{1/2}\{\widehat{\Lambda}^*(t) - \Lambda^*(t)\} \rightarrow_{\mathcal{D}} W_t$. W_t is a Gaussian process with mean zero and variance σ_t^2, where $\sigma_t^2 = \int_t^\tau \{\lambda^*(u)du/v(u)\}$ and $v(u) = E[n^{-1}\bar{Y}(u)]$.

The optional variation process of martingale leads to a variance estimator of $\widehat{\Lambda}^*(t)$,

$$\hat{\text{var}}^{(1)}[\widehat{\Lambda}^*(t)] = \sum_{i=1}^n \int_t^\tau \frac{dN_i(s)}{\bar{Y}(s)^2}. \tag{20.4}$$

In the following context, this estimator is referred to as the naive variance estimator. Klein (1991) suggested an alternative variance estimator, by assuming a binomial distribution for a jump in the event counting process. It can be also explained by the predictable variation process of a martingale (Andersen et al. 1993). This alternative variance estimator has the form

$$\hat{\text{var}}^{(2)}[\widehat{\Lambda}^*(t)] = \sum_{i=1}^n \int_t^\tau \frac{(\bar{Y}(s) - \Delta N_i(s))dN_i(s)}{\bar{Y}(s)^3}. \tag{20.5}$$

20.3 Nonparametric Inference for the Cumulative Hazard Function

20.3.1 Estimation of the Cumulative Hazard Function

Definition of $\Lambda(t)$ (Eq. (20.1)) suggests a plug-in estimator if an estimator of $G(t)$ is inserted in the formula. Since $G(t) = P(L \le t) = P(L^* \ge \tau - t)$, $G(t)$ can be viewed as the survival function of L^* on the reversed time axis. Consequently, the truncated version Kaplan-Meier estimator is utilized for estimating $G(t)$ (Woodroofe 1985; Keiding and Gill 1990),

$$\widehat{G}(t) = \prod_{s>t}\left(1 - \frac{d\bar{N}(s)}{\bar{Y}(s)}\right). \tag{20.6}$$

Then the plug-in estimator of $\Lambda(t)$ (Kalbfleisch and Lawless 1989) is given by

$$\widehat{\Lambda}(t) = \int_0^t \frac{d\widehat{G}(s)}{1 - \widehat{G}(s^-)}, \tag{20.7}$$

where $\widehat{G}(s^-)$ is the Kaplan-Meier estimate of distribution probability prior to time s. The regular Nelson-Aalen estimator is not feasible for $\Lambda(t)$ because proper weights should be assigned to observations to correct for biased selection. A weighted version Nelson-Aalen estimator of $\Lambda(t)$ can be found in Shen (2010).

Lagakos et al. (1988) noted the relation between the reverse-time and forward-time hazard functions, when investigating the weight of the weighted log-rank test between two independent truncated samples. Using another subscript to indicate sample 1 or 2, the following relation exists:

$$\frac{\lambda_1^*(t)}{\lambda_2^*(t)} = \frac{\lambda_1(t)\{1 - G_1(t)\}G_2(t)}{\lambda_2(t)\{1 - G_2(t)\}G_1(t)}.$$

One can easily conclude that, when two forward-time hazards have a constant ratio, the proportionality on the reverse-time hazards does not hold.

Assuming that $G(t)$ is differentiable, we further clarify the relation between $\Lambda(t)$ and $\Lambda^*(t)$ as follows:

$$\Lambda(t) = -\log(1 - \exp[-\Lambda^*(t)]). \tag{20.8}$$

The above equation is true because $G(t) = \exp(-\Lambda^*(t))$ and $1 - G(t) = \exp(-\Lambda(t))$. Using weak convergence of $\sqrt{n}\{\widehat{\Lambda}^*(t) - \Lambda^*(t)\}$ and applying the generalized delta method, we have the result,

$$n^{1/2}\{\widehat{\Lambda}(t) - \Lambda(t)\} \rightarrow_{\mathcal{D}} \kappa(\Lambda^*(t))W_t \tag{20.9}$$

where

$$\kappa(\Lambda^*(t)) = -\frac{\exp(-\Lambda^*(t))}{1-\exp(-\Lambda^*(t))} = -\frac{G(t)}{1-G(t)}.$$

It follows that the asymptotic variance of $\widehat{\Lambda}(t)$ is given by

$$\mathrm{var}[\widehat{\Lambda}(t)] \approx \left[\frac{G(t)}{1-G(t)}\right]^2 \mathrm{var}[\widehat{\Lambda}^*(t)]. \tag{20.10}$$

The variance estimators of $\widehat{\Lambda}^*(t)$ given in Eqs. (20.4) and (20.5) can be plugged into the above equation, leading to two variance estimators of $\widehat{\Lambda}(t)$. The naive variance estimator will be

$$\widehat{\mathrm{var}}^{(1)}[\widehat{\Lambda}(t)] = \left[\frac{\widehat{G}(t)}{1-\widehat{G}(t^-)}\right]^2 \sum_{i=1}^{n} \int_t^{\tau} \frac{dN_i(s)}{\bar{Y}(s)^2}. \tag{20.11}$$

The alternative variance estimator is given by

$$\widehat{\mathrm{var}}^{(2)}[\widehat{\Lambda}(t)] = \left[\frac{\widehat{G}(t)}{1-\widehat{G}(t^-)}\right]^2 \sum_{i=1}^{n} \int_t^{\tau} \frac{(\bar{Y}(s)-\Delta N_i(s))dN_i(s)}{\bar{Y}(s)^3}. \tag{20.12}$$

In above two formulas, we particularly chose $1-\widehat{G}(t^-)$, instead of $1-\widehat{G}(t)$, for estimating $1-G(t)$ in Eq. (20.10), because the same form is used in $\widehat{\Lambda}(t)$ (Eq. (20.7)).

20.3.2 One-Sample Log-Rank Test

In survival analysis, one practical question is to compare the mortality rate of the target population to a standard population. The hypothesis needs to be tested is given by

$$H_0 : \lambda(s) = \lambda_0(s), \ \forall s \le t,$$

where $\lambda_0(t)$ is the known hazard rate function. Let $\Lambda_0(t) = \int_0^t \lambda_0(u)du$. Let $W(t)$ be a weight process. Here we consider a general test statistic, which is an integrated process,

$$U(t) = \int_0^t W(u)d[\widehat{\Lambda}(u) - \Lambda_0(u)].$$

A common choice of $W(t)$ is the risk set process. For the complete data or for the context of right censoring and/or left truncation, the statistic is exactly the difference between observed and expected number of events. Consequently, the statistic is equivalent to the one-sample log-rank statistic. Under the context of right truncation, usage of $\bar{Y}(t)$ for $W(t)$ does not lead to the interpretation of observed or expected number of events. The test studied in this section is indeed a closed-form log-rank test.

Under the null hypothesis, we show in the appendix that a variance estimator of $U(t)$ is given by

$$\widehat{\sigma}(t)^2 = \int_0^t \left[W(s) \frac{\widehat{G}(s)}{1 - \widehat{G}(s-)} - \int_0^s W(u) d\left(\frac{\widehat{G}(u)}{1 - \widehat{G}(u-)} \right) \right]^2 \frac{d\bar{N}(s)}{\bar{Y}(s)^2}$$

$$+ \int_t^\tau \left[\int_0^t W(u) d\left(\frac{\widehat{G}(u)}{1 - \widehat{G}(u-)} \right) \right]^2 \frac{d\bar{N}(s)}{\bar{Y}(s)^2}.$$

The appendix sketches the asymptotic distribution of the test statistic $U(t)$. Therefore, the statistic $Z(t) = U(t)/\widehat{\sigma}(t)$ asymptotically follows a standard normal distribution when the null hypothesis is true.

20.4 Two-Sample Weighted Tests

In this section, we introduce a family of weighted tests comparing the hazard rate function between two independent samples. Two truncated samples can be summarized as $\{L_{i1}^0, T_{i1}^0\}(i = 1, \cdots, n_1)$ and $\{L_{i2}^0, T_{i2}^0\}(i = 1, \cdots, n_2)$, where $L_{i1}^0 \leq T_{i1}^0$ and $L_{i2}^0 \leq T_{i2}^0$. The following notations are needed for the two-sample tests: $\bar{N}_1(t) = \sum_{i=1}^{n_1} I(L_{i1}^0 \leq t)$, $\bar{N}_2(t) = \sum_{i=1}^{n_2} I(L_{i2}^0 \leq t)$, $\bar{N}_\bullet(t) = \bar{N}_1(t) + \bar{N}_2(t)$, $\bar{Y}_1(t) = \sum_{i=1}^{n_1} I(L_{i1}^0 \leq t \leq T_{i1}^0)$, $\bar{Y}_2(t) = \sum_{i=1}^{n_2} I(L_{i2}^0 \leq t \leq T_{i2}^0)$, and $\bar{Y}_\bullet(t) = \bar{Y}_1(t) + \bar{Y}_2(t)$.

Let $\lambda_1(t)$ and $\lambda_2(t)$ be the hazard functions of L_1 and L_2, respectively. Tests should be developed for the hypothesis, $H_0 : \lambda_1(s) = \lambda_2(s),\ \forall s \leq t$. The standard integrated process for testing such hypothesis has the form

$$U^*(t) = \int_0^t L(s) d\widehat{\Lambda}_1(s) - \int_0^t L(s) d\widehat{\Lambda}_2(s).$$

According to Andersen et al. (1993, V.2), the weight process $L(s)$ is expressed as

$$L(s) = K(s)\bar{Y}_1(s)\bar{Y}_2(s)\{\bar{Y}_1(s) + \bar{Y}_2(s)\}^{-1}.$$

Under the general context, $K(t)$ should be a non-negative process depending on $(\bar{N}_\bullet, \bar{Y}_\bullet)$ only, and it is defined to be zero whenever $\bar{Y}_\bullet$ is zero. Several options for $K(t)$ lead to a few standard tests. For example, $K(t) = I(\bar{Y}_\bullet(t) > 0)$ leads to the two-sample log-rank test. A few other choices include $\bar{Y}_\bullet(t)$ and $\sqrt{\bar{Y}_\bullet(t)}$, leading to Gehan test and Tarone and Ware test, respectively (Andersen et al. 1993, V.2). However, one should note that, for the right truncated data, these tests do not have the usual interpretation about observed and expected number of events.

Following similar derivation developed for the one-sample test, we can establish the asymptotic distribution of $U^*(t)$ at time t. It is mean zero normal distribution with variance

$$\sqrt{\frac{n_1 n_2}{n}}\left[\int_0^t L(s)d\widehat{\Lambda}_1(s) - \int_0^t L(s)d\widehat{\Lambda}_2(s) - \left(\int_0^t L(s)d\Lambda_1(s) - \int_0^t L(s)d\Lambda_2(s)\right)\right].$$

The variance of statistic $U^*(t)$ can be estimated by

$$\widehat{\sigma}^*(t)^2 = \int_0^t \left[L(s)\frac{\widehat{G}_\bullet(s)}{1-\widehat{G}_\bullet(s-)} - \int_0^s L(u)d\left(\frac{\widehat{G}_\bullet(u)}{1-\widehat{G}_\bullet(u-)}\right)\right]^2 \frac{d\left[\bar{N}_1(s)+\bar{N}_2(s)\right]}{\bar{Y}_1(s)\bar{Y}_2(s)}$$

$$+\int_t^\tau \left[\int_0^t L(u)d\left(\frac{\widehat{G}_\bullet(u)}{1-\widehat{G}_\bullet(u-)}\right)\right]^2 \frac{d\left[\bar{N}_1(s)+\bar{N}_2(s)\right]}{\bar{Y}_1(s)\bar{Y}_2(s)}.$$

The test statistics $Z^*(t) = U^*(t)/\widehat{\sigma}^*(t)$ asymptotically follows a standard normal distribution.

20.5 Simulation Studies

20.5.1 Study I

We preferred to choose distribution of L to be defined on a bounded interval. Two such distributions were considered in Study I, uniform[0, 1] and exponential distribution truncated at 1.2, with the respective cumulative hazard functions

$$\Lambda(t) = -\log(1-t), \quad 0 \le t \le 1$$

and

$$\Lambda(t) = -\log\left(1 - \frac{1-e^{-t}}{1-e^{-1.2}}\right), \quad 0 < t < 1.2.$$

The truncation variable T was generated from an exponential distribution with the rate parameter γ. The value of γ was searched to yield the predetermined truncation rates, 25% and 50%. The truncation rate for one sample is defined as $(N - n)/N$, where N is the size of the pool from which the truncated sample of size n is selected. Two levels of sample size, 200 and 400, were considered. Each simulated setting contained 1000 replicates. Let $\widehat{\Lambda}^{(i)}(t)$ denote the cumulative hazard estimate for the ith replicate at t. Let $\bar{\widehat{\Lambda}}(t)$ be the average cumulative hazard estimate across 1000 replicates, where $\bar{\widehat{\Lambda}}(t) = \sum_{i=1}^{1000} \widehat{\Lambda}^{(i)}(t)$. The bias was defined as the deviation between average cumulative hazard estimate and the true value, that is, Bias $= \bar{\widehat{\Lambda}}(t) - \Lambda(t)$. The variation among 1000 cumulative hazard estimates was evaluated by the sample variance,

$$\text{Sample variance} = \frac{1}{1000-1} \sum_{i=1}^{1000} \left(\widehat{\Lambda}^{(i)}(t) - \bar{\widehat{\Lambda}}(t) \right)^2 .$$

We evaluated two variance estimators, Eqs. (20.11) and (20.12), for individual sample, and further obtained the averages,

$$\text{Estimated variance} = \frac{1}{1000} \sum_{i=1}^{1000} \widehat{\text{var}}^{(k)}[\widehat{\Lambda}^{(i)}(t)], \quad k = 1, 2.$$

For each variance estimator, 95% confidence interval was calculated for each replicate and actual coverage rate across 1000 replicates was obtained.

For the settings that L followed uniform distribution, we report the estimation result at $t = 0.2, 0.5, 0.8$, corresponding to 0.2, 0.5, 0.8 in $G(t)$ (Table 20.1). For the settings that L followed the truncated exponential distribution, we evaluated at $t = 0.15, 0.43, 0.82$, still relating to 0.2, 0.5, 0.8 in $G(t)$ (Table 20.2). In both tables, biases are about zero across all settings. Both variance estimators were evaluated very close to each other, and the averages match the variation existing among the cumulative hazard estimates. The coverage percentages are close to the nominal level, with the exception for small t and heavy truncation, in which slight undercoverage is observed.

20.5.2 *Study II*

The performances of the weighted tests were investigated in this study. We still used the uniform and truncated exponential distributions for L and exponential distribution for T. For the first set of the simulated settings, uniform[0, 1] was consistently used as the underlying distribution of L for sample 1, while distribution of L for sample 2 varied among uniform[0, 1], uniform[0, 1.2], uniform[0, 1.3] (see Table 20.3). Variable T in samples 1 and 2 was generated from exponential

Table 20.1 Simulation results for variance estimation of $\widehat{\Lambda}(t)$ when the underlying distribution is uniform[0, 1]

					Naive variance estimator		Alternative variance estimator	
n	$L\%$	t	Bias	SVar	EVar	Coverage	EVar	Coverage
200	25	0.20	0.000	0.0012	0.0012	0.943	0.0012	0.944
		0.50	−0.003	0.0059	0.0056	0.941	0.0056	0.941
		0.80	−0.005	0.0269	0.0253	0.950	0.0253	0.950
	50	0.20	−0.002	0.0014	0.0012	0.929	0.0012	0.933
		0.50	0.001	0.0081	0.0077	0.949	0.0076	0.950
		0.80	0.002	0.0445	0.0408	0.937	0.0404	0.937
400	25	0.20	0.000	0.0006	0.0006	0.950	0.0006	0.951
		0.50	0.002	0.0027	0.0029	0.952	0.0029	0.952
		0.80	0.000	0.0135	0.0128	0.939	0.0128	0.939
	50	0.20	0.000	0.0006	0.0006	0.947	0.0006	0.947
		0.50	0.003	0.0038	0.0038	0.959	0.0038	0.959
		0.80	0.004	0.0207	0.0207	0.947	0.0207	0.947

SVar: sample variance; EVar: estimated variance

Table 20.2 Simulation results for variance estimation of $\widehat{\Lambda}(t)$ when the underlying distribution is truncated exponential

					Naive variance estimator		Alternative variance estimator	
n	$L\%$	t	Bias	SVar	EVar	Coverage	EVar	Coverage
200	25	0.15	0.000	0.0012	0.0012	0.947	0.0012	0.945
		0.43	−0.003	0.0059	0.0058	0.941	0.0058	0.941
		0.82	−0.005	0.0286	0.0279	0.942	0.0276	0.942
	50	0.15	−0.002	0.0012	0.0012	0.935	0.0012	0.933
		0.43	0.001	0.0090	0.0088	0.941	0.0086	0.937
		0.82	0.008	0.0620	0.0562	0.945	0.0552	0.944
400	25	0.15	0.000	0.0006	0.0006	0.953	0.0006	0.953
		0.43	0.002	0.0031	0.0029	0.954	0.0029	0.954
		0.82	0.000	0.0144	0.0142	0.933	0.0142	0.933
	50	0.15	0.000	0.0006	0.0006	0.948	0.0006	0.947
		0.43	0.002	0.0048	0.0045	0.951	0.0045	0.948
		0.82	0.004	0.0306	0.0286	0.948	0.0282	0.948

SVar: sample variance; EVar: estimated variance

distributions with different rate parameters, to produce the same level of truncation rate between two samples. The exponential distributions truncated at 1.2 was the underlying distribution of L for the second set of the settings. The rate parameter values for two samples are provided in Table 20.4.

The null hypothesis, $H_0 : \lambda_1(s) = \lambda_2(s),\ \forall s \leq t$, was rejected at the significance level 0.05 using the log-rank, Gehan, Tarone, and Ware tests discussed in Sect. 20.4. The proportions of rejection among 1000 replicates at the selected time points are shown in Tables 20.3 and 20.4. In both tables, when the distributions of L for two samples are identical, the observed rejection rates are close to the significance level 0.05. When the distributions vary between two samples, in Table 20.3, the observed power increases by time, while Table 20.4 shows a different trend that the observed power increases first but declines when t gets large. We depicted the underlying cumulative hazard functions, to explore the reason of different trends between these two tables. When the underlying distributions are uniform, the difference between two cumulative hazard functions monotonically increases by time. When the underlying distributions are exponential, the difference increases first and then declines when t becomes larger.

The observed power levels look similar between three tests. There is a trend that the log-rank test has a higher level of power when t is small. It is known that the log-rank test is most powerful under the context of proportional hazards. For the simulated settings, the hazards are close to proportional for small t, leading to a better power result in the log-rank test.

20.6 The Blood Transfusion Infected AIDS Data

We analyzed the AIDS data set described in Sect. 20.1, focusing on the cumulative hazard function of the incubation time. Let L denote the incubation time. The truncation time T is the time from infection to the study closing date, July 1, 1986. This data set was conventionally divided into three subgroups, children (age range 1–4), adults (age range 5–59), and elderly patients (age $\geq$ 60), with the sizes 34, 120 and 141, respectively. The largest incubation times were, respectively, 43, 89, and 83 months in children, adults, and elderly patients.

Figure 20.1 depicts the cumulative hazard estimates for the three subgroups. Children was clearly associated with a greatly higher intensity of AIDS onset, while the adults and elderly patients had similar level of intensity. We further conducted the weighted tests discussed in Sect. 20.4 between any two subgroups. Table 20.5 shows the test results for comparing hazards up to 12, 24, and 36 months. There are strong evidences supporting different hazards for children versus adults (log-rank test $P < 0.001$, up to 36 month), and children versus elderly patients (log-rank test $P < 0.001$, up to 36 month). For adults versus elderly patients, these two groups are not significantly different on the hazard (log-rank test $P = 0.66$, up to 36 month).

20.7 Discussion

With right truncation, earlier researches about the hazard function are limited to the proportional hazards regression model by Finkelstein et al. (1993) and semi-parametric two-sample tests by Shen (2010). This paper focuses on nonparametric inferences of the forward-time hazard function. The two-sample tests studied here

Table 20.3 The proportions of rejection for the hypothesis $H_0 : \lambda_1(s) = \lambda_2(s),\ \forall s \leq t$, when the underlying distributions are uniform

			Proportion rejecting H_0 at level 0.05								
			uniform[0, 1], uniform[0, 1]			uniform[0, 1], uniform[0, 1.2]			uniform[0, 1], uniform[0, 1.3]		
n	$L\%$	t	Log-rank	Gehan	Tarone&Ware	Log-rank	Gehan	Tarone&Ware	Log-rank	Gehan	Tarone&Ware
200	25%	0.20	0.046	0.050	0.050	0.131	0.115	0.118	0.200	0.164	0.172
		0.50	0.036	0.037	0.036	0.311	0.278	0.286	0.544	0.494	0.522
		0.80	0.036	0.040	0.037	0.773	0.745	0.756	0.950	0.932	0.942
	50%	0.20	0.049	0.052	0.048	0.128	0.109	0.114	0.178	0.174	0.172
		0.50	0.056	0.052	0.053	0.269	0.263	0.260	0.435	0.404	0.418
		0.80	0.051	0.053	0.052	0.585	0.579	0.586	0.826	0.823	0.822
400	25%	0.20	0.057	0.047	0.052	0.186	0.157	0.173	0.342	0.300	0.310
		0.50	0.047	0.049	0.052	0.535	0.488	0.505	0.823	0.768	0.807
		0.80	0.055	0.054	0.054	0.966	0.948	0.957	0.997	0.997	0.997
	50%	0.20	0.046	0.052	0.050	0.193	0.182	0.189	0.373	0.319	0.345
		0.50	0.052	0.052	0.054	0.469	0.442	0.456	0.745	0.715	0.730
		0.80	0.038	0.037	0.038	0.856	0.851	0.853	0.985	0.983	0.985

Table 20.4 The proportions of rejection for the hypothesis $H_0 : \lambda_1(s) = \lambda_2(s),\ \forall s \le t$, when the underlying distributions are truncated exponential

			Proportion rejecting H_0 at level 0.05								
			exp(1.0), exp(1.0)			exp(1.0), exp(1.5)			exp(1.0), exp(2.0)		
n	$L\%$	t	Log-rank	Gehan	Tarone&Ware	Log-rank	Gehan	Tarone&Ware	Log-rank	Gehan	Tarone&Ware
200	25	0.15	0.043	0.049	0.048	0.142	0.133	0.136	0.389	0.327	0.354
		0.43	0.042	0.037	0.039	0.189	0.150	0.173	0.509	0.412	0.457
		0.82	0.044	0.048	0.044	0.130	0.113	0.118	0.333	0.294	0.317
	50	0.15	0.054	0.053	0.053	0.157	0.144	0.150	0.413	0.364	0.380
		0.43	0.052	0.054	0.052	0.169	0.155	0.160	0.404	0.368	0.385
		0.82	0.056	0.055	0.055	0.119	0.123	0.121	0.273	0.274	0.269
400	25	0.15	0.045	0.046	0.046	0.255	0.212	0.235	0.667	0.545	0.611
		0.43	0.054	0.056	0.052	0.309	0.250	0.275	0.830	0.730	0.776
		0.82	0.052	0.053	0.050	0.194	0.155	0.171	0.583	0.485	0.522
	50	0.15	0.050	0.062	0.053	0.243	0.197	0.217	0.642	0.573	0.604
		0.43	0.046	0.045	0.046	0.248	0.224	0.233	0.642	0.597	0.615
		0.82	0.039	0.039	0.040	0.163	0.157	0.158	0.429	0.457	0.446

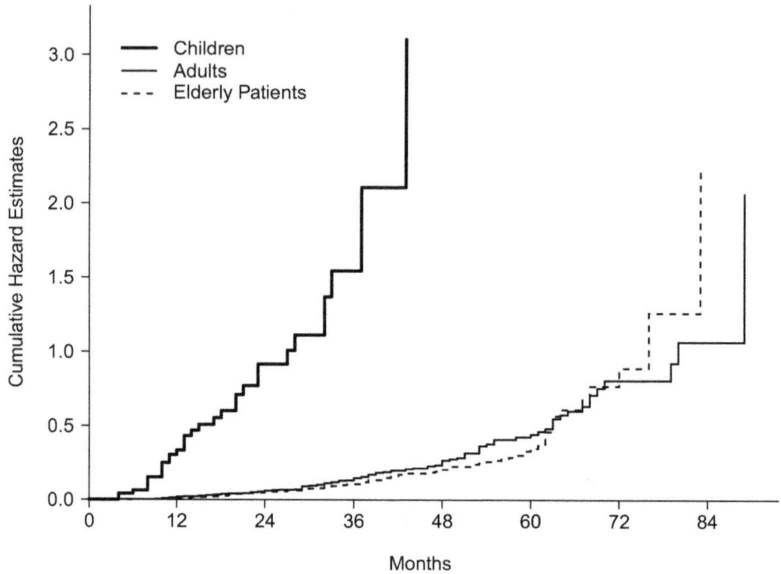

Fig. 20.1 Cumulative hazard estimates of AIDS incubation time for (a) children, (b) adults and (c) elderly patients

provide a useful tool for analyzing right truncated data. A future research direction is the Kolmogorov-Smirnov type of test for the cumulative hazard function. Since difference in cumulative hazard function between two samples can be expressed as a function of martingale process, the simulation method (Lin et al. 1994) can be employed to produce a confidence band for the differences over a time interval. The supremum test over an interval $[t_1, t_2]$ can be implemented using the realized processes. This test is expected to be more powerful in detecting a difference when two cumulative hazard functions cross at some time point.

Appendix: Asymptotic Distribution of $U(t)$

Suppose that $\lambda_0(t)$ is the true hazard function. Let $\lambda_0^*(t)$ and $\Lambda_0^*(t)$ denote the corresponding reverse-time hazard and cumulative hazard functions. Let $G_0(t)$ be the true distribution function. The following equation specifies the relation between $\Lambda_0(t)$ and $\Lambda_0^*(t)$,

$$\int_0^t W(u)d\Lambda_0(u) = -\int_0^t W(u)\frac{G_0(u)}{1-G_0(u)}d\Lambda_0^*(u).$$

The first step is to decompose the test statistic into two components,

Table 20.5 The weighted log-rank tests for comparing the hazard rates between subgroups in the AIDS blood transfusion data set

t	Subgroup	Log-rank		Gehan		Tarone&Ware	
		Z^*	P	Z^*	P	Z^*	P
12 month	Children vs. adults	9.97	<0.001	8.94	<0.001	9.41	<0.001
	Children vs. elderly patients	16.02	<0.001	15.25	<0.001	15.62	<0.001
	Adults vs. elderly patients	0.92	0.356	0.99	0.321	0.96	0.337
24 month	Children vs. adults	11.57	<0.001	11.11	<0.001	11.32	<0.001
	Children vs. elderly patients	11.02	<0.001	10.40	<0.001	10.63	<0.001
	Adults vs. elderly patients	0.09	0.926	0.06	0.949	0.07	0.944
36 month	Children vs. adults	9.66	<0.001	9.10	<0.001	9.35	<0.001
	Children vs. elderly patients	10.57	<0.001	10.30	<0.001	10.38	<0.001
	Adults vs. elderly patients	0.45	0.656	0.48	0.630	0.46	0.643

$$\sqrt{n}\int_0^t W(u)d[\widehat{\Lambda}(u)-\Lambda_0(u)]$$

$$=\sqrt{n}\int_t^0 W(u)\frac{\widehat{G}(u)}{1-\widehat{G}(u)}d\left[\widehat{\Lambda}^*(u)-\Lambda_0^*(u)\right]$$

$$+\sqrt{n}\int_t^0 W(u)\left[\frac{\widehat{G}(u)}{1-\widehat{G}(u)}-\frac{G_0(u)}{1-G_0(u)}\right]d\Lambda_0^*(u).$$

In the following context $\approx$ indicates asymptotic equivalence. The first component is asymptotically equivalent to the sum of martingales

$$\sqrt{n}\int_t^0 W(u)\frac{\widehat{G}(u)}{1-\widehat{G}(u)}d\left[\widehat{\Lambda}^*(u)-\Lambda_0^*(u)\right]\approx\sqrt{n}\int_t^0 W(u)\frac{G_0(u)}{1-G_0(u)}\frac{d\bar{M}(u)}{\bar{Y}(u)}.$$

Applying the delta method on the second component, we have

$$\sqrt{n}\int_t^0 W(u)\left[\frac{\widehat{G}(u)}{1-\widehat{G}(u)}-\frac{G_0(u)}{1-G_0(u)}\right]d\Lambda_0^*(u)$$

$$\approx\sqrt{n}\int_t^0 W(u)\frac{-G(u)}{[1-G(u)]^2}\left[\widehat{\Lambda}^*(u)-\Lambda_0^*(u)\right]d\Lambda_0^*(u)$$

$$=\sqrt{n}\int_t^0 W(u)\left[\int_\infty^u\frac{d\bar{M}(s)}{\bar{Y}(s)}\right]d\left(\frac{G_0(u)}{1-G_0(u)}\right).$$

For the above double integrals, changing order of integration leads to

$$\sqrt{n}\left\{\int_0^t\left[\int_0^s W(u)d\left(\frac{G_0(u)}{1-G_0(u)}\right)\right]\frac{d\bar{M}(s)}{\bar{Y}(s)}\right.$$

$$\left.+\int_t^\tau\left[\int_0^t W(u)d\left(\frac{G_0(u)}{1-G_0(u)}\right)\right]\frac{d\bar{M}(s)}{\bar{Y}(s)}\right\}.$$

Combining the derived results, one can show that

$$\sqrt{n}\int_0^t W(u)d[\widehat{\Lambda}(u)-\Lambda_0(u)]\approx$$

$$\sqrt{n}\left\{\int_t^0\left[W(s)\frac{G_0(s)}{1-G_0(s)}-\int_0^s W(u)d\left(\frac{G_0(u)}{1-G_0(u)}\right)\right]\frac{d\bar{M}(s)}{\bar{Y}(s)}\right.$$

$$\left.-\int_\tau^t\left[\int_0^t W(u)d\left(\frac{G_0(u)}{1-G_0(u)}\right)\right]\frac{d\bar{M}(s)}{\bar{Y}(s)}\right\}.$$

Using the martingale central limit theorem, $\sqrt{n}\int_0^t W(u)d[\widehat{\Lambda}(u)-\Lambda_0(u)]$ converges in distribution to zero-mean normal random variable, with the variance

$$\int_0^t \left[W(s)\frac{G_0(s)}{1-G_0(s)} - \int_0^s W(u)d\left(\frac{G_0(u)}{1-G_0(u)}\right)\right]^2 \frac{\lambda_0^*(s)ds}{y(s)}$$

$$+\int_t^\tau \left[\int_0^t W(u)d\left(\frac{G_0(u)}{1-G_0(u)}\right)\right]^2 \frac{\lambda_0^*(s)ds}{y(s)}.$$

References

Aalen, O. O. (1978). Nonparametric inference for a family of counting processes. *The Annals of Statistics, 6*, 701–726.

Andersen, P. K., Borgan, Ø., Gill, R. D., & Keiding, N. (1993). *Statistical models based on counting processes*. New York: Springer.

Bilker, W. B., & Wang, M. C. (1996). A semiparametric extension of the Mann-Whitney test for randomly truncated data. *Biometrika, 52*, 10–20.

Chao, M. T., & Lo, S. H. (1988). Some representations of the non-parametric maximum likelihood estimators with truncated data. *The Annals of Statistics, 16*, 661–668.

Chi, Y., Tsai, W. Y., & Chiang, C. L. (2007). Testing the equality of two survival functions with right truncated data. *Statistics in Medicine, 26*, 812–827.

Finkelstein, D. M., Moore, D. F., & Schoenfeld, D. A. (1993). A proportional hazards model for truncated AIDS data. *Biometrics, 49*, 731–740.

Gross, S. T., & Huber-Carol, C. (1992). Regression models for truncated survival data. *Scandinavian Journal of Statistics, 19*, 193–213.

Kalbfleisch, J. D., & Lawless, J. F. (1989). Inference based on retrospective ascertainment: An analysis of the data on transfusion-related AIDS. *Journal of the American Statistical Association, 84*, 360–372.

Kalbfleisch, J. D., & Lawless, J. F. (1991). Regression models for right truncated data with applications to AIDS incubation times and reporting lags. *Statistica Sinica, 1*, 19–32.

Kaplan, E. L., & Meier, P. (1958). Nonparametric estimation from incomplete observations. *Journal of the American Statistical Association, 53*, 457–481.

Keiding, N., & Gill, R. D. (1990). Random truncation models and Markov process. *The Annals of Statistics 18*, 582–602.

Klein, J. P. (1991). Small sample moments of some estimators of the variance of the Kaplan-Meier and Nelson-Aalen estimators. *Scandinavian Journal of Statistics, 18*, 333–340.

Klein, J. P., & Moeschberger, M. L. (2003). *Survival analysis: Techniques for censored and truncated data*. New York: Springer.

Lagakos, S. W., Barraj, L. M., & Gruttola, V. (1988). Nonparametric analysis of truncated survival data with applications to AIDS. *Biometrika, 75*, 515–523.

Lin, D. Y., Fleming, T. R., & Wei, L. J. (1994). Confidence bands for survival curves under the proportional hazards model. *Biometrika, 81*, 73–81.

Lui, K. J., Lawrence, D. L., Morgan, W. M., Peterman, T. A., Haverkos, H. W., & Bregman, D. J. (1986). A model-based approach for estimating the mean incubation period of transfusion-associated acquired immunodeficiency syndrome. *Proceedings of National Academy of Sciences USA, 83*, 3051–3055.

Medley, G. F., Anderson, R. M., Cox, D.R., & Billiard, L. (1987). Incubation period of AIDS in patients infected via blood transfusion. *Nature, 328*, 719–721.

Nelson, W. (1969). Hazard plotting for incomplete failure data. *Journal of Quality Technology, 1*, 27–52.

Shen, P. (2010). A class of semiparametric rank-based tests for right-truncated data. *Statistics and Probability Letters, 80*, 1459–1466.

Tsai, W. Y. (1990). Testing the assumption of independence between truncated time and failure time. *Biometrika, 77*, 169–178.

Wang, M. C. (1989). A semiparametric model for randomly truncated data. *Journal of the American Statistical Association, 84*, 742–748.

Wang, M. C., Jewell, N. P., & Tsai, W. Y. (1986). Asymptotic properties of the product limit estimate under random truncation. *The Annals of Statistics, 14*, 1597–1605

Woodroofe, M. (1985). Estimating a distribution function with truncated data. *The Annals of Statistics, 13*, 163–177.

Chapter 21
Empirical Study on High-Dimensional Variable Selection and Prediction Under Competing Risks

Jiayi Hou and Ronghui Xu

21.1 Introduction

Competing risks occur when multiple types of failures co-exist and the occurrence of one type of failure may prevent the observation of the other types of failure. In addition the failure times may be subject to right-censoring. In the regression settings the Cox proportional hazards model can be used to model the so-called cause-specific hazards, and existing software for fitting the Cox model for classical survival data without competing risks can be used to fit the proportional cause-specific hazards model (Kalbfleisch and Prentice 2011). Under this model, however, the dependence of the cumulative incidence function of a particular failure type on the covariates involves also the effects of the covariates on the cause-specific hazards of all other types of failures. Beyersmann et al. (2007) showed as an example in patients receiving peripheral blood stem-cell transplantation, while the cause-specific hazard ratio for certain baseline risk factors of bloodstream infection (competing with the event of neutropenia) might be similar, the corresponding cumulative incidence functions can be quite different. In order to link the covariates directly to the cumulative incidence functions (CIF), Fine and Gray (1999) proposed to model the subdistribution hazards.The proportional hazards modeling of the subdistribution hazards, also known as Fine-Gray model, has gained popularity in

J. Hou (✉)
Altman Clinical and Translational Research Institute, University of California, San Diego, La Jolla, CA, USA

R. Xu
Department of Family Medicine and Public Health, University of California, San Diego, La Jolla, CA, USA

Department of Mathematics, University of California, San Diego, La Jolla, CA, USA
e-mail: rxu@ucsd.edu

Y. Zhao, D.-G. Chen (eds.), *New Frontiers of Biostatistics and Bioinformatics*, ICSA Book Series in Statistics, https://doi.org/10.1007/978-3-319-99389-8_21

recent years. The proportional cause-specific hazards model, and the proportional subdistribution hazards model are typically not valid at the same time, and limited empirical experiences seem to indicate that in real data applications the two models can lead to similar conclusions (Geskus 2016).

Researchers have studied different approaches to analyze survival data with high dimensional covariates. Notably, Tibshirani (1997) proposed the least absolute shrinkage and selection operator (LASSO) under the Cox proportional hazards model. Zhang and Lu (2007) investigated the statistical properties of adaptive LASSO for the Cox proportional hazards model. Hothorn et al. (2006) introduced a random forest algorithm and a generic gradient boosting algorithm for right-censoring data. When considering theoretical aspects, Bradic et al. (2011) studied a group of penalty functions and established strong oracle properties of non-concave penalized methods for ultra high dimensional covariates in the presence of right-censoring. In comparison, very few high-dimensional methods have been developed in the presence of competing risks. Binder et al. (2009) first proposed a boosting approach for fitting the proportional subdistribution hazards model. Very recently, Fu et al. (2016) considered penalized approaches under the same model. In this chapter, we will consider both the proportional cause-specific hazards (PCSH) model and the proportional subdistribution hazards (PSDH) model, and empirically investigate the accuracy of variable selection and prediction using existing computational software under either model. This leads to the Binder et al. approach under the PSDH model, and LASSO approach under the PCSH model, both being readily implemented. Both approaches rely critically on the selection of a "penalty" parameter, and there are different ways to select this parameter. We will empirically evaluate these different methods using Monte Carlo simulations.

21.2 Competing Risk Models

Let $\epsilon = 1, \ldots, J$ be the cause or type (we use the two words interchangeably in the following) of failure. Let $T = \min_{j=1}^{J} \tilde{T}_j$ denote the observed failure time if there is no censoring which is due to one of the causes, while failures from other types or causes are latent. Let $X_i = \min(T_i, C_i)$, $\delta_i = I(T_i \le C_i)$, where C_i is the potential censoring time, and is assumed non-informative. Denote $S(t) = P(T > t)$ the survival function of T. The cumulative incidence function (CIF) for failure type j is $F_j(t) = P(T \le t, \epsilon = j)$. Obviously $S(t) = 1 - \sum_{j=1}^{J} F_j(t)$, and $\sum_{j=1}^{J} F_j(\infty) = 1$. Denote the cause-specific hazard function of type j as $\lambda_j(t) = \lim_{\Delta t \to 0+} Pr(t \le T < t + \Delta t, \epsilon = j | T \ge t)/\Delta t$. Then one can also show that

$$F_j(t) = \int_0^t \lambda_j(u) S(u) du, \tag{21.1}$$

leading to a nonparametric estimate of the CIF (Fleming and Harrington 2011):

$$\hat{\lambda}_j(t_i) = \frac{d_{ji}}{n_i} \tag{21.2}$$

where d_{ji} denotes the number of failures from cause j at ordered time t_i, where $t_1 < t_2 \cdots t_i \cdots < \infty$, and n_i the number of subjects at risk at t_i, and

$$\hat{S}(t) = \prod_{i:t_i \le t} \left(1 - \sum_{j=1}^{J} \hat{\lambda}_j(t_i)\right). \tag{21.3}$$

Then we have $\hat{F}_j(t) = \sum_{i:t_i \le t} \hat{p}_j(t_i)$, where $\hat{p}_j(t_i) = \hat{\lambda}_j(t_i)\hat{S}(t_i^-)$.

21.2.1 The PCSH Model

Given a vector of covariates Z, under the proportional hazards assumption of the cause-specific hazard function we have

$$\lambda_j(t|Z) = \lambda_{0j}(t)\exp(\beta_j' Z), \tag{21.4}$$

where β_j is a vector of coefficients associated with cause j, for $j = 1, \ldots, J$.

To estimate β_j, we can use any software for the regular Cox model to model one type of event at a time and treat all other types of event as if censored. This is because the (partial) likelihood for all event types factors into a separate likelihood function for each event type, and the likelihood function for each event type treats all other types of events as if censored.

To estimate the cumulative incidence function given $Z = z_0$, we have similar to the above nonparametric estimation:

$$\begin{aligned}\hat{F}_j(t; z_0) &= \int_0^t \hat{S}(u; z_0) d\hat{\Lambda}_j(u; z_0) \\ &= \sum_{i=1}^{n} \frac{\hat{S}(X_i; z_0)\delta_{ji} I(X_i \le t)\exp(\hat{\beta}_j' z_0)}{\sum_{i'=1}^{n} I(X_i \le X_{i'})\exp(\hat{\beta}_j' Z_{i'})},\end{aligned} \tag{21.5}$$

where $\hat{S}(u; z_0) = \exp\{-\sum_{j=1}^{J} \hat{\Lambda}_j(u; z_0)\}$, $\hat{\Lambda}_j(u; z_0) = \hat{\Lambda}_{0j}(u)\exp(\hat{\beta}_j{}' z_0)$, the baseline cumulative hazard $\hat{\Lambda}_{0j}(u)$ is a Breslow-type estimator (Breslow 1974), and $\delta_{ji} = I(\epsilon_i = j)\delta_i$ indicates if an event occurs at time X_i due to cause j.

Notice that in estimating the overall survival function $\hat{S}$ we need to fit the models for all event types, even if we are only interested in the CIF of type j.

21.2.2 *The PSDH Model*

Gray (1988) introduced the subdistribution hazard function as $\tilde{\lambda}_j(t) = -\frac{d}{dt}\log\{1 - F_j(t)\}$. Under the proportional hazards assumption of the subdistribution hazard function for cause 1 we have (Fine and Gray 1999)

$$\tilde{\lambda}_1(t|Z) = \tilde{\lambda}_0(t)\exp(\beta' Z). \tag{21.6}$$

It is easy to see that model (21.6) provides a direct way to estimate the CIF of cause 1, so that there is no need to fit models for the other causes in order to estimate CIF_1, which is the cumulative incidence function due to cause 1. Fine and Gray (1999) proposed estimating equations for β. Geskus (2011) further showed that these estimating equations can be solved using weighted Cox regression, i.e. software for the regular Cox model incorporating weights. The baseline subdistribution hazard is again estimated using a modified version of Breslow's estimator.

21.3 Regularization

Classical statistical methods, such as stepwise regression, have been known to suffer from inconsistency and are computationally infeasible when the number of covariates is equal to or greater than the (effective) sample size. A group of statistical learning methods, in particular supervised learning has shown good performance empirically when the data is of high dimensionality (Fan and Li 2001). The goals of these methods are (Bühlmann and van de Geer 2011) (1) prediction: to find a set of covariates which results in minimal prediction error in independent test data; (2) variable selection: estimate the true sparsity pattern with low false positive rate for each covariate. In theory, consistent variable selection requires stronger assumptions, which are more difficult to meet in practice. In this paper, we will study the performance of statistical learning methods in estimating the true cumulative incidence function F_j. These statistical learning methods often involve the selection of a tuning parameter, based on the minimal estimated prediction error. There are two ways to estimate this prediction error: cross-validation which is computationally intensive, or approximation methods such as the C_p type statistics. When a log-likelihood loss function is used, the latter leads to the well-known Akaike information criterion (AIC). Another commonly used information based criterion is Bayesian information criterion (BIC), which imposes a larger penalty than the AIC.

21.3.1 LASSO

LASSO is an L_1 penalization method proposed by Tibshirani (1996) for building parsimonious models when the performance of classical methods such as stepwise regression or best subset selection is not satisfactory. For linear regression LASSO solves a penalized least squares problem along the regularization path, where the regression coefficients associated with unimportant covariates shrink to exactly zero while granting non-zero coefficients for important covariates. The theoretical properties of LASSO have been extensively studied under the linear regression model. Meinshausen and Bühlmann (2006) showed consistency of LASSO under the neighborhood stability condition, when the true non-zero coefficients are sufficiently large in absolute value. This condition is equivalent to the irrepresentable condition used by Zhao and Yu (2006). Although some of these theoretical conditions might be difficult to achieve in practice, LASSO has gained numerous attention as a technique to reduce dimensionality and construct predictive models. One of the main reasons for its popularity is its computational simplicity, involving convex optimization only. Alternative versions of LASSO have been proposed to handle grouped and categorical data. For example, Yuan and Lin (2006) introduced group LASSO to include or exclude the grouped variable by replacing the L_1 penalty with $\|\boldsymbol{\beta}\|_K = (\boldsymbol{\beta}^T \boldsymbol{K} \boldsymbol{\beta})^{1/2}$, where $\boldsymbol{K}$ is a symmetric positive definite matrix. In a more recent paper, Gertheiss and Tutz (2010) introduced a different penalty function $J(\boldsymbol{\beta}) = \sum_{i>j} w_{ij}|\beta_i - \beta_j|$, which is similar to the adaptive LASSO Zou (2006).

Tibshirani (1997) extended LASSO to the Cox regression model, where the log partial likelihood is penalized by $\lambda\|\boldsymbol{\beta}\|_1$. In fitting the PCSH model, the Cox regression software is used, and we apply the same LASSO algorithm as proposed in Tibshirani (1997). The penalty parameter λ can be determined by different methods, and in the following we consider:

- CV10: λ associated with the minimum tenfold cross-validated (CV) negative predictive log partial likelihood (referred to as "error" in the following);
- CV+1SE: λ associated with the minimum tenfold CV error plus one standard error of the CV estimated errors;
- min AIC: λ associated with the minimum AIC criteria;
- elbow AIC: λ associated with the largest descent in AIC.

In the above under the Cox model, the AIC is defined as $-2\log(L) + 2s$, where L is the partial likelihood and $s = |S(\hat{\boldsymbol{\beta}})|$ is the number of non-zero regression coefficients, i.e. the size of the active set $S(\hat{\boldsymbol{\beta}})$ (Verweij and Van Houwelingen 1993; Xu et al. 2009). We apply these definitions to the PCSH model, where k would be the number of observed events from the cause of interest. The "elbow" criteria are described in Tibshirani et al. (2001) as a way to avoid over-selection in practice.

21.3.2 Boosting

Freund and Schapire (1997) introduced the AdaBoost algorithm to solve classification problems by combining rough and moderately accurate "rules of thumb" repeatedly. Later, Friedman (2001) developed boosting methods for linear regression as a numerical optimization method to minimize the squared error loss function. Boosting can be viewed as a gradient descent optimization algorithm in function space, and is essentially the same as the matching pursuit algorithm in signal processing (Mallat and Zhang 1993). Bühlmann (2006) proved that boosting with the squared error loss is consistent in high-dimensional linear models, where the number of predictors is allowed to grow as fast as exponential to the sample size.

For the PSDH model with high-dimensional data, Binder et al. (2009) proposed a likelihood based boosting approach, where the likelihood is the same as the partial likelihood in Fine and Gray (1999) for complete (i.e., no censoring) data, but otherwise with weights in the risk sets to account for censoring:

$$L(\beta) = \prod_{i=1}^{n} \left[\frac{\exp(\beta' Z_i)}{\sum_{l \in R_i} w_l(X_i) \exp(\beta' Z_l)} \right]^{I(\delta_i \epsilon_i = 1)}, \tag{21.7}$$

where $R_i = \{l : X_l \geq X_i \text{ or } \delta_l \epsilon_l > 1\}$ is the risk set consisting of individuals who have not had any event or who have had an event of other causes, and $w_l(t) = \hat{G}(t) I(t \geq X_l) \delta_l / \hat{G}(X_l) + I(t < X_l)$ (Binder et al. missed the second summand) where $\hat{G}$ is the Kaplan-Meier estimate of $P(C > t)$. This boosting procedure incorporates the gradient descent in function space (Friedman 2001) to maximize the partial likelihood from PSDH model. This procedure has been implemented in R package 'CoxBoost'.

The number of boosting steps γ, which is the main tuning parameter for this approach, can be determined by the following criteria:

- CV10: γ associated with the minimum tenfold CV negative predictive log partial likelihood;
- min AIC: γ associated with the minimum AIC criteria;
- elbow AIC: γ associated with the largest descent in AIC.

21.4 Simulations

21.4.1 Setup

To investigate the performance of LASSO and boosting under the PCSH and PSDH models, respectively, we conducted comprehensive simulation studies with both continuous and dichotomized covariates in competing risks data. We assumed

$J = 2$, and we considered sample size $n = 500$ and number of covariates $p = 20$, 500, and 1000. We repeated each simulation setting 100 times.

For continuous covariates, the covariate vector for each subject was generated for the following correlation structures:

1. Independent: each covariate was independently generated from $N(0, 1)$;
2. Exchangeable: the covariate vector was generated from a multivariate normal distribution with mean zero, marginal variance of one, and a block diagonal covariance matrix—each block of size 10 and within a block the pairwise correlation $\rho(i, i') = 0.5$.
3. AR(1): the covariate vector was generated from a multivariate normal distribution with mean zero, marginal variance of one, and a block diagonal covariance matrix—each block of size 10 and within a block the pairwise correlation $\rho(i, i') = 0.5^{|i-i'|}$.

For binary covariates, the covariate vector was first generated the same as in the above, then dichotomized at threshold 0, which results in a balanced binary distribution. We set the number of non-zero regression coefficients, i.e. the size of the active set, to be $s_1 = 5$ and $s_2 = 3$ for causes 1 and 2, respectively. We let $\boldsymbol{\beta}_{1,1\cdots,5} = (1.96, -0.79, -0.5, -1.35, 1.29)$, $\boldsymbol{\beta}_{2,11\cdots,13} = (-1.16, -0.86, 0.5)$ and the rest of the $\boldsymbol{\beta}_1$ and $\boldsymbol{\beta}_2$ values were zero. These $\boldsymbol{\beta}$ values were used under both the PCSH and the PSDH models.

To simulate survival outcomes under the PCSH model we followed the approach described in Beyersmann et al. (2009); that is, we simulated the event time T first, then we simulated the cause ϵ given T. We assumed the baseline hazard functions for type 1 and 2 failures to be $\lambda_{01}(t) = 0.15$ and $\lambda_{02}(t) = 0.10$, respectively. The overall (not cause-specific) cumulative hazard function for T was then $\Lambda(t|z) = t\{\lambda_{01} \exp(\boldsymbol{\beta}_1{}'z) + \lambda_{02} \exp(\boldsymbol{\beta}_2{}'z)\}$, and T was generated using the fact that $U = \exp(-\Lambda(T)) \sim U(0, 1)$ given z. The cause ϵ was generated proportional to the cause-specific hazard function, i.e. $P(\epsilon = 1|z) = \lambda_{01} \exp(\boldsymbol{\beta}_1{}'z)/\{\lambda_{01} \exp(\boldsymbol{\beta}_1{}'z) + \lambda_{02} \exp(\boldsymbol{\beta}_2{}'z)\}$. Under this model, the true CIF for cause j was

$$\mathrm{CIF}_j(t|z) = \int_0^t S(u|z)\lambda_{0j} \exp(\boldsymbol{\beta}_j{}'z)du = \lambda_{0j} \exp(\boldsymbol{\beta}_j{}'z)\frac{e^{tM}}{M}, \tag{21.8}$$

where $M = -\{\lambda_{01} \exp(\boldsymbol{\beta}_1{}'z) + \lambda_{02} \exp(\boldsymbol{\beta}_2{}'z)\}$. The censoring times were generated from $U(0, 20)$, which resulted in an average event rate of 45.8% for cause 1 and 33.6% for cause 2 with continuous covariates, an average event rate of 51.8% for cause 1 and 27.2% for cause 2 with balanced binary covariates.

To simulate under the PSDH model we followed the approach described in Fine and Gray (1999). The CIF for failure from cause 1 was given by

$$\mathrm{CIF}_1(t|z) = P(T \le t, \epsilon = 1|z) = 1 - \{1 - p(1 - e^{-t})\}^{\exp(\boldsymbol{\beta}_1' z)}, \tag{21.9}$$

where we used $p = 0.6$. As this was a subdistribution function, with a point mass $1 - \mathrm{CIF}_1(\infty|z)$ at infinity, the proper distribution function that was used to generate T was $F(t|z) = \mathrm{CIF}_1(t|z)/\mathrm{CIF}_1(\infty|z)$, so that $F(T) \sim U(0, 1)$ given z. Note that $P(\epsilon = 1|z) = \mathrm{CIF}_1(\infty|z)$, and $P(\epsilon = 2|z) = 1 - P(\epsilon = 1|z)$. Finally the event times for failure from cause 2 were generated according to an exponential distribution with rate $\exp(\boldsymbol{\beta}_2' z)$. The censoring times were generated from $U(0, 20)$, resulting in an average event rate of 53.5% for cause 1 and 35.1% for cause 2 with continuous covariates, an average event rate of 55.8% for cause 1 and 33.4% for cause 2 with binary covariates.

21.4.2 Results

We evaluate the performance of prediction at a given covariate vector value $\mathbf{z_0}$. We set $z_0 = (0.5, \cdots, 0.5)_{1\times p}$ for the continuous case; and for all the binary cases each element of $\mathbf{z_0}$ was independently drawn with a fixed seed from Bernoulli distribution with $p = 0.5$. Figures 21.1, 21.2, 21.3, and 21.4 show the empirical distributions of the estimated $CIF_1(2)$ over the 100 simulation runs, where the vertical line marks the true $CIF_1(2)$; the empirical distributions were plotted using the R function "density()". The PCSH model with LASSO was used to estimate $CIF_1(2)$ in Figs. 21.1 and 21.2, and the PSDH model with boosting was used in Figs. 21.3 and 21.4.

In the figures the blue dashed lines are for the oracle estimator, which fits the exact true active set $S(\beta)$. The oracle estimator varied extremely slightly when the three correlation structures for Z were generated separately, which appeared to be due to Monte Carlo variation, and the one under the AR(1) structure is plotted here. The solid lines are the estimated $CIF_1(2)$ under each model after regularization using LASSO or boosting, with different colors representing different correlation structures of Z.

Under the PSDH model using LASSO to regularize, the performances were generally not satisfactory as compared to the oracle estimator. The worst performances were seen when using minimum AIC to choose the penalty parameter; some of these results were so extreme that "density()" failed to work. Elbow AIC criteria is slightly better. CV10 had the best performance for binary covariates for $p = 20$, but it too deteriorated for $p = 500$ and 1000.

Under the PSDH model using boosting, in Fig. 21.3 we see that for continuous covariates, the estimators performed reasonably well when CV10 or minimum AIC was used to choose the number of boosting steps; with CV10 the estimation was perhaps the best. The performance deteriorated with binary covariates for $p = 500$ and 1000. We note that in Bühlmann (2006) simulation studies (Table 1) the mean squared error for boosting with correlated design was also smaller than that with uncorrelated design, and their Figure 1 showed that boosting tended to select more

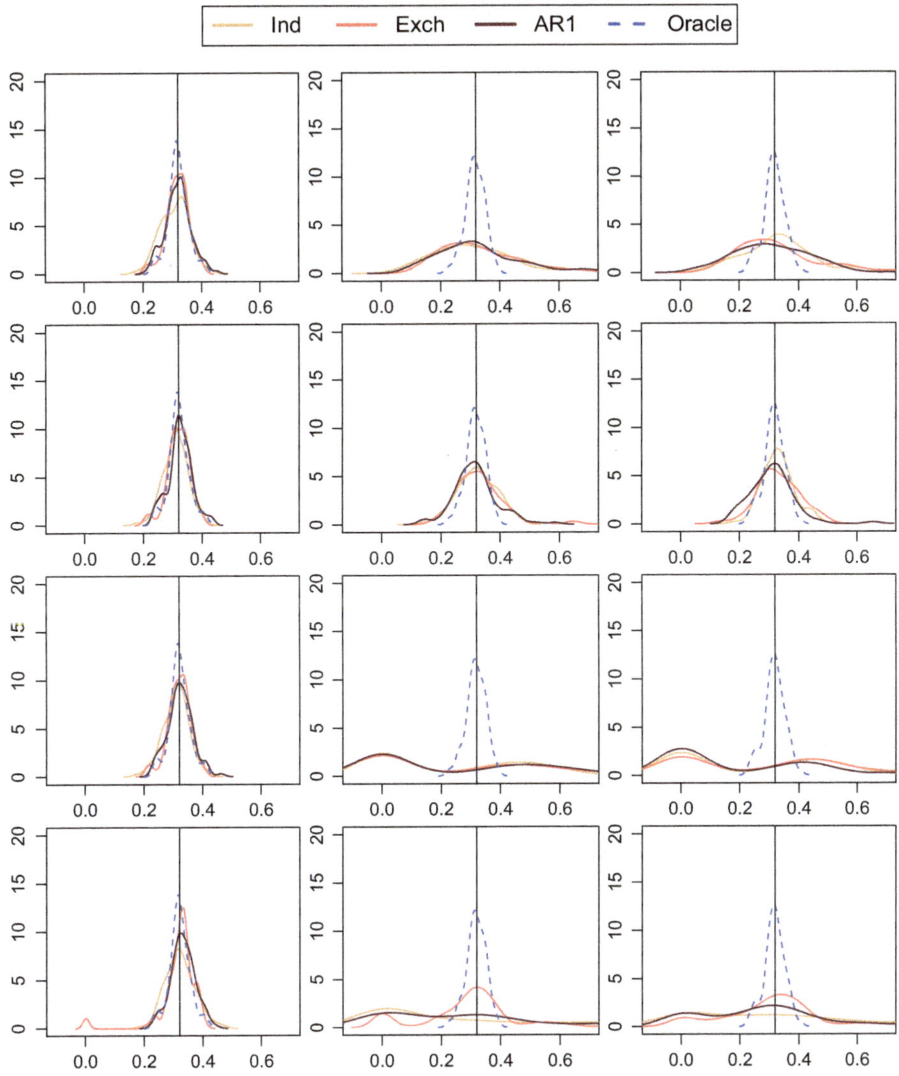

Fig. 21.1 The (smoothed) empirical distribution of $C\hat{I}F_1(2)$, estimated under the PCSH model with LASSO, for continuous covariates. The three columns correspond to $p = 20$, 500, and 1000. The rows correspond to different ways of selecting λ, from top to bottom: (1) CV10, (2) CV+1SE, (3) minimum AIC and (4) elbow AIC. The true $CIF_1(2|z_0) = 0.32$

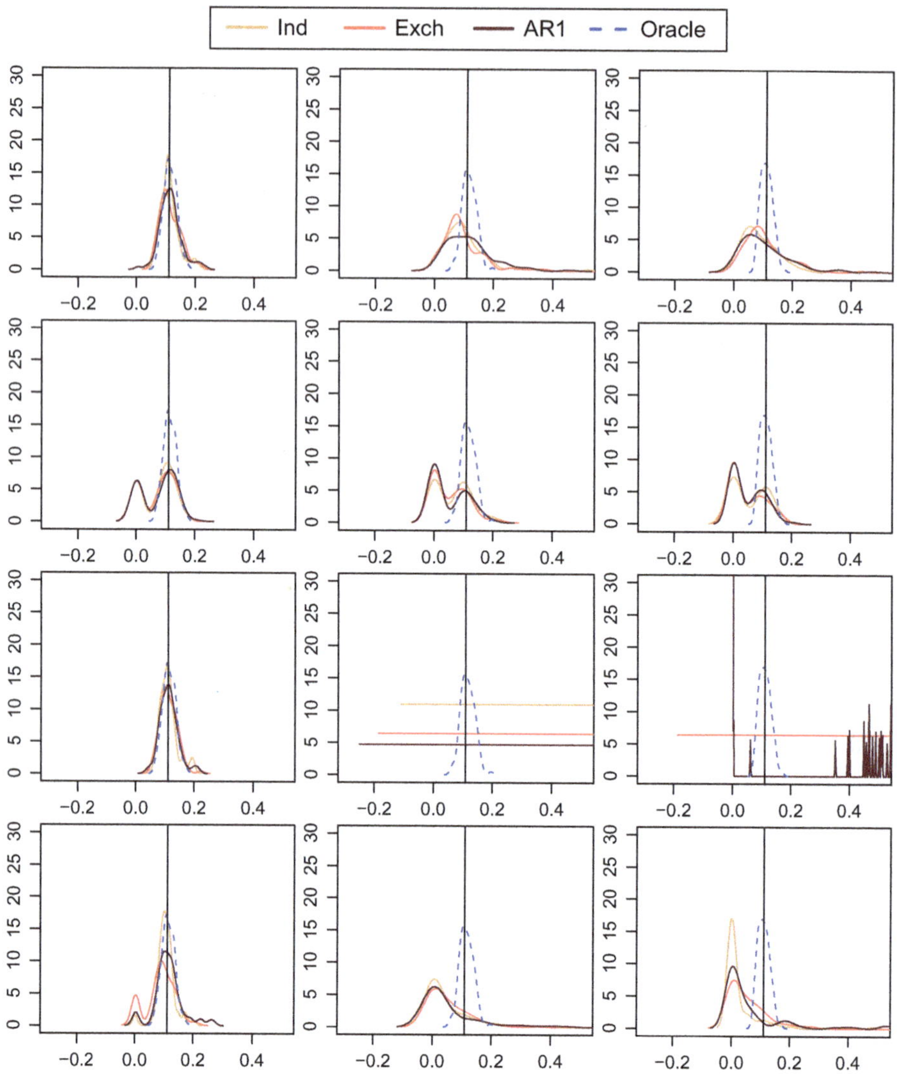

Fig. 21.2 The (smoothed) empirical distribution of $C\hat{I}F_1(2)$, estimated under the PCSH model with LASSO, for balanced binary covariates. The three columns correspond to $p = 20$, 500, and 1000. The rows correspond to different ways of selecting λ, from top to bottom: (1) CV10, (2) CV+1SE, (3) minimum AIC and (4) elbow AIC. The true $CIF_1(2|z_0) = 0.11$

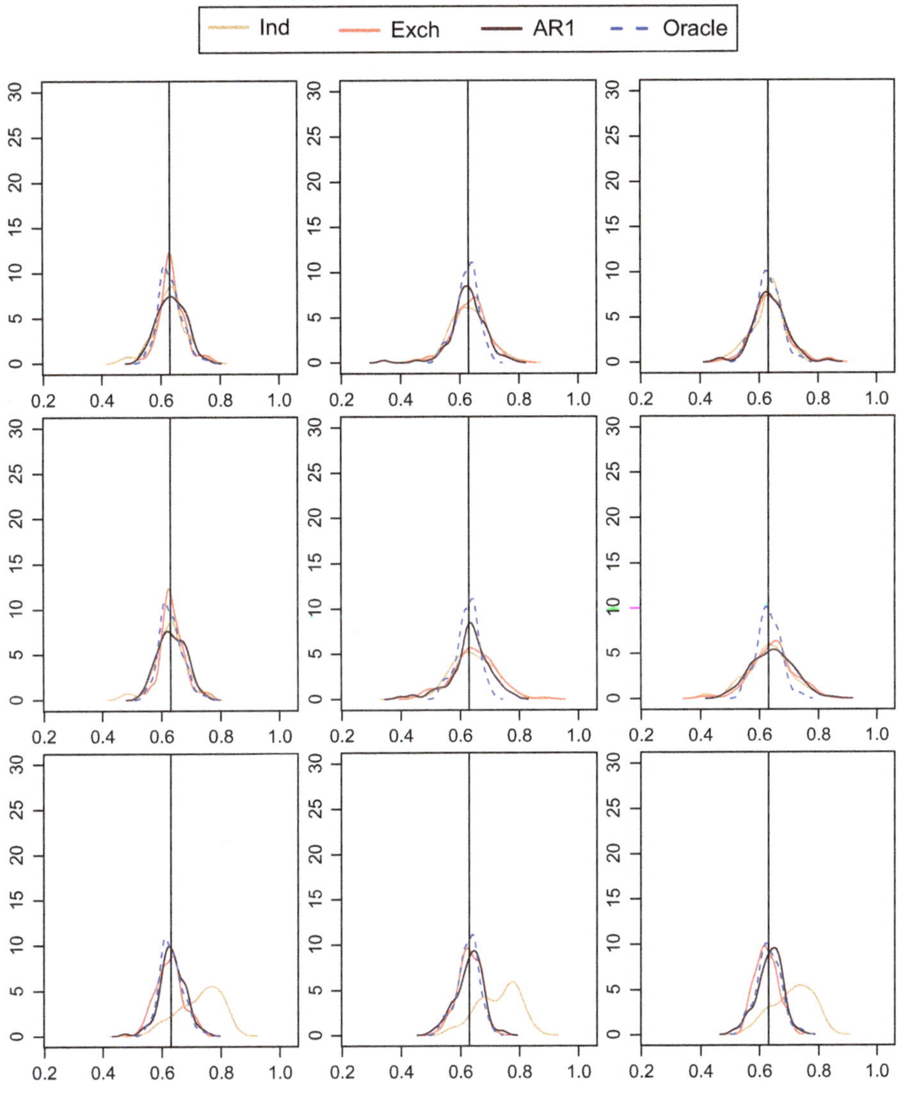

Fig. 21.3 The (smoothed) empirical distribution of $C\hat{I}F_1(2)$, estimated under the PSDH model with boosting, for continuous covariates. The three columns correspond to $p = 20, 500$, and 1000. The rows correspond to different ways of selecting γ, from top to bottom: (1) CV10, (2) minimum AIC and (3) elbow AIC. The true $CIF_1(2|z_0) = 0.63$

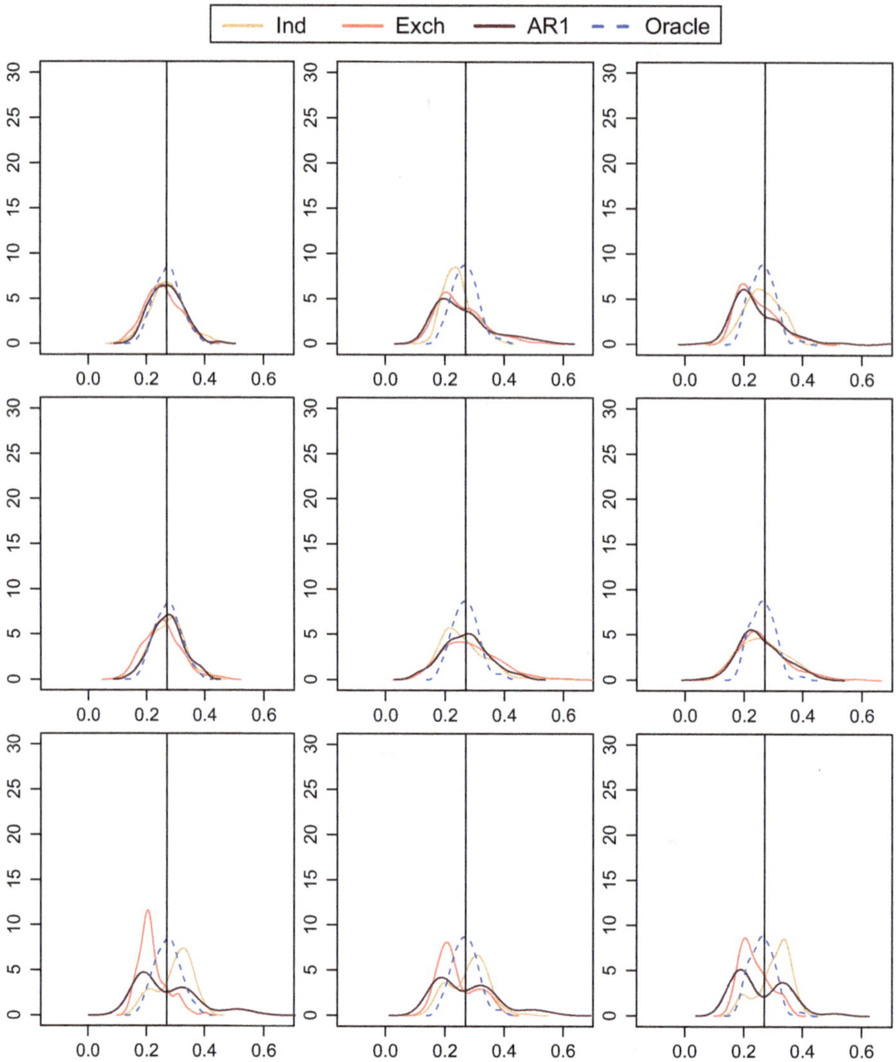

Fig. 21.4 The (smoothed) empirical distribution of $C\hat{I}F_1(2)$, estimated under the PSDH model with boosting, for balanced binary covariates. The three columns correspond to $p = 20$, 500, and 1000. The rows correspond to different ways of selecting γ, from top to bottom: (1) CV10, (2) minimum AIC and (3) elbow AIC. The true $CIF_1(2|z_0) = 0.27$

covariates in the uncorrelated design than the correlated design. This may explain the bias introduced due to under-selection in the uncorrelated design (Fig. 21.3).

The results of variable selection are presented in Tables 21.1, 21.2, 21.3, 21.4, 21.5, 21.6, and 21.7. In the following tables, $|S(\hat{\beta})|$ is the size of the estimated active set, i.e. number of non-zero estimated regression coefficients. The median number of selected variables are reported, and in () are the median absolute deviation (MAD) of selection. One can see it was difficult to achieve good model consistency (i.e., selection). When the selection is extremely poor, for example a couple of hundred false positives, then the prediction results were very poor as well. Boosting had no more than five false positives in all cases.

Table 21.1 The median (MAD) number of selected variables by LASSO under the PCSH model

Continuous	$\lvert S(\hat{\beta})\rvert$	#True positives	#False positives
$p = 20$			
Independence	12(1)	5(0)	7(1)
Exchangeable	12(2)	5(0)	7(2)
AR1	12(1)	5(0)	7(1)
$p = 500$			
Independence	35(7)	5(0)	30(7)
Exchangeable	35(5)	5(0)	30(5)
AR1	37(6)	5(0)	32(6)
$p = 1000$			
Independence	41(8)	5(0)	36(8)
Exchangeable	44(7)	5(0)	39(7)
AR1	41(7.5)	5(0)	36(7.5)
Binary (balanced)	$\lvert S(\hat{\beta})\rvert$	#True positives	#False positives
$p = 20$			
Independence	11(2)	5(0)	6(2)
Exchangeable	12(2)	5(0)	7(2)
AR1	12(2)	5(0)	7(2)
$p = 500$			
Independence	22(6)	5(0)	17(6)
Exchangeable	27(6)	5(0)	22(6)
AR1	25(6)	5(0)	20(6)
$p = 1000$			
Independence	23(6)	5(0)	18(6)
Exchangeable	29(5)	5(0)	24(5)
AR1	27(8)	5(0)	22(8)

The penalty parameter is chosen using CV10

Table 21.2 The median (MAD) number of selected variables by LASSO under the PCSH model

Continuous	$\lvert S(\hat{\beta})\rvert$	#True positives	#False positives
$p = 20$			
Independence	6(1)	5(0)	1(1)
Exchangeable	6(1)	5(0)	1(1)
AR1	6(1)	5(0)	1(1)
$p = 500$			
Independence	9(2)	5(0)	4(3)
Exchangeable	12(3)	5(0)	7(3)
AR1	10(3)	5(0)	5(3)
$p = 1000$			
Independence	9(3)	5(0)	4(3)
Exchangeable	13(4)	5(0)	8(4)
AR1	12(4)	5(0)	7(4)
Binary (balanced)	$\lvert S(\hat{\beta})\rvert$	#True positives	#False positives
$p = 20$			
Independence	5(0)	5(0)	0(0)
Exchangeable	5(1)	5(0)	0(0)
AR1	5(0)	5(0)	0(0)
$p = 500$			
Independence	5(1)	5(0)	1(1)
Exchangeable	6(1)	4(0)	2(1)
AR1	7(2)	5(0)	2(2)
$p = 1000$			
Independence	5(1)	5(0)	0(0)
Exchangeable	6(2)	4(0)	1(1)
AR1	6(1)	5(0)	1(1)

The penalty parameter is chosen using CV+1SE

21.5 Discussion

The rapid accumulation of data across many fields, medicine in particular, has created unique challenges in statistics. The distinct issues with high-dimensional data have come to be recognized recently, including for example, the rapid noise accumulation, the unrealistic independence assumption, and the necessity for novel robust data analysis methods (Fan et al. 2014). While researchers work to meet these challenges, some of the methods proposed in the literature do not necessarily scale well to large data sets. In this paper, we considered the feasible implementations of statistical learning methods under the PCSH and PSDH models. We empirically

Table 21.3 The median (MAD) number of selected variables by LASSO under the PCSH model

Continuous	$\lvert S(\hat{\beta})\rvert$	#True positives	#False positives
$p = 20$			
Independence	7(1)	5(0)	2(1)
Exchangeable	7(1)	5(0)	2(1)
AR1	7(1)	5(0)	2(1)
$p = 500$			
Independence	280(16.5)	5(0)	275(16.5)
Exchangeable	303.5(13)	5(0)	298.5(13)
AR1	295.5(11.5)	5(0)	290.5(11.5)
$p = 1000$			
Independence	260(15)	5(0)	255(15)
Exchangeable	292(16.5)	5(0)	287(16.5)
AR1	282(17)	5(0)	277(17)
Binary (balanced)	$\lvert S(\hat{\beta})\rvert$	#True positives	#False positives
$p = 20$			
Independence	7(1)	5(0)	2(1)
Exchangeable	7(1)	5(0)	2(1)
AR1	7(1)	5(0)	2(1)
$p = 500$			
Independence	364.5(16.5)	5(0)	359.5(16.5)
Exchangeable	372(17)	5(0)	367(17)
AR1	369(15)	5(0)	364(15)
$p = 1000$			
Independence	344(17.5)	5(0)	339.5(17.5)
Exchangeable	352(14.5)	5(0)	347(14.5)
AR1	362(16.5)	5(0)	357(16.5)

The penalty parameter is chosen using min AIC

studied their performance in variable selection and prediction through comprehensive simulations in both low- and high-dimensional settings with different covariate structures.

In the limited comparisons that we are aware of in the literature, the two models seem to give somewhat comparable results (Geskus 2016). While the PSDH model was proposed in order to associate the CIF due to one cause directly with the covariates without having to specifically model the other causes, the PCSH model might be more flexible precisely due to the fact that it allows different modeling of different causes in the CIF. This is certainly worth future investigation. We also note that while the proportional hazards assumption is used in both models, there

Table 21.4 The median (MAD) number of selected variables by LASSO under the PCSH model

Continuous	$\lvert S(\hat{\beta})\rvert$	#True positives	#False positives
$p = 20$			
Independence	5.5(0.5)	5(0)	0.5(0.5)
Exchangeable	5(1)	5(0)	0(0)
AR1	5(1)	5(0)	0(0)
$p = 500$			
Independence	225(21)	5(0)	220(21)
Exchangeable	27.5(18.5)	5(0)	22.5(18.5)
AR1	200(51)	5(0)	195(51)
$p = 1000$			
Independence	162.5(72)	5(0)	157.5(72)
Exchangeable	31.5(22)	5(0)	26.5(22)
AR1	61(49)	5(0)	56(49)
Binary (balanced)	$\lvert S(\hat{\beta})\rvert$	#True positives	#False positives
$p = 20$			
Independence	6(1)	5(0)	1(1)
Exchangeable	5(1)	5(0)	0(0)
AR1	5(1)	5(0)	0(0)
$p = 500$			
Independence	271.5(41)	5(0)	266.5(41)
Exchangeable	171.5(127.5)	5(0)	166.5(127.5)
AR1	277(40)	5(0)	272(40)
$p = 1000$			
Independence	271(26)	5(0)	266(26)
Exchangeable	70.5(60)	5(0)	65.5(59.5)
AR1	247(68)	5(0)	242(68)

The penalty parameter is chosen using elbow AIC

has been recent work considering other modeling approaches such as the additive hazards in the presence of competing risks (Zheng et al. 2017).

Finally, we note that in the high-dimensional context methods developed for continuous data may behave differently for binary data especially if sparsity presents. In a recent paper Mukherjee et al. (2015) showed that when a binary design matrix is sufficiently sparse, no signal can be detected irrespective of its strength. This finding echoes the challenges that we have observed in our simulation studies.

Table 21.5 The median (MAD) number of selected variables by Boosting under the PSDH model

Continuous	$\|S(\hat{\beta})\|$	#True positives	#False positives
$p = 20$			
Independence	7(2)	5(0)	2(2)
Exchangeable	10(3)	5(0)	5(3)
AR1	9(2)	5(0)	4(2)
$p = 500$			
Independence	5(0)	5(0)	0(0)
Exchangeable	6(1)	5(0)	1(1)
AR1	6(1)	5(0)	1(1)
$p = 1000$			
Independence	5(0)	5(0)	0(0)
Exchangeable	5(0)	5(0)	0(0)
AR1	5(0)	5(0)	0(0)
Binary (balanced)	$\|S(\hat{\beta})\|$	#True positives	#False positives
$p = 20$			
Independence	6(1)	5(0)	1(1)
Exchangeable	7(2)	5(0)	2(2)
AR1	6(1)	5(0)	1(1)
$p = 500$			
Independence	5(1)	5(0)	0(0)
Exchangeable	4(1)	4(0)	0(0)
AR1	4.5(0.5)	4(0)	0(0)
$p = 1000$			
Independence	5(1)	4(1)	0(0)
Exchangeable	4(0)	4(0)	0(0)
AR1	4(0)	4(0)	0(0)

The number of steps is chosen using CV10

Table 21.6 The median (MAD) number of selected variables by Boosting under the PSDH model

Continuous	$\|S(\hat{\beta})\|$	#True positives	#False positives
$p = 20$			
Independence	7(1)	5(0)	2(1)
Exchangeable	7(1)	5(0)	2(1)
AR1	7(1)	5(0)	2(1)
$p = 500$			
Independence	7(1)	5(0)	2(1)
Exchangeable	8(1)	5(0)	3(1)
AR1	7.5(1.5)	5(0)	2.5(1.5)
$p = 1000$			
Independence	7(1)	5(0)	2(1)
Exchangeable	7(1)	5(0)	2(1)
AR1	7(1)	5(0)	2(1)

(continued)

Table 21.6 (continued)

Continuous	$\lvert S(\hat{\beta})\rvert$	#True positives	#False positives
Binary (balanced)	$\lvert S(\hat{\beta})\rvert$	#True positives	#False positives
$p = 20$			
Independence	7(1)	5(0)	2(1)
Exchangeable	7(1)	5(0)	2.5(1.5)
AR1	7(1)	5(0)	2(1)
$p = 500$			
Independence	7(1)	5(0)	2(0)
Exchangeable	7(1)	5(0)	2(1)
AR1	7(0)	5(0)	2(0)
$p = 1000$			
Independence	7(1)	5(0)	2(1)
Exchangeable	6(1)	5(0)	2(1)
AR1	7(0)	5(0)	2(0)

The number of steps is chosen using min AIC

Table 21.7 The median (MAD) number of selected variables by Boosting under the PSDH model

Continuous	$\lvert S(\hat{\beta})\rvert$	#True positives	#False positives
$p = 20$			
Independence	3(0)	3(0)	0(0)
Exchangeable	4(1)	4(1)	0(0)
AR1	4.5(0.5)	4.5(0.5)	0(0)
$p = 500$			
Independence	3(0)	3(0)	0(0)
Exchangeable	4(0)	4(0)	0(0)
AR1	4(1)	4(1)	0(0)
$p = 1000$			
Independence	4(1)	4(1)	0(0)
Exchangeable	4(1)	4(1)	0(0)
AR1	4.5(0.5)	4.5(0.5)	0(0)
Binary (balanced)	$\lvert S(\hat{\beta})\rvert$	#True positives	#False positives
$p = 20$			
Independence	3(0)	3(0)	0(0)
Exchangeable	3.5(0.5)	3.5(0.5)	0(0)
AR1	4(0)	4(0)	0(0)
$p = 500$			
Independence	3(0)	3(0)	0(0)
Exchangeable	3(1)	3(1)	0(0)
AR1	4(0)	4(0)	0(0)
$p = 1000$			
Independence	3(0)	3(0)	0(0)
Exchangeable	3(1)	3(1)	0(0)
AR1	4(0)	4(0)	0(0)

The number of steps is chosen using elbow AIC

References

Beyersmann, J., Dettenkofer, M., Bertz, H., & Schumacher, M. (2007). A competing risks analysis of bloodstream infection after stem-cell transplantation using subdistribution hazards and cause-specific hazards. *Statistics in Medicine, 26*(30), 5360–5369.

Beyersmann, J., Latouche, A., Buchholz, A., & Schumacher, M. (2009). Simulating competing risks data in survival analysis. *Statistics in Medicine, 28*(6), 956–971.

Binder, H., Allignol, A., Schumacher, M., & Beyersmann, J. (2009). Boosting for high-dimensional time-to-event data with competing risks. *Bioinformatics, 25*(7), 890–896.

Bradic, J., Fan, J., & Jiang, J. (2011). Regularization for Cox's proportional hazards model with np-dimensionality. *Annals of Statistics, 39*(6), 3092.

Breslow, N. (1974). Covariance analysis of censored survival data. *Biometrics, 30*, 89–99.

Bühlmann, P. (2006). Boosting for high-dimensional linear models. *The Annals of Statistics, 34*(2), 559–583.

Bühlmann, P., & van de Geer, S. (2011). *Statistics for high-dimensional data: Methods, theory and applications*. Berlin: Springer.

Fan, J., Han, F., & Liu, H. (2014). Challenges of big data analysis. *National Science Review, 1*(2), 293–314.

Fan, J., & Li, R. (2001). Variable selection via nonconcave penalized likelihood and its oracle properties. *Journal of the American Statistical Association, 96*(456), 1348–1360.

Fine, J. P., & Gray, R. J. (1999). A proportional hazards model for the subdistribution of a competing risk. *Journal of the American Statistical Association, 94*(446), 496–509.

Fleming, T. R., & Harrington, D. P. (2011). *Counting processes and survival analysis* (Vol. 169). Hoboken: John Wiley & Sons.

Freund, Y., & Schapire, R. E. (1997). A desicion-theoretic generalization of on-line learning and an application to boosting. *Journal of Computer and System Sciences, 55*, 119–139.

Friedman, J. H. (2001). Greedy function approximation: A gradient boosting machine. *Annals of Statistics, 29*(5), 1189–1232.

Fu, Z., Parikh, C. R., & Zhou, B. (2016). Penalized variable selection in competing risks regression. *Lifetime Data Analysis, 23*, 353–376. https://doi.org/10.1007/s10985-016-9362-3.

Gertheiss, J., & Tutz, G. (2010). Sparse modeling of categorial explanatory variables. *The Annals of Applied Statistics, 4*(4), 2150–2180.

Geskus, R. B. (2011). Cause-specific cumulative incidence estimation and the Fine-Gray model under both left truncation and right censoring. *Biometrics, 67*(1), 39–49.

Geskus, R. B. (2016). *Data analysis with competing risks and intermediate states*. Boca Raton, FL: Taylor & Francis Group, LLC.

Gray, R. J. (1988). A class of K-sample tests for comparing the cumulative incidence of a competing risk. *The Annals of Statistics, 16*(3), 1141–1154.

Hothorn, T., Bühlmann, P., Dudoit, S., Molinaro, A., & Van Der Laan, M. J. (2006). Survival ensembles. *Biostatistics, 7*(3):355–373.

Kalbfleisch, J. D., & Prentice, R. L. (2011). *The statistical analysis of failure time data* (Vol. 360). Hoboken: John Wiley & Sons.

Mallat, S. G., & Zhang, Z. (1993). Matching pursuits with time-frequency dictionaries. *IEEE Transactions on Signal Processing, 41*(12), 3397–3415.

Meinshausen, N., & Bühlmann, P. (2006). High-dimensional graphs and variable selection with the lasso. *Annals of Statistics, 34*, 1436–1462.

Mukherjee, R., Pillai, N. S., & Lin, X. (2015). Hypothesis testing for high-dimensional sparse binary regression. *Annals of Statistics, 43*(1), 352.

Tibshirani, R. (1996). Regression shrinkage and selection via the Lasso. *Journal of the Royal Statistical Society. Series B (Methodological), 58*, 267–288.

Tibshirani, R. (1997). The lasso method for variable selection in the Cox model. *Statistics in Medicine, 16*(4), 385–395.

Tibshirani, R., Walther, G., & Hastie, T. (2001). Estimating the number of clusters in a dataset via the gap statistic. *Journal of the Royal Statistical Society, Series B, 63*(2), 411–423.

Verweij, P. J. M., & Van Houwelingen, H. C. (1993). Cross-validation in survival analysis. *Statistics in Medicine, 12*(24), 2305–2314.

Xu, R., Vaida, F., & Harrington, D. P. (2009). Using profile likelihood for semiparametric model selection with application to proportional hazards mixed models. *Statistica Sinica, 19*, 819–842.

Yuan, M., & Lin, Y. (2006). Model selection and estimation in regression with grouped variables. *Journal of the Royal Statistical Society: Series B (Statistical Methodology), 68*(1), 49–67.

Zhang, H. H., & Lu, W. (2007). Adaptive lasso for Cox's proportional hazards model. *Biometrika, 94*(3), 691–703.

Zhao, P., & Yu, B. (2006). On model selection consistency of lasso. *Journal of Machine Learning Research, 7*(Nov), 2541–2563.

Zheng, C., Dai, R., Hari, P. N., & Zhang, M.-J. (2017). Instrumental variable with competing risk model. *Statistics in Medicine, 36*, 1240–1255.

Zou, H. (2006). The adaptive lasso and its oracle properties. *Journal of the American Statistical Association, 101*(476), 1418–1429.

Chapter 22
Nonparametric Estimation of a Hazard Rate Function with Right Truncated Data

Haci Akcin, Xu Zhang, and Yichuan Zhao

22.1 Introduction

Truncation is one type of incompleteness often appearing in time-to-event data. A truncated sample contains replicates of the variables (L, T) subject to the constraint $L < T$. For a truncated sample, T is left truncated by L and L is right truncated by T. Truncation is closely related to biased sampling, where the probability of selection depends on the length of the variable.

Analysis of left truncated survival data has received much attention. The survival function of T can be estimated by the truncated version of Kaplan-Meier estimator (1958). Woodroofe (1985), Wang et al. (1986), Keiding and Gill (1990), and Chen et al. (1995) studied asymptotic properties of the left truncated version of the Kaplan-Meier estimator. Lai and Ying (1991) and Gurler et al. (1993) focused on the nonparametric inference of survival function with left truncated and right censored data. Uzunogullari and Wang (1992) studied the kernel estimators of the hazard rate function with left truncated and right censored data. They particularly considered adaptive bandwidth to get smoother curves and more precise estimation result. Regression analysis in the context of left truncation and right censoring

H. Akcin (✉)
Department of Risk Management and Insurance, Georgia State University, Atlanta, GA, USA
e-mail: hakcin1@gsu.edu

X. Zhang
Center for Clinical and Translational Sciences, University of Texas Health Science Center, Houston, TX, USA
e-mail: Xu.Zhang@uth.tmc.edu

Y. Zhao
Department of Mathematics and Statistics, Georgia State University, Atlanta, GA, USA
e-mail: yichuan@gsu.edu

Y. Zhao, D.-G. Chen (eds.), *New Frontiers of Biostatistics and Bioinformatics*, ICSA Book Series in Statistics, https://doi.org/10.1007/978-3-319-99389-8_22

was studied by Klein and Zhang for its application in evaluating outcome of bone marrow transplantation (1996).

Right truncation has been routinely tackled by transforming it to left truncation. Let τ be a large constant. The transformed variable $\tau - L$ is left truncated by $\tau - T$. Using this relationship, the distribution function of L coincides with the survival function of the transformed variable $\tau - L$, and the truncated version of the Kaplan-Meier estimator became the natural estimation method (Lagakos et al. 1988; Woodroofe 1985; Keiding and Gill 1990). In recent years, Chi et al. (2007) developed a test to compare integrated weighted differences between two survival functions. An important survival quantity related to the transformed variable $\tau - L$ is its hazard rate function. This function is commonly interpreted as a hazard rate function with τ as the origin and counting backwards along the time axis. Therefore, it is known as reverse-time hazard or retro-hazard. Lagakos et al. (1988) proposed a weighted log-rank test to compare the reverse-time hazard rate functions. Gross and Huber-Carol (1992) and Kalbfleisch and Lawless (1991) studied Cox regression modeling the reverse-time hazard rate function. The Nelson-Aalen estimator is applicable for estimating the cumulative reverse-time hazard.

Natural interpretation of reverse-time hazard does not exist. A few statisticians noted this drawback and developed the inferences about the forward-time hazard rate function. Finkelstein et al. (1993) considered Cox regression modeling the hazard rate function and proposed the full likelihood approach to estimate the regression coefficients. Shen (2010) utilized the inverse probability technique to estimate the cumulative hazard function and proposed a semiparametric test to compare weighted cumulative hazard functions. In general, inference about the hazard rate function under right truncation is scarce. In this study we directly estimate the hazard rate function and develop the nonparametric inferences. Our motivation for estimating the hazard rate function was based on the dynamic feature of this function.

The remainder of this book chapter is organized as follows. Section 22.2 describes non-parametric inference of reverse-time hazard rate function and its kernel smoothing estimate. Subsequently, kernel smoothing of forward-time hazard rate function is provided in Sect. 22.3. Results of simulation study for forward-time hazard rate are presented in Sect. 22.4. As an example, AIDS blood transfusion data set is analyzed using forward-time hazard rate in Sect. 22.5. A brief discussion is given in Sect. 22.6.

22.2 Nonparametric Inference of Reverse-Time Hazard Rate Function

In survival analysis, one often needs to find the risk of an individual at a certain time. Estimation becomes cumbersome under random truncation. Formally, let (L, T) be random variables with the constraint $L < T$. A truncated sample can be described as $\{L_i, T_i\}$, $i = 1, 2, \ldots, n$, and $L_i < T_i$. The variable of study interest, L, is right truncated by the truncation variable, T. Let F and G be distribution functions for T and L, respectively. Let (a_C, b_C) be the inner support for any distribution function C and defined as $a_C = \inf\{z > 0 : C(z) > 0\}$ and $b_C = \sup\{z > 0 : C(z) < 1\}$. Consequently the inner supports for F and G are (a_F, b_F) and (a_G, b_G), respectively. F and G are estimable only if $a_G < b_F$. Another issue with truncated sample is indentifiability. In general one can choose $a^* = \min(L_1, \ldots, L_n)$ and $b^* = \max(T_1, \ldots, T_n)$, then the conditional distribution functions $F^*(t) = P(T \leq t|T \geq a^*)$ and $G^*(t) = P(L \leq t|L \leq b^*)$ are identifiable (Klein and Moeschberger 2003). For simplicity $a_G = 0$ and $b_G < b^*$ are assumed so that G agrees with the conditional distribution function G^* and becomes identifiable.

Let $\alpha(t)$ and $A(t)$ be the hazard rate function and cumulative hazard function of L, respectively, with the definitions

$$\alpha(t) = \lim_{\Delta t \to 0} \frac{P\,[t \leq L < t + \Delta t|L \geq t]}{\Delta t} \tag{22.1}$$

and

$$A(t) = \int_0^t \alpha(u)du = \int_0^t \frac{dG(u)}{P(L \geq u)}, \tag{22.2}$$

where $G(t)$ is the distribution function of L, $G(t) = P(L \leq t)$.

To estimate $\alpha(t)$ one should first obtain estimates of $A(t)$, and then apply some smoothing technique to find smoothed slopes of $A(t)$. Different smoothing methods were proposed to estimate hazard rate. Kernel smoothing, spline method, and local polynomial method are the most common techniques. Kernel smoothing and local polynomial method are theoretically more tractable than spline method (Wang 2005). Kernel smoothing is used here to estimate hazard rate function.

With a truncated sample, right truncation can be easily transformed to become left truncation. Let τ be a large constant greater than $\max\{T_1, \cdots, T_n\}$ and consider the transformed variables $L^* = \tau - L$, $T^* = \tau - T$. For the newly constructed sample $\{L_i^*, T_i^*\}$, $i = 1, \cdots, n$, there is the constraint $L_i^* > T_i^*$. Therefore, the variable L^* is left truncated by the variable T^*. The hazard rate function of L^* is a quantity with τ as its origin and counting backwards along the time axis towards zero. As a result, such a quantity is called as "reverse-time hazard" by Lagakos et al. (1988) or "retro hazard" by Keiding and Gill (1990). Let $\alpha^*(t)$ and $A^*(t)$, respectively, denote the reverse-time hazard rate function and cumulative hazard function, with the definitions

$$\alpha^*(t) = \lim_{\Delta t \to 0} \frac{P[t \geq L > t - \Delta t|L \leq t]}{\Delta t} \tag{22.3}$$

and

$$A^*(t) = \int_t^\tau \alpha^*(u)du = \int_t^\tau \frac{dG(u)}{P(L \le u)}. \tag{22.4}$$

The function $A^*(t)$ can be estimated by the Nelson-Aalen estimator. A definition about the reverse-time martingale is needed in order to establish the inference of the Nelson-Aalen estimator. For a truncated sample, we define the following counting processes $N_i^L(t) = I(L_i \ge t)$, $\bar{N}^L(t) = \sum_{i=1}^n N_i^L(t)$ and $Y_i(t) = I(L_i \le t \le T_i)$, $\bar{Y}(t) = \sum_{i=1}^n Y_i(t)$. Also define $N_i(t) = I(L_i \le t)$ and $\bar{N}(t) = \sum_{i=1}^n N_i(t)$. The counting process $N_i^L(t)$ is defined to count an event via the reversed-time axis. The corresponding martingale is given by

$$M_i^*(t) = N_i^L(t) - \int_\tau^t Y_i(u)d\Lambda^*(u)$$

It is the standard result that $M^*(t)$ is a local square integrable martingale (Keiding and Gill 1990). Let $\bar{M}^*(t) = \sum_{i=1}^n M_i^*(t)$, the Nelson-Aalen type estimator of $A^*(t)$ is given by

$$\widehat{A}^*(t) = \int_\tau^t \frac{J(u)}{\bar{Y}(u)} d\bar{N}^L(u) = \sum_{i=1}^n \int_t^\tau \frac{J(u)}{\bar{Y}(u)} dI(L_i \le u),$$

where $J(t) = I(\bar{Y}(t) > 0)$, if $\bar{Y}(t) = 0$, then $J(t)/\bar{Y}(t)$ is defined as 0. Consider $A^{*s}(t) = \int_\tau^t \alpha^*(u)J(u)du$. It follows that

$$\widehat{A}^*(t) - A^{*s}(t) = \int_\tau^t \frac{J(u)}{\bar{Y}(u)} d\bar{M}^*(u)$$

and $\widehat{A}^*(t) - A^{*s}(t)$ has the predictable variation process $\int_\tau^t \{J(u)\alpha^*(u)/\bar{Y}(u)\}du$.

It can be proven that $\sqrt{n}\{\widehat{A}^*(t) - A^{*s}(t)\}$ converges in distribution to a zero-mean Gaussian process with the predictable variation process $\int_\tau^t \{J(u)\alpha^*(u)/y(u)\}du$, where $y(t) = E[n^{-1}\bar{Y}(t)]$. Watson and Leadbetter (1964) defined the kernel function estimator of hazard rate. Andersen et al. (1993) generalized the kernel function estimator based on counting process proposed by Ramlau-Hansen (1983).

Using the similar approach, we can estimate reverse-time hazard $\alpha^*(t)$ by

$$\widehat{\alpha}^*(t) = \frac{1}{b}\int_\tau^0 K\left(\frac{t-u}{b}\right) d\widehat{A}^*(u) \tag{22.5}$$

The kernel function is a bounded function between $[-1, 1]$ and should be integrated to 1. The bandwidth b is a positive parameter. Estimation of the reverse-time hazard rate function has not been studied before, because its interpretation is not natural. Our interest centered on the inference of the forward-time hazard rate function because it is the natural and basic function for a time-to-event variable. We sketched the result for $\widehat{\alpha}^*(t)$ as follows. Let $\alpha^{*s}(t)$ be the smoothed version of $\alpha^*(t)$,

$$\alpha^{*s}(t) = \frac{1}{b}\int_{\tau}^{0} K\left(\frac{t-u}{b}\right) dA^{*s}(u). \tag{22.6}$$

Regarding the smoothed hazard rate estimate, it can be shown that

$$\begin{aligned}\widehat{\alpha}^*(t) - \alpha^{*s}(t) &= \frac{1}{b}\int_{\tau}^{0} K\left(\frac{t-u}{b}\right) d(\widehat{A}^* - A^{*s})(u) \\ &= \frac{1}{b}\int_{\tau}^{0} K\left(\frac{t-u}{b}\right) \frac{J(u)}{\bar{Y}(u)} d\bar{M}^*(u).\end{aligned} \tag{22.7}$$

$\widehat{\alpha}^*(t) - \alpha^{*s}(t)$ is a stochastic integral with respect to local martingale $\bar{M}^*(t)$. Asymptotic normality follows the martingale central limit theorem. A naive variance estimator of $\widehat{\alpha}^*(t)$ is given by

$$\frac{1}{b^2}\int_{\tau}^{0} J(u)\left\{\frac{K(\frac{t-u}{b})}{\bar{Y}(u)}\right\}^2 d\bar{N}^L(u) = \frac{1}{b^2}\int_{0}^{\tau} J(u)\left\{\frac{K(\frac{t-u}{b})}{\bar{Y}(u)}\right\}^2 d\bar{N}(u). \tag{22.8}$$

22.3 Nonparametric Inference of Hazard Rate Function

Estimation of the distribution function of L has been well studied. $G(t)$ can be estimated by the right truncated version of the Kaplan-Meier estimator (Woodroofe 1985; Keiding and Gill 1990),

$$\widehat{G}(t) = \prod_{u>t}\left(1 - \frac{d[\sum_{i=1}^{n} I(L_i \le u)]}{\bar{Y}(u)}\right). \tag{22.9}$$

Under the context of right truncation, Nelson-Aalen estimator of the cumulative hazard function is not applicable. Instead, one has to consider a plug-in estimator,

$$\widehat{A}(t) = \int_0^t \frac{d\widehat{G}(u)}{1-\widehat{G}(u-)}. \tag{22.10}$$

The forward-time hazard rate can be estimated by

$$\widehat{\alpha}(t) = \frac{1}{b}\int_0^{\tau} K\left(\frac{t-u}{b}\right) d\widehat{A}(u). \tag{22.11}$$

Based on the definitions of $A^*(t)$ and $A(t)$, we can have

$$dA(t) = \frac{-G(t)}{1 - G(t-)} dA^*(t).$$

Using this relationship and estimators given by (22.9) and (22.10), $\alpha(t)$ can be identically estimated by

$$\widehat{\alpha}(t) = \frac{1}{b} \int_0^\tau K\left(\frac{t-u}{b}\right) \frac{-\widehat{G}(u)}{1 - \widehat{G}(u-)} d\widehat{A}^*(u). \tag{22.12}$$

Define $A^s(t) = \int_0^t \alpha(u) J(u) du$. The smoothed function $\alpha^s(t)$ can be written as

$$\begin{aligned}\alpha^s(t) &= \frac{1}{b} \int_0^\tau K\left(\frac{t-u}{b}\right) dA^s(u) \\ &= \frac{1}{b} \int_0^\tau K\left(\frac{t-u}{b}\right) \frac{-G(u)}{1 - G(u-)} dA^{*s}(u). \end{aligned} \tag{22.13}$$

We showed in the appendix that $(nb)^{1/2}[\widehat{\alpha}(t) - \alpha^s(t)]$ is asymptotically equivalent to the sum of functions of martingales, and through the martingale central limit theorem, $(nb)^{1/2}[\widehat{\alpha}(t) - \alpha^s(t)]$ converges in distribution to a normal random variable with mean zero. Based on the results given in the appendix, we will estimate the variance of $\widehat{\alpha}(t)$ by the formula

$$\hat{\text{var}}[\widehat{\alpha}(t)] = \frac{1}{b^2} \int_\tau^0 \left[K\left(\frac{t-u}{b}\right) \frac{\widehat{G}(u)}{1 - \widehat{G}(u-)} \right.$$

$$\left. - \int_0^u K\left(\frac{t-x}{b}\right) d\left(\frac{\widehat{G}(x)}{1 - \widehat{G}(x-)}\right)\right]^2 J(u) \frac{d\bar{N}^L(u)}{\bar{Y}(u)^2}.$$

The kernel smoothed estimator of $\alpha(t)$ is a weighted average of crude hazard estimates over event times close to t. Most kernel functions allow the event times closer to t to have more weight than those farther from t. Bandwidth, b, controls the width of window. b is chosen to include those events that fall in the interval $[t - b, t + b]$. Symmetric kernel functions are commonly used such as uniform, Epanechnikov and biweight, with the following expressions:

$$K(x) = 1/2, \qquad -1 \le x \le 1 \quad \text{(uniform kernel)},$$

$$K(x) = 3(1 - x^2)/4, \qquad -1 \le x \le 1 \quad \text{(Epanecnikov kernel)},$$

$$K(x) = 15(1 - x^2)^2/16, \qquad -1 \le x \le 1 \quad \text{(biweight kernel)}.$$

The above kernels are applicable if $b \leq t \leq t_n - b$, where t_n is the largest event time. When $t < b$, adjustment is necessary because $t - b$ is less than zero. In this case symmetric kernels need to be modified and asymmetric kernels should be used. Gasser and Muller (1979) suggested the boundary kernel method to modify kernels. The boundary kernel method uses linear multiples of the kernel function around the boundary.

The main question is to find the best bandwidth for kernel smoothed estimates of hazard rate. There is a trade off between bias and variance in terms of choosing the bandwidth b. Generally speaking, small bandwidth will result in less smooth curve. Consequently, there will be smaller bias but larger variance. One way to choose the optimum bandwidth is to use mean integrated squared error (MISE) to see what value of b minimizes such error (Klein and Moeschberger 2003). MISE of $\widehat{\alpha}$ is defined by

$$\text{MISE}(b) = E\left[\int_0^{\tau} [\widehat{\alpha}(u) - \alpha(u)]^2 du\right] = E\left[\int_0^{\tau} \widehat{\alpha}^2(u) du\right]$$

$$-2E\left[\int_0^{\tau} \hat{\alpha}(u)\alpha(u) du\right] + E\left[\int_0^{\tau} \alpha^2(u) du\right]$$

MISE(b) depends both on the kernel used to estimate α and on the bandwidth b. Since the last term is independent from both kernel and bandwidth, it can be ignored. Let $t_1 < t_2 < \ldots < t_n$ be distinct event times, first term can be estimated by using trapezoidal rule, and the second term can be estimated by using cross-validation estimate given by Ramlau-Hansen (1983). Optimum bandwidth, b, minimizes the following function (Klein and Moeschberger 2003);

$$g(b) = \sum_{i=1}^{n-1} \left(\frac{t_{i+1} - t_i}{2}\right) [\widehat{\alpha}^2(t_i) + \widehat{\alpha}^2(t_{i+1})]$$

$$-\frac{2}{b} \sum_{i \neq j} K\left(\frac{t_i - t_j}{b}\right) \Delta\widehat{A}(t_i)\Delta\widehat{A}(t_j). \tag{22.14}$$

22.4 Simulation Study

We conducted a simulation study to assess the performance of the kernel smoothed hazard function. Random variables (L, T) were generated with constraint of $L < T$. Two settings were considered for distribution of L, uniform [0,1] and exponential(1) truncated at 1.2. The truncation variable T was generated from exponential(λ) for both settings. The following steps were taken to get a sample with size n: first, random variables (L, T) were generated; second, if $L > T$, we regenerate the pair until getting a pair satisfying $L < T$. Let size of truncated sample be n_t, then the truncation rate is calculated by $n_t/(n + n_t)$. In our simulation study truncation rates were chosen to be 25% and 50%. In order to obtain these truncation rates, we searched for appropriate λ values.

Each simulated setting contained 1000 replicates. For simplicity, uniform kernel was used in estimation. In order to obtain optimum bandwidth, we have searched for b, which minimized $g(b)$ given in (22.14) for each replicate. Searching for optimum bandwidth can be computationally challenging when sample size is large. Due to this limitation, the sample sizes of simulation settings were chosen to be 100 and 200. Let $\bar{\widehat{\alpha}}(t)$ be the average of kernel smoothed hazard estimates of 1000 replicates and $\widehat{\alpha}^{(i)}(t)$ be the kernel smoothed hazard estimate for the ith replicate, then $\bar{\widehat{\alpha}}(t) = \frac{1}{1000}\sum_{i=1}^{1000}\widehat{\alpha}^{(i)}(t)$.

The relative bias provides a measure of the magnitude for the bias,

$$\text{Relative bias} = \frac{B[\bar{\widehat{\alpha}}(t)]}{\alpha(t)} = \frac{\bar{\widehat{\alpha}}(t) - \alpha(t)}{\alpha(t)},$$

where the bias, $B[\bar{\widehat{\alpha}}(t)]$, was defined as the deviation between the average kernel smoothed hazard estimate and the true value.

The variance estimator $\widehat{\text{var}}[\widehat{\alpha}(t)]$ evaluated for each replicate and the average of these variance estimates was calculated as

$$\text{Estimated variance} = \frac{1}{1000}\sum_{i=1}^{1000}\widehat{\text{var}}[\widehat{\alpha}^{(i)}(t)].$$

Sample variances were evaluated by the formula

$$\text{Sample variance} = \frac{1}{1000-1}\sum_{i=1}^{1000}\left(\widehat{\alpha}^{(i)}(t) - \bar{\widehat{\alpha}}(t)\right)^2.$$

The 95% confidence intervals of variance estimators and coverage probabilities for each replicate were calculated. The estimation results were reported at time points that corresponds to 0.2, 0.5, 0.8 in $G(t)$. For this reason, results were evaluated at $t = 0.2, 0.5, 0.8$ for uniform distribution and at $t = 0.15, 0.43, 0.82$ for truncated-exponential distribution (see Table 22.1). True hazard rates for Uniform[0,1] at 0.2, 0.5, and 0.8 are 1.25, 2.0, and 5.0, respectively. Similarly, true hazard rates for exponential(1) truncated at 1.2 are 1.53, 1.85, and 3.13, respectively. Relative biases are very small for both distributions. Estimated variances have very close values to sample variances. There is an optimum bandwidth search for each setting, so it is not possible to observe a clear trend for relative bias and variance. There is obvious under-coverage for sample size 100. The coverage probabilities improved when the sample size increased to 200 though they are still slightly below 0.95. We anticipate improvement should a larger size be employed.

Table 22.1 The simulation results for estimating $\alpha^s(t)$

Sample size	Distribution of L	Truncation rate	t	Relative bias (%)	Sample variance	Estimated variance	95% CI coverage
$n = 100$	U[0,1]	25	0.20	0.33	0.497	0.499	0.924
			0.50	1.05	0.905	0.868	0.913
			0.80	0.90	2.255	2.296	0.913
		50	0.20	−0.96	0.414	0.415	0.920
			0.50	0.26	0.881	0.803	0.897
			0.80	−1.07	2.525	2.330	0.897
	Exp(1) truncated at 1.2	25	0.15	−0.24	0.427	0.413	0.923
			0.43	−0.21	0.662	0.609	0.900
			0.82	2.64	1.409	1.371	0.920
		50	0.15	−0.81	0.377	0.370	0.939
			0.43	0.34	0.683	0.597	0.898
			0.82	0.76	1.711	1.501	0.888
$n = 200$	U[0,1]	25	0.20	1.23	0.492	0.484	0.930
			0.50	1.84	0.860	0.844	0.926
			0.80	1.24	2.194	2.249	0.926
		50	0.20	−0.93	0.389	0.377	0.920
			0.50	−0.80	0.735	0.742	0.937
			0.80	0.36	2.304	2.262	0.919
	Exp(1) truncated at 1.2	25	0.15	0.74	0.391	0.380	0.937
			0.43	−1.30	0.570	0.559	0.918
			0.82	1.39	1.352	1.278	0.908
		50	0.15	0.35	0.314	0.301	0.924
			0.43	0.67	0.488	0.498	0.942
			0.82	1.03	1.385	1.322	0.911

22.5 The Blood Transfusion Infected AIDS Data

For illustration purpose we analyzed the blood transfusion infected AIDS data set. The data were collected by Centers for Disease Control and Prevention (CDC) which required reporting of all AIDS onsets up to July 1, 1986. Study of interest is the AIDS incubation time, which is the duration between infection with HIV and the onset of AIDS. If blood transfusion was the cause of HIV infection, then the infection date can be determined retrospectively. The study population was defined to be all blood transfusion infected HIV subjects by the closing date, July,1 1986. Since only subjects whose AIDS onsets occurred earlier than the closing date could possibly be included in the sample, the AIDS incubation time in the data set would be shorter than the duration between infection date and the closing date. In other words, the AIDS incubation time was right truncated by the duration between infection date and the closing date.

This AIDS data set can be found in Kalbfleisch and Lawless (1989). The data set contained 295 AIDS cases diagnosed between January 1, 1978 and July 1, 1986, with the earliest infection date in April 1978 and the latest one in February 1986. The data set included the following variables, AIDS incubation time in months, infection time in months starting from January 1, 1978 and age at blood transfusion.

Let L denote the incubation time. The truncation time T is the time from infection to end of study which is July 1, 1986. Our goal was to obtain kernel smoothed hazard rate estimates and compare them between different age groups. Similar to previous researches, we considered three age groups for analysis, children (1–4 years), adults (5–59 years), and elderly patients ($\geq$60). Sample sizes were 34 for children 120 for adults and 141 for elderly people. The largest incubation times recorded were respectively 43, 89, and 83 months for children, adults, and elderly patients.

We used kernel smoothing to estimate smoothed hazard rate function for right truncated data. We looked for optimum bandwidths for three kernels in each age group. Optimum bandwidth selected for adults were $b = 5, 12, 11$ with respective of using uniform, Epanechnikov, and biweight kernels. We chose $b = 8, 61, 31$ for elderly and $b = 8, 32, 8$ for children for these three kernels, respectively. Figure 22.1 depicts smoothed hazard functions using three kernels for each age group. Epanechnikov and biweight kernels assign higher weight in the middle and less weight towards the tails where uniform kernel assigns homogeneous weight. There is a great similarity in the estimation results of three kernels. Therefore, we only present the results regarding precision evaluation and two-sample comparison based on the uniform kernel. Figure 22.2 shows the smoothed hazard rate curves and pointwise 95% confidence intervals for each group. Figure 22.3 shows the differences between two kernel smoothed hazard rate functions and 95% pointwise confidence intervals. Comparisons between adults versus children and elderly versus children have been depicted until 40 months as the largest incubation time for children was 43 months. Due to the similar reason, comparison between adults versus elderly has been depicted up to 80 months.

The hazard rate curves using Epanechnikov and biweight kernels for adults increase by time for all three kernels. There is a sudden decrease at about 65 months for uniform kernel smoothed curve. In elderly patients, Epanechikov kernel results in a much flat curve. For hazard rate smoothed curves using biweight and uniform kernels, rapid increase occurs after 50 months. Children had much higher smoothed hazard rate curves compared to the other two groups. The hazard rate curves using uniform and biweight kernels increase steadily up to 30 months and then increase rapidly. The Epanechnikov kernel smoothed hazard rate curve remains at the almost the same level after 20 months.

In Fig. 22.3, the pointwise confidence intervals for differences in hazard rate function between adults and elderly contain value zero for almost whole study period, indicating highly similar risk level between these two populations. For comparison between children and each of the other two groups, the pointwise confidence intervals for differences do not include value zero starting from early

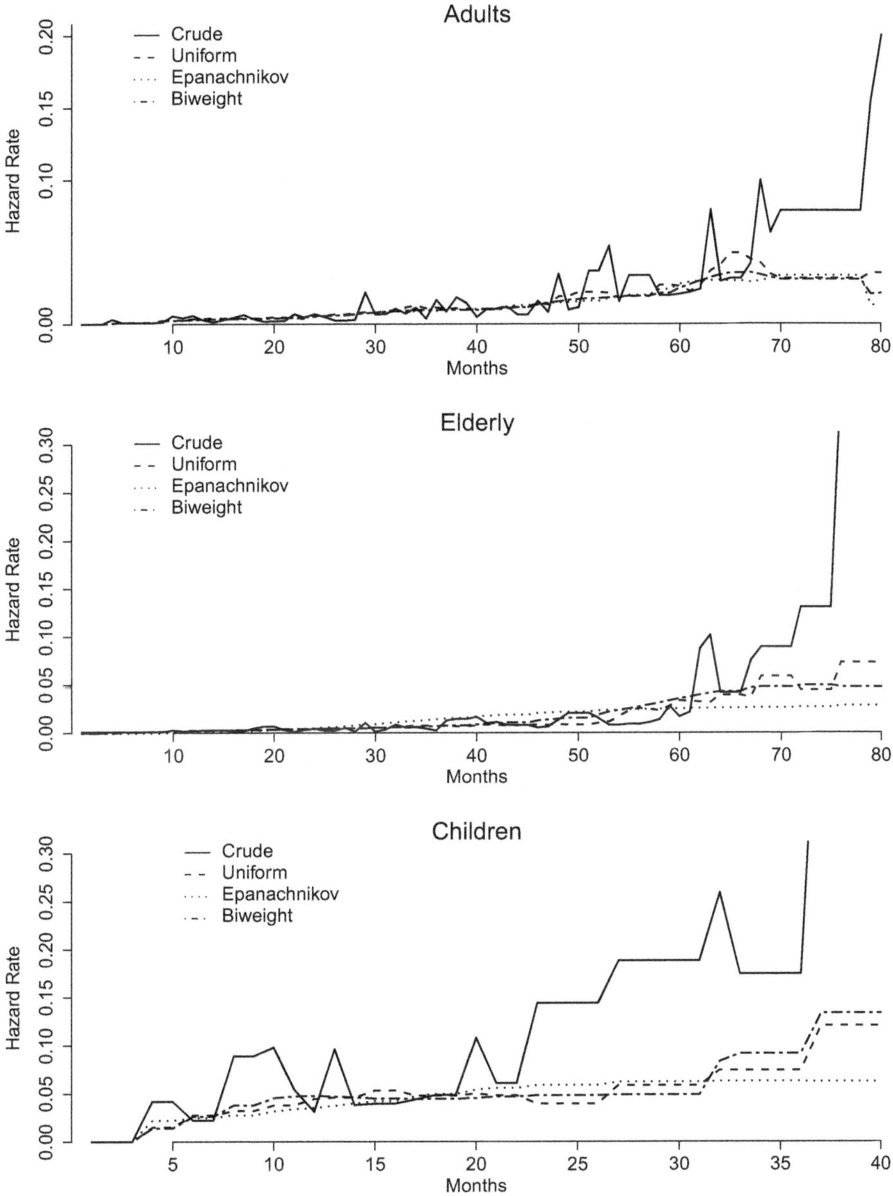

Fig. 22.1 Crude and smoothed hazard rate functions

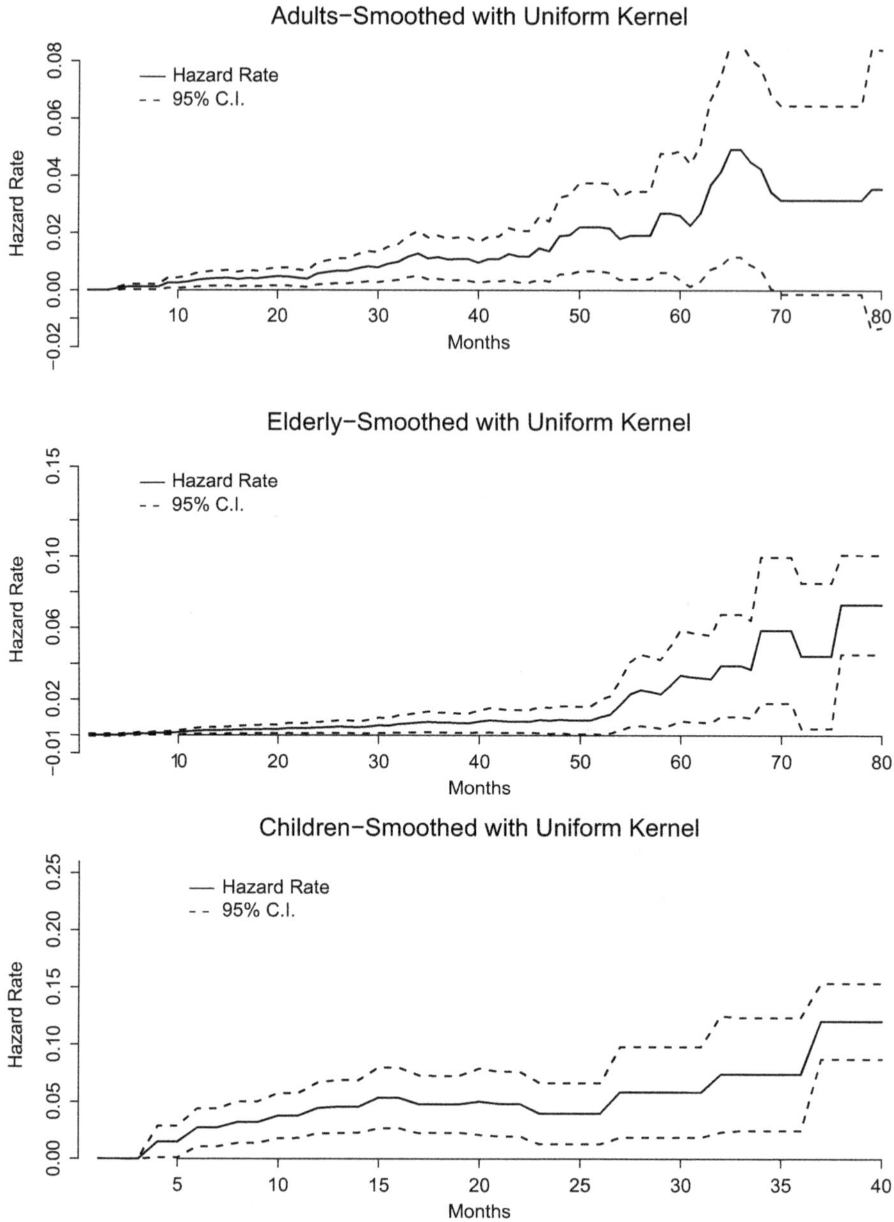

Fig. 22.2 Uniform-kernel smoothed hazard rate functions and 95% confidence intervals

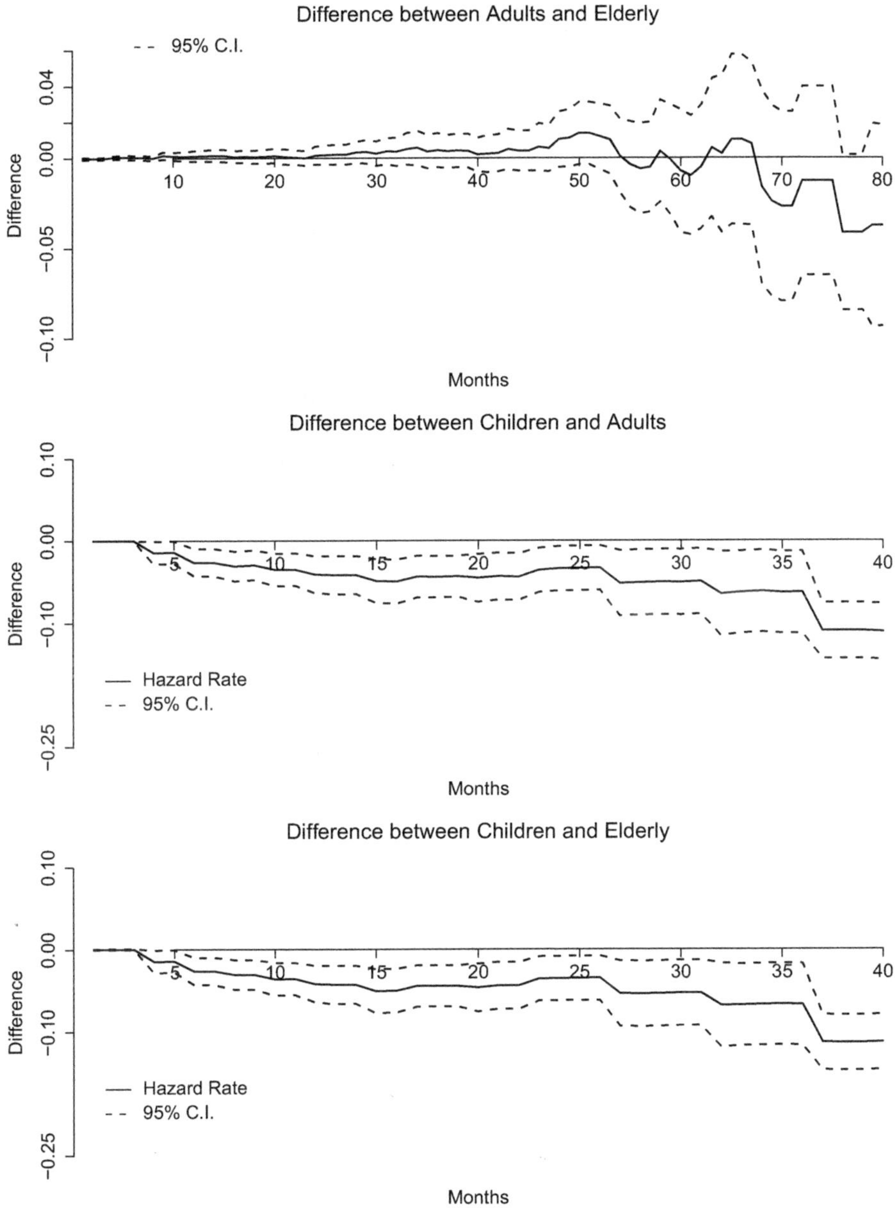

Fig. 22.3 Differences of hazard rate functions and 95% confidence intervals

study period. The results suggest that children were consistently subject to elevated risk of infection compared to people of senior ages.

22.6 Discussion

Our motivation was to study one important survival quantity, the hazard rate function, for right-truncated data. The reverse-time hazard has been studied by many researchers but the forward-time hazard didn't receive the same degree of attention. One of the earliest researches on forward-time hazard was the Cox model studied by Finkelstein et al. (1993). Our study is useful in examining the shape of hazard rate function and can provide direct assessment of proportional hazards assumption. The nonparametric inference provided here permits comparison of hazard rate functions in forward-time between two samples. The result is easily interpretable in real-life applications.

Additional researches can be conducted for hazard rate function with right truncated data. A test about proportional hazards is practically needed for justifying inclusion of a covariate in a Cox model. Sometime one may be interested in testing whether a hazard rate function monotonically increase or decrease over a time interval.

Acknowledgements This book chapter has been greatly improved following the comments of two referees. The authors appreciate referees' insightful suggestions on the contents of this book chapter.

Appendix

Asymptotic properties of kernel estimator of intensity were established by Ramlau-Hansen (1983). In this study we exploratively investigate the limiting distribution of $\widehat{\alpha}(t)$. In the following context, "$\approx$" indicates asymptotic equivalence. Note that $(nb)^{1/2}[\widehat{\alpha}(t) - \alpha^s(t)]$ can be expressed as

$$(nb)^{1/2}[\widehat{\alpha}(t) - \alpha^s(t)] = \frac{(nb)^{1/2}}{b}\int_0^\tau K\left(\frac{t-u}{b}\right)\left[\frac{-\widehat{G}(u)}{1-\widehat{G}(u)}d(\widehat{A}^* - A^*)(u)\right]$$

$$-\frac{(nb)^{1/2}}{b}\int_0^\tau K\left(\frac{t-u}{b}\right)\left(\frac{\widehat{G}(u)}{1-\widehat{G}(u-)} - \frac{G(u)}{1-G(u-)}\right)dA^*(u)$$

For the first term on the right-hand side of the above equation, it can be shown that

$$\frac{(nb)^{1/2}}{b}\int_0^\tau K\left(\frac{t-u}{b}\right)\frac{-\widehat{G}(u)}{1-\widehat{G}(u-)}d(\widehat{A}^* - A^*)(u)$$

$$\approx \sqrt{\frac{n}{b}} \int_{\tau}^{0} K\left(\frac{t-u}{b}\right) \frac{G(u)}{1-G(u-)} d(\widehat{A}^{*} - A^{*})(u)$$

$$= \sqrt{\frac{n}{b}} \int_{\tau}^{0} K\left(\frac{t-u}{b}\right) \frac{G(u)}{1-G(u-)} J(u) \frac{d\bar{M}^{*}(u)}{\bar{Y}(u)}$$

To investigate the second term on the right-hand side, we first consider the Taylor series expansion,

$$\frac{\widehat{G}(u)}{1-\widehat{G}(u^{-})} - \frac{G(u)}{1-G(u-)} \approx \frac{d}{dA^{*}(u)}\left(\frac{G(u)}{1-G(u-)}\right)(\widehat{A}^{*} - A^{*})(u).$$

Then we will have

$$\frac{(nb)^{1/2}}{b} \int_{0}^{\tau} K\left(\frac{t-u}{b}\right)\left[-\left(\frac{\widehat{G}(u)}{1-\widehat{G}(u-)} - \frac{G(u)}{1-G(u-)}\right) dA^{*}(u)\right]$$

$$\approx \sqrt{\frac{n}{b}} \int_{0}^{\tau} K\left(\frac{t-u}{b}\right)\left[-d\left(\frac{G(u)}{1-G(u-)}\right)(\widehat{A}^{*} - A^{*})(u)\right]$$

$$= \sqrt{\frac{n}{b}} \int_{0}^{\tau} K\left(\frac{t-u}{b}\right)\left[-d\left(\frac{G(u)}{1-G(u-)}\right) \int_{\infty}^{u} J(x) \frac{d\bar{M}^{*}(x)}{\bar{Y}(x)}\right]$$

$$= \sqrt{\frac{n}{b}} \int_{\tau}^{0} \left[-\int_{0}^{u} K\left(\frac{t-y}{b}\right) d\left(\frac{G(y)}{1-G(y-)}\right)\right] J(x) \frac{d\bar{M}^{*}(x)}{\bar{Y}(x)}.$$

Combining the above results, $(nb)^{1/2}[\widehat{\alpha}_n(t) - \alpha_n^{s}(t)]$ is asymptotically equal to

$$\sqrt{\frac{1}{nb}} \int_{\tau}^{0} \left[K\left(\frac{t-u}{b}\right) \frac{G(u)}{1-G(u-)} - \int_{0}^{u} K\left(\frac{t-y}{b}\right) d\left(\frac{G(y)}{1-G(y-)}\right)\right] J(u) \frac{d\bar{M}^{*}(u)}{\bar{Y}(u)/n}. \qquad (22.15)$$

Through the martingale central limit theorem, when $n \to \infty, b \to 0, nb \to \infty$, $(nb)^{1/2}[\widehat{\alpha}(t) - \alpha^{s}(t)]$ converges in distribution to a normal random variable with mean zero and the following variance function,

$$\frac{1}{b} \int_{\tau}^{0} \left[K\left(\frac{t-u}{b}\right) \frac{G(u)}{1-G(u-)} - \int_{0}^{u} K\left(\frac{t-x}{b}\right) d\left(\frac{G(x)}{1-G(x-)}\right)\right]^{2} \frac{\alpha(u)du}{y(u)}.$$

In addition, it needs to prove that $(nb)^{1/2}[\alpha_n^{s}(t) - \alpha(t)]$ is asymptotically negligible. Some regularity conditions for establishing such a result can be found in Ramlau-Hansen (1983, §4). We do not investigate this topic here.

References

Andersen, P. K., Borgan, Ø., Gill, R. D., & Keiding. N. (1993). *Statistical models based on counting processes*. New York: Springer.

Chen, K., Chao, M. T., & Lo, S. H. (1995). On strong uniform consistency of the Lynden-Bell estimator for truncated Data. *The Annals of Statistics,23*, 440–449.

Chi, Y., Tsai, W. Y., & Chiang, C. L. (2007). Testing the equality of two survival functions with right truncated data. *Statistics in Medicine, 26*, 812–827.

Finkelstein, D. M., Moore, D. F., & Schoenfeld, D. A. (1993). A proportional hazards model for truncated AIDS data. *Biometrics, 49*, 731–740.

Gasser, T. H., & Muller, H. G. (1979). Kernel estimation of regression functions. In: T. Gasser & M. Rosenblatt (Eds.), *Smoothing techniques for curve estimation: Vol. 757. Lecture notes in mathematics* (pp. 23–68). Heidelberg: Springer-Verlag.

Gross, S. T., & Huber-Carol, C. (1992). Regression models for truncated survival data. *Scandinavian Journal of Statistics, 19*, 193–213.

Gurler, U., Stute, W., & Wang, J. L. (1993). Weak and strong quantile representations for randomly truncated data with applications. *Statistics and Probability Letters, 17*, 139–148.

Kalbfleisch, J. D., & Lawless, J. F. (1989). Inference based on retrospective ascertainment: An analysis of the data on transfusion-related AIDS. *Journal of the American Statistical Association, 84*, 360–372.

Kalbfleisch, J. D., & Lawless, J. F. (1991). Regression models for right truncated data with applications to AIDS incubation times and reporting lags. *Statistica Sinica, 1*, 19–32.

Kaplan, E. L., & Meier, P. (1958). Nonparametric estimation from incomplete observations. *Journal of the American Statistical Association, 53*, 457–481.

Keiding, N., & Gill, R. D. (1990). Random truncation models and Markov process. *The Annals of Statistics, 18*, 582–602.

Klein, J. P., & Moeschberger, M. L. (2003). *Survival analysis: Techniques for censored and truncated data*. New York: Springer.

Klein, J. P., & Zhang, M. J. (1996). Statistical challenges in comparing chemotherapy and bone-marrow transplantation as a treatment for leukemia. In N. P. Jewell et al. (Eds.) *Life data: Models in reliability and survival analysis* (pp. 175–185). New York: Springer.

Lagakos, S. W., Barraj, L. M., & Gruttola, V. (1988). Nonparametric analysis of truncated survival data with applications to AIDS. *Biometrika, 75*, 515–523.

Lai, T. L., & Ying, Z. (1991). Estimating a distribution function with truncated and censored data. *The Annals of Statistics, 19*, 417–442.

Ramlau-Hansen, H. (1983). Smoothing counting process intensities by means of kernel functions. *The Annals of Statistics, 11*, 453–466.

Shen, P. (2010). A class of semiparametric rank-based tests for right-truncated data. *Statistics and Probability Letters, 80*, 1459–1466.

Uzunogullari, U., & Wang, J. L. (1992). A comparison of hazard rate estimators for left truncated and right censored data. *Biometrika, 79*, 297–310.

Wang, J. L. (2005). Smoothing hazard rate. In *Encyclopedia of biostatistics* (2nd ed., pp. 4986–4997). New York: Wiley.

Wang, M. C., Jewell, N. P., & Tsai, W. Y. (1986). Asymptotic properties of the product limit estimate under random truncation. *The Annals of Statistics, 14*, 1597–1605.

Watson, G. S., & Leadbetter, M. R. (1964). Hazard analysis I. *Biometrika, 51*, 175–184.

Woodroofe, M. (1985). Estimating a distribution function with truncated data. *The Annals of Statistics, 13*, 163–177.

Index

Y. Zhao, D.-G. Chen (eds.), *New Frontiers of Biostatistics and Bioinformatics*, ICSA Book Series in Statistics, https://doi.org/10.1007/978-3-319-99389-8